Applied Electronic Math with Calculators

John W. Tontsch
Los Angeles Pierce College

SCIENCE RESEARCH ASSOCIATES, INC.
Chicago, Henley-on-Thames, Sydney, Toronto
A Subsidiary of IBM

Acquisition Editor	Alan W. Lowe
Project Editor	Ronald Q. Lewton
Copy Editor	Eva Marie Strock
Text and Cover Designer	Judith A. Olson
Illustrators	Jean Foster, John F. Foster
Compositor	Interactive Composition Corporation

Library of Congress Cataloging in Publication Data

Tontsch, John W.
    Applied electronic math, with calculators.

    Includes index.
    1. Electronics—Mathematics.   2. Calculating-
machines.   I. Title.
TK7835.T66   512'.1'02854   81-9166
ISBN 0-574-21550-6   AACR2

10 9 8 7 6 5 4 3

The author explaining the 9s complement arithmetic method to one of his classes.

About the Author

John W. Tontsch received his degree in electronic engineering from the School of Engineering at Northwestern University, Evanston, Illinois. He joined IBM, serving as a customer engineer, then in 1963 joined the Rocketdyne Division of what is currently Rockwell International in Canoga Park, California. While at Rocketdyne he developed and taught in-house training programs in numerical control. In 1964 he was invited to set up a program in this field at Los Angeles Pierce College; he has been on the staff there ever since, as a full-time electronics instructor for five years and then as the first instructor in the newly established computer technology program. He wrote and coordinated an extensive, multiyear grant for the National Science Foundation to establish an electronic/computer curriculum and then taught for five years in this field. Tontsch is now Coordinator of Instruction within the college administration at Pierce. He continues to teach courses at Pierce College in the mathematics of electronics and electronic circuits. He is coauthor of a textbook titled *Fundamental Circuit Analysis,* also published by SRA.

Preface

This text presents both the traditional mathematics required for electronics technician education and instructions for using the hand-held calculator to solve problems. Considering the enormous time-saving advantages the calculator offers, students must understand how to use this powerful and inexpensive tool. It seems logical to assume that all students with a calculator (and almost all students have one) can get the correct answer to a problem if they understand the theory involved. However, in practice many students get wrong answers despite their understanding of the theory because they do not understand that the calculator executes functions in a predetermined order. This text presents the concepts and skills needed by students to accurately perform routine computations with a hand-held calculator, thus providing more time for considering basic electronic theory.

At no point is a compromise made in developing the theory behind mathematics. It is the author's contention that students must understand what they are doing during a calculation, not merely pushing memorized sequences of buttons to find a numerical answer. For this reason all chapters fully develop the mathematical concepts involved; the concluding sections of each chapter (when applicable) thoroughly illustrate how these concepts are applied when using a calculator.

It is quite difficult, if not totally impractical, to develop a text about the mathematics of electronics and not discuss simple electronic concepts. Therefore it is assumed that students have some fundamental DC and/or AC theory background, or at least are currently taking a basic electronics course. This prerequisite is particularly important for Chapters 9, 13, and 18, which deal with problems in electronics. These chapters can be temporarily skipped in classes with little or no electronics background. Regardless of which condition prevails, this text either gives students a healthy head start in electronics theory or helps reinforce the theory if they are also taking a theory course.

This book incorporates many important concepts that other mathematical books neglect. In addition to the calculator concepts, such topics as matrix analysis, electronic network theorems, and simple rules of Boolean algebra used in constructing desired logic effects are included. Examples of problem solutions prevail throughout each chapter. Numerous problems are then presented both at the end of each chapter section and as a review of each major subdivision of the text. Calculator key-in functions are boxed for easy identification. Additionally, calculator solutions are presented in both the algebraic hierarchy format and the RPN format. Only up-to-date topics are covered; for example, meter movements, which are being replaced by digital readout devices, are not discussed. In those chapters dedicated to nondecimal numbering systems as used in computers, examples show how the calculator can convert from one numbering system into another.

In perspective, this text lays the mathematical foundations necessary for students to comprehend and apply modern electronics theory.

The author and publisher would like to thank the following educators for their valued review comments:

Robert L. Benne	Linn Technical College
Richard T. Burchell	Riverside City College
Charles G. King	Fresno City College
Edwin E. Pollock	Cabrillo College

glossary of arithmetic definitions _____

Terms used in the four basic arithmetic operations are:

Addition   augend + addend = sum

Subtraction   minuend − subtrahend = difference

Multiplication   multiplicand × multiplier = product

Division   dividend ÷ divisor = quotient

Absolute value   The value of any number regardless of its sign; symbolized by one vertical line on either side of the number. For example, $|+5| = 5$ $|-5| = 5$.

Addend   A number or quantity to be added to another number or quantity (the *augend*).

Augend   A number having another number (the *addend*) added to it.

Base   The number or variable to be raised to a *power*.

Binomial   Two terms separated by a plus (+) or minus (−) sign.

Coefficient   Any number, especially when associated with variables; for example, 5 is the coefficient of the term $5ab$.

Denominator   The term under the division symbol: numerator/denominator. That number which shows into how many equal parts the whole is divided.

Difference   The resultant after one term (the *subtrahend*) is subtracted from another term (the *minuend*).

Dividend   A number into which another number (the *divisor*) is to be divided.

Divisor   The number being divided into another number (the *dividend*).

Exponent   A number indicating how many times another (*base*) number is to be multiplied by itself.

Expression   The collection of all the terms and operations constituting an evaluation.

Integer   A number not having any fractional or decimal parts associated with it. On the calculator, only that portion of a number to the left of the decimal point.

Irrational number   A number that cannot be completely expressed as either an *integer* or a *quotient* of an integer, such as $\sqrt{2}$ or π.

Logarithm   The *exponent* to which some base number must be raised to equal some quantity. For example, the base number 10 must be raised to the power 2 to equal 100; therefore the logarithm of 100 to the base 10 is 2.

Minuend   The number from which another number (the *subtrahend*) is subtracted.

Monomial   A single collection of a coefficient and/or variables not separated by a + or − sign, such as $6.25ab$.

Multiplicand   A number to be multiplied by another number (the *multiplier*).

Multiplier   The number by which another number (the *miltiplicand*) is to be multiplied.

Numerator   The term above the division symbol: numerator/denominator. That term of a fraction which shows how many of the specified parts of a unit are taken.

Operand   Any symbol or quantity upon which an operation is performed.

Operator   A symbol signifying a function to be performed.

Polynomial   A general definition describing an expression containing more than one term. However, because *binomial* and *trinomial* are reserved for describing expressions of two and three terms, respectively, *polynomial* is sometimes reserved for describing an expression of four or more terms.

Power   Used synonymously with *exponent*. A general term that describes how many times a number is multiplied by itself.

Product   The resultant of any two terms multiplied.

Quotient   The number resulting from the division of one quantity (the *dividend*) by another quantity (the *divisor*).

Radical sign   The symbol $\sqrt{}$ over any quantity (the *radicand*) signifying that its root is to be extracted.

Radicand   The term under the *radical sign*.

Radix   A number made the *base* of a system of numbering.

Rational number   A number without a *radical sign* expressible as an *integer* or a *quotient* of an integer.

Reciprocal   The quantity resulting from the division of 1 by a given quantity.

Root   A quantity that when multiplied by itself a specified number of times produces a given quantity.

Sign   A symbol specifying a given operator or operation to be performed.

Subscript   A number written below a variable designator. Usually used to distinguish one designator from another when the same symbolic designator is used several times, such as R_1, R_2, R_3, and so on.

Subtrahend   The quantity subtracted from another quantity (the *minuend*).

Sum   The resultant of the addition of two or more quantities.

Term   A collection of *coefficients* and/or variables within an *expression* separated from other collections by a $+$ or $-$ sign. For example, $5a^2 + 3a - 6$ contains three terms.

Trinomial   A *polynomial* of only three terms.

fundamental concepts of the calculator _____

THE CHOICE

The calculator is universally accepted as a classroom device, just as indispensable today as the slide rule was years ago. But with so many calculators on the market, which calculator should be used in conjunction with this text?

First, it should be a *scientific* calculator. Calculators manufactured for statistics, real estate analysis, telling time, and so on are not as capable of performing scientific calculations as are calculators designed specifically for that purpose.

Second, the calculator should be reasonably priced for what it is intended to do. Many advanced, expensive calculators can be programmed, read magnetic strips, be connected to an external printer, have extensive memory storage, and accept plug-in modules with sophisticated built-in mathematical algorithms. These features are certainly attractive but not necessary to solve electronic problems. The author has found three very adequate scientific calculators costing $20 to $35 each and capable of handling all the problems in this textbook. The two major differences between the three calculators (which probably account for the price variation) are the number of nested parentheses capable of being handled and the number of memory storage locations. The three scientific calculators used to verify all the calculator examples are the Casio *fx*-80, Texas Instruments' TI-55, and the Sharp EL-5813.

ALGEBRAIC HIERARCHY VERSUS RPN

The calculators use one of two built-in methodologies. The first format requires entry of data via the *algebraic hierarchy* method. For example, if 5 is to be added to 6, the key-in sequence is

$$5 \quad \boxed{+} \quad 6 \quad \boxed{=}$$

The second format requires entry of data via the *Reverse Polish Notation* (RPN) method. The previous problem is keyed in as

$$5 \quad \boxed{\text{ENTER}} \quad 6 \quad \boxed{+}$$

On the surface, the sample problem is not that complicated. However, it is the author's opinion that more complex problems, like those involving nested paren-

Photo 1 Casio *fx*-80. Costs less than $20, has limited memory and parentheses capabilities. (*Courtesy of Casio, Inc. and Doremus & Company*)

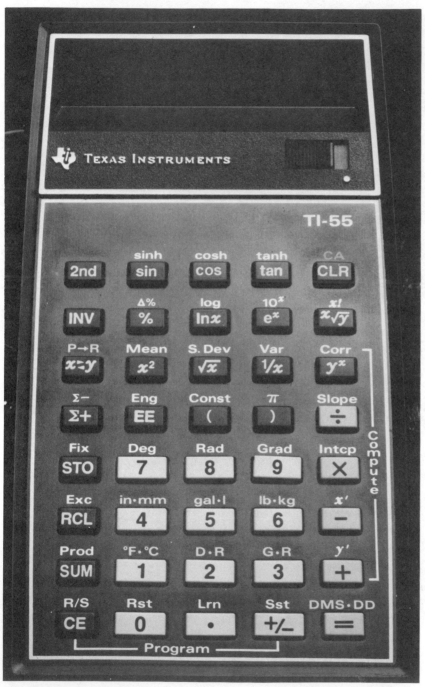

Photo 2 Texas Instruments TI-55. Costs less than $30, has medium memory and parentheses capabilities. (*Courtesy of Texas Instruments, Incorporated*)

Photo 3   Sharp EL-5813. Costs less than $35. Of the three scientific calculators shown here, this one has more capability for nested parentheses calculations and medium memory storage, including a working storage memory. (*Courtesy of Sharp Electronics Corporation*)

theses, require a preknowledge of how to approach the solution before the RPN method can be efficiently used. In short, if readers already know several fundamental principles of algebra, then they can use an RPN-type calculator. However, in this book it is assumed that readers do not thoroughly know the fundamentals of algebra. This text does not engage in the pros and cons of algebraic hierarchy versus the RPN format; however, it is the author's observation that it is easier for students in the formative learning stage to key in a problem in a straightforward, left-to-right sequence, just as it appears. Thus the book emphasizes the algebraic hierarchy approach when discussing calculator applications.

However, it is also the author's experience that even in fundamental mathematics classes about 5 to 10 percent of the students do have calculators that utilize the RPN format. Therefore, the Appendix is included to illustrate *every* calculator problem using the RPN method that has a corresponding illustration using the algebraic hierarchy method. Photo 4 shows the Hewlett Packard 31E, the calculator chosen for RPN format comparison with the *fx*-80, TI-55, and EL-5813. The 31E, which was purchased for under $42, uses an ENTER key rather than parentheses keys. Besides the basic scientific keys, it has four memory locations and a LAST *x* register, which automatically stores the last value present before performing any calculation.

THE GROUND RULES

To establish a simple procedure for keying the same data into three different calculators, it is necessary to recognize the differences in the layouts of the keyboards. For example, the *fx*-80 INV e^x key to raise *e* (epsilon, explained later in the text) to the power *x*; the TI-55 and the EL-5813 use the standard e^x key. Furthermore, the *fx*-80 uses the INV key to facilitate the second function of a key, the EL-5813 uses 2nd F , and the TI-55 uses either the 2nd or INV 2nd keys. In other words, various manufacturers locate their somewhat different scientific keys at different places on their respective keyboards. In this book, when displaying a problem's solution, only the function required is illustrated; we assume that readers will be able to initiate that function on their particular calculators.

Two of the calculators discussed may have a scientific function that the third calculator does not. For example, the *fx*-80 does not have a key for the hyperbolic function, whereas the TI-55 and the EL-5813 do. When a scientific function is illustrated in the text and one of the calculators does not have this function key, this is so stated. However, a student should not immediately throw away an otherwise good calculator; alternate methods of calculating the problem are demonstrated, even if they require extra keying-in strokes.

Finally, at any point in a sequence of steps, the display on one manufacturer's calculator may vary slightly from the display on the other two calculators. For example, on the TI-55, opening a parenthesis does not change the display, but on the *fx*-80 and EL-5813 the level of the parenthesis is displayed. For example,

$$2 \times ((6+2) \div (1+3))$$

should be keyed in exactly as the sequence is written, from left to right. What happens on each of the three calculators is shown in the table at the top of the next page.

The intermediate results differ only at the points where a parenthesis is being opened, that is, when the [(key is depressed. The TI-55 continues to display the value just prior to the depression of the [(key; the *fx*-80 and EL-5813 delete the preceding display and temporarily indicate the level of nesting parentheses. This is merely a technique used by each individual manufacturer.

Press (P)	Display (D)			Intermediate Points of Difference
	fx-80	TI-55	EL-5813	
2	2	2	2	
×	2	2	2	
⟦(	(1 0	2	(	*
⟦(	(2 0	2	((	*
6	6	6	6	
+	6	6	6	
2	2	2	2	
)	8	8	8	
÷	8	8	8	
⟦(	(2 0	8	((	*
1	1	1	1	
+	1	1	1	
3	3	3	3	
)	4	4	4	
)	2	2	2	
=	4	4	4	

Calculator terminology, such as "nesting parentheses," is adequately explained throughout the text. The important concept to remember is that at any given intermediate point (especially when opening a parenthesis), one manufacturer's display may differ from another's, but the end results are always identical. All examples in the text end at a point where the three calculators yield identical results, unless expressly indicated to the contrary to stress major points.

SIGNIFICANT DIGITS AND THE Fix KEY

Most calculators have either an 8- or a 10-digit display, but in the practical world this degree of accuracy is much more than necessary. For example, suppose we want to calculate the value of a resistor required to produce a given effect. Say that the exact mathematical value is 3272.125 ohms (Ω, the electrical unit of resistance). This number has seven significant digits (SDs). (Incidentally, 3272.1250 has eight significant digits, even though the last digit is zero, because it specifies a value for the fourth position beyond the decimal point.) The number of significant digits is thus the number of digits the number itself contains, including trailing zeros but excepting leading zeros: 003272.1250 still contains only eight significant digits.

In our problem, a 3272.125-Ω resistor would be very expensive to manufacture, and maintaining such a precise value would be very difficult because of the changes

Photo 4 Hewlett-Packard 31E. (*Courtesy of Hewlett-Packard Company*)

in resistance caused by temperature variations. In our practical world, the higher levels of accuracy are deleted and the number is *rounded off*. In this case, the round off is to the nearest 100 Ω, giving us a 3300-Ω resistor, which has a ± 10 percent tolerance.

To round off a number, we must identify critical points within the number. The highest weighted digit position (except leading zeros)—the digit(s) before the decimal—is called the *most significant digit* (MSD). The lowest weighted digit position (including trailing zeros)—the digit(s) after the decimal—is called the *least significant digit* (LSD):

$$\overbrace{003272.1250}^{8 \text{ SDs}}$$
$$\underset{\text{M S D} \qquad \text{L S D}}{\uparrow \qquad \uparrow}$$

To round off a number, we have to consider whether the round off is confined to the right or the left of the decimal point.

Rules for the *right* of the decimal point are:

1. Determine the digit position that is to become the new LSD of the rounded-off number.

2. Call the digit adjacent to and to the right of this new LSD *LSD − 1*.

3. If LSD − 1 is less than 5 (<5), delete it and all subsequent *lesser* position digits when forming the new number.

4. If LSD − 1 is greater than or equal to 5 ($\geq$5), increase the LSD by 1 before deleting LSD − 1 and all other *lesser* position digits when forming the new number.

Example

Round off the following to the indicated position.

a. 143.6219 to the thousandths place

b. 27.8432 to the hundredths place

c. 123.495 to the hundredths place

Solutions

a. 143.621$\cancel{9}$ = 143.622

b. 27.84$\cancel{3}\,\cancel{2}$ = 27.84

c. 123.49$\cancel{5}$ = 123.49 + .01 = 123.50   Note carryover from hundredths to tenths place

Rules for the *left* of the decimal point are:

1. Determine the digit position to which the rounding off is to occur.

2. If the digit immediately to the right of this position is <5, replace it and any other *lesser* digit positions, up to the decimal point, with a zero. Delete all digits to the right of and including the decimal point.

3. If the digit immediately to the right is ≥5, increment the digit to which the round off is to occur by 1 before applying the actions of rule 2. Watch for carryovers.

Example

Round off the following to the indicated position.

a. 131.260 to the tenths position

b. 1632.840 to the units position

c. 3799.6 to the units position

Solutions

a. 13X.Z 6 $\emptyset$ = 130

b. 1632.8 A $\emptyset$ = 1633

c. 3799.6 = 3800   Note carryover from units position all the way to hundredths position

   Some calculators, such as the TI-55, have a [Fix] key that rounds off the display to a fixed number of digits after the decimal point. For example, [Fix 2] rounds out the display to two places to the right of the decimal point. [Fix 0] rounds out the display to the nearest whole unit integer. To further clarify the Fix function, consider an eight-digit display of 123.52739. To display the effects of Fix 3, 2, 1, and 0, respectively:

[Fix]	Display
3	123.527
2	123.53
1	123.5
0	124

Note that the Fix function *only* rounds off the displayed number; the calculator *internally* continues to calculate at full register accuracy. The benefit of the [Fix] key is that you can observe only the accuracy desired. How many significant digits should be carried out when determining the answer to a problem? Three factors that influence this decision are:

1. Is it really that meaningful to express a number such as 2/3 as .66666666 rather than .667?

2. The calculator always displays all the digits within its capacity, regardless of where we choose to round off the number.

3. For practical electronic problems, we can round off to the closest commercial value regardless of the number of significant digits mathematically displayed.

Therefore, to display reasonable accuracy consistent with reasonable brevity, the answers in the text are usually displayed with a Fix number much less than the full capacity of the calculator. We use Fix numbers mainly in the initial chapters; as a reader develops a sixth sense as to how many digits are practical, we stop mentioning the Fix number and display answers to reasonable limits. Thus, if an answer (especially to an electronic problem) is shown as 123.5 but the calculator displays 123.45678, readers should not panic: the calculator is OK and the answer is correct. Additionally, to conform with the way a calculator displays a number, trailing zeros are not shown.

ontents _____

Chapter 16.   Complex Algebra   362

Chapter 17.   Generation of the Periodic Sine Function   389

Chapter 18.   Applications 3: AC Electronic Circuits   414

PART IV   LOGARITHMS, GRAPHS, AND EXPONENTS

Chapter 19.   Logarithms   437

CALCULATORS, FRACTIONS, AND UNIT SYSTEMS

1 PRIORITIES, SIGNS, AND PARENTHESES

1.1 INTRODUCTION

Mathematics is an exact science. Our economy, the flight of *Voyager* past Jupiter and Saturn, and even the orbits of the planets themselves are all structured and explainable by mathematics. The list of phenomena that can be expressed mathematically is endless. This science has to be exact; consider the consequences of each person interpreting differently a specific problem. For example, *without* using your calculator, evaluate the following simple expression:

$$10 - 3 + 5 \times 8 \div 4$$

Did you get 24 as an answer? Pretty simple, right? Wrong! The answer is 17. To prove it, key the problem into your calculator. Satisfied now that the answer is 17? To find out why the answer is 17 and not 24, we must consider the concept of priorities.

1.2 + AND − PRIORITIES

The basic arithmetic functions plus (also called addition and denoted by the symbol +) and minus (also called subtraction and denoted by the symbol −) are considered of equal priority and assigned the lowest priority position. If a series of terms has only plus and minus functions performed on the terms, operations proceed left to right, just like reading this page.

Example 1

a. $\underbrace{11 - 7} + 3$
   $= \quad 4 \quad + 3 = 7$

b. $\underbrace{-13 + 7.5} - 6 + 3$
   $= \quad \underbrace{-5.5 \quad - 6} + 3$
   $= \quad \quad \underbrace{-11.5 \quad + 3} = -8.5$

c. $\underbrace{-22.9 - 13.6} - 7.2 + 21.8$
   $= \quad \underbrace{-36.5 \quad - 7.2} + 21.8$
   $= \quad \quad \underbrace{-43.7 \quad + 21.8} = -21.9$

Note that only one sign precedes each term, which indicates what function is to be performed on that term. If the first term of a series has no sign preceding it (such as the 11 in Example 1a), that term is assumed to be positive.

Sometimes two operator signs precede a term. When this happens, the second operator and any associated term can be set off in parentheses for clarity, although this is not mandatory (and is not done in this book). The two general rules are:

1. If both operators (signs) are alike—both positive or both negative—they may be replaced by a single positive sign.

2. If one operator is positive and the other negative, regardless of the order, they may be replaced by a single negative sign.

Generally, when two plus and/or minus operators separate operands (the terms on which the function is performed), the first sign is the operation to be performed and the second sign is the positive or negative value of the second operand.

Example 2

a. $18 + 6 - -4$
    $= \underline{18 + 6} + 4$
    $= 24 + 4 = 28$

b. $17 + {}^-3 - {}^-4$
    $= \underline{17 - 3} + 4$
    $= 14 + 4 = 18$

c. $-13.2 - +7.1 + {}^-6.2$
    $= \underline{-13.2 - 7.1} - 6.2$
    $= -20.3 - 6.2 = -26.5$

d. $-7.1 - -1.2 - -9.3$
    $= \underline{-7.1 + 1.2} + 9.3$
    $= -5.9 + 9.3 = 3.4$

e. $9.32 - +6.1 - -5.72 + {}^-0.09 + +1.3$
    $= \underline{9.32 - 6.1} + 5.72 - 0.09 + 1.3$
    $= \underline{3.22 + 5.72} - 0.09 + 1.3$
    $= \underline{8.94 - 0.09} + 1.3$
    $= 8.85 + 1.3 = 10.15$

PROBLEMS

Where there are multiple operator signs between the operands, manually reduce them to their appropriate positive or negative value. Then, using your calculator, determine the final answers.

1. $13 - 2 + 6 =$

2. $7.2 + 1.9 - 8.3 =$

3. $16.2 - 8.4 - 8.9 =$

4. $54.7 + 132.6 - 146.2 =$

5. $112.6 - 321.2 + 412.7 - 31.2 =$

6. $31.6 + -35.2 - 16.9 =$

7. $152.7 - -13.2 - -78.1 =$

8. $38.9 + -1.72 + -11.83 =$

9. $254.8 - 182.6 - -172.2 - -32.5 =$

10. $142.32 - +161.7 - -182.4 + -132.6 =$

11. $0.09 - +0.16 - -0.84 + -1.32 =$

12. $5.7 - 3.2 + 62 - 347 =$

13. $579 - -64 - 72 + 97 =$

14. $6.21 + -3.42 - -6.3 + 5.86 =$

15. $123.7 + -642.5 - 861.7 =$

In Problems 16 to 25, first determine the solution manually and then check your results using your calculator.

16. $1.5 + -3.2 - 6.9 =$

17. $21 - -13 + 17 + 10 =$

18. $12 + -18 - 4 - -7 =$

19. $18 + -15 + 9 =$

20. $36 + -49 + -31 - -10 =$

21. $9.6 + -7.7 + -5.2 + 3.4 =$

22. $19 + -8 + -5 + -16 =$

23. $1.5 - -1.9 - -3.1 =$

24. $25 + -10 - 14 =$

25. $2.7 + -0.9 - 1.2 + -3.1 =$

1.3 × AND ÷ PRIORITIES

Within the mathematical hierarchy of priorities, times (also called multiplication and denoted by the symbols × or ·) and divide (also called division and denoted by the symbol ÷) operators are considered equally important but are one level higher in priority than the plus and minus operators. Thus, if only times and divide operators connect a string of terms, operations are performed from left to right regardless of priority because both operations have equal priority. But in a string of functions that also includes plus and minus signs, multiplication and division functions are performed first (from left to right), before the plus and minus functions are performed (again from left to right). A good mnemonic for remembering this is *my dear aunt Sally* (*m* for multiply, *d* for divide, *a* for addition, and *s* for subtraction).

Because the times and divide operators can yield answers much longer than the original operands, all intermediate and final stages in the following examples are set at $\boxed{\text{Fix 3}}$ (the numbers are rounded off to three places after the decimal). Remember, however, that this is done only for reasonable brevity; the calculator continues to operate internally at full register capacity.

Example 1 (only $\times$ and $\div$ operators)

a. $\quad 3 \times 6 \div 9 \times 4$

$\quad = \quad 18 \div 9 \times 4$

$\quad = \quad\quad 2 \quad \times 4 = 8$

b. $\quad 4.2 \div 3 \times 5.1 \div 4$

$\quad = \quad\quad 1.4 \times 5.1 \div 4$

$\quad = \quad\quad\quad 7.14 \quad \div 4 = 1.785$

c. $\quad 1.7 \div 6.2 \div 3.1 \times 4.8$

$\quad = \quad\quad 0.274 \div 3.1 \times 4.8$

$\quad = \quad\quad\quad 0.088 \quad \times 4.8 = 0.425$

Example 2 (intermixing $+$, $-$, $\times$, and $\div$ operators)

a. $\quad 10 - 3 + 5 \times 8 \div 4$ step 1

$\quad = 10 - 3 + \quad 40 \quad \div 4$ step 2

$\quad = 10 - 3 + \quad\quad 10$ step 3

$\quad = \quad 7 \quad + \quad\quad 10 \quad = 17$ step 4

Note that this is the same problem we used at the beginning of the chapter. The explanation as to why the answer is 17 should now be apparent. First, of the four operations to be performed, the times and divide must be executed before the plus and minus because they have higher priority. Second, the times function must be executed first because it is encountered first when proceeding from left to right. This means that $5 \times 8 = 40$ (step 1) is then divided by 4 (step 2), resulting in 10. Proceeding again from left to right, the problem becomes $10 - 3 = 7$ (step 3) $+ 10$ (step 4), for a resultant of 17.

b. $4 + \underbrace{5 \times 3} \times 2 - 6 \div 1.5$

   $= 4 + \underbrace{15 \times 2} - 6 \div 1.5$

   $= 4 + 30 - \underbrace{6 \div 1.5}$

   $= \underbrace{4 + 30} - 4$

   $= 34 - 4 = 30$

c. $\underbrace{3.5 \div 1.75} - 6.2 \times 3.6 + 2.7$

   $= 2 - \underbrace{6.2 \times 3.6} + 2.7$

   $= \underbrace{2 - 22.32} + 2.7$

   $= -20.32 + 2.7 = -17.62$

d. $\underbrace{3.1 \times 4.05} - 7.2 \div 3.6 + 2.4$

   $= 12.555 - \underbrace{7.2 \div 3.6} + 2.4$

   $= \underbrace{12.555 - 2} + 2.4$

   $= 10.555 + 2.4 = 12.955$

e. $4.75 - \underbrace{3.21 \times 1.67} + 2.5 \div 1.4$

   $= 4.75 - 5.361 + \underbrace{2.5 \div 1.4}$

   $= \underbrace{4.75 - 5.361} + 1.786$

   $= -0.611 + 1.786 = 1.175$

You may also encounter multiple operators between terms, where at least one operator is the $\times$ or $\div$ function. When this occurs, the first operator must be a $\times$ or $\div$ (higher priority) and the second operator must be a $+$ or $-$ (lower priority). This restriction does not exist when both operators are the $+$ and/or $-$ functions. For example, the following setups are *not* permissible combinations:

1. $3.85 \times \div 1.7$

2. $-3.85 - \times 1.7$

Example 1 is invalid because if the first operator is a $\times$ or $\div$, the second operator must be of lower priority ($+$ or $-$). Example 2 is invalid because if the first operator is a $+$ or $-$, the second must be of at least equal priority, which in this case would be another $+$ or $-$.

Here are two examples of permissible combinations of mixed priorities between terms:

1. 3.85×-1.7 multiply positive 3.85 by negative 1.7

2. $-3.85 \div +1.7$ divide negative 3.85 by positive 1.7

When you encounter sequences such as these, remember that the higher priority operators (times or divide) specify what is to be done, and the lower priority operators (plus or minus) specify the positive or negative value of the operands on which the function is performed. The following two rules govern multiple times and division:

1. If the two operands on which a times or divide function is performed have the same polarity—both positive or both negative—the resultant is positive.

2. If the two operands on which a times or divide function is performed have unlike signs—one positive and one negative—the resultant is negative, regardless of the order.

One final comment. The / or — symbol is used interchangeably with ÷ to signify division; all three symbols are acceptable in mathematics.

Example 3

a. 1.32×-1.67

$= -2.2044$

b. $-16.72 \div +3.84$

$= -4.354$

c. $-3.41/2.56 \times -8.71$

$= \quad -1.332 \quad \times -8.71 = 11.602$

d. $-1.38 \times -2.42 + 7.81 \div 2.6$

$= \qquad 3.340 \qquad + 7.81 \div 2.6$

$= \qquad 3.340 \qquad + \quad 3.004 \quad = 6.344$

e. $17.1 - -2.4 \times 3.1 \div 4.7 \times 0.09 + 2$

$= 17.1 - \quad -7.440 \quad \div 4.7 \times 0.09 + 2$

$= 17.1 - \qquad -1.583 \quad \times 0.09 + 2$

$= 17.1 - \qquad\qquad -0.142 \qquad + 2$

$= \qquad\qquad 17.242 \qquad\qquad + 2 = 19.242$

f. $-2.5 \times -1.4 \div -3.7 + 2.1 \div -1.6$

$= \qquad 3.5 \qquad \div -3.7 + 2.1 \div -1.6$

$= \qquad\qquad -0.946 \qquad + 2.1 \div -1.6$

$= \qquad\qquad -0.946 \qquad + \quad -1.313$

$= \qquad\qquad -0.946 \qquad - \quad 1.313 \quad = -2.259$

g.
$$
\begin{aligned}
& \underbrace{1.72 \times -3.1} + 6.4 \div -3.7 - 1.5 \div -0.67 \\
=\ & -5.332 + \underbrace{6.4 \div -3.7} - 1.5 \div -0.67 \\
=\ & -5.332 + -1.730 - \underbrace{1.5 \div -0.67} \\
=\ & -5.332 + \underbrace{-1.730} \underbrace{-} -2.239 \\
=\ & \underbrace{-5.332 - 1.730} + 2.239 \\
=\ & -7.062 + 2.239 = -4.823
\end{aligned}
$$

PROBLEMS

Using Example 3 as a guide, *manually* analyze (step by step) Problems 1 to 10 by determining the:

- Order of priorities
- Polarity of the resultant after any two operations

Use the calculator only when necessary to obtain individual numerical resultants.

1. $4 \times 9 \div 2 \times 3 =$

2. $6.1 \div 2.4 \div 1.6 =$

3. $13.2 \times -6.1 - 3.4 =$

4. $24.6 \div 7.2 + 1.3 \times 5.35 =$

5. $24.2 \div 6.3 \times -1.2 \div 2.4 + 6 =$

6. $32.4 \times -0.17 + -13.2 \div 7.5 =$

7. $5.25 \div -17.2 \times -3.6 - 2.4 =$

8. $0.076 \times -31.7 - 6.2 \times -5.1 =$

9. $3.54 + 0.57 \times 6.2 + -3.1 =$

10. $0.0035 \div -0.072 \times -0.52 =$

1.4 PARENTHESES

Why use parentheses? To explain, evaluate the following problem:

$$36/3 \times 4$$

According to the rules discussed so far, we first determine that $36/3 = 12$ and then, multiplying by 4, obtain a correct resultant of 48. Now suppose we want the product of 3×4 (which is 12) and then have 36 divided by this product, for a resultant of

3. How can we alter the normal left-to-right order of priorities to obtain the desired results? The answer is to establish a priority level even higher than that of the times and divide functions.

This new higher priority is the parentheses. More specifically, parentheses indicate that all the operations within them are to be performed before any other operation begins. To obtain the results discussed earlier, rewrite the problem as:

$$36/(3 \times 4)$$

The normal left-to-right operation of the $/$ and $\times$ is altered because the parentheses force an evaluation of $3 \times 4 = 12$ before the division process assumes priority. In effect, the problem becomes $36/12 = 3$. Parentheses thus override the normal priority execution sequence. To illustrate, look at the following example, where the three problems in Example 1, Section 1.2 have been reworked using parentheses to alter the normal evaluation process. Note the difference in the solutions.

Example 1

a. $11 - \underbrace{(7 + 3)}$
   $= 11 - \quad 10 \quad = 1$

b. $-13 + 7.5 - \underbrace{(6 + 3)}$
   $= \underbrace{-13 + 7.5} - \quad 9$
   $= \quad -5.5 \quad - \quad 9 \quad = -14.5$

c. $-22.9 - \underbrace{(13.6 - 7.2)} + 21.8$
   $= \underbrace{-22.9 - \quad 6.4} \quad + 21.8$
   $= \quad -29.3 \quad + 21.8 = -7.5$

Parentheses have the highest (first) priority in evaluating problems. Multiplication and division have third priority, and addition and subtraction have the lowest (fourth) priority. (The second priority is omitted because it is reserved for *special functions*, a topic covered in Chapter 2. To summarize:

Priority	Function
1	Parentheses ( )
2	Special functions
3	$\times$, $\div$
4	$+$, $-$

Any sign preceding a parenthesis affects the entire evaluation of the parenthesis. There may be any number of parentheses throughout an expression, and parentheses may be nested; that is, one set may exist within another set. If nesting occurs, evaluations begin within the innermost parentheses and progress to the outermost parentheses.

Example 2

a. $\underbrace{(1.3 \div 2.4)}$ $\div$ $\underbrace{(5.6 \times -1.3)}$

=   0.542 $\div$ -7.28 = -0.074

b. $3.6 \times \underbrace{(1.7 - 0.62)}$ $\div$ $\underbrace{(3.9 - 2.4)}$

= $\underbrace{3.6 \times \quad\quad 1.08}$ $\div$ 1.5

=   3.888 $\div$ 1.5   = 2.592

c. $23.1 \times (\underbrace{(2.4 - 1.6)} \div \underbrace{(1.2 \times 5.7)})$

= $23.1 \times (\underbrace{\quad 0.8 \quad \div \quad 6.84 \quad})$

= $23.1 \times \quad\quad\quad 0.117 \quad\quad\quad$ = 2.702

d. $\underbrace{(12.2 \div -1.4)} \div (\underbrace{(3.7 - 1.6)} + 2.4 \times \underbrace{(1.3 \div 0.7)})$

= -8.714 $\div$ (2.1 $+ \underbrace{2.4 \times \quad\quad 1.857}$)

= -8.714 $\div$ (2.1 $+$ $\underbrace{4.457 \quad}$)

= -8.714 $\div$ 6.557   = -1.329

e. $(4.6 \times \underbrace{(-4.1 \div 2.3)} + 4.2) \div 1.4$

= $(\underbrace{4.6 \times \quad -1.783}$ $+ 4.2) \div 1.4$

= $(\underbrace{\quad -8.2 \quad\quad\quad + 4.2)} \div 1.4$

= -4 $\div 1.4 = -2.857$

PROBLEMS

As in Example 2, manually reduce Problems 1 to 7 to a single solution by:

- Applying the priority levels
- Observing polarities after each operation is performed
  between any two operands

Use your calculator only to evaluate numeric results between two operands. Later sections illustrate how the entire problem can be keyed in at once.

1. $(4.25 \times 3.1) \div (4.6 \times -1.3)$

2. $(5.2 - 1.1 \times 2.6) + (4.7 \div 2.3 - 1)$

3. $(3.1 \times (4.5 \div 2.1 + 4) \times 5.6)$

4. $4.75 + (3.1 + (4.2 + 6 \times (3.5 \div 2)))$

5. $17.8 \times ((1.4 \div 2.65) \times (5.7 \div 2.8) \div 6)$

6. $8.1 \div (3.6 \times -1.2 \div 4.5) + (1.6 \div -0.35)$

7. $20 + (1.1 \times (2.2 + 3.3 \times (4.4 \div 5.5 \div 6.6))) \div 7.7$

1.5 CALCULATOR BASICS

Today the electronic calculator has completely replaced the slide rule in the class-room. Not only is the calculator more versatile, it provides an accuracy virtually unattainable by the slide rule. However, as with the slide rule, you have to know how to use it to get accurate results. Unlike the slide rule, the calculator usually has storage facilities for later recall when solving problems. Also, inexpensive program-mable calculators can virtually duplicate the calculating ability of small computers. For now we emphasize the basic functions and how to correctly *key in* data for problems. Our previous discussions about priorities, signs, and parentheses will help you quickly and accurately evaluate problem solutions.

We assume that most scientific calculators used for standard mathematical prob-lem solving (as opposed to statistical problem solving) have at least the functions illustrated by the photographs of the Casio *fx*-80, TI-55, Sharp 5813, and HP 31E calculators in "Fundamental Concepts of the Calculator." The minimum functions are the left and right parentheses (not necessary for RPN calculators); the four basic arithmetic operators ($+$, $-$, $\times$, $\div$); roots and powers ($\sqrt[x]{y}$, y^x); the trigonometric functions (sin, cos, tan); the log functions (ln x, log x), and the inverse function (the $\boxed{\text{INV}}$ key). Additionally, most calculators have certain special-function keys such as $\boxed{\pi}$ (pi), $\boxed{\sqrt{x}}$, $\boxed{x^2}$, $\boxed{1/x}$. For electronic application, the calculator must have an $\boxed{\text{EE}}$ key (enter exponent) for quickly inserting powers of 10. All these functions are discussed in later chapters.

The ability to store ($\boxed{\text{STO}}$ key) and recall ($\boxed{\text{RCL}}$ key) data is also neces-sary for more advanced problem solving. Of the four calculators shown, only the *fx*-80 has limited storage (one location). If your calculator does not have store and recall ability, you have to manually record and key in all storage values beyond one number when needed; you cannot recall those excess values from memory. This concept is illustrated in subsequent sections requiring more than one storage location.

Before attempting problems like those in Sections 1.2 to 1.4, you must recog-nize the difference between the minus $\boxed{-}$ and the change sign $\boxed{+/-}$ keys. If you actually want to substract any quantity (regardless of sign) from another, use the minus key between the quantities:

$$17.3 \quad - \quad 6.7$$
$$\text{Minuend} \quad - \quad \text{Subtrahend}$$

PRESS	DISPLAY
17.3	17.3
$\boxed{-}$	17.3
6.7	6.7
$\boxed{=}$	10.6

In this example, both the minuend and subtrahend are positive, so the keying-in process merely subtracts one positive number from another. (*Remember*: with the RPN calculator you key in 17.3, ENTER , 6.7, −. See the Appendix for RPN examples corresponding to text examples.)

Now suppose that the subtrahend is negative:

Operator (subtract)   Sign of operand (negative)

$$17.3 \quad \underset{\downarrow}{-} \quad \overset{\downarrow}{-}6.7$$

PRESS	DISPLAY	COMMENT
17.3	17.3	Minuend
−	17.3	Minuend positive; subtract
6.7	6.7	Subtrahend
+ / −	−6.7	Subtrahend negative
=	24	Solution

Notice that you must depress + / − key after keying in 6.7. You *cannot* simply key in the following:

PRESS	DISPLAY
17.3	17.3
−	17.3
−	Foul, flashing display, abort launch, or whatever

There is no such thing as

Minuend, subtract, subtract, . . .

but there is a way of telling the calculator the following:

Minuend, subtract, subtrahend (which is negative)

This is like coming in by the back door, but it works. In short, the − sign defines the function (and priority), and the + / − defines a change in the sign of the displayed number. On RPN calculators the change sign key is often CHS .

Note too that the calculator displays the resultant accrued up to that point in the problem. This implies that calculator priorities are being followed. Implicit in this statement is that once a left parenthesis (is set, the highest priority mode is entered and the cumulating results are accruing within the parenthesis. They are terminated by a right parenthesis. The equal = key completes all operations that are pending up to that point. Example 1 illustrates these concepts.

Example 1 (see Appendix for RPN methods)

a. $9.34 \times -1.6 \div 3.2$

PRESS	DISPLAY	COMMENTS
9.34	9.34	The first term
$\boxed{\times}$	9.34	Above term now identified as multiplicand
1.6	1.6	Multiplier of above problem
$\boxed{+/-}$	−1.6	Changes sign of multiplier
$\boxed{\div}$	−14.944	Resultant of 9.34×-1.6
3.2	3.2	Divisor of above resultant
$\boxed{=}$	−4.67	Solution of entire problem

b. $-13.6 \times (3.2 + -1.65) + 2$

PRESS	DISPLAY	COMMENTS
13.6	13.6	Value of first term
$\boxed{+/-}$	−13.6	Changes sign of first term
$\boxed{\times}$	−13.6	First term becomes multiplicand
$\boxed{(}$ 3.2 $\boxed{+}$		
1.65 $\boxed{+/-}$ $\boxed{)}$	1.55	Multiplier evaluated
$\boxed{+}$	−21.08	Pending product completed; display becomes augend (a number having another number added to it)
2	2	Addend of pending sum
$\boxed{=}$	−19.08	Pending sum and problem completed

c. $\dfrac{7 \times (3 + 5) + 42}{(5 + 6 \div 3) \times 7}$

PRESS	DISPLAY	COMMENTS
$\boxed{(}$ 7 $\boxed{\times}$ $\boxed{(}$ 3 $\boxed{+}$ 5 $\boxed{)}$	8	(3 + 5) is evaluated
$\boxed{+}$	56	$7 \times (3 + 5)$ is evaluated
42 $\boxed{)}$	98	Numerator is evaluated

PRESS	DISPLAY	COMMENTS
$\div$ ((5 +		
6 $\div$ 3)	7	(5 + 6 ÷ 3) is evaluated
× 7)	49	Denominator is evaluated
=	2	Problem is completed

Note: Because the entire numerator is to be divided by the entire denominator, both the numerator and denominator must be enclosed within parentheses, although the problem does not indicate this.

d. $37.6 \div 4.19 \times (16.3 \div -1.6 + 7.8)$

PRESS	DISPLAY	COMMENTS
37.6 $\div$ 4.19 ×	8.974	Division completed; quotient, which is now the display, becomes multiplicand of a new problem
(16.3 $\div$ 1.6 +/−		
+ 7.8)	−2.388	Parentheses evaluated; display is now the multiplier
=	−21.425	Problem completed

e. $2 \times (0.2 \times (2 \times (0.2 + 2)))$

Paren 1
Paren 2
Paren 3

PRESS	DISPLAY	COMMENTS
2 × (	2	Sets up product of 2 times parentheses 3
.2 × (2 × (		
.2 + 2)	2.2	Evaluates parentheses 1
)	4.4	Evaluates parentheses 2
)	0.88	Evaluates parentheses 3
=	1.76	Completes problem, that is, 2 multiplied by parentheses 3

f. $(2.1 + (6.3 \div 1.7)) \div (3.21 \div (1.6 \times 2.3))$

PRESS	DISPLAY	COMMENTS
$[($ 2.1 $[+]$	2.1	
$[($ 6.3 $[\div]$ 1.7 $[)]$	3.706	Innermost of first nested parentheses evaluated
$[)]$	5.806	First group of nested parentheses evaluated
$[\div]$	5.806	Display now dividend of problem
$[($ 3.21 $[\div]$	3.21	
$[($ 1.6 $[\times]$ 2.3 $[)]$	3.68	Innermost of second nested parentheses evaluated
$[)]$	0.872	Second group of nested parentheses evaluated; this is now the divisor
$[=]$	6.656	Pending division completed

PROBLEMS

Using your calculator, compute the correct resultants of Problems 1 to 10.

1. $12 - (3 \times 6.7) =$

2. $(-12 + 3.6) \div (7.9 \times -3.2) =$

3. $3.1 + (2.7 \times -6.7) - (1.2 \times 3.5) =$

4. $(7.1 + 1.2 -(1.5 \times 3.2)) + 1.6 =$

5. $2.25 - ((1.6 \div 1.7) \times (3.1 + 1.6)) =$

6. $-1.17 \times ((1.23 - 4.6) \div (1.9 + 2.6)) =$

7. $-3.68 \div ((3.24 \div -1.6) + (2.84 \times -1.68)) =$

8. $(4.81 + (3.21 \div -1.67) \div 2.4) - 1.4 =$

9. $((3.57 + 1.23 \div 6.1) + 2.4) \times 3.61 - 2.7 =$

10. $-6.42 + ((-3.21 \times -4.27) \times (2.31 + 6.24)) \div 5.1 =$

1.6 SAMPLE ELECTRONIC PROBLEMS

To give you a feeling for how the preceding rules should be applied, we now perform simple exercises with calculators; these exercises will be useful when you apply them to electronic theory. It is not critical at this time to understand the theory—it is more important that you know how to apply the mathematics, so that when you later encounter the theory you will be able to easily obtain the numerical solutions.

One electronic theory concept involves parallel resistors, or, more specifically, the equivalent resistance of parallel resistors; Figure 1.1 illustrates this concept. Although the general solution involves the $\boxed{1/x}$ (reciprocal) key, we defer this approach until we discuss special-function keys. Now we consider the *product-over-sum* approach. The theorem is

> Two resistors in parallel can be replaced by one resistor whose value is equal to the product of the original two resistors divided by the sum of the original two resistors $(R_1 \times R_2)/(R_1 + R_2)$.

For example, if one resistor has a value of 3300 ohms (Ω) and the other a value of 2200 Ω, the equivalent resistance is

$$R = (3300 \times 2200)/(3300 + 2200) = 7260000/5500 = 1320 \ \Omega$$

Note that writing the problem as

$$3300 \times 2200/3300 + 2200$$

results in 4400 Ω, which is incorrect. Once again, note the value of the parentheses. In the first example the entire denominator is divided into the numerator product. In the second example, only 3300 is incorrectly divided into the numerator product and

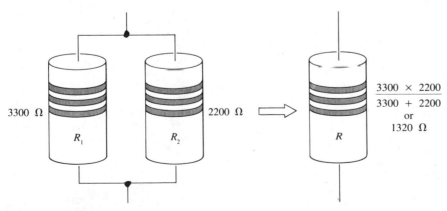

Figure 1.1 Calculations for replacing two resistors with one resistor.

then 2200 is incorrectly added to the resultant. Using your calculator, verify the following by the product-over-sum method, that is, $(R_1 \times R_2)/(R_1 + R_2)$:

R_1	in parallel with	R_2	=	R
1500		2200		891.89
4700		5600		2555.34
3300		6800		2221.78
1000		4700		824.56
10,000		6800		4047.62

Now let us expand this concept to three resistors in parallel. Figure 1.2 illustrates that one approach is to first find the equivalent resistance of any two of the parallel resistors and then consider it as one new resistance in parallel with the third resistor. As shown in Chapter 7, when algebraic manipulation and simplification are discussed, this is the same as

$$R = \frac{R_1 \times R_2 \times R_3}{R_1 \times R_2 + R_1 \times R_3 + R_2 \times R_3} \tag{1.1}$$

To illustrate the first method, consider resistor one (R_1) equal to 30 Ω, resistor two (R_2) 60 Ω, and resistor three (R_3) 80 Ω. As shown in Figure 1.2, first form a new resistance (R_4), which is the product-over-sum combination of R_1 and R_2:

$$R_4 = \frac{R_1 \times R_2}{R_1 + R_2} = \frac{30 \times 60}{30 + 60} = \frac{1800}{90} = 20$$

Next, combine the new R_4 with the old R_3 using a new product-over-sum combination; let us call this resultant R:

$$R = \frac{R_4 \times R_3}{R_4 + R_3} = \frac{20 \times 80}{20 + 80} = \frac{1600}{100} = 16 \ \Omega$$

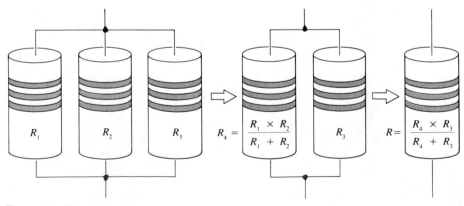

Figure 1.2 Calculations for replacing three resistors with one resistor.

Now try plugging the original three values of R_1, R_2, and R_3 into Equation (1.1):

$$R = \frac{30 \times 60 \times 80}{(30 \times 60 + 30 \times 80 + 60 \times 80)} = \frac{144,000}{9000} = 16 \ \Omega$$

What is the purpose of the parenthesis in the divisor (denominator)?

Did you get 16 Ω for the resistance when you tried the problem? Now that you are satisfied that the simple application of Equation (1.1) produces the same resultant as the two separate product-over-sum approaches, use Equation (1.1) and your calculator to verify the following (Fix 1):

R_1	R_2	R_3	Resultant
100	220	330	56.9
220	470	680	122.8
330	680	100	69.0
100	330	470	66.0
220	470	150	75.0

PROBLEMS

Using your calculator in Problems 1 to 10, compute the equivalent resistance of the following parallel combinations:

1. 1500; 3300

2. 4700; 2200

3. 2200; 1500

4. 4700; 560

5. 5600; 1500

6. 4700; 10,000

7. 330; 150; 670

8. 220; 100; 470

9. 560; 220; 330

10. 330; 1000; 220

SUMMARY

When analyzing problems that involve the four basic arithmetic functions ($+$, $-$, $\times$, and $\div$), you must assign a priority to the operator according to the following rules:

1. You must execute the times and divide operations before the plus and minus operations.
2. Within any given string of operations of equal priority, the execution sequence is from left to right.

Frequently, in a string of functions two operator signs are adjacent. The first operator sign indicates the function to be performed, and the second operator sign indicates the positive or negative value of the adjacent operand. Two adjacent signs of the same polarity can be replaced with a positive sign; two adjacent signs of opposite polarity can be replaced with a negative sign.

Parentheses have the highest priority when executing a sequence of mathematical functions. That is, the resultant of all operations enclosed within a set of parentheses is first evaluated before the remainder of the problem is undertaken. If parentheses are nested—one set of parentheses enclosed within another—the innermost sets are evaluated first. Future evaluations progress toward the outermost set of parentheses.

2 FRACTIONS AND THE RECIPROCAL FUNCTION

2.1 ADDITION AND SUBTRACTION OF FRACTIONS

There are two ways of expressing a number as a combination of an integer and some proportionate part of an integer. The first, and the one most directly used with the calculator, is the decimal format, such as 3.125. The second format is fractional notation, such as 3 1/8. We start with the traditional decimal format; later in the chapter we show how the calculator can be used to represent a number in fractional form.

The first concept to understand when adding or subtracting fractions is that of the *least common denominator* (LCD). The LCD is the simplest or smallest integer into which the denominators can be evenly divided. For example, if 1/2 is to be added to 1/3, the smallest number into which both the 2 (first denominator) and the 3 (second denominator) can be evenly divided is 6. In a simple case such as this, the LCD is merely the product of the two denominators. But in more complex cases each denominator has to be broken down into its smallest multiples; then selected portions of each of these multiples are used to find the LCD. For example, what is the sum of 5/12 and 9/28? Although a common denominator could be 336 (12 × 28), this is not the LCD because it is not the smallest or simplest term into which the denominators can be rewritten. To obtain the LCD, first break down each denominator into the product of its smallest integer parts as follows:

$$12 = 2 \times 2 \times 3$$
$$28 = 2 \times 2 \times 7$$

Now multiply the maximum number of times any integer is used in each denominator; the product is the LCD. Because the integer 2 is used twice and the integers 3 and 7 are each used once, the LCD is

$$LCD = 2 \times 2 \times 3 \times 7 = 84$$

This is certainly much smaller than the denominator of 336.

Before proceeding, note that the previous expressions can also be written as

$$12 = 2 \cdot 2 \cdot 3$$
$$28 = 2 \cdot 2 \cdot 7$$

The center dot · represents multiplication and is often used rather than the times sign to avoid confusion with the letter X. To separate the product of a series of numbers, such as $12 = 2 \cdot 2 \cdot 3$, you must use the dot or times sign so that the resultant will not look like the number 223. However, any two terms that are not both integers or fractions can be written without the center dot or times sign because no confusion will result. For example, $2 \times a$ can be written as $2a$.

The second concept is that of converting all the terms into the same common fractional unit. In other words, all terms must be changed so that they have the same denominator. When changing fractions to the same denominator, such as 1/2 or 1/3 into sixths, each term is merely being multiplied by 1 (unity). Multiplying any term by 1 does not change its value. For example:

$$\frac{3}{4} = 0.75$$

$$\frac{3}{4} \cdot \frac{5}{5} = \frac{15}{20} = 0.75$$

Notice that 5/5 is the same as 1 or unity. Multiplying the numerator and denominator of any fraction by a combination of terms that represents unity does not alter the equality of the fraction; it does make it easier to change fractions into the same denominator so that they can be added or subtracted. To illustrate, consider adding 1/2 and 1/3; 1/2 may be changed as follows:

$$\frac{1}{2} \cdot \frac{3}{3} = \frac{3}{6}$$

To change the denominator (2) into sixths, it has to be multiplied by 3. Therefore the numerator (1) also has to be multiplied by 3 to retain the equality. Similarly:

$$\frac{1}{3} \cdot \frac{2}{2} = \frac{2}{6}$$

To change the denominator (3) into sixths, it has to be multiplied by 2. Therefore the numerator (1) also has to be multiplied by 2 to retain the equality. Now that both terms have been converted into the same denominator, they can be added as follows:

$$\frac{3}{6} + \frac{2}{6} = \frac{5}{6}$$

Note that only the numerators were added, to produce a total of 5; the denominators were not added to produce a total of 12. Many students initially erroneously add both the numerators and denominators. Once all terms have been converted into the same denominator, only the numerators are added, and their total has the same denominator.

Example 1

Find the LCD of the following fractional combinations.

a. $\dfrac{5}{12}, \dfrac{5}{18}$

$12 = 2 \cdot 2 \cdot 3$

$18 = 2 \cdot 3 \cdot 3$

$LCD = 2 \cdot 2 \cdot 3 \cdot 3 = 36$

b. $\dfrac{1}{9}, \dfrac{7}{15}$

$9 = 3 \cdot 3$

$15 = 3 \cdot 5$

$LCD = 3 \cdot 3 \cdot 5 = 45$

c. $\dfrac{2}{3}, \dfrac{3}{12}, \dfrac{8}{21}$

$3 = 1 \cdot 3$

$12 = 2 \cdot 2 \cdot 3$

$21 = 3 \cdot 7$

$LCD = 1 \cdot 2 \cdot 2 \cdot 3 \cdot 7 = 84$

d. $\dfrac{5}{8}, \dfrac{4}{9}, \dfrac{7}{30}$

$8 = 2 \cdot 2 \cdot 2$

$9 = 3 \cdot 3$

$30 = 2 \cdot 3 \cdot 5$

$LCD = 2 \cdot 2 \cdot 2 \cdot 3 \cdot 3 \cdot 5 = 360$

e. $\dfrac{1}{4}, \dfrac{7}{36}, \dfrac{2}{15}, \dfrac{13}{45}$

$4 = 2 \cdot 2$

$36 = 2 \cdot 2 \cdot 3 \cdot 3$

$15 = 3 \cdot 5$

$45 = 3 \cdot 3 \cdot 5$

$LCD = 2 \cdot 2 \cdot 3 \cdot 3 \cdot 5 = 180$

Example 2

Using the LCDs of Example 1, perform the indicated operations.

a. $\dfrac{5}{12} + \dfrac{5}{18}$

To change 12ths into 36ths, both numerator and denominator must be multiplied by 3; to change 18ths into 36ths, both numerator and denominator must be multiplied by 2.

$$\therefore \frac{5}{12} \cdot \frac{3}{3} + \frac{5}{18} \cdot \frac{2}{2} = \frac{15}{36} + \frac{10}{36} = \frac{25}{36}$$

Note that the symbol $\therefore$ represents the word "therefore."

b. $\dfrac{1}{9} - \dfrac{7}{15}$

$$\frac{1}{9} \cdot \frac{5}{5} - \frac{7}{15} \cdot \frac{3}{3} = \frac{5}{45} - \frac{21}{45} = \frac{-16}{45}$$

c. $\dfrac{2}{3} + \dfrac{3}{12} - \dfrac{8}{21}$

$\dfrac{2}{3} \cdot \dfrac{28}{28} + \dfrac{3}{12} \cdot \dfrac{7}{7} - \dfrac{8}{21} \cdot \dfrac{4}{4}$

$= \dfrac{56}{84} + \dfrac{21}{84} - \dfrac{32}{84} = \dfrac{45}{84}$

d. $\dfrac{5}{8} - \dfrac{4}{9} + \dfrac{7}{30}$

$\dfrac{5}{8} \cdot \dfrac{45}{45} - \dfrac{4}{9} \cdot \dfrac{40}{40} - \dfrac{7}{30} \cdot \dfrac{12}{12}$

$= \dfrac{225}{360} - \dfrac{160}{360} + \dfrac{84}{360} = \dfrac{149}{360}$

e. $\dfrac{1}{4} - \dfrac{7}{36} - \dfrac{2}{15} + \dfrac{13}{45}$

$\dfrac{1}{4} \cdot \dfrac{45}{45} - \dfrac{7}{36} \cdot \dfrac{5}{5} - \dfrac{2}{15} \cdot \dfrac{12}{12} + \dfrac{13}{45} \cdot \dfrac{4}{4}$

$= \dfrac{45}{180} - \dfrac{35}{180} - \dfrac{24}{180} + \dfrac{52}{180} = \dfrac{38}{180}$

PROBLEMS

In Problems 1 to 10 find the LCD:

1. 12, 15

2. 6, 18, 40

3. 3, 25, 80

4. 10, 15, 45

5. 4, 7, 56

6. 12, 20, 30

7. 20, 45, 85

8. 16, 21, 72

9. 4, 12, 36, 42

10. 5, 21, 35, 52

In Problems 11 to 21, manually perform the following operations and give your answers in fractional format.

11. $\dfrac{2}{3} + \dfrac{1}{2}$

12. $\dfrac{4}{5} + \dfrac{2}{10}$

13. $\dfrac{21}{27} + \dfrac{3}{9}$

14. $\dfrac{7}{8} - \dfrac{2}{3}$

15. $\dfrac{4}{5} - \dfrac{9}{10}$

16. $\dfrac{4}{9} + \dfrac{2}{3} - \dfrac{1}{6}$

17. $\dfrac{5}{7} - \dfrac{4}{5} + \dfrac{24}{70}$ **19.** $\dfrac{13}{15} - \dfrac{5}{12} - \dfrac{2}{8}$ **21.** $\dfrac{1}{2} + \dfrac{1}{3} + \dfrac{1}{4} + \dfrac{1}{5}$

18. $\dfrac{19}{42} - \dfrac{7}{8} + \dfrac{2}{6}$ **20.** $\dfrac{6}{7} + \dfrac{4}{5} - \dfrac{26}{30}$

2.2 USING THE CALCULATOR TO DISPLAY FRACTIONS

We mentioned that the calculator displays resultants in a decimal format, which in many cases is entirely satisfactory. However, it is possible to obtain resultants in a fractional format with only a minimal amount of paperwork. All you have to know are the integers that comprise the LCD. To illustrate, find the solution to the following in both decimal and fractional formats:

$$\frac{2}{5} + \frac{7}{27} - \frac{5}{9} =$$

$$5 = 5$$
$$27 = 3 \cdot 3 \cdot 3$$
$$9 = 3 \cdot 3$$
$$\therefore \text{LCD} = 3 \cdot 3 \cdot 3 \cdot 5$$

PRESS	DISPLAY	COMMENTS
2 ÷ 5 + 7 ÷		
27 − 5 ÷ 9 =	0.1037	Answer in decimal format (Fix 4)
× [(3 ×		
3 × 3 × 5)]	135	LCD, which is the denominator of the fractional answer
=	14	Numerator of fractional answer

To test the results, divide 14 by 135 and see if 0.1037 appears. To derive the fractional answer, multiply the decimal answer (0.1037) by the LCD (135). Note that parentheses were used to separate the numerator and denominator of the answers.

Example 1

Using your calculator, find the decimal and fractional answers to the following problems. You may find it helpful to use paper and pencil to determine the integers of the LCD.

a. $\dfrac{3}{7} - \dfrac{1}{6} + \dfrac{3}{35}$

$7 = 7$

$6 = 2 \cdot 3$

$35 = 5 \cdot 7$

$\therefore \text{LCD} = 7 \cdot 2 \cdot 3 \cdot 5$

PRESS	DISPLAY	COMMENTS
3 ÷ 7 − 1 ÷		
6 + 3 ÷ 35 =	0.3476	Decimal answer
× (| 7 × 2 ×		
3 × 5 |)	210	Denominator
=	73	Numerator

b. $\dfrac{2}{21} + \dfrac{7}{9} - \dfrac{3}{45}$

$21 = 3 \cdot 7$

$9 = 3 \cdot 3$

$45 = 3 \cdot 3 \cdot 5$

$\text{LCD} = 3 \cdot 3 \cdot 5 \cdot 7$

PRESS	DISPLAY	COMMENTS
2 ÷ 21 + 7 ÷		
9 − 3 ÷ 45 =	0.8063	Decimal answer
× (| 3 × 3 ×		
5 × 7 |)	315	Denominator
=	254	Numerator

c. $\dfrac{7}{32} + \dfrac{23}{24} - \dfrac{7}{8}$

$32 = 2 \cdot 2 \cdot 2 \cdot 2 \cdot 2$

$24 = 2 \cdot 2 \cdot 2 \cdot 3$

$8 = 2 \cdot 2 \cdot 2$

$\text{LCD} = 2 \cdot 2 \cdot 2 \cdot 2 \cdot 2 \cdot 3$

PRESS	DISPLAY	COMMENTS
7 ÷ 32 + 23 ÷		
24 − 7 ÷ 8 =	0.3021	Decimal answer
× ((2 × 2 × 2		
× 2 × 2 × 3))	96	Denominator
=	29	Numerator

PROBLEMS

Using your calculator, give your answers to Problems 1 to 11 in decimal and fractional formats.

1. $\dfrac{2}{3} + \dfrac{1}{2}$ **4.** $\dfrac{7}{8} - \dfrac{2}{3}$ **7.** $\dfrac{5}{7} - \dfrac{4}{5} + \dfrac{24}{70}$ **10.** $\dfrac{6}{7} + \dfrac{4}{5} - \dfrac{26}{30}$

2. $\dfrac{4}{5} + \dfrac{2}{10}$ **5.** $\dfrac{4}{5} - \dfrac{9}{10}$ **8.** $\dfrac{19}{42} - \dfrac{7}{8} + \dfrac{2}{6}$ **11.** $\dfrac{1}{2} + \dfrac{1}{3} + \dfrac{1}{4} + \dfrac{1}{5}$

3. $\dfrac{21}{27} + \dfrac{3}{9}$ **6.** $\dfrac{4}{9} + \dfrac{2}{3} - \dfrac{1}{6}$ **9.** $\dfrac{13}{15} - \dfrac{5}{12} - \dfrac{2}{8}$

2.3 CALCULATOR RECIPROCAL KEY $\boxed{1/x}$

Frequently the numerator of a fraction has a value of 1. Two examples from electronics are

$$\frac{1}{2\pi f C}$$

or

$$\frac{1}{2\pi\sqrt{LC}}$$

The format 1/denominator occurs so frequently that it has a special name: *reciprocal*. This is also one of the special-function keys on the calculator; as such, "reciprocal" has priority two in the execution sequence. To determine the reciprocal of any number, such as 12, merely press

12 $\boxed{1/x}$ 2 keystrokes (*Note*: an entire number is considered here as 1 keystroke)

The alternate method is

$$1 \boxed{\div} \; 12 \boxed{=} \qquad \text{4 keystrokes}$$

The second method requires twice as many keystrokes, and using the divide function places the priority at level three, as opposed to priority level two when the reciprocal key is used. To illustrate, consider the parallel resistors in Chapter 1. The general formula for a number (N) of resistors is

$$R = \cfrac{1}{\cfrac{1}{R_1} + \cfrac{1}{R_2} + \; \cdots \; \cfrac{1}{R_N}} \tag{2.1}$$

The equivalent resistance (R) is equal to the reciprocal of the sum of the reciprocals of each resistor. If only two resistors are used in the calculation (the minimum number possible to be in parallel), two approaches are possible. Assume 20- and 30-Ω resistors.

Approach 1: Use the divide key.

Press: $1 \boxed{\div} \; \boxed{(} \; 1 \boxed{\div} 20 \boxed{+} \; 1 \boxed{\div} \; 30 \boxed{)} \; \boxed{=}$

Answer: 12

Keystrokes: 12

Approach 2: Use the reciprocal key.

Press: $20 \boxed{1/x} \; \boxed{+} \; 30 \; \boxed{1/x} \; \boxed{=} \; \boxed{1/x}$

Answer: 12

Keystrokes: 7

Obviously it is much easier to use the reciprocal key than to key in the "one, divided by, denominator, equal" sequence.

Example 1 ($\boxed{\text{Fix 4}}$)

a.   Evaluate $\dfrac{1}{6} + \dfrac{1}{13} - \dfrac{1}{8}$

PRESS   **DISPLAY**

$6 \boxed{1/x} \quad \boxed{+} \; 13 \; \boxed{1/x}$

$\boxed{-} \; 8 \; \boxed{1/x} \quad \boxed{=}$ 0.1186

b. Evaluate $\dfrac{1}{2 \times \left(\dfrac{1}{8} - \dfrac{1}{12}\right)}$

PRESS	DISPLAY	COMMENT
2 $\boxed{\times}$ $\boxed{(}$ 8 $\boxed{1/x}$		
$\boxed{-}$ 12 $\boxed{1/x}$ $\boxed{)}$	0.0417	Parentheses evaluated
$\boxed{=}$	0.0833	Denominator evaluated
$\boxed{1/x}$	12	1/denominator evaluated

c. Evaluate $\dfrac{\dfrac{1}{8}}{\dfrac{1}{3} + \dfrac{1}{4} - \dfrac{1}{7}}$

PRESS	DISPLAY	COMMENT
8 $\boxed{1/x}$ $\boxed{\div}$	0.125	Calculates numerator; sets up division
$\boxed{(}$ 3 $\boxed{1/x}$ $\boxed{+}$ 4		
$\boxed{1/x}$ $\boxed{-}$ 7 $\boxed{1/x}$ $\boxed{)}$	0.4405	Calculates denominator
$\boxed{=}$	0.2838	Completes pending division

d. Evaluate $\dfrac{\dfrac{1}{12} + \dfrac{1}{8}}{2 \times \left(\dfrac{1}{6} - \dfrac{1}{15}\right)}$

PRESS	DISPLAY	COMMENT
$\boxed{(}$ 12 $\boxed{1/x}$ $\boxed{+}$		
8 $\boxed{1/x}$ $\boxed{)}$ $\boxed{\div}$	0.2083	Calculates numerator; sets up division
$\boxed{(}$ 2 $\boxed{\times}$ $\boxed{(}$ 6 $\boxed{1/x}$		
$\boxed{-}$ 15 $\boxed{1/x}$ $\boxed{)}$ $\boxed{)}$	0.2	Calculates denominator
$\boxed{=}$	1.0417	Completes pending division

e. Calculate the equivalent resistance of the four parallel resistors 3300, 5600, 4700, 6800. Use the format of Equation (2.1).

PRESS   **DISPLAY**

3300 $\boxed{1/x}$ $\boxed{+}$ 5600 $\boxed{1/x}$ $\boxed{+}$

   4700 $\boxed{1/x}$ $\boxed{+}$ 6800 $\boxed{1/x}$ $\boxed{=}$ $\boxed{1/x}$ 1188.4579

PROBLEMS

Using your calculator, determine in decimal format the resultant of the following nine reciprocals:

1. $\dfrac{1}{2} + \dfrac{1}{5} - \dfrac{1}{9}$

2. $\dfrac{\dfrac{1}{3} + \dfrac{1}{7}}{\dfrac{1}{6}}$

3. $\dfrac{\dfrac{1}{9} \times \dfrac{1}{23}}{\dfrac{1}{4}}$

4. $\dfrac{\dfrac{1}{5} - \dfrac{1}{2}}{\dfrac{1}{7} + \dfrac{1}{4}}$

5. $\dfrac{\dfrac{1}{8} \times \dfrac{1}{9} - \dfrac{1}{6}}{\dfrac{1}{7}}$

6. $\dfrac{\dfrac{1}{24} + \dfrac{1}{17} - \dfrac{1}{6}}{\dfrac{1}{4} \times \dfrac{1}{5}}$

7. $\dfrac{\dfrac{1}{17} \times \dfrac{1}{19} - \dfrac{1}{15}}{\dfrac{1}{4} - \dfrac{1}{5}}$

8. $\dfrac{\dfrac{1}{4} \times \left(\dfrac{1}{6} - \dfrac{1}{3}\right)}{\left(\dfrac{1}{2} \times \dfrac{1}{3}\right) - \dfrac{1}{4}}$

9. $\dfrac{\left(\dfrac{1}{5} + \dfrac{1}{4}\right) \times \left(\dfrac{1}{6} - \dfrac{1}{7}\right)}{\left(\dfrac{1}{3} - \dfrac{1}{9}\right)}$

2.4 MULTIPLICATION OF FRACTIONS

Unlike adding and subtracting fractions, multiplying fractions does not require finding the LCD.

To multiply fractions, multiply all numerators and divide the resultant by the products of all denominators.

As examples:

$$\frac{2}{7} \cdot \frac{3}{5} = \frac{2 \cdot 3}{7 \cdot 5} = \frac{6}{35}$$

$$\frac{1}{4} \cdot \frac{3}{7} \cdot \frac{5}{9} = \frac{1 \cdot 3 \cdot 5}{4 \cdot 7 \cdot 9} = \frac{15}{252}$$

It is frequently helpful to break down any numerator or denominator into its components, such as breaking down 15 into $3 \cdot 5$. If any component of any numerator equals any component of any denominator, the equal parts can be canceled because any number divided by itself equals 1 (unity). From the previous example:

$$\frac{1}{4} \cdot \frac{3}{7} \cdot \frac{5}{9}$$

$$= \frac{1}{2 \cdot 2} \cdot \frac{\cancel{3}}{7} \cdot \frac{5}{3 \cdot \cancel{3}} = \frac{1 \cdot 5}{2 \cdot 2 \cdot 7 \cdot 3} = \frac{5}{84}$$

You can also get the same result by breaking down the final answer into components and then canceling:

$$\frac{1}{4} \cdot \frac{3}{7} \cdot \frac{5}{9} = \frac{15}{252} = \frac{\cancel{3} \cdot 5}{\cancel{3} \cdot 3 \cdot 4 \cdot 7} = \frac{5}{84}$$

The advantage of performing cancellations first and then finding the resultant is that you can work with much smaller numbers and therefore see breakdowns more easily.

Example 1

a. $\dfrac{21}{16} \cdot \dfrac{32}{7}$

$$\frac{\cancel{7} \cdot 3}{\cancel{2} \cdot \cancel{2} \cdot \cancel{2} \cdot \cancel{2}} \cdot \frac{\cancel{2} \cdot \cancel{2} \cdot \cancel{2} \cdot \cancel{2} \cdot 2}{\cancel{7}} = 3 \cdot 2 = 6$$

b. $\dfrac{4}{27} \cdot \dfrac{6}{35} \cdot \dfrac{7}{20}$

$$\frac{\cancel{2} \cdot \cancel{2}}{3 \cdot 3 \cdot \cancel{3}} \cdot \frac{\cancel{3} \cdot 2}{\cancel{7} \cdot 5} \cdot \frac{\cancel{7}}{\cancel{2} \cdot \cancel{2} \cdot 5} = \frac{2}{3 \cdot 3 \cdot 5 \cdot 5} = \frac{2}{225}$$

c. $\dfrac{6}{9} \cdot \dfrac{25}{36} \cdot \dfrac{63}{15}$

$$\frac{\cancel{6}}{\cancel{9}} \cdot \frac{5 \cdot \cancel{5}}{\cancel{6} \cdot 6} \cdot \frac{7 \cdot \cancel{9}}{3 \cdot \cancel{5}} = \frac{5 \cdot 7}{6 \cdot 3} = \frac{35}{18}$$

You can cancel without breaking down terms into their *most* elementary parts. For example, if there is a 9 in the denominator and a 63 in the numerator, then you should write 63 as 9 · 7, not 3 · 3 · 7.

d. $\dfrac{25}{8} \cdot \dfrac{3}{75} \cdot \dfrac{24}{9}$

$$\frac{\cancel{25}}{\cancel{8}} \cdot \frac{\cancel{3}}{\cancel{25} \cdot 3} \cdot \frac{\cancel{8} \cdot \cancel{3}}{\cancel{3} \cdot \cancel{3}} = \frac{1}{3}$$

A final point: When all terms of the numerators or denominators are canceled, as in d, their collective product is 1, not zero. Remember that all cancellations are replaced by 1; therefore the product of all 1s is 1.

PROBLEMS

In Problems 1 to 10, what are the products in fractional format? Simplify during the process whenever possible.

1. $\dfrac{1}{3} \cdot \dfrac{2}{6}$

2. $\dfrac{7}{8} \cdot \dfrac{16}{4}$

3. $\dfrac{3}{4} \cdot \dfrac{9}{6} \cdot \dfrac{1}{36}$

4. $\dfrac{2}{3} \cdot \dfrac{12}{6} \cdot \dfrac{18}{4}$

5. $\dfrac{8}{9} \cdot \dfrac{18}{11} \cdot \dfrac{33}{36}$

6. $\dfrac{4}{5} \cdot \dfrac{15}{9} \cdot \dfrac{10}{8}$

7. $\dfrac{1}{7} \cdot \dfrac{49}{16} \cdot \dfrac{30}{21}$

8. $\dfrac{20}{81} \cdot \dfrac{9}{14} \cdot \dfrac{21}{3}$

9. $\dfrac{6}{7} \cdot \dfrac{42}{3} \cdot \dfrac{21}{8}$

10. $\dfrac{1}{9} \cdot \dfrac{20}{42} \cdot \dfrac{30}{4}$

2.5 DIVISION OF FRACTIONS

A fraction consists of a numerator and a denominator. In our discussion so far, each numerator and denominator has been a whole number. But what happens if the numerator or denominator of a *simple* fraction is itself a fraction, consisting of its own numerator and denominator? Such fractions are called *complex* fractions. One of the simplest forms of a complex fraction is the reciprocal of a reciprocal, such as the format used in the parallel resistor equation. One example is

$$\frac{1}{1/4}$$

To avoid confusion when examples like this occur, both the slash and horizontal line can be used to represent the various layers of division. In a fairly simple complex fraction as just shown, the horizontal line separates the major numerator 1 from the major denominator 1/4. The 1/4 is denoted as a fraction by the *slash*. When the layers of fractions become extremely complex, the terms are separated by lengthening the horizontal line. For example:

$$\frac{1/7}{3/8 + 1 / (5/7)}$$

The major numerator 1/7 is divided by a major denominator that itself consists of the sum of 3/8 plus the reciprocal of 5/7. As you can see, you have to develop some intuitive reasoning to combine multilayered complex fractions. For now we consider the solution of complex fractions in their simpler form.

Let us redefine the first complex fraction into a question, such as, How many times can a quarter (one-fourth of $1) divide into $1? The answer is obviously 4. How does

$$\frac{1}{1/4}$$

translate into 4? First, consider the major numerator (an integer) as itself divided by 1. Such an assumption produces the result of 1/1 and does not change the value of the numerator. The fraction now looks like

$$\frac{1/1}{1/4}$$

Second, the denominator 1/4 is multiplied by its inverse, namely 4/1. However, any number multiplied by its inverse is unity:

$$\frac{1}{4} \cdot \frac{4}{1} = \frac{4}{4} = 1$$

Remember, to keep the equality of a fraction, the numerator must also be multiplied by 4/1. The fraction now takes the following form:

$$\frac{1/1}{1/4} \cdot \frac{4/1}{4/1} = \frac{1/1 \cdot 4/1}{1/4 \cdot 4/1} = \frac{4/1}{1} = \frac{4}{1} = 4$$

To divide a fraction by a fraction, invert the denominator fraction (divisor) and multiply it by the numerator fraction (dividend).

If either the numerator or denominator of a complex fraction is a whole number, make it a fraction by dividing it by 1. Example 1 clarifies these points.

Example 1

a. $\dfrac{3}{2/5}$

$$\frac{3/1}{2/5} = \frac{3}{1} \cdot \frac{5}{2} = \frac{15}{2}$$

b. $\dfrac{2/5}{3}$

$$\frac{2/5}{3/1} = \frac{2}{5} \cdot \frac{1}{3} = \frac{2}{15}$$

c. $\dfrac{4/7}{5/6}$

$$\frac{4}{7} \cdot \frac{6}{5} = \frac{24}{35}$$

d. $\dfrac{25/7}{15/49}$

$$\frac{25}{7} \cdot \frac{49}{15} = \frac{\cancel{5} \cdot 5}{\cancel{7}} \cdot \frac{7 \cdot \cancel{7}}{3 \cdot \cancel{5}} = \frac{35}{3}$$

e. $\dfrac{24/5}{8/35}$

$$\frac{24}{5} \cdot \frac{35}{8} = \frac{\cancel{8} \cdot 3}{\cancel{5}} \cdot \frac{7 \cdot \cancel{5}}{\cancel{8}} = 21$$

PROBLEMS

In Problems 1 to 10 determine the resultant of the divisions in fractional format.

1. $\dfrac{2}{1/5}$

2. $\dfrac{3}{6/7}$

3. $\dfrac{2/5}{4}$

4. $\dfrac{1/9}{7}$

5. $\dfrac{1/5}{2/7}$

6. $\dfrac{11/9}{3/11}$

7. $\dfrac{24/5}{15/8}$

8. $\dfrac{1/2 \cdot 1/3}{1/4}$

9. $\dfrac{1/4 \cdot 1/5}{1/6 \cdot 1/7}$

10. $\dfrac{1/4 - 1/5}{1/6 - 1/7}$

2.6 SIGNS AND MIXED NUMBERS

Just as there are rules of signs for the addition, subtraction, and multiplication of numbers, there are simple rules for fractions.

A single sign preceding a fraction can be associated with either the numerator or denominator without changing its meaning.

$$+\frac{3}{4} = \frac{+3}{4} = \frac{3}{+4}$$

$$-\frac{5}{6} = \frac{-5}{6} = \frac{5}{-6}$$

Any two negative signs associated with a fraction can be replaced with a single positive sign.

$$-\frac{-7}{8} = \frac{7}{8} \text{ is the same as } -\frac{7}{-8} = \frac{7}{8} \text{ is the same as } \frac{-7}{-8} = \frac{7}{8}$$

Example 1

a. $\dfrac{-\dfrac{3}{4}}{-\dfrac{5}{8}}$

b. $-\dfrac{\dfrac{-5}{6}}{\dfrac{3}{-4}}$

$$\frac{3}{4} \cdot \frac{8}{5} = \frac{3}{\cancel{4}} \cdot \frac{\cancel{4} \cdot 2}{5} = \frac{6}{5}$$

$$-\frac{5}{6} \cdot \frac{4}{3} = -\frac{5}{\cancel{2} \cdot 3} \cdot \frac{\cancel{2} \cdot 2}{3} = -\frac{10}{9}$$

c. $\dfrac{\dfrac{-7}{-8}}{\dfrac{-3}{4}}$

d. $-\dfrac{\dfrac{-9}{7}}{\dfrac{-3}{14}}$

$$-\frac{\dfrac{7}{8}}{\dfrac{3}{4}} = -\frac{7}{\cancel{4} \cdot 2} \cdot \frac{\cancel{4}}{3} = -\frac{7}{6}$$

$$\overset{3}{\underset{1}{\cancel{9}}} \cdot \overset{2}{\underset{1}{\cancel{14}}} \cdot \frac{6}{1} = 6$$

Note from d that you can take shortcuts. For instance, the numerator 9 was not broken into 3 · 3 because obviously the denominator 3 divides evenly into the 9. Similarly, for the numerator 14 and the denominator 7, the 7 was canceled and re-placed with a 1 because it divides evenly into the 14 twice.

Also note from Example 1 that an odd number of negative signs with a fraction results in a negative answer, but an even number of negative signs results in a posi-tive answer:

e. $\dfrac{-\dfrac{-10}{16}}{\dfrac{5}{-8}}$

f. $\dfrac{\dfrac{7}{-36}}{\dfrac{14}{-9}}$

$$-\overset{2}{\underset{2}{\cancel{10}}} \cdot \overset{1}{\underset{1}{\cancel{8}}} = -\frac{2}{2} = -1$$

$$\overset{1}{\underset{4}{\cancel{7}}} \cdot \overset{1}{\underset{2}{\cancel{9}}} = \frac{1}{8}$$

Mixed numbers such as 3 1/8 contain both an integer and a fraction. It is often useful to change a mixed number into a pure fraction. To do this, recognize that just as a whole number such as 23 means two 10s plus three 1s, 3 1/8 means 3 (which will be written as 24/8) plus 1/8, for a total of 25/8. To get the numerator 25 in

25/8, multiply the denominator 8 of the fraction 1/8 by the whole number 3. Then add the numerator 1 of the fraction to this product, for the final resultant:

$$3 \underset{\times}{\overset{+}{\longleftarrow}} \frac{1}{8} = \frac{8 \times 3 + 1}{8} = \frac{25}{8}$$

Example 2

Convert the following mixed numbers into fractions.

a. $2\dfrac{2}{7}$

$$\frac{7 \times 2 + 2}{7} = \frac{16}{7}$$

c. $4\dfrac{5}{8}$

$$\frac{8 \times 4 + 5}{8} = \frac{37}{8}$$

b. $5\dfrac{4}{9}$

$$\frac{9 \times 5 + 4}{9} = \frac{49}{9}$$

d. $11\dfrac{2}{7}$

$$\frac{7 \times 11 + 2}{7} = \frac{79}{7}$$

Example 3

Reduce the following complex fractions to simple fractions.

a. $\dfrac{3\frac{5}{8}}{\frac{2}{7}}$

$$\frac{\frac{29}{8}}{\frac{2}{7}} = \frac{29}{8} \cdot \frac{7}{2} = \frac{203}{16}$$

c. $\dfrac{4\frac{5}{8}}{7\frac{1}{4}}$

$$\frac{\frac{37}{8}}{\frac{29}{4}} = \frac{37}{\cancel{8}_2} \cdot \frac{\cancel{4}^1}{29} = \frac{37}{58}$$

e. $\dfrac{-1\frac{1}{5}}{\frac{-3}{5}}$

$$\frac{-\frac{6}{5}}{-\frac{3}{5}} = \frac{\cancel{6}^2}{\cancel{5}_1} \cdot \frac{\cancel{5}^1}{\cancel{3}_1} = \frac{2}{1} = 2$$

b. $\dfrac{\frac{7}{9}}{2\frac{1}{9}}$

$$\frac{\frac{7}{9}}{\frac{19}{9}} = \frac{7}{\cancel{9}} \cdot \frac{\cancel{9}^1}{19} = \frac{7}{19}$$

d. $\dfrac{\frac{-5}{3}}{4\frac{5}{8}}$

$$\frac{\frac{-5}{3}}{\frac{37}{8}} = -\frac{5}{3} \cdot \frac{8}{37} = -\frac{40}{111}$$

One final rule concerning fractions:

The numerator and denominator of a fraction can be multiplied or divided by the same number without the value of the fraction changing.

For example:

$$\frac{3}{5} = 0.6 \qquad \text{or} \qquad \frac{3 \cdot 4}{5 \cdot 4} = \frac{12}{20} = 0.6$$

$$\frac{25}{35} = 0.7143 \qquad \text{or} \qquad \frac{\dfrac{25}{5}}{\dfrac{35}{5}} = \frac{5}{7} = 0.7143$$

However, the same number *cannot* be *added to or subtracted from* the numerator and denominator of a fraction without changing the equality of the original fraction. For example:

$$\frac{3}{5} = 0.6 \qquad \frac{3+1}{5+1} = \frac{4}{6} = 0.6667 \neq 0.6$$

$$\frac{6}{11} = 0.5455 \qquad \frac{6-1}{11-1} = \frac{5}{10} = 0.5 \neq 0.5455$$

The symbol $\neq$ means "is not equal to."

This principle of retaining equality when multiplying or dividing both the numerator and denominator of a simple fraction by the same number permits easier manipulation of mixed numbers and/or complex fractions.

Example 4

Perform the indicated operations, and simplify wherever possible.

a.
$$\frac{3\dfrac{5}{8}}{2\dfrac{1}{4}}$$

b.
$$\frac{-5\dfrac{5}{7}}{\dfrac{15}{21}}$$

$$\frac{\dfrac{29}{8}}{\dfrac{9}{4}} = \frac{29}{\overset{}{\underset{2}{\cancel{8}}}} \cdot \frac{\overset{1}{\cancel{4}}}{9} = \frac{29}{18}$$

$$\frac{-\dfrac{40}{7}}{\dfrac{5}{7}} = -\frac{\overset{8}{\cancel{40}}}{\cancel{7}} \cdot \frac{\overset{1}{\cancel{7}}}{\underset{1}{\cancel{5}}} = -8$$

c. $\dfrac{1\frac{2}{3} \times \frac{7}{8}}{\frac{6}{8}}$

e. $\dfrac{\frac{2}{5} - 5\frac{1}{3}}{3\frac{1}{4} \times 4\frac{1}{3}}$

$$\dfrac{\frac{5}{3} \times \frac{7}{8}}{\frac{3}{4}} = \dfrac{\frac{35}{24}}{\frac{3}{4}} = \frac{35}{\underset{6}{24}} \times \frac{\overset{1}{4}}{3} = \frac{35}{18}$$

$$\dfrac{\frac{2}{5} - \frac{16}{3}}{\frac{13}{4} \times \frac{13}{3}} = \dfrac{\frac{6}{15} - \frac{80}{15}}{\frac{169}{12}} = -\frac{74}{\underset{5}{15}} \times \frac{\overset{4}{12}}{169} = \frac{-296}{845}$$

d. $\dfrac{6\frac{1}{2} - \frac{3}{4}}{2\frac{1}{3} \times \frac{1}{6}}$

$$\dfrac{\frac{13}{2} - \frac{3}{4}}{\frac{7}{3} \times \frac{1}{6}} = \dfrac{\frac{26}{4} - \frac{3}{4}}{\frac{7}{18}} = \frac{23}{\underset{2}{4}} \times \frac{\overset{9}{18}}{7} = \frac{207}{14}$$

PROBLEMS

Determine the resultants of Problems 1 to 10 in fractional formats:

1. $\dfrac{-\frac{1}{5}}{\frac{-3}{7}}$

4. $\dfrac{\frac{-7}{4}}{6\frac{1}{3}}$

2. $\dfrac{-\frac{1}{4}}{\frac{-2}{-3}}$

5. $\dfrac{-4\frac{1}{5}}{7\frac{1}{8}}$

3. $\dfrac{-3\frac{1}{8}}{\frac{-2}{3}}$

6. $\dfrac{2\frac{1}{4} + \frac{3}{2}}{\frac{3}{7}}$

7. $\dfrac{3\frac{1}{8} \times 4\frac{1}{6}}{\frac{-2}{5}}$

9. $\dfrac{8\frac{1}{4} \times 3\frac{9}{5}}{7\frac{1}{2} \times 1\frac{1}{3}}$

8. $\dfrac{\frac{-1}{3} \times 4\frac{1}{5}}{3\frac{2}{7} \times \frac{1}{2}}$

10. $\dfrac{8\frac{1}{4} - 3\frac{9}{5}}{7\frac{1}{2} + 1\frac{1}{3}}$

2.7 CALCULATOR EXAMPLES OF COMPLEX FRACTIONS

The calculator readily solves all forms of complex fractions. Be sure to use parentheses to properly group terms. Example 1 illustrates this concept.

Example 1 (⌷ Fix 4 ⌷)

Solve the following five problems using your calculator.

a. $\dfrac{2\frac{5}{8}}{1\frac{9}{11}}$

PRESS	DISPLAY	COMMENTS
⟦(2 ⌷+⌷ 5 ⌷÷⌷ 8 ⟦) ⌷÷⌷	2.625	Calculates numerator; sets up division
⟦(1 ⌷+⌷ 9 ⌷÷⌷ 11 ⟦)	1.8182	Calculates denominator
⌷=⌷	1.4438	Completes problem

b. $\dfrac{1\frac{7}{8} \times 2\frac{5}{6}}{4\frac{1}{3}}$

PRESS	DISPLAY	COMMENTS
⟦(1 ⌷+⌷ 7 ⌷÷⌷ 8 ⟦) ⌷×⌷	1.875	Calculates multiplicand of numerator
⟦(2 ⌷+⌷ 5 ⌷÷⌷ 6 ⟦)	2.8333	Calculates multiplier of numerator

PRESS	DISPLAY	COMMENTS
$\boxed{\div}$	5.3125	Completes pending product; sets up division
$\boxed{(}$ 4 $\boxed{+}$ 3 $\boxed{1/x}$ $\boxed{)}$	4.333	Calculates divisor
$\boxed{=}$	1.226	Completes problem

c. $\dfrac{\dfrac{-5}{6} \times 4\dfrac{1}{3}}{\dfrac{2}{5} \times \dfrac{-3}{4}}$

PRESS	DISPLAY	COMMENTS
$\boxed{(}$ $\boxed{(}$ 5 $\boxed{+/-}$		
$\boxed{\div}$ 6 $\boxed{)}$ $\boxed{\times}$	−0.8333	Multiplicand of numerator
$\boxed{(}$ 4 $\boxed{+}$ 3 $\boxed{1/x}$ $\boxed{)}$	4.3333	Multiplier of numerator
$\boxed{)}$ $\boxed{\div}$	−3.6111	Determines numerator product; sets up division
$\boxed{(}$ $\boxed{(}$ 2 $\boxed{\div}$ 5 $\boxed{)}$ $\boxed{\times}$	0.4	Multiplicand of denominator
$\boxed{(}$ 3 $\boxed{+/-}$ $\boxed{\div}$ 4 $\boxed{)}$	−0.75	Multiplier of denominator
$\boxed{)}$	−0.3	Determines product of denominator
$\boxed{=}$	12.037	Completes pending division

d. $\dfrac{3.17 \times \left(2\dfrac{2}{5} - 1\dfrac{7}{8}\right)}{3\dfrac{7}{9}} - 3\dfrac{1}{5}$

PRESS	DISPLAY	COMMENTS
$\boxed{(}$ 3.17 $\boxed{\times}$ $\boxed{(}$ $\boxed{(}$		
2 $\boxed{+}$ 2 $\boxed{\div}$ 5 $\boxed{)}$	2.4	
$\boxed{-}$ $\boxed{(}$ 1 $\boxed{+}$ 7 $\boxed{\div}$		
8 $\boxed{)}$ $\boxed{)}$ $\boxed{)}$ $\boxed{\div}$	1.6643	Completes calculation of numerator; sets up division
$\boxed{(}$ 3 $\boxed{+}$ 7 $\boxed{\div}$ 9 $\boxed{)}$ $\boxed{=}$	0.4405	Completes division
$\boxed{-}$ $\boxed{(}$ 3 $\boxed{+}$		
5 $\boxed{1/x}$ $\boxed{)}$ $\boxed{=}$	−2.7595	Completes problem

$$\text{e.} \quad \frac{\overbrace{\left(4\frac{1}{9} \times 3\frac{1}{3}\right)}^{\text{Paren 1}} \overbrace{\left(2\frac{1}{2} - 1\frac{1}{3}\right)}^{\text{Paren 2}}}{\underbrace{\left(5\frac{1}{6} + 3\frac{1}{7}\right)}_{\text{Paren 3}}}$$

PRESS	DISPLAY	COMMENTS
⦅ ⦅ 4 + 9 1/x ⦆		
× ⦅ 3 +		
3 1/x ⦆ ⦆ ⋏	13.7037	Paren 1; sets up product
⦅ ⦅ 2 + 2 1/x ⦆		
− ⦅ 1 +		
3 1/x ⦆ ⦆	1.1667	Paren 2
= ÷	15.9877	Completes product; sets up division
⦅ ⦅ 5 + 6 1/x ⦆		
+ ⦅ 3 +		
7 1/x ⦆ ⦆	8.3095	Calculates paren 3, the divisor
=	1.924	Completes problem

PROBLEMS

Using your calculator, determine in decimal format the answers to Problems 1 to 20.

1. $\dfrac{-\dfrac{1}{5}}{\dfrac{-3}{7}}$

2. $\dfrac{-\dfrac{1}{4}}{\dfrac{-2}{-3}}$

3. $\dfrac{-3\dfrac{1}{8}}{\dfrac{-2}{3}}$

4. $\dfrac{\dfrac{-7}{4}}{6\dfrac{1}{3}}$

5. $\dfrac{-4\frac{1}{5}}{7\frac{1}{8}}$

6. $\dfrac{2\frac{1}{4} + \frac{3}{2}}{\frac{3}{7}}$

7. $\dfrac{3\frac{1}{8} \times 4\frac{1}{6}}{\frac{-2}{5}}$

8. $\dfrac{\frac{-1}{3} \times 4\frac{1}{5}}{3\frac{2}{7} \times \frac{1}{2}}$

9. $\dfrac{8\frac{1}{4} \times 3\frac{9}{5}}{7\frac{1}{2} + 1\frac{1}{3}}$

10. $\dfrac{8\frac{1}{4} - 3\frac{9}{5}}{7\frac{1}{2} + 1\frac{1}{3}}$

11. $2.53\left(1\frac{2}{7} - \frac{3}{5}\right) + 2\frac{1}{8}$

12. $\dfrac{\left(1\frac{1}{6} \times 3\frac{1}{5}\right) - \left(2\frac{1}{7} + 5\frac{1}{8}\right)}{3\frac{7}{4}}$

13. $8.29\left(\frac{9}{5} + \frac{2}{3}\right) - 3.17\left(\frac{4}{5} - \frac{1}{7}\right)$

14. $2.55\left(\frac{1}{2} - \frac{1}{3} + \frac{1}{4}\right) + 7.15\left(2\frac{1}{3}\right)$

15. $\dfrac{\left(3\frac{1}{5} \times 2\frac{1}{7}\right) + \left(5\frac{1}{8} - 3\frac{1}{3}\right)}{3\frac{3}{4}}$

16. $\dfrac{\left(-1\frac{1}{5} + \frac{-5}{9}\right) - 3\frac{1}{2}}{-\frac{1}{3}}$

17. $\dfrac{6\frac{1}{2} - \left(7.2 \times -2\frac{1}{4}\right)}{-3\frac{1}{5}}$

18. $\dfrac{\frac{2}{7} \div 2\frac{1}{4}}{\left(5\frac{1}{2}\right)\left(1\frac{3}{5} + \frac{5}{7}\right)}$

19. $\dfrac{\left(\frac{5}{6} - 2\frac{4}{5}\right) \div 5}{1\frac{1}{5}}$

20. $\dfrac{2.1\left(\frac{5}{8} - \frac{3}{4}\right) + \frac{7}{8}}{5\frac{1}{8} - 6\frac{3}{16}}$

SUMMARY

To add or subtract fractions, first determine the least common denominator (LCD). Then convert all the fractions to this LCD. Next, add or subtract the numerators of the fractions. Simplify the resultant further if necessary.

To multiply fractions, multiply all numerators and then divide the resultant by the product of all denominators. To divide a fraction by another fraction, multiply the numerator by the inverse of the denominator and then multiply the fractions. Convert any mixed numbers into fractions before performing any of the four basic arithmetic functions on them.

The reciprocal key on a calculator, identified as $\boxed{1/x}$, implies a ratio whose numerator is 1 and whose denominator is the number for which the reciprocal is sought. To find the reciprocal of a number, it is easier to use the reciprocal key rather than the "one, divided by, number, equal" sequence.

You can use the calculator to display a sequence of operations on fractions in fractional rather than decimal form by

1.   Keying in the fractional problem as presented
2.   Multiplying this decimal resultant by the LCD

If the LCD is within parentheses, the fractional denominator is displayed when the parentheses are closed; the fractional numerator is displayed when the $\boxed{=}$ key is depressed.

3 POWERS AND ROOTS

3.1 THE FOUR LAWS OF EXPONENTS

It is more convenient to represent a number repeatedly multiplied by itself as a number raised to a power (also called exponent or superscript) rather than as a series of individual multiplications. For example, if the integer 2 is raised to the fifth power, which means four discrete multiplication processes, either of the following representations is appropriate:

$$2^5 \equiv 2 \cdot 2 \cdot 2 \cdot 2 \cdot 2 \qquad \equiv \text{ means "is the same as," "identical with"}$$

Other examples include

$$4^6 \equiv 4 \cdot 4 \cdot 4 \cdot 4 \cdot 4 \cdot 4$$
$$10^3 \equiv 10 \cdot 10 \cdot 10$$
$$9.6^2 \equiv 9.6 \cdot 9.6$$

Thus, the superscript next to the base number indicates the number of times the number is being multiplied by itself. When we discuss the $\boxed{y^x}$ function of a calculator, the power does not have to be an integer. The generalized form of an exponential representation is

$$a^m$$

where a = base number
   m = exponent to which a is raised

To multiply or divide the *same* base numbers with different exponents, raise a number with a power to another power, or find the root of a number raised to a power, use the four *laws of exponents*. To analyze the first law—multiplication—consider the following problem:

$$2^3 \cdot 2^2 = 8 \cdot 4 = 32$$

But 32 is also 2^5; therefore

$$2^3 \cdot 2^2 = 2^{3+2} = 2^5 = 32$$

To multiply the same base numbers raised to different powers, add the exponents.

$$a^m \cdot a^n = a^{m+n} \tag{3.1}$$

This rule applies only to numbers with the *same* base. Thus the following is an incorrect application of the rule:

$$a^m \cdot b^n \neq (a \cdot b)^{m+n}$$

Example 1

Perform the indicated operations.

a. $4^5 \cdot 4^2$
    $4^{5+2} = 4^7$

b. $3^2 \cdot 3^1 \cdot 3^4$
    $3^{2+1+4} = 3^7$

c. $6^{1.1} \cdot 6^{2.4}$
    $6^{1.1+2.4} = 6^{3.5}$

d. $4.2^{1.8} \cdot 4.2^{3.6}$
    $4.2^{1.8+3.6} = 4.2^{5.4}$

e. $\pi^{1.5} \cdot \pi^{2.3} \cdot \pi^{0.6}$
    $\pi^{1.5+2.3+0.6} = \pi^{4.4}$

There is a similar analogy for dividing the same base numbers with different exponents:

$$\frac{3^5}{3^2} = \frac{243}{9} = 27$$

But 27 is also 3^3; therefore

$$\frac{3^5}{3^2} = 3^5 \times 3^{-2} = 3^{5-2} = 3^3 = 27$$

To divide the same base numbers raised to powers, subtract the exponent of the denominator from the exponent of the numerator.

$$\frac{a^m}{a^n} = a^{m-n} \tag{3.2}$$

Example 2

Perform the indicated operations.

a. $\dfrac{5^3}{5^2}$
    $5^{3-2} = 5^1$

b. $\dfrac{4.1^7}{4.1^3}$
    $4.1^{7-3} = 4.1^4$

c. $\dfrac{3^{6.2}}{3^{1.5}}$
    $3^{6.2-1.5} = 3^{4.7}$

d. $\dfrac{8.21^{1.67}}{8.21^{0.93}}$
    $8.21^{1.67-0.93} = 8.21^{0.74}$

e. $\dfrac{\pi^{3.17}}{\pi^{1.62}}$

$\pi^{3.17-1.62} = \pi^{1.55}$

g. $\dfrac{6.7^{2.8}}{6.7^{-3.2}}$

$6.7^{2.8-(-3.2)} = 6.7^6$

f. $\dfrac{4.5^{-1.2}}{4.5^{2.6}}$

$4.5^{-1.2-(2.6)} = 4.5^{-3.8}$

The third law involves the power of a base number with an exponent raised to another power. For example,

$$(2^3)^2 = (8)^2 = 64$$

Because 64 is also 2^6, it follows that

$$(2^3)^2 = 2^{3 \cdot 2} = 2^6 = 64$$

To raise a base number with an exponent to another power, multiply the exponents.

$$(a^m)^n = a^{m \cdot n} \tag{3.3}$$

Example 3

Perform the indicated operations.

a. $(3^2)^5$
$3^{2 \cdot 5} = 3^{10}$

d. $((3^2)^3)^4$
$3^{2 \cdot 3 \cdot 4} = 3^{24}$

b. $(4^{1.5})^3$
$4^{(1.5) \cdot 3} = 4^{4.5}$

e. $((\pi^{1.1})^{0.6})^{2.3}$
$\pi^{(1.1) \cdot (0.6) \cdot (2.3)} = \pi^{1.518}$

c. $(1.6^{2.4})^{1.5}$
$1.6^{(2.4) \cdot (1.5)} = 1.6^{3.6}$

The fourth law of exponents concerns roots, but first let us investigate the exponent 1. In the following decreasing progression of powers, note that each resultant is one-half its former value.

$$2^4 = 16$$

Decrease power by 1	$2^3 = 8$	one-half previous value
Decrease power by 1	$2^2 = 4$	one-half previous value
Decrease power by 1	$2^1 = 2$	one-half previous value

(If the base number is 3, each resultant is one-third its former value; if the base number is 4, each resultant is one-fourth its former value; and so on.)

$$3^4 = 81 \qquad 4^4 = 256$$
$$3^3 = 27 \qquad 4^3 = 64$$
$$3^2 = 9 \qquad 4^2 = 16$$
$$3^1 = 3 \qquad 4^1 = 4$$

Thus, any number with an exponent of 1 ultimately equals itself. This concept is important when dealing with roots, as explained in a moment.

The root of a number is a number that when multiplied by itself the number of times indicated by the root is equal to the original number. For example, the *square* root of a number is a number that when multiplied by itself equals the original number. The *cube* root of a number is a number that when multiplied by itself three times equals the original number. The *fourth* root requires four multiplications, and so on. In mathematical notation,

$$\sqrt[n]{\text{number}}$$

where n indicates the degree (index) of the root; that is, it indicates the number of times the resultant number must be multiplied by itself to equal the original number. The $\sqrt{}$ symbol is called the radical sign, and the number within the radical sign is often called the radicand.

Example 4

a. $3 \cdot 3 = 9$

$\therefore \sqrt[2]{9} = 3$

b. $2 \cdot 2 \cdot 2 = 8$

$\therefore \sqrt[3]{8} = 2$

c. $5 \cdot 5 \cdot 5 \cdot 5 = 625$

$\therefore \sqrt[4]{625} = 5$

If the $\sqrt{}$ symbol does *not* specify a value for n, n is assumed to be the positive square root, or $\sqrt[2]{}$.

How do you extract the square root of an exponential number such as $\sqrt{2^6}$?

$$\sqrt{2^6} = \sqrt{64} = 8$$

Because 8 is really 2^3, we can use a simpler format:

$$\sqrt{2^6} = 2^{6/2} = 2^3 = 8$$

To obtain the root of a base number raised to a power, divide the exponent under the radical sign by the index of the root.

$$\sqrt[n]{a^m} = a^{m/n} \qquad\qquad (3.4)$$

Example 5

Perform the indicated operations.

a. $\sqrt{3^5}$

   $3^{5/2} = 3^{2.5}$

b. $\sqrt[3]{4^{1.5}}$

   $4^{1.5/3} = 4^{0.5}$ Note that $4^{0.5} = 4^{1/2} = \sqrt[2]{4^1} = 2$

c. $\sqrt[9]{\pi^{8.2}}$

   $\pi^{8.2/9} = \pi^{0.911}$ Exponent to $\boxed{\text{Fix 3}}$

In Example 4a, how do you apply Equation (3.4) when the radicand does not have a power? Remember that any number with an exponent 1 is merely itself; Example 4a can thus be rewritten as

$$\sqrt[2]{9^1}$$

When applying Equation (3.4), the resultant is

$$9^{1/2}$$

Thus,

   To obtain the root of a number that has no exponent, write the number with a power that is the reciprocal of the root.

Example 6

Express the following radical formats as fractional and decimal exponents.

a. $\sqrt[3]{10}$

   $10^{1/3} = 10^{0.333}$ Exponent to $\boxed{\text{Fix 3}}$

b. $\sqrt[4]{7}$

   $7^{1/4} = 7^{0.25}$

c. $\sqrt[2]{5^3}$

   $5^{3/2} = 5^{1.5}$

d. $\sqrt[3]{4^{2.4}}$

   $4^{2.4/3} = 4^{0.8}$

e. $\sqrt[5]{\pi^3}$

   $\pi^{3/5} = \pi^{0.6}$

Example 7

Evaluate the following exponential formats in radical sign format.

a. $6^{1/2}$

$\sqrt[2]{6}$

b. $3^{1/4}$

$\sqrt[4]{3}$

c. $7^{2/7}$

$\sqrt[7]{7^2}$

d. $9^{1.5/4}$

$\sqrt[4]{9^{1.5}}$

e. $\pi^{6/7}$

$\sqrt[7]{\pi^6}$

Example 8

Applying all four laws of exponents, evaluate the following (where appropriate, exponents are set at $\boxed{\text{Fix 3}}$):

a. $\dfrac{2^3 \cdot 2^5}{2^2}$

$2^{3+5-2} = 2^6$

b. $\dfrac{3^{11.6} \cdot 3^2}{(3^2)^3}$

$\dfrac{3^{11.6+2}}{3^6} = 3^{11.6+2-6} = 3^{7.6}$

c. $\dfrac{(4.3^{1.5})^{2.4}}{\sqrt[3]{4.3^2}}$

$\dfrac{4.3^{(1.5) \cdot (2.4)}}{4.3^{.667}} = 4.3^{3.6-0.667} = 4.3^{2.933}$

d. $\dfrac{(\pi^{2.4})^{1.9}}{\pi^{4.2} \cdot \pi^{0.15}}$

$\dfrac{\pi^{(2.4) \cdot (1.9)}}{\pi^{4.2+0.15}} = \dfrac{\pi^{4.56}}{\pi^{4.35}} = \pi^{4.56-4.35} = \pi^{0.21}$

PROBLEMS

Solve Problems 1 to 14 by applying the four laws of exponents.

1. $3^2 \cdot 3^3 \cdot 3$

2. $\dfrac{4^6}{4^3}$

3. $\dfrac{5^4 \cdot 5^2}{5}$

4. $(2.3^{1.6})^2$

5. $(\pi^{2.1} \cdot \pi^{3.9})^2$

6. $\sqrt[3]{1.8^2} \cdot 1.8^2$

7. $\sqrt{2.4^{1.5}} \cdot \sqrt{2.4^3}$

8. $\dfrac{(8.9^{1/2})^3}{\sqrt{8.9}}$

9. $\left(\dfrac{1.5^{1.5}}{1.5}\right)^2$

10. $\sqrt[4]{2.8^3} \cdot \sqrt{2.8^3} \cdot \sqrt{2.8}$

11. $((1.25^2)^3)^4$

12. $\dfrac{(\sqrt{7.75})^3}{(\sqrt[3]{7.75})^2}$

13. $\dfrac{6.24^{2/3} \cdot 6.24^{5/2}}{\sqrt{6.24^3}}$

14. $\sqrt{\dfrac{\pi^{2.1}}{\pi^{1.3}}}$

In Problems 15 to 24 replace the question marks by a power that will give the indicated solution.

15. $3^? \cdot 3^6 = 3^{12}$

16. $\dfrac{4^5 \cdot 4^3}{4^?} = 4^6$

17. $(3^5 \cdot 3^2)^? = 3^{14}$

18. $(2^5 \cdot 2^7 \cdot 2^?)^2 = 2^{26}$

19. $(\sqrt{0.02})(0.02)^3(0.02)^? = (0.02)^4$

20. $\left(\dfrac{9.8^5}{9.8^?}\right)^3 = 9.8^9$

21. $\left(\sqrt{\dfrac{\pi^5}{\pi^3}}\right)^? = \pi^3$

22. $\dfrac{(\sqrt{26})^3}{(\sqrt[4]{26})^?} = 26$

23. $\sqrt{3.4} \cdot \sqrt[3]{3.4} \cdot \sqrt[?]{(3.4)^2} = 3.4$

24. $((0.5)^?)^4 = 0.5^{24}$

3.2 THE EXPONENT ZERO

We discussed the exponent 1 in Section 3.1. To show the end result of progressively decreasing exponents until they reach a value of zero, observe the following (recall that each decreasing evaluation is 1/base number of its former value).

$$
\begin{array}{lll}
2^4 = 16 & 3^4 = 81 & 4^4 = 256 \\
2^3 = 8 & 3^3 = 27 & 4^3 = 64 \\
2^2 = 4 & 3^2 = 9 & 4^2 - 16 \\
2^1 - 2 & 3^1 = 3 & 4^1 = 4 \\
2^0 = 1 & 3^0 = 1 & 4^0 = 1
\end{array}
$$

Regardless of the base, all evaluations ultimately reach the same value—1—when the exponent is zero. Thus,

Any number other than 0 raised to the power of zero is 1.

3.3 NEGATIVE POWERS

So far, all powers have been expressed as nonnegative numbers. What happens to evaluations when the decreasing progression of powers continues past zero, that is, to a negative number? If each decreasing evaluation equals 1/base number of its former value, a continuation of the chart in Section 3.2 is

$$
\begin{array}{lll}
2^2 = 4 & 3^2 = 9 & 4^2 = 16 \\
2^1 = 2 & 3^1 = 3 & 4^1 = 4 \\
2^0 = 1 & 3^0 = 1 & 4^0 = 1 \\
2^{-1} = 1/2 & 3^{-1} = 1/3 & 4^{-1} = 1/4
\end{array}
$$

All evaluate as 1
All evaluate as 1/base

Generally, the following results if any base number (a) is raised to any negative power (m):

$$a^{-m} = \frac{1}{a^m} \tag{3.5}$$

To prove this, rewrite a^{-m} as $a^{-m}/1$. Dividing by 1 does not change its value. Now multiply both the numerator and denominator by the same base (a) raised to the same positive exponent (m):

$$\frac{a^{-m}}{1} \cdot \frac{a^m}{a^m} = \frac{a^{m+(-m)}}{a^m} = \frac{a^0}{a^m} = \frac{1}{a^m}$$

The same reasoning applies if the exponent occurs in the denominator rather than the numerator:

$$\frac{1}{a^{-m}} = \frac{1}{a^{-m}} \cdot \frac{a^m}{a^m} = \frac{a^m}{a^{m+(-m)}} = \frac{a^m}{a^0} = \frac{a^m}{1} = a^m$$

We now have a simple rule for negative exponents:

Make a negative exponent positive (or vice versa) by moving it down from the numerator to the denominator (or up from the denominator to the numerator).

Example 1

Change all negative exponents to positive exponents. Use the four laws of exponents to combine terms where appropriate.

a. $3^{-2} \cdot 4^3$

$$\frac{4^3}{3^2}$$

b. $10^{-4} \cdot 6^{-2}$

$$\frac{1}{10^4 \cdot 6^2}$$

c. $\dfrac{5^{-3}}{2^{-6}}$

$$\frac{2^6}{5^3}$$

d. $\dfrac{\sqrt[3]{7^{-2}}}{6^{-1} \cdot 6^{-4}}$

$$\frac{7^{-2/3}}{6^{-5}} = \frac{6^5}{7^{2/3}}$$

e. $\dfrac{2^{1/2} \cdot 3^{-1/4}}{2^{-3/2}}$

$$\frac{2^{1/2} \cdot 2^{3/2}}{3^{1/4}} = \frac{2^{1/2+3/2}}{3^{1/4}} = \frac{2^2}{3^{1/4}}$$

f. $\dfrac{(4^{-1/3})^{4/2} \cdot 2^{-1/5}}{4^{-3}}$

$$\frac{4^{-2/3} \cdot 4^3}{2^{1/5}} = \frac{4^{3+(-2/3)}}{2^{1/5}} = \frac{4^{7/3}}{2^{1/5}}$$

PROBLEMS

Simplify Problems 1 to 8 where appropriate, and change all negative exponents to positive exponents.

1. 7.1^{-1}

2. $3.4^{-2} \cdot 3.4^{-1.5}$

3. $\dfrac{1}{6.7^{-2}}$

4. $\dfrac{5^{-2} \cdot 5^{-4}}{5^{-1}}$

5. $\left(\dfrac{6.5}{6.5^{-1}}\right)^{-2}$

6. $(2.5^{-1})^{-2} \cdot (2.5^{-2})^{-1}$

7. $\left(\dfrac{1.75^{-2}}{1.75^3}\right)^{-1.5}$

8. $\dfrac{3.6^{-2} \cdot 4.7^{-3}}{4.7^2 \cdot 3.6^{-3}}$

3.4 ADDITION AND SUBTRACTION OF EXPONENTS

We discussed combining exponents of the same base number when the operators separating them were of priority three (multiplication and division) or two (special functions like powers and roots). But what happens when the operator is priority four (addition or subtraction)? Are there rules that apply to the following types of problems?

1. $4^2 + 4^3$
2. $3^4 - 3^{-2}$
3. $5^2 + 4^{-3}$

None of the four laws of exponents can be applied to the first two examples because the laws do not relate to addition or subtraction, and they certainly cannot relate to the third example because the base numbers are different. The only way to handle such problems is to first evaluate each individual term and then perform the indicated operation.

Example 1

Determine the numerical evaluation of the following ($\boxed{\text{Fix 3}}$).

a. $4^2 + 4^3$

$$16 + 64 = 80$$

b. $3^2 - 3^{-2}$

$$9 - \frac{1}{3^2} = 9 - \frac{1}{9} = \frac{81}{9} - \frac{1}{9} = \frac{80}{9} = 8.889$$

c. $5^2 + 4^{-3}$

$$25 + \frac{1}{4^3} = 25 + \frac{1}{64} = \frac{1601}{64} = 25.016$$

d. $7^{-2} \cdot 3^2$

$$\frac{1}{7^2} \cdot 9 = \frac{1}{49} \cdot \frac{9}{1} = \frac{9}{49} = 0.184$$

e. $\dfrac{3^{-2}}{4^{-4}}$

$$\frac{4^4}{3^2} = \frac{256}{9} = 28.444$$

PROBLEMS

Evaluate Problems 1 to 10.

1. $3^2 + 3^5 - 3^4$

2. $5^{-2} + 5$

3. $4^{-2} - 4^{-1}$

4. $1.5^2 + 3^{-1}$

5. $6.5^3 + 5^{-1}$

6. $4^{-2} \cdot 3^2$

7. $6^{-2} \cdot 5^3$

8. $\left(\dfrac{4^3}{3^3}\right)^2$

9. $\dfrac{2.5^{-2}}{1.5^{-2}}$

10. $\dfrac{3^{-2} - 3^{-3}}{3}$

3.5 $\boxed{y^x}$ AND $\boxed{\sqrt[x]{y}}$ KEYS AND THEIR FUNCTIONS

These two keys generally raise any base number to any power (y^x) or find the nth root ($\sqrt[x]{y}$) of any base number. Depending upon the manufacturer, $\boxed{y^x}$ might be designated as $\boxed{x^y}$, and $\boxed{\sqrt[x]{y}}$ might be designated as $\boxed{x^{1/y}}$. For simplicity, $\boxed{y^x}$ used in the book also implies $\boxed{x^y}$; similarly, $\boxed{\sqrt[x]{y}}$ used here also implies $\boxed{x^{1/y}}$.

To avoid confusion, note that the key $\boxed{y^x}$ represents a generalized base raised to any power; the $\boxed{\sqrt[x]{y}}$ key represents the root to which that base is extracted. However, most scientific calculators have certain keys that operate on special values of bases: the $\boxed{e^x}$ and $\boxed{10^x}$ keys. The ϵ of the $\boxed{e^x}$ key is epsilon (which to $\boxed{\text{Fix 3}}$ has a value of 2.718). The 10 in the $\boxed{10^x}$ key is the base of the decimal numbering system; the $\boxed{10^x}$ key raises 10 to any power x. Both the keys (and their inverse) are discussed in more detail in Chapter 19, "Logarithms." In the following examples we learn how to manipulate the calculator keys $\boxed{y^x}$ (or $\boxed{x^y}$) and $\boxed{\sqrt[x]{y}}$ (or $\boxed{x^{1/y}}$) for many types of problems involving base numbers raised to powers or the extraction of the roots of base numbers.

The $\boxed{y^x}$ and $\boxed{\sqrt[x]{y}}$ keys are of priority two and require two entries: the values of both x and y. Remember, these special-function entries react immediately upon the number in the display register and are different from single-entry special functions like $\boxed{\sqrt{x}}$.

Example 1 ($\boxed{\text{Fix 3}}$)

Determine the resultant of the following problems using the calculator.

a. $3.2^{1.6}$

PRESS	DISPLAY
3.2 $\boxed{y^x}$ 1.6 $\boxed{=}$	6.430

b. $\sqrt[3]{1.84}$

PRESS	DISPLAY
1.84 $\boxed{\sqrt[x]{y}}$ 3 $\boxed{=}$	1.225

c. $\sqrt[4]{3.84^{2.61}}$

PRESS	DISPLAY
3.84 $\boxed{y^x}$ 2.61 $\boxed{\sqrt[x]{y}}$ 4 $\boxed{=}$	2.406

d. $8.1^{1.62} + \sqrt{3.4}$

PRESS	DISPLAY
8.1 $\boxed{y^x}$ 1.62 $\boxed{+}$ 3.4 $\boxed{\sqrt{}}$ $\boxed{=}$	31.475

e. $(2.32^{1.2} - 2.64^{0.87})^{1.35}$

PRESS	DISPLAY
$\boxed{(}$ 2.32 $\boxed{y^x}$ 1.2 $\boxed{-}$ 2.64 $\boxed{y^x}$ .87 $\boxed{)}$ $\boxed{y^x}$ 1.35 $\boxed{=}$	0.308

$$\text{f.} \quad \overbrace{\frac{6.4^{3/7}}{1.6^{2/5}}}^{\text{Paren 1}} \underset{\text{Paren 2}}{} - \overbrace{\frac{1.4^{1/3}}{3.7^{1/4}}}^{\text{Paren 3}}_{\text{Paren 4}}$$

PRESS	DISPLAY	COMMENTS
$($ 6.4 y^x $($		
3 $\div$ 7 $)$ $)$ $\div$	2.216	Paren 1; sets up division
$($ 1.6 y^x $($ 2		
$\div$ 5 $)$ $)$	1.207	Paren 2
$-$	1.836	Completes division; sets up display as minuend
$($ 1.4 y^x $($ 3		
$1/x$ $)$ $)$ $\div$	1.119	Paren 3; sets up division
$($ 3.7 y^x $($ 4		
$1/x$ $)$ $)$	1.387	Paren 4
$=$	1.029	Completes pending division and problem

Example 1f illustrates fractional exponents. To evaluate these exponents, an extra set of parentheses was used to isolate this evaluation from the pending y^x function. Consider a value such as

$$6.4^{3/7} \equiv \sqrt[7]{6.4^3}$$

This can also be evaluated by pressing

6.4 $\boxed{y^x}$ 3 $\boxed{\sqrt[x]{y}}$ 7

This is a much shorter sequence. Therefore, if we are not concerned with the value of *each* parenthesis, Example 1f can be shortened to the following form:

PRESS	DISPLAY	COMMENT
6.4 y^x 3 $\sqrt[x]{y}$ 7 $\div$		
1.6 x^2 $\sqrt[x]{y}$ 5 $-$	1.836	Entire minuend evaluated
1.4 y^x 3 $1/x$ $\div$		
3.7 y^x 4 $1/x$ $=$	1.029	Completes problem

PROBLEMS

Using your calculator, evaluate Problems 1 to 18.

1. $1.45^{3.2}$

2. $2.81^{1.7} + 1.2^{3.4}$

3. $\sqrt[4]{4.25}$

4. $\sqrt[3]{8.42^2}$

5. $\dfrac{2.86^{1.5}}{2.1^{0.4}}$

6. $(6.14^{-1.2})^2$

7. $\dfrac{14.7^{1/3}}{6.4^{-1}}$

8. $\sqrt[3]{10.2^2} - \sqrt[2]{10.2^3}$

9. $\sqrt{42.3^{-1}} + \sqrt[3]{16.1^{-2}}$

10. $\left(\dfrac{17.4^{1.2}}{13.3^{2.1}}\right)^2$

11. $\dfrac{3.1^{2/3}}{4.6^{3/4}} - \sqrt{6.8}$

12. $\dfrac{1}{3.4^{1.3}} - \dfrac{1}{4.6^{1.5}}$

13. $\dfrac{1}{\dfrac{1}{4.6^{1.2}}} + \sqrt[3]{7.1}$

14. $\dfrac{3.8^{-1.2}}{0.76^{1.5}} + \dfrac{2.5^{3.1}}{4^{-2}}$

15. $\left(\dfrac{1.52^{1.1}}{0.43^{-2}}\right)^{2.1} - \sqrt[4]{6.8}$

16. $\left(\dfrac{2.5^{1.5}}{3.6^{-2.1}}\right)^{3.4} + \dfrac{1.2^{-2}}{0.056}$

17. $\left(\dfrac{\sqrt{4.7^3}}{2.5^{1.4}}\right)^{-0.85}$

18. $\left(\dfrac{1}{\dfrac{\sqrt{16.5}}{3.5^{1.2}}}\right)^{2.1} - \left(\dfrac{1}{\dfrac{\sqrt[3]{13.2}}{4.2}}\right)^{1.5}$

SUMMARY

A superscript indicates the number of times a base number is used when multiplied by itself. There are four laws for exponential operations.

Rule 1. To multiply the same base numbers with different exponents, add the exponents.

Rule 2. To divide the same base numbers with different exponents, subtract the exponent of the denominator from the exponent of the numerator.

Rule 3. To raise a base number with a superscript to another power, multiply the exponents.

Rule 4. To find the root of a base number with a superscript, divide the exponent under the radical sign by the index of the root.

The root of a number is a number that when multiplied by itself the number of times indicated by the index is the original number. In fractional power, the numerator indicates the power to which the base number is raised, and the denominator indicates the root of the base number.

Any number raised to the zero power is 1. The exception is zero raised to the zero power, which is mathematically undefined. Any number raised to the power 1 is itself.

A negative exponent can be changed to a positive exponent (or vice versa) by changing its location between numerator and denominator.

To add or subtract exponentials, evaluate each term and then perform the indicated operation(s).

The primary keys used on a calculator to determine the power and root of a number are $\boxed{y^x}$ and $\boxed{\sqrt[x]{y}}$, respectively.

4 SCIENTIFIC NOTATION, SYSTEMS, AND CONVERSIONS

4.1 SCIENTIFIC NOTATION

Daily we manipulate numbers, as we average test scores, balance our checking accounts, pay for and receive change from purchases, and perform various other mathematical manipulations. Most of these activities involve a range of numbers that we can easily work with. However, in many fields the magnitude of the numbers being handled is beyond the average. For example, a typical number in astronomy is 5,879,000,000,000 miles (mi) which is the approximate distance light travels in 1 year (a light-year). Electronics is another field that involves large (or small) numbers. Frequency, which is measured in units called *hertz* (Hz), might look like 160,250,000 Hz.

A system of notation—called *scientific notation*—has been developed to simplify the reading and writing of numbers of very large magnitudes. Scientific notation involves the base 10 and reduces a very large number to a more manageable one. To convert a number greater than 1 to scientific notation:

Move the decimal point to the left until the value of the number is between 1 and 10. Multiply this resultant by 10 raised to a positive power equal to the number of places the decimal point was shifted left.

Thus, for 5,879,000,000,000 mi, move the decimal point (consider the decimal point the end of a number) to the left until the value represents a number between 1 and 10. In our light-year example, we move left 12 places to get a number between 1 and 10: 5.879:

$$\overleftarrow{5{,}879{,}000{,}000{,}000.}$$

Now multiply the resultant 5.879 by 10 raised to a power that is the number of places the decimal point was shifted. Because the decimal point moved 12 positions to the left, the power is 12: 10^{12}. Scientific notation is therefore 5.879×10^{12}. Similarly, scientific notation for the frequency example 160,250,000 Hz is 1.6025×10^{8}.

Example 1

Rewrite each of the following numbers in scientific notation.

a. 3,256
    3.256×10^{3}

b. 1,750,000
    1.75×10^{6}

c. 2,115,000,000 d. 42.7
   2.115×10^9 4.27×10^1

Numbers can also be extremely minute. For example, *capacitance* is measured in in a unit called the *farad* (F). A typical capacitor value is

$$0.00000000015 \text{ F}$$

Another example is the charge of an electron expressed in *coulombs* (C). The electron charge is written as, for example,

$$0.00000000000000000016 \text{ C}$$

Very small numbers such as these can be just as difficult to handle as very large numbers. To write a number whose value is less than 1 in scientific notation:

Move the decimal point to the right until the value of the number is between 1 and 10. Multiply this resultant by 10 raised to a negative power equal to the number of places the decimal point was shifted right.

In the 0.00000000015 F example, move the decimal to the right until the value represents a number between 1 and 10; the decimal point thus must be shifted 10 places. To retain the original value of the number, now multiply the resultant 1.5 times 10 raised to a negative power that is the number of places the decimal point was shifted. Because the decimal point moved 10 positions to the right, the power is -10: 10^{-10}. Scientific notation is thus 1.5×10^{-10}. Likewise, the charge of 0.00000000000000000016 C is 1.6×10^{-19} in scientific notation.

Example 2

Rewrite each of the following numbers in scientific notation.

a. 0.00126 c. 0.0000025
   1.26×10^{-3} 2.5×10^{-6}

b. 0.169 d. 0.000200
   1.69×10^{-1} 2×10^{-4}

To illustrate further how much easier it is to read scientific notation, can you recognize which of the following two numbers is larger:

$$0.00000000000123 \quad \text{or} \quad 0.000000000000567$$

At first it appears that the significant digits of the second number, 567, are larger than the significant digits of the former number, 123. However, when expressed in scientific notation, the two numbers are 1.23×10^{-12} and 5.67×10^{-13}. The power of 10 of the second number indicates that the decimal point is actually farther to the left of the first significant digit than is the decimal point in the second number; hence the first number is actually the larger of the two.

PROBLEMS

Rewrite Problems 1 to 12 as numbers between 1 and 10 times the appropriate power of 10.

1.	0.000156	**7.**	12
2.	0.000000235	**8.**	12.4
3.	1685	**9.**	0.1
4.	0.152	**10.**	10
5.	3,650,000	**11.**	100
6.	0.0123	**12.**	10.1

4.2 PREFIXES AND ENGINEERING NOTATION

A prefix is frequently used with scientific notation to represent powers of 10 measured in units of 3 (just as commas are used after every third zero in very large numbers). Refer to Table 4.1; note that, starting with the smallest prefix (pico) and progressing up toward the largest (tera), each successive prefix is 1000 (10^3) times larger than its predecessor. Because 1000 represents three 0s, it follows from Table 4.1 that all exponents of 10 assigned a prefix are divisible by 3. (There are four prefixes assigned to powers of 10 not divisible by 3, namely, 10^{-2}, 10^{-1}, 10^1, and 10^2. The prefixes for 10 raised to the power ± 1 are called deca and deci, respectively, and hecto and centi, respectively, for the power ± 2.) Thus powers such as 10^{-5} and 10^8 do not have prefixes.

Numbers with the prefixes in Table 4.1 are written in *engineering notation,* which differs from scientific notation in that the power of 10 in scientific notation is the number of places the decimal is shifted to convert a number to between 1 and 10,

TABLE 4.1   DECIMAL MULTIPLES

Number	Power of 10	Prefix	Abbreviation
1,000,000,000,000	10^{12}	tera	T
1,000,000,000	10^9	giga	G
1,000,000	10^6	mega	M
1,000	10^3	kilo	k or K
1	10^0	. . .	. . .
0.001	10^{-3}	milli	m
0.000001	10^{-6}	micro	μ
0.000000001	10^{-9}	nano	n
0.000000000001	10^{-12}	pico	p

whereas in engineering notation the decimal is adjusted to between 1 and 1000 so that the power of 10 is divisible by 3. In other words, to convert to engineering notation:

Move the decimal so the resulting number lies between 1 and 1000 and the number of places the decimal moves is divisible by 3.

We stress this difference between the notations because calculators have specific keys for setting the display into the desired notation. These features are investigated in Section 4.4.

Example 1

Convert the following numbers into engineering notation, that is, into a number between 1 and 1000 with the appropriate prefix. (*Note:* The units of measurement are electronic terms.)

a. 1,625,000 ohms (Ω)
$1.625 \times 10^6 \ \Omega$
= 1.625 megohms (MΩ)

b. 32,700 Hz
32.7×10^3 Hz
= 32.7 kilohertz (kHz)

c. 0.0152 henry (H)
15.2×10^{-3} H
= 15.2 millihenries (mH)

d. 0.000000000125 F
125×10^{-12} F
= 125 picofarad (pF)

PROBLEMS

Express Problems 1 to 10 as numbers between 1 and 1000 and use the appropriate prefixes.

1. 1,875,000 Hz

2. 0.000050 F

3. 0.0025 H

4. 0.00000007 ampere (A)

5. 1400 watts (W)

6. 1000 Ω

7. 1,000,000 Ω **9.** 11,100 W

8. 0.1 A   **10.** 0.0001 H

4.3 APPLYING THE LAWS OF EXPONENTS TO SCIENTIFIC NOTATION

Chapter 3 illustrated the laws of exponents when base numbers raised to powers were separated by operators. In scientific notation, coefficients are normally intermixed with base 10 numbers that have exponents. For example, a common electronics expression is

$$2 \cdot \pi \cdot f \cdot L$$

If f is $5 \cdot 10^6$ and L is $0.2 \cdot 10^{-3}$, the expression becomes

$$2 \cdot \pi \cdot (5 \cdot 10^6)(0.2 \cdot 10^{-3})$$

Here is the rule for solving such a problem:

To perform operations on coefficients and base 10 numbers with exponents, separate the coefficients from the powers of 10. Apply the indicated operations to the coefficients and the laws of exponents to the powers of 10.

Example 1

Perform the indicated operations on the following powers of 10. The operators in Example 1 n to p have priority four ($+$ or $-$), so the four laws of exponents are not applicable. Express all your answers in scientific notation.

a. $3 \cdot 10^3 \cdot 4 \cdot 10^{-1}$
$(3 \cdot 4)(10^3 \cdot 10^{-1})$
$= 12 \cdot 10^2 = 1.2 \cdot 10^3$

b. $4.1 \cdot 10^9 \cdot 3 \cdot 10^{-9}$
$(4.1 \cdot 3)(10^9 \cdot 10^{-9})$
$= 12.3 \cdot 10^0 = 12.3 = 1.23 \cdot 10^1$

c. $8 \cdot 10^{-5} \cdot 4 \cdot 10^{-4}$
$(8 \cdot 4)(10^{-5} \cdot 10^{-4})$
$= 32 \cdot 10^{-9} = 3.2 \cdot 10^{-8}$

d. $\dfrac{4 \cdot 10^{-2}}{2 \cdot 10^3}$
$\left(\dfrac{4}{2}\right)\left(\dfrac{10^{-2}}{10^3}\right) = 2 \cdot 10^{-2-3} = 2 \cdot 10^{-5}$

e. $(5 \cdot 10^2)^3$
$(5)^3 (10^2)^3 = 125 \cdot 10^{2 \cdot 3} = 125 \cdot 10^6$
$= 1.25 \cdot 10^8$

f. $\sqrt[2]{64 \cdot 10^{-8}}$
$\sqrt[2]{64} \cdot 10^{-8/2} = 8 \cdot 10^{-4}$

g. $\dfrac{12 \cdot 10^3 \cdot 6 \cdot 10^{-2}}{3 \cdot 10^4}$
$\left(\dfrac{12 \cdot 6}{3}\right)\left(\dfrac{10^3 \cdot 10^{-2}}{10^4}\right)$
$= 24 \cdot 10^{3+(-2)+(-4)} = 24 \cdot 10^{-3} = 2.4 \cdot 10^{-2}$

h. $\sqrt[2]{3^3 \cdot 10^4}$
$(3^3)^{1/2} \cdot 10^{4/2} = 27^{1/2} \cdot 10^2$
$= 100 \sqrt{27} = 519.615 = 5.19615 \cdot 10^2$

i. $(2^2 \cdot 2^4)^3$
$(2^6)^3 = 2^{18} = 2.62144 \cdot 10^5$

j. $\sqrt[2]{(1000)^2 \cdot 10^2}$
$\sqrt[2]{10^6 \cdot 10^2} = \sqrt[2]{10^8}$
$= 10^4 = 1 \cdot 10^4$

k. $(3 \cdot 10^2 \cdot 4 \cdot 10^3)^2$
$(3 \cdot 4)^2 (10^2 \cdot 10^3)^2$
$= (12)^2 \times (10^5)^2 = 144 \cdot 10^{10} = 1.44 \cdot 10^{12}$

l. $\sqrt[4]{17^2 \cdot 10^3}$
$17^{2/4} \cdot 10^{3/4} = 23.1859 = 2.31859 \cdot 10^1$

m. $\sqrt[4]{0.125^3 \cdot 10^8}$
$0.125^{3/4} \cdot 10^{8/4}$
$= 0.125^{3/4} \cdot 10^2 = 21.0224 = 2.10224 \cdot 10^1$

n. $14 \cdot 10^2 + 3 \cdot 10^3$
$1.4 \cdot 10^3 + 3 \cdot 10^3 = 4.2 \cdot 10^3$

o. $6 \cdot 10^{-3} - 241 \cdot 10^{-5}$
$6 \cdot 10^{-3} - 2.41 \cdot 10^{-3} = 3.59 \cdot 10^{-3}$

p. $5.21 \cdot 10^4 - 17,800 + 0.63 \cdot 10^5$
$5.21 \cdot 10^4 - 1.78 \cdot 10^4 + 6.3 \cdot 10^4 = 9.73 \cdot 10^4$

PROBLEMS

In Problems 1 to 16, perform the indicated operations and express your answers in scientific notation.

1. $2 \cdot 10^2 \cdot 6.1 \cdot 10^4$

2. $3.2 \cdot 10^{-2} \cdot 4.6 \cdot 10^{-5} \cdot 1.7$

3. $\dfrac{3.94 \cdot 10^4}{1.76 \cdot 10^{-2}}$

4. $\dfrac{(4.78 \cdot 10^6)(8.62 \cdot 10^{-3})}{3.1 \cdot 10^{-4}}$

5. $\dfrac{10{,}000 \cdot 6.2 \cdot 10^{-2}}{3.1 \cdot 10^6}$

6. $\left(\dfrac{0.0025 \cdot 10^3}{1.61 \cdot 10^{-2}}\right)^2$

7. $(89.6 \cdot 10^6)^{1/2}$

8. $\left(\dfrac{23.4 \cdot 10^{-2}}{4.6}\right)^{1/2}$

9. $(0.000008)^3$

10. $((0.000064)^{1/2})^{1/3}$

11. $((1000)^3 \cdot 10^{-9})^{1/2}$

12. $\dfrac{0.125^{1/3} \cdot 60 \cdot 10^{-3}}{30 \cdot 10^{-3}}$

13. $((0.0002)^3)^2$

14. $2 \cdot 10^3 + 5 \cdot 10^4$

15. $4100 \cdot 10^{-4} \cdot 0.0034 \cdot 10^3$

16. $13.1 \cdot 10^2 + 1432 + 98{,}000 \cdot 10^{-2}$

4.4 CALCULATOR EE AND ENG KEYS

Many calculators do not display more than 8 (sometimes 10) digits, which is a problem because electronic terms are often larger or smaller. For example, suppose you are keying in values for an equation whose magnitudes are similar to those of an electron's charge. After the first eight zeros, the display is filled, but there are still many more digits to go. To avoid this situation, the enter exponent (EE) key enters and displays in scientific notation. [On some calculators this is the exponent (EXP) key.]

Example 1

Key in the following values in scientific notation. Note the second set of double digits following the EE function.

a. 5.879×10^{12}

PRESS	DISPLAY	COMMENT
5.879	5.879	Significant digits of number
EE	5.879 00	Second set of digits for powers of 10
12	5.879 12	Represents 5.879×10^{12}

b. -32.4×10^{14}

PRESS	DISPLAY	COMMENT
32.4	32.4	Significant digits of number
+/−	−32.4	Changes sign
EE	−32.4 00	Second set of digits for power of 10
14	−32.4 14	Represents -32.4×10^{14}

c. 1.6×10^{-19}

PRESS	DISPLAY	COMMENT
1.6	1.6	Significant digits of number
EE	1.6 00	Second set of digits for power of 10
+/−	1.6 −00	Exponent is negative
19	1.6 −19	Represents 1.6×10^{-19}

d. Repeat part c and observe the order of entry.

PRESS	DISPLAY	COMMENT
1.6	1.6	Significant digits of number
$\boxed{EE}$	1.6 00	Second set of digits for power of 10
19	1.6 19	Sets number as 1.6×10^{19}
$\boxed{+/-}$	1.6 −19	Changes number to 1.6×10^{-19}

The significance of parts c and d is that once the $\boxed{EE}$ key is depressed, it does not matter whether the negative sign of the exponent is declared before or after the actual value of the exponent is entered. However, it is recommended that you acquire the habit of following format d; that is, enter negative exponents into the calculator in this order.

Significant digits, $\boxed{EE}$ *exponent, make exponent negative*
rather than
Significant digits, $\boxed{EE}$ *make exponent negative, exponent*

The reason is that many calculators, when using the $\boxed{y^x}$ function, do not calculate the correct result if the power is expressed in the latter sequence. Therefore, when exponents are negative, enter the sign *after* entering the value, to ensure less chance of error. To test this premise, try this simple power problem:

Evaluate 2^{-3}

Method 1

Press: 2 $\boxed{y^x}$ $\boxed{+/-}$ 3 $\boxed{=}$
Result: 8 (incorrect)

Method 2

Press: 2 $\boxed{y^x}$ 3 $\boxed{+/-}$ $\boxed{=}$
Result: 0.125 (correct)

It should now be apparent why the second method is preferable.

Example 2

Perform the following operations using the $\boxed{EE}$ or $\boxed{EXP}$ key ($\boxed{Fix\ 4}$).

a. $\dfrac{8.9 \times 6.02 \times 10^{23}}{63.5}$ calculates number of free electrons in 1 cubic centimeter (cm^3) of copper

PRESS	DISPLAY	COMMENT
8.9 $\boxed{\times}$ 6.02 $\boxed{\text{EE}}$ 23 $\boxed{\div}$	5.3578 24	Calculates numerator; sets up division
63.5 $\boxed{=}$	8.4375 22	Completes problem; answer = 8.4375×10^{22}

b. $\dfrac{9 \times 10^9 (1.6 \times 10^{-19})^2}{(0.529 \times 10^{-10})^2}$ calculates force of attraction between electron and proton of hydrogen atom

PRESS	DISPLAY	COMMENT
9 $\boxed{\text{EE}}$ 9 $\boxed{\times}$ 1.6 $\boxed{\text{EE}}$		
19 $\boxed{+/-}$ $\boxed{x^2}$ $\boxed{\div}$	2.3040 −28	Calculates numerator; sets up division
.529 $\boxed{\text{EE}}$ 10 $\boxed{+/-}$		
$\boxed{x^2}$ $\boxed{=}$	8.2332 −08	Answer = 8.2332×10^{-8}

c. $1.6 \times 10^{-19} \left(\dfrac{9 \times 10^9}{125 \times 10^{-9}} \right)^{1/2}$ calculates distance in meters (m) between proton and electron for attractive force of 125×10^{-9} newtons (N)

PRESS	DISPLAY
1.6 $\boxed{\text{EE}}$ 19 $\boxed{+/-}$ $\boxed{\times}$ $\boxed{(}$ 9 $\boxed{\text{EE}}$ 9 $\boxed{\div}$	
125 $\boxed{\text{EE}}$ 9 $\boxed{+/-}$ $\boxed{)}$ $\boxed{\sqrt{}}$ $\boxed{=}$	4.2933 −11

The $\boxed{\text{ENG}}$ key is a modification of the $\boxed{\text{EE}}$; its only function is to change scientific notation into engineering notation. The $\boxed{\text{ENG}}$ key automatically shifts the decimal point so the resultant power of 10 is divisible by 3. This in turn usually permits either replacing the power of 10 with a new unit or replacing the base unit with one containing a prefix. For example, it is easier to say 124 dollars rather than 124,000 mils.

Example 3

Rework the problems in Exercise 2, with the following exception. After you key in the problem, depress the $\boxed{\text{ENG}}$ key to shift into engineering notation. Notice the display differences.

a. $\dfrac{8.9 \times 6.02 \times 10^{23}}{63.5}$

PRESS	DISPLAY	COMMENT
8.9 $\boxed{\times}$ 6.02 $\boxed{EE}$ 23 $\boxed{\div}$		
63.5 $\boxed{=}$ $\boxed{ENG}$	84.375 21	Answer in engineering notation

b. $\dfrac{9 \times 10^9 (1.6 \times 10^{-19})^2}{(0.529 \times 10^{-10})^2}$

PRESS	DISPLAY	COMMENT
9 $\boxed{EE}$ 9 $\boxed{\times}$ 1.6 $\boxed{EE}$ 19		
$\boxed{+/-}$ $\boxed{x^2}$ $\boxed{\div}$ .529 $\boxed{EE}$		
10 $\boxed{+/-}$ $\boxed{x^2}$ $\boxed{=}$ $\boxed{ENG}$	82.3325 −09	Answer in engineering notation

c. $1.6 \times 10^{-19} \times \left(\dfrac{9 \times 10^9}{125 \times 10^{-9}} \right)^{1/2}$

PRESS	DISPLAY	COMMENT
1.6 $\boxed{EE}$ 19 $\boxed{+/-}$ $\boxed{\times}$ $\boxed{(}$ 9		
$\boxed{EE}$ 9 $\boxed{\div}$ 125 $\boxed{EE}$ 9		
$\boxed{+/-}$ $\boxed{)}$ $\boxed{\sqrt{}}$ $\boxed{=}$ $\boxed{ENG}$	42.9325 −12	Answer in engineering notation

PROBLEMS

Using your calculator, work out the following power of 10 problems. Set to engineering notation and $\boxed{Fix\ 3}$. If your calculator does not have engineering notation, work out the problem in enter exponent notation and manually change to engineering notation.

1. $\dfrac{2.45 \times 10^{-3}}{4.68 \times 10^{12}}$

2. $(13.7 \times 10^{-4})^2 - (63.2 \times 10^{-5})^2$

3. $\dfrac{(18.4 \times 10^{-3})(16.3 \times 10^2)}{(6.49 \times 10^{-12})^{1/2}}$

4. $\dfrac{(\sqrt{21.45} \times 10^{13})^{1/2}}{3.7 \times 10^7 (12.4 \times 10^{-2})^2}$

5. $\dfrac{(3.1^{1.5} \times 10^{-4})^2}{1.7^{2.5} \times 10^3} \times 10^{-5}$

6. $\dfrac{(2.5 \times 10^{-3})^{1.2}}{(4.7 \times 10^{-5})^{2.4}}$

7. $\dfrac{(13.2^3 \times 10^{11})^{1/7}}{3.4 \times 10^{-5} - 4.2 \times 10^{-6}}$

9. $\dfrac{(6.32 \times 10^{12})^{1.3}}{(7.1 \times 10^{10})^{1.4}} - \dfrac{(13.3 \times 10^{13})^{1.5}}{(8.9 \times 10^{11})^{1.6}}$

8. $\dfrac{12.32^{1.5} \times 10^{-11}}{2.16^{2.5} \times 10^{-13}} - (10^{14})^{1/8}$

4.5 UNIT SYSTEMS

You have encountered units in systems of measurement other than the English system. For example, the speedometer of many foreign cars is calibrated in kilometers per hour as opposed to American cars which are calibrated in miles per hour. The name of the former system of measure is the *metric system,* of which there are two versions: MKS (meter, kilogram, and second) and CGS (centimeter, gram, and second). The major difference between the two versions is only the scale of magnitude of each measurable unit. The MKS system's values of units of measurement are more closely aligned to the English system than are those in the CGS system. Generally, the CGS version is better adapted to the measurement of very small quantities than is the MKS version. However, the CGS system possesses the same advantages as MKS in that its units can be easily converted to multiples of 10.

Table 4.2 is the basic measurable units in the three systems. The only unit common to the three systems is the unit of time, namely, the second. Table 4.3 is a relative comparison of the magnitudes of different measurable units in the MKS and English systems.

The rationale for the growing popularity of the metric system is obvious if you consider the awkwardness of numbers in the English system versus those in the metric system. For example, as you advance up the scale of length in the English system, it takes 12 inches to equal a foot, 3 feet to equal a yard, and 1760 yards to equal a mile. Such conversion numbers are not at all convenient. But in the metric system, 10 millimeters equals 1 centimeter, 10 centimeters equals 1 decimeter, 10 decimeters equals 1 meter and so on. And even the names in the metric system are more rational. The prefix *milli* implies $1/1000$; the prefix *centi* implies $1/100$. The same reasoning applies to the other prefixes in the metric system, such as deci and kilo. No wonder most of the major industrial countries have already gone to the metric system; current legislation in the United States calls for a gradual transition to the metric system over the next decade(s).

TABLE 4.2 UNIT SYSTEMS

Dimension	English	MKS	CGS
Length	foot (ft)	meter (m)	centimeter (cm)
Mass	slug (32.3 lb)	kilogram (kg)	gram (g)
Time	second (s)	second (s)	second (s)
Force	pound (lb)	newton (N)	dyne (dyn)
Energy		joule (J)	erg
(force × distance)	(ft · lb)	(N · m)	(dyn · cm)

TABLE 4.3 COMPARISON OF METRIC AND ENGLISH UNITS

English Unit	=	MKS Unit
1 in		0.0254 m or 2.54 cm
1 ft		0.3048 m or 30.48 cm
1 yd		0.9144 m or 91.44 cm
1 mi		1.6094 km
1 lb		$\begin{cases} 0.4536 \text{ kg} \\ 4.45 \text{ N} \end{cases}$
1 slug (32.2 lb)		14.6 kg
1 ft^2		0.0929 m^2
1 ft^3		0.0283 m^3
1 gal		3.7853 L
1 qt		0.9463 L

MKS Unit	=	English Unit
1 m		$\begin{cases} 39.37 \text{ in} \\ 3.2808 \text{ ft} \\ 1.0936 \text{ yd} \end{cases}$
1 km		0.62137 mi
1 kg		2.2046 lb
1 m^2		10.7629 ft^2
1 m^3		35.3144 ft^3
1 L		$\begin{cases} 0.2642 \text{ gal} \\ 1.0567 \text{ qt} \end{cases}$

Miscellaneous Conversions
1 N (MKS force) = 10^5 dyn (CGS force)
1 J (MKS energy) = 10^7 erg (CGS energy)
Temperature °F = (%)°C + 32°   °F = degrees Fahrenheit
Temperature °C = (%)(°F − 32°)   °C = degrees Celsius

Note: Prior to 1948, the Celsius temperature scale was called "centigrade." In 1948, the Ninth General Conference on Weights and Measurements changed the name to Celsius, in honor of the scientist instrumental in developing the scale.

4.6 UNIT CONVERSIONS

Here we consider two types of conversions. The first involves changes of prefix. For example, it is easier to read and write 65 pF rather than 0.000065 μF, yet in both cases, capacitance is still expressed within the basic unit of the farad. Only the prefix preceding farad is changed, to make the number easier to read and express verbally (65 picofarads as opposed to 65 millionths of a microfarad).

The key to conversions is so simple that it is frequently overlooked. It involves nothing more than multiplying the unit to be converted by 1, that is, by the equivalent of 1 (unity) rather than numeric 1. Any ratio in which the numerator and denominator are equal equates to 1. This basic truth is also valid for measurable units such as length and time and purely dimensionless numbers. For example, because there are 1 million pF (the smaller unit) in 1 μF (the larger unit), then the ratio of 10^6 pF to 1 μF must be 1:

$$\frac{10^6 \text{ pF}}{1 \ \mu\text{F}} = 1$$

Expanding this fact to change $0.000065 \ \mu F$ to pF produces the following:

$$0.000065 \ \mu F \times 1$$

is the same as
$$0.000065 \ \mu F \times \frac{10^6 \ pF}{1 \ \mu F} = 65 \ pF$$

The key to determining the equivalent of 1, that is, the conversion ratio, is to place the unit desired in the numerator and the unit to be changed in the denominator. As an example, to change milliamperes to microamperes, the conversion factor is

$$\frac{10^3 \ \mu A}{1 \ mA} \quad \begin{array}{l} \text{unit desired in numerator} \\ \text{unit to be changed in denominator} \end{array}$$

Table 4.4 lists several common conversions that illustrate this basic approach.

TABLE 4.4   CONVERSIONS OF PREFIXES

Changing from	to	Multiply by
micro (10^{-6})	milli (10^{-3})	$\dfrac{1 \ milli}{10^3 \ micro}$
micro (10^{-6})	unity	$\dfrac{1}{10^6 \ micro}$
milli (10^{-3})	micro (10^{-6})	$\dfrac{10^3 \ micro}{1 \ milli}$
milli (10^{-3})	unity	$\dfrac{1}{10^3 \ milli}$
milli (10^{-3})	kilo (10^3)	$\dfrac{1 \ kilo}{10^6 \ milli}$
unity	mega	$\dfrac{1 \ mega}{10^6}$
unity	kilo	$\dfrac{1 \ kilo}{10^3}$
unity	milli	$\dfrac{10^3 \ milli}{1}$
unity	micro	$\dfrac{10^6 \ micro}{1}$

Example 1

a.   Convert $0.0176 \ mA$ to μA.
$$0.0176 \ mA \times \frac{10^3 \ \mu A}{1 \ mA} = 0.0176 \times 10^3 \ \mu A = 17.6 \ \mu A$$

b.   Convert 1,560,000 Hz to kHz.
$$1{,}560{,}000 \ Hz \times \frac{1 \ kHz}{10^3 \ Hz} = 1560 \ kHz$$

c. Convert 0.000024 H to μH.

$$0.000024 \text{ H} \times \frac{10^6 \ \mu\text{H}}{1 \text{ H}} = 24 \ \mu\text{H}$$

d. Convert 1,500,000 Ω to MΩ.

$$1,500,000 \ \Omega \times \frac{1 \text{ M}\Omega}{10^6 \ \Omega} = 1.5 \text{ M}\Omega$$

e. Convert 0.000000150 A to nanoamperes (nA).

$$0.000000150 \text{ A} \times \frac{10^9 \text{ nA}}{1 \text{ A}} = 150 \text{ nA}$$

The second type of conversion involves changes between systems, such as changing the unit of length in the English system (mille) to the unit of length in the MKS system (kilometer). At this point it is important to understand that all units in the final answer must exist within the *same* system. For example, to measure energy in a unit called newton-mile is not acceptable because a newton is associated with the MKS system and the mile is the basic unit of length in the English system.

As with prefix conversions, the coefficients of the units to be changed are multiplied by the equivalent of 1. Table 4.3 is useful for obtaining conversions between systems.

Example 2

a. Convert 60 feet per second (ft/s) to millimeters per minute (mm/min).

$$\frac{60 \text{ ft}}{\text{s}} \times \frac{60 \text{ s}}{1 \text{ min}} = \frac{3600 \text{ ft}}{\text{min}} \qquad \text{step 1}$$

$$\frac{3600 \text{ ft}}{\text{min}} \times \frac{12 \text{ in}}{\text{ft}} = \frac{43,200 \text{ in}}{\text{min}} \qquad \text{step 2}$$

$$\frac{43,200 \text{ in}}{\text{min}} \times \frac{2.54 \text{ cm}}{\text{in}} = \frac{109,728 \text{ cm}}{\text{min}} \qquad \text{step 3}$$

$$\frac{109,728 \text{ cm}}{\text{min}} \times \frac{10 \text{ mm}}{1 \text{ cm}} = \frac{1.09728 \times 10^6 \text{ mm}}{\text{min}} \qquad \text{step 4}$$

In actual practice, all these steps are performed as one instead of four distinct operations:

$$\frac{60 \text{ ft}}{\text{s}} \times \frac{60 \text{ s}}{\text{min}} \times \frac{12 \text{ in}}{\text{ft}} \times \frac{2.54 \text{ cm}}{\text{in}} \times \frac{10 \text{ mm}}{\text{cm}} = \frac{1.09728 \times 10^6 \text{ mm}}{\text{min}}$$

$$\approx \frac{1.1 \times 10^6 \text{ mm}}{\text{min}} \qquad \approx \text{ means "approximately equal to"}$$

b. Convert 65 miles per hour (mi/h) to kilometers per hour (km/h).

$$\frac{65 \text{ mi}}{h} \times \frac{1.609 \text{ km}}{\text{mi}} = \frac{104.6 \text{ km}}{h}$$

c. Convert 0.85 gallon (gal) to liters (L).

$$0.85 \text{ gal} \times \frac{3.7853 \text{ L}}{1 \text{ gal}} = 3.218 \text{ L}$$

d. Convert 2.85 cubic yards (yd³) to cubic meters (m³).

$$2.85 \text{ yd}^3 \times \frac{27 \text{ ft}^3}{1 \text{ yd}^3} \times \frac{0.0283 \text{ m}^3}{\text{ft}^3} = 2.178 \text{ m}^3$$

PROBLEMS

In Problems 1 to 6, perform prefix conversions.

1. 0.000015 F = _____ μF = _____ pF

2. 1,650,000 Ω _____ kΩ = _____ MΩ

3. 85.6 kHz = _____ Hz = _____ MHz

4. 1635 MHz = _____ GHz = _____ kHz

5. 4650 nA = _____ μA = _____ mA

6. 0.015 mH = _____ H = _____ μH

In Problems 7 to 13, perform system conversions.

7. 60 mi/h = _____ km/h

8. 42 yd/min = _____ cm/s

9. 3 ft² = _____ cm²

10. 2.5 ft³ = _____ m³

11. 2.5 gal = _____ L

12. 8.4 yd = _____ decimeters (dm)

13.   245 yd/min = _____ cm/h

14.   The basic equation for the linear distance transversed by a moving object is $s = vt$, where s = distance, v = velocity, and t = time. What is the distance traveled in meters in 0.15 s if an object travels 10 mi/min?

15.   A rectangular object covers 6.25 ft^2. If one side is 0.85 m long, what is the length of the other side in meters?

SUMMARY

Scientific notation simplifies the writing and recognition of very large or very small numbers. To express a large number as a value between 1 and 10, move the decimal to the left until this value is reached. Multiply the resultant by 10 raised to a positive power equal to the number of places the decimal shifted left. To express a small number as a value between 1 and 10, move the decimal to the right until this value is reached. Multiply the resultant by 10 raised to a negative power equal to the number of places the decimal shifted right.

Prefixes also express numerical magnitudes. If a number can be written with a power of 10 equal to a prefix notation, this notation may be substituted for the power of 10.

When mathematically manipulating numbers expressed in scientific notation, the laws of exponents are utilized, most frequently when the base number is 10, although any base numbering system can be used.

Calculators are ideally suited for working with powers of 10 via the EE key. The ENG key changes the answer from scientific notation into engineering notation.

It is often necessary to convert unit prefixes or units between different systems of measurement. Multiply the undesired unit by the "equivalent" of unity, which should be such that when forming the product, the unwanted terms cancel out and are replaced with the desired terms.

REVIEW PROBLEMS, PART I

The following problems are a review of Part I of the text and directly relate to the numbered sections.

Section 1.2
Calculate Review Problems 1 to 5.

1. $7.3 + 2.5 - 6.7$

2. $25.6 - 906.4 + -125.9$

3. $6,251 - 42,352 + 73,005 - 2,599$

4. $0.952 - +0.062 - -2.06 + -1.321$

5. $56.1 - +43.1 + -66.07 - 1.033$

Section 1.3
Calculate Review Problems 6 to 10.

6. $36.2 \div 17.1 + 12.9 \times 5.6$

7. $-42.5 \times -1.24 + -34.9 \div 6.7$

8. $-6.21 + 0.62 \times 3.1 + -4.2$

9. $0.0061 - 0.3426 \times -0.9231$

10. $45,672 + -59,842 \div 140 + 5,807$

Sections 1.4, 1.5
Calculate Review Problems 11 to 15.

11. $(-25 + 2.6) \div (3.9 \times -36.1)$

12. $(9.2 + 4.7 - (3.7 \times 6.8)) + 19.3$

13. $-6.02 \div ((3.47 \div -1.5) + (3.67 \times -2.98))$

14. $((2.62 + 1.98 \div 3.2) + 6.2) \times 2.71 - 3.4$

15. $(721 - 53 \times 15) - (245 \times 26 - 153)$

Section 1.6

In Review Problems 16 to 21 compute the equivalent resistance of the parallel combinations.

16. 1700; 3600

17. 2500; 18,000

18. 3300; 10,000

19. 220; 3300; 1000

20. 2200; 4700; 1500

21. 420; 500; 560

Sections 2.1, 2.2

In Problems 22 to 29 find the LCD for the denominators.

22.   9, 16, 12

23.   3, 5, 20

24.   4, 9, 12, 15

25.   25, 45, 15

26.   3, 8, 20

27.   63, 14, 28

28.   6, 15, 7

29.   4, 5, 7

Evaluate Review Problems 30 to 36 and give your answers in fractional form.

30. $\dfrac{1}{5} + \dfrac{3}{4}$

31. $\dfrac{14}{39} + \dfrac{2}{3}$

32. $\dfrac{4}{5} + \dfrac{1}{3} - \dfrac{2}{9}$

33. $\dfrac{5}{7} + \dfrac{3}{5} - \dfrac{7}{15}$

34. $\dfrac{1}{3} + \dfrac{2}{7} + \dfrac{3}{4} + \dfrac{1}{2}$

35. $\dfrac{13}{42} - \dfrac{3}{8} + \dfrac{1}{4}$

36. $\dfrac{3}{7} + \dfrac{4}{9} + \dfrac{17}{42}$

Section 2.3

In Review Problems 37 to 43 use your calculator to determine the resultant of the reciprocals.

37. $\dfrac{1}{2/3}$

38. $\dfrac{1}{3} + \dfrac{1}{5} - \dfrac{1}{8}$

39. $\dfrac{\dfrac{1}{7} \times \dfrac{1}{16}}{\dfrac{1}{5}}$

40. $\dfrac{\dfrac{1}{6} + \dfrac{1}{13} - \dfrac{1}{7}}{\dfrac{1}{9}}$

41. $\dfrac{\dfrac{1}{13} + \dfrac{1}{15} \times \dfrac{1}{3}}{\dfrac{1}{6} - \dfrac{1}{9}}$

42. $\dfrac{\dfrac{1}{3}\left(\dfrac{1}{5} - \dfrac{1}{7}\right)}{\left(\dfrac{1}{6} \times \dfrac{1}{3}\right) - \dfrac{1}{2}}$

43. $\dfrac{\left(\dfrac{1}{7} + \dfrac{1}{3}\right)\left(\dfrac{1}{4} - \dfrac{1}{3}\right)}{\dfrac{1}{4} \times \dfrac{1}{10}}$

Section 2.4

In Review Problems 44 to 49 evaluate the products and express your answers in fractional form.

44. $\dfrac{2}{5} \cdot \dfrac{3}{4}$

45. $\dfrac{7}{9} \cdot \dfrac{18}{5}$

46. $\dfrac{4}{5} \cdot \dfrac{12}{7} \cdot \dfrac{15}{4}$

47. $\dfrac{7}{9} \cdot \dfrac{13}{64} \cdot \dfrac{8}{39}$

48. $\dfrac{5}{21} \cdot \dfrac{7}{36} \cdot \dfrac{42}{45}$

49. $\dfrac{20}{33} \cdot \dfrac{24}{35} \cdot \dfrac{11}{12}$

Section 2.5

In Review Problems 50 to 55 determine the resultant of the division and express your answers in fractional form.

50. $\dfrac{5}{4/7}$

52. $\dfrac{\text{¾}}{\text{⅖}}$

54. $\dfrac{\text{¾} \cdot \text{⅙}}{\text{⅓} \cdot \text{⅝}}$

51. $\dfrac{3/7}{21}$

53. $\dfrac{2\text{⅕}}{\text{¹⁴⁄₁₅}}$

55. $\dfrac{\text{¾} - \text{⅖}}{\text{⅙} - \text{⅔}}$

Section 2.6

Evaluate Review Problems 56 to 61 and express your answers in fractional form.

56. $\dfrac{6\frac{1}{3}}{2\frac{1}{6}}$

58. $\dfrac{2\frac{1}{5} + \frac{1}{2}}{\frac{9}{2}}$

60. $\dfrac{\frac{-10}{11} \times 2\frac{4}{5}}{-1\frac{1}{2} \times -3\frac{1}{3}}$

57. $\dfrac{-1\frac{1}{15}}{6\frac{2}{3}}$

59. $\dfrac{2\frac{1}{6} - 3\frac{3}{4}}{-\frac{7}{12}}$

61. $\dfrac{3\frac{1}{5} - 4\frac{1}{3}}{6\frac{1}{5} - 2\frac{5}{6}}$

Section 2.7

Using your calculator, evaluate Review Problems 62 to 65 and leave your answers in decimal form.

62. $3.41\left(1\frac{3}{7} - \frac{2}{5}\right) + 3\frac{1}{4}$

64. $3.6\left(\frac{2}{3} + \frac{1}{5} - \frac{3}{7}\right) + 2.1\left(5\frac{4}{9}\right)$

63. $\dfrac{\left(1\frac{1}{5} \times 2\frac{1}{4}\right) - \left(3\frac{1}{3} + 6\frac{3}{8}\right)}{2\frac{5}{9}}$

65. $\dfrac{\left(4\frac{1}{4} \times 3\frac{2}{5}\right) - \left(4\frac{1}{7} + 2\frac{1}{6}\right)}{4\frac{3}{7}}$

Section 3.1

Solve Review Problems 66 to 77 using the laws of exponents.

66. $16^2 \cdot 16^3 \cdot 16$

67. $\dfrac{15^5}{15^3}$

68. $\dfrac{3^6 \cdot 3^7}{3^9}$

69. $(2^4 \cdot 2^5)^3$

70. $\sqrt[2]{3.2} \cdot 3.2^5$

71. $\dfrac{\sqrt[3]{15^2} \cdot 15^4}{15^2}$

72. $\sqrt{0.02} \cdot \sqrt[3]{0.02} \cdot 0.02$

73. $\dfrac{(5.6^{3/4})^4}{\sqrt{5.6}}$

74. $\left(\dfrac{42^{1.2}}{42^{0.8}}\right)^3$

75. $(\sqrt[3]{0.4})(\sqrt{0.4})(\sqrt{0.4})$

76. $(((2)^2)^4)^{1/2}$

77. $\dfrac{3.02^{8/5} \cdot 3.02^{7/4}}{(\sqrt{3.02})^2}$

Section 3.3

Simplify Review Problems 78 to 83 and leave all your answers with nonnegative exponents.

78. 3.4^{-2}

79. $\dfrac{1}{8^{-5} \cdot 8^3}$

80. $4.02^{-5} \cdot 4.02^3$

81. $\dfrac{7^{-2} \cdot 7^3}{7^4}$

82. $\left(\dfrac{6.2^{-3}}{6.2^{-2}}\right)^6$

83. $\left(\dfrac{0.41^{-3} \cdot 25^{-2}}{0.41^{-2}}\right) 25^3$

Section 3.4

Evaluate Review Problems 84 to 88 and express your answers in decimal form.

84. $5^{-3} + 5^2$

85. $8^{-2} \cdot 3^3$

86. $(0.02)^{-4} \cdot 4^2$

87. $\dfrac{4^{-2} - 4^{-3}}{4^2}$

88. $\left(\dfrac{3^3}{6^2}\right)^2$

Section 3.5

Using your calculator, evaluate Review Problems 89 to 94.

89. $25.4^{1.2}$

90. $(8.24^{-0.25})^3$

91. $(\sqrt{111} - \sqrt[3]{14})^{2.01}$

92. $\dfrac{1}{\left(\dfrac{1}{3.2}\right)^5} + (\sqrt{6.7})^5$

93. $(3\sqrt[3]{17} + 4\sqrt{21})^{-2/3}$

94. $\dfrac{4.6^{-2.1}}{0.34^{1.5}} - \dfrac{5^{6.2}}{3^{-3}}$

Section 4.1

Write the numbers in Review Problems 95 to 101 in scientific notation.

95. 0.0000785   **98.** 130.5   **101.** 0.000139

96. 56,721,000   **99.** 0.001

97. 0.2300   **100.** 1,000,000

Section 4.2

Express Review Problems 102 to 107 in engineering notation.

102. 0.000002 F   **104.** 0.00001 A   **106.** 0.00010 H

103. 10,000 Ω **105.** 1500 W   **107.** 1,750,000 Hz

Sections 4.3, 4.4

In Review Problems 108 to 118 perform the indicated operations and express your answers in both engineering and scientific notation.

108. $(3.24 \times 10^4)(6.98 \times 10^{-2})(2.6 \times 10^3)$

109. $(421 \times 10^4)^{1/3}$

110. $4 \times 10^3 \times 5.9 \times 10^{15}$

111. $9.8 \times 10^{-2} \times 3.1 \times 10^{-4} \times 2.1 \times 10$

112. $(0.0002)^5$

113. $((2.56 \times 10^{-4}) \times 10^{15})^{1/2}$

114. $22 \times 10^6 + 5 \times 10^7$

115. $6.3 \times 10^4 - 0.021 \times 10^6$

116. $9 \times 10^{-2} + 0.0004 \times 10^3$

117. $\dfrac{(3.2 \times 10^{-4})^{2.3}}{(4.1 \times 10^{-6})^{3.2}}$

118. $\dfrac{(7.61 \times 10^5)^{3.2}}{(9.8 \times 10^8)^{1.1}} - \dfrac{(4.1 \times 10^6)^{1.9}}{(3.8 \times 10^{-4})^{4.1}}$

Sections 4.5, 4.6

In Review Problems 119 to 123 perform prefix conversions.

119. 120 kHz = _____ Hz = _____ MHz

120. 4.398 μA = _____ nA = _____ mA

121. 20 H = _____ mH = _____ μH

122. 1800 MHz = _____ GHz = _____ kHz

123. 18 μF = _____ F = _____ pF

In Review Problems 124 to 128 perform system conversions.

124. 4.9 ft^2 = _____ cm^2

125. 10 gal = _____ L

126. 143 yd = _____ m

127. 45 mi/h = _____ km/min

128. 20 gal/day = _____ L/h

ALGEBRA

FUNDAMENTALS OF ALGEBRA

5.1 VARIABLES

Albegra involves the use of variables. A variable is a designator of any value, usually in the form of a letter, that represents some quantity within an expression; for example, in Chapter 1, R_1 and R_2 represented different resistors. Variables can be combined to form a resultant, such as the equivalent resistance (R_T). Recall that when R_1 and R_2 were in parallel, their equivalency was evaluated as

$$R_T = \frac{1}{\dfrac{1}{R_1} + \dfrac{1}{R_2}} \qquad (5.1)$$

You can substitute the values of R_1 and R_2 into Equation (5.1) to find the resistance. For example, if R_1 is 20 Ω and R_2 is 30 Ω and these values are substituted into Equation (5.1), the resultant is

$$R_T = \frac{1}{\dfrac{1}{20} + \dfrac{1}{30}} = 12 \ \Omega$$

If the variables are 30 and 60 Ω, the resultant is

$$R_T = \frac{1}{\dfrac{1}{30} + \dfrac{1}{60}} = 20 \ \Omega$$

Thus the general rule is evaluate expressions that contain variables by substituting the value of the variables into the expression and then performing the operations indicated.

We mentioned in Chapter 2 that a refinement is permitted to indicate multiplication. Because the center dot can be confusing when intermixed with a sequence of decimal points and the letter X is often used to designate an unknown variable, the absence of any operator symbol implies multiplication. For example, the expression

$$2\pi f L$$

indicates that each of the four items are multiplied. If f equals 1.25 MHz and L equals 0.35 mH, then the expression is

$$2\pi(1.25 \times 10^6)(0.35 \times 10^{-3}) = 2.749 \times 10^3$$

Example 1

Evaluate the following commonly encountered electrical expressions by substituting the values indicated for each variable.

a. $\dfrac{1}{2\pi f C}$ $f = 7.15$ MHz

$C = 0.0005 \ \mu F = 500$ pF

$$\dfrac{1}{2\pi(7.15 \times 10^6)(500 \times 10^{-12})} = 44.52$$

b. $\dfrac{1}{2\pi\sqrt{LC}}$ $L = 0.045$ mH

$C = 2000$ pF

$$\dfrac{1}{2\pi(0.045 \times 10^{-3} \times 2000 \times 10^{-12})^{1/2}} = 530.52 \times 10^3$$

c. $\omega L - \dfrac{1}{\omega C}$ $\omega = 10\pi \times 10^3$

$L = 20$ mH

$C = 0.022 \ \mu F$

$$10\pi \times 10^3 \times 20 \times 10^{-3} - \dfrac{1}{10\pi \times 10^3 \times 0.022 \times 10^{-6}} = -818.54$$

d. $\sqrt{R^2 + (\omega L)^2}$ $R = 250$
$\omega = 5\pi \times 10^3$
$L = 35 \ mH$
$$(250^2 + (5\pi \times 10^3 \times 35 \times 10^{-3})^2)^{1/2} = 603.95$$

PROBLEMS

Evaluate Problems 1 to 7. Note that ϵ is set at Fix 3 but can be more fully expanded by keying in 1 $\boxed{e^x}$.

1. $\dfrac{R_1 R_2}{R_1 + R_2}$ $R_1 = 4.7$ kΩ

$R_2 = 22$ kΩ

3. $\dfrac{1}{2\pi\sqrt{LC}}\sqrt{1 - \dfrac{1}{Q^2}}$ $L = 12.5 \ \mu H$
$C = 3300$ pF
$Q = 20$

2. $\dfrac{R_1 R_2 R_3}{R_1 R_2 + R_1 R_3 + R_2 R_3}$ $R_1 = 6.8$ kΩ
$R_2 = 10$ kΩ
$R_3 = 3.3$ kΩ

4. $\dfrac{1}{2\pi RC}$ $R = 5.6$ kΩ
$C = 0.0033 \ \mu F$

5. $25\epsilon^{-t/2.5 \times 10^{-3}}$ $\epsilon = 2.718$
$t = 3.4 \times 10^{-3}$

7. $\dfrac{1}{2\pi}\sqrt{\dfrac{L - R^2C}{L^2C}}$ $L = 125\mu\text{H}$
$C = 0.01\mu\text{F}$
$R = 12.5\Omega$

6. $15\left(1 - \epsilon^{(-3.5\times 10^{-3}/RC)}\right)$ $\epsilon = 2.718$
$R = 2.2 \text{ k}\Omega$
$C = 1\mu\text{F}$

5.2 ADDITION OF ALGEBRAIC TERMS

To study how to apply basic arithmetic operations to algebraic expressions, the easiest starting point is *monomials*, that is, singular terms. Suppose you are supplying electronic products and one product is the 8080 microprocessor chip. If you have five chips in stock and this chip is designated item x, a way to express this inventory is $5x$. If you sell three of the 8080s, there are two 8080s remaining in stock, which can be expressed as $2x$. Notice that when reducing the stock inventory one coefficient is subtracted from the other, but you do not have to subtract one x from the other x because x is the designator that merely names the item (in this case, an 8080 chip). The item name must still be retained.

$$5x - 3x = 2x$$

The addition process involves working with singular terms (monomials) one at a time.

Multiple terms are called *polynomials*. If you have another type of microprocessor chip in your inventory (the Z80) and it is designated y, 10 such chips can be written as $10y$. Prior to the sale mentioned earlier, the total inventory of these two different chips can be expressed as $5x + 10y$. After you sell the three 8080s, the total inventory can be mathematically expressed as

$$\underbrace{5x + 10y}_{\text{Beginning inventory}} - \underbrace{3x}_{\text{Sales}} = \underbrace{2x + 10y}_{\text{Ending inventory}} \qquad (5.2)$$

Two points are immediately apparent from the total transactions represented by Equation (5.2):

1. You must keep different designators separate; that is, you cannot sum the original inventory of $5x$ and $10y$ as $15xy$ because there is no inventory item designated xy, and certainly there are not 15 of these new items.

2. When adding or subtracting terms with the same designation, operate only upon the coefficients and retain the same designator name.

If a designator contains more than one letter, it is still considered a single entity. Only when another term contains the exact same name can you perform an addition or subtraction operation. For example, suppose you expand your inventory and buy fourteen 6800 microprocessor chips. Designate this chip xy. When added to the inventory left in Equation (5.2), the total stock is now

$$2x + 10y + 14xy \qquad (5.3)$$

Now, the following week you buy six more 8080s and sell three of the Z80s and five of the 6800s. To bring the inventory in equation (5.3) up to date, combine Equation (5.3) and the week's transactions as follows:

$$\underbrace{2x + 10y + 14xy}_{\text{Beginning inventory}} + \underbrace{6x - 3y - 5xy}_{\text{Week's transactions}} = \underbrace{8x + 7y + 9xy}_{\text{Ending inventory}}$$

A more convenient way to write these expressions is to align like terms into vertical columns and then perform the operations indicated:

$$
\begin{array}{ll}
2x + 10y + 14xy & \text{augend} \\
\underline{6x - 3y - 5xy} & \text{addend} \\
8x + 7y + 9xy & \text{sum}
\end{array}
$$

Although the augend, addend, and sum represent three terms each (and hence are referred to as polynomials), each term is operated upon as though it is an individual monomial.

Example 1

Add the following, using your calculator where appropriate.

a.
$$
\begin{array}{l}
2x + 3y \\
\underline{x - y} \\
3x + 2y
\end{array}
$$

b.
$$
\begin{array}{l}
14e - 2f + 3g \\
\underline{-5e + f + 2g} \\
9e - f + 5g
\end{array}
$$

c.
$$
\begin{array}{l}
14.1u + 13.6v - 7.1w \\
\underline{-5.8u - 7.1v + 5.2w} \\
8.3u + 6.5v - 1.9w
\end{array}
$$

d.
$$
\begin{array}{l}
\dfrac{1}{8}x + \dfrac{2}{3}y - \dfrac{4}{5}z \\[2mm]
\dfrac{2}{3}x - y + \dfrac{1}{6}z
\end{array}
\quad \text{implies} \quad
\begin{array}{l}
\dfrac{3}{24}x + \dfrac{2}{3}y - \dfrac{24}{30}z \\[2mm]
\dfrac{16}{24}x - \dfrac{3}{3}y + \dfrac{5}{30}z \\[2mm]
\hline
\dfrac{19}{24}x - \dfrac{1}{3}y - \dfrac{19}{30}z
\end{array}
$$

e. $\begin{array}{r} 3\frac{1}{4}s - \frac{3}{4}t \\ -2\frac{1}{5}s + 2\frac{1}{8}t \\ \hline \end{array}$ implies $\begin{array}{r} \frac{65}{20}s - \frac{6}{8}t \\ -\frac{44}{20}s + \frac{17}{8}t \\ \hline \frac{21}{20}s + \frac{11}{8}t \end{array}$

PROBLEMS

Perform the additions in Problems 1 to 10.

1. $(3x - 4.7y) + (1.6x - 3.1y)$

2. $(-4.7s - 6.1t) + (-1.1s + 3.4i)$

3. $(3.24u - 4.8w + 6.1v) + (-8.1u + 3.6v)$

4. $\left(3\frac{5}{8}a + 6\frac{1}{2}b\right) + \left(4\frac{1}{5}a - 3\frac{1}{6}b\right)$

5. $\left(\frac{1}{7}x + \frac{1}{6}y - \frac{1}{5}z\right) + \left(\frac{3}{14}x - \frac{4}{11}y - \frac{5}{6}z\right)$

6. $\left(\frac{3}{8}f - \frac{4}{7}g - \frac{5}{6}h\right) + \left(4\frac{1}{2}f - 2\frac{1}{6}h\right)$

7. $\begin{array}{l} 3.1i + 2.6j \\ 4.1i \qquad\quad - 3.6k \\ \hline \qquad\quad 6.1j - 3.2k \end{array}$

8. $\begin{array}{l} 0.067l + 0.123m - 0.056n \\ 0.137l \qquad\qquad\quad + 1.19\ n \\ \hline 0.998l - 0.346m - \qquad n \end{array}$

9. $\begin{array}{l} \frac{1}{8}p + \frac{4}{5}q - \frac{1}{6}r \\[4pt] \frac{1}{2}p - \frac{3}{6}q + \frac{1}{2}r \\[4pt] \frac{5}{6}p + \frac{1}{2}q - \frac{1}{3}r \\ \hline \end{array}$

10. $2\frac{1}{2}s + 4\frac{1}{3}t$

$3\frac{1}{6}s - 2\frac{1}{2}t - 6\frac{1}{7}u$

$\frac{7}{8}s - t + u$

5.3 GROUPING AND SUBTRACTION OF ALGEBRAIC TERMS

Before beginning the study of subtraction, it is necessary to expand upon the concept of grouping terms by parentheses. Problems taking the format

$$\text{Augend} \ + \ \text{addend} = \text{sum}$$

present no difficulty because each term within the parentheses assumes the sign preceding it: the *implicit* positive sign before the augend and the *explicit* positive sign before the addend. Thus, the first rule of parentheses is

> You can remove parentheses implicitly or explicitly preceded by a positive sign without any change to the sign of each term within the parentheses.

Problems of the form

$$\text{Augend} \ + \ \text{addend} = \text{sum}$$

can therefore be added term by term according to the individual sign preceding each term. Even problems that involve subtraction and contain only monomials can be handled this way.

$$(5a) - (+3a)$$
$$\text{Minuend} - \text{subtrahend}$$

can be rewritten as

$$(5a) + (-3a)$$
$$\text{Augend} + \text{addend}$$

and the laws of addition applied. This is valid because, as explained in Chapter 1, two mixed signs of priority four ($+$ and $-$) can be replaced with one negative sign, resulting in

$$5a - 3a = 2a$$

It is when *two* or more terms are contained within parentheses of either the minuend or subtrahend that care and some readjustment of signs are required before you can execute the operations. For example:

$$25 - (12)$$

is the same as $25 - (7 + 8 - 3)$

which results in 13. However, this latter form can also be written as

$$25 - 7 - 8 + 3 = 13$$

Notice that when the parentheses of the subtrahend $(7 + 8 - 3)$ preceded by the minus sign was removed, each term within the parentheses changes sign. The second rule of parentheses is therefore

> If a set of parentheses is preceded by a minus sign you can remove the parentheses and drop the minus sign if the sign of each term within the parentheses is changed.

Example 1

Remove the parentheses and solve the following problems.

a. $(3a + 2b) + (a - 4b)$

$3a + 2b + a - 4b = 4a - 2b$

b. $7x - 4y - (3x + 4y)$

$7x - 4y - 3x - 4y = 4x - 8y$

c. $-(2p + 3.1q - 2.7r) + (4.1p + r)$

$-2p - 3.1q + 2.7r + 4.1p + r$

$= 2.1p - 3.1q + 3.7r$

d. $\left(\dfrac{7}{8}m + \dfrac{2}{3}v\right) - \left(-\dfrac{1}{8}m + \dfrac{1}{6}v\right)$

$\dfrac{7}{8}m + \dfrac{2}{3}v + \dfrac{1}{8}m - \dfrac{1}{6}v$

$= \dfrac{8}{8}m + \dfrac{3}{6}v$

$= m + \dfrac{1}{2}v$

e. $-\left(2\dfrac{1}{5}g - 3\dfrac{1}{4}h\right) - \left(-1\dfrac{1}{2}g + 2\dfrac{1}{5}h\right)$

$-\dfrac{11}{5}g + \dfrac{13}{4}h + \dfrac{3}{2}g - \dfrac{11}{5}h$

$= -\dfrac{22}{10}g + \dfrac{65}{20}h + \dfrac{15}{10}g - \dfrac{44}{20}h$

$= -\dfrac{7}{10}g + \dfrac{21}{20}h$

The third rule for grouping involves nested parentheses and is the same as the rule developed for calculators:

When one group of parentheses is within another group, apply the rules of parentheses to the innermost group first, and then extend the rules to the outermost set of parentheses.

Example 2

Evaluate the following problems by first removing all parentheses.

a. $7 + (3 - 2 + (4 - 1) + 2)$
   $= 7 + (3 - 2 + 4 - 1 + 2)$
   $= 7 + 3 - 2 + 4 - 1 + 2 = 13$

b. $(2a + 3b) - (4a + (a - b) + 3b)$
   $= 2a + 3b - (4a + a - b + 3b)$
   $= 2a + 3b - 4a - a + b - 3b$
   $= -3a + b$

c. $x - (y + z - (2x + y) + 3z)$
   $= x - (y + z - 2x - y + 3z)$
   $= x - y - z + 2x + y - 3z$
   $= 3x - 4z$

d. $u - (v - w - (2u - (2v - 2w)))$
   $= u - (v - w - (2u - 2v + 2w))$
   $= u - (v - w - 2u + 2v - 2w)$
   $= u - v + w + 2u - 2v + 2w$
   $= 3u - 3v + 3w$

e. $-(3.1a - (4.5a + 2.6b - (1.7a - 2.9b + 3.2c) - a) - b)$
   $= -(3.1a - (4.5a + 2.6b - 1.7a + 2.9b - 3.2c - a) - b)$
   $= -(3.1a - 4.5a - 2.6b + 1.7a - 2.9b + 3.2c + a - b)$
   $= -3.1a + 4.5a + 2.6b - 1.7a + 2.9b - 3.2c - a + b$
   $= -1.3a + 6.5b - 3.2c$

PROBLEMS

In Problems 1 to 10 simplify and perform the indicated operations.

1. $(7.16a - 7) - (3.1a + 2)$

2. $2 + (2.2 - (0.2 + 2.2) + 2) - 2.2$

3. $(3.1a - 4.6b - 7.2) - (4.75b - 2.1)$

4. $-(3.62x - 4.1y + 3.82z) - (1.2x + 1.72y - 6.8z)$

5. $-(3.1a - (2.6b - 1.7c) - 4.2) - (3.4a - (2.1b + c))$

6. $-1.6r - (6.1s + 4.7t - (3.1r - s + (-2.6t + r)))$

7. $\frac{1}{8}d - \left(\frac{1}{6}e + \frac{2}{5}f - \left(\frac{2}{4}e - \frac{1}{6}f\right) - \frac{2}{14}d\right)$

8. $-\left(\frac{2}{3}h + \frac{4}{5}j - \left(\frac{2}{7}k - \frac{1}{8}j + \left(\frac{1}{5}h - j\right)\right)\right)$

9. $\frac{3}{4}u + \left(3\frac{1}{8}v - 2\frac{1}{5}w - \left(3\frac{1}{5}u - v\right) - w\right)$

10. $-a - (b - c - (-a - b - (-c - a - b - (b - c))) - a)$

5.4 CALCULATOR VERIFICATION—STORAGE AND RECALL

There is no analytical way to find the values of several variables within a single equation. However, a calculator can be useful for verifying that the reduction of terms is correct, especially when nested parentheses are involved. When there are at least as many equations as variables, calculators can be very helpful for solving each variable. This is discussed in Chapters 11 and 12, "Equations: Simultaneous," and "Determinants and Matrix Analysis," respectively. Here we discuss how to verify the correct reduction of a single complex expression into a simpler expression.

Recall that variables can be of any value. When you substitute this value into an expression and perform the indicated operations, the total expression assumes a unique numeric value. If you substitute the same value into both the original and the simplified expressions, both expressions should yield the same numeric answer when evaluated; this indicates that the simplification was done correctly. Here is an extremely simple illustration of this principle:

$$2a - 3b + 6a + 5b = ? \tag{5.4}$$
$$8a + 2b = ? \tag{5.5}$$

Suppose you arbitrarily assign a the value of 1.5 and b the value of 2.3. Equation (5.4) is thus

$$2(1.5) - 3(2.3) + 6(1.5) + 5(2.3) = 16.6$$

Equation (5.5) becomes

$$8(1.5) + 2(2.3) = 16.6$$

We can quote a geometry theorem: "Things equal to the same thing are equal to each other." Certainly, within the constraints of reducing an expression of one level of complexity into an expression of less complexity, if both expressions equate to the same value when the same numeric substitutions are made for each variable, then it may reasonably be concluded that the *reduction* is correct.

Because we want to store numerical values to verify the reduction of equations, we now consider the *storage* (STO) and *recall* (RCL) functions of various calculators. For the four scientific calculators used in this text, the following conclusions apply:

- TI-55   Capacity to store STO and recall RCL numeric values into 10 addressable locations: locations 0 through 9.

- HP 31E   Capacity to store STO and recall RCL numeric values into four addressable locations: locations 0 through 3.

- SHARP EL-5813   Capacity to store STO and recall numeric values into six addressable locations: locations K1 through K6 . Additionally, has a working memory storage designated by $x{\rightarrow}M$ and its recall RM .

- CASIO fx-80   Limited to only one memory location—memory in M in —and memory recall MR .

In all the following examples that involve storing more than one value, you will have to manually record and key in the extra values on those calculators which have less storage and recall capacity than required, usually a maximum of three values. Because most of the scientific calculators represented in this book have more storage capacity than one location, we use the store and recall functions throughout the examples. That is, rather than keying in the values of 1.5 for *a* and 2.3 for *b* every time they are used in Equation (5.4), place these values into memory using the STO key. For convenience, place the 1.5 in memory location 1 and the 2.3 in memory location 2. Note that for those calculators with more than 10 addressable locations, you may have to key in 01 instead of 1 to store in memory location 1, and so on. To store the 1.5 and 2.3 values, execute the following simple sequence:

PRESS	DISPLAY	COMMENT
1.5 STO 1 [1]	1.5	Stores 1.5 in memory location 1
2.3 STO 2 [2]	2.3	Stores 2.3 in memory location 2

---

[1]This is K1 on a calculator like the Sharp EL-5813.
[2]This value has to be manually recorded if your calculator has only one memory storage location.

Now any time you need these values in an evaluation, you can substitute [RCL 1] or [RCL 2] for the values during the execution of the operators within the expression. These *commands* will recall the previously stored values exactly as they were placed in memory, thereby ensuring that no error occurs when you make variable substitutions during the calculations. An advantage of the [RCL] key is that you cannot make a simple mistake such as accidentally keying in 1.4 instead of 1.5 for one of the *a* variable. Second, even if you do make an error when placing an arbitrary value for a variable into the memory location, that same value is consistently substituted for the variable when you use the [RCL] key, ensuring that a correct simplification can *still* be verified. Equation (5.4) is now illustrated from the beginning. Verify that

$$2a - 3b + 6a + 5b = 8a + 2b$$

PRESS	DISPLAY	COMMENTS
1.5 [STO 1]	1.5	Stores *a* in memory location 1
2.3 [STO 2]	2.3	Stores *b* in memory location 2
[Fix 3]	2.300	Sets display to round off at third decimal position
[CLR]	0	Clears display
2 [×] [RCL 1] [−] 3		
[×] [RCL 2] [+] 6		
[×] [RCL 1] [+] 5		
[×] [RCL 2] [=]	16.600	Evaluates original expression
[CLR]	0	Clears display for evaluation of reduced expression
8 [×] [RCL 1] [+] 2		
[×] [RCL 2] [=]	16.600	Evaluation of reduced expression verified

Because both expressions evaluated to the same numeric answer, the algebraic simplification must have been correct.

Example 1

Verify the simplification of Example 1a, c, and e, Section 5.3 by using [Fix 3] and substituting 1.75 for the first variable [STO 1] , 2.5 for the second variable [STO 2] , and 3.25 for the third variable if used [STO 3] . If you have a calculator with limited storage, you will have to key in the second and/or third variable.

1a: $(3a + 2b) + (a - 4b) = 4a - 2b$

PRESS   **DISPLAY**

⌈(3 ⌈×⌉ ⌈RCL 1⌉ ⌈+⌉ 2

⌈×⌉ ⌈RCL 2⌉ ⌈)⌉ ⌈+⌉

⌈(⌈RCL 1⌉ ⌈−⌉ 4 ⌈×⌉

⌈RCL 2⌉ ⌈)⌉ ⌈=⌉   2.000

⌈CLR⌉   0

4 ⌈×⌉ ⌈RCL 1⌉ ⌈−⌉ 2

⌈×⌉ ⌈RCL 2⌉ ⌈=⌉   2.000

1c: $-(2p + 3.1q - 2.7r) + (4.1p + r) = 2.1p - 3.1q + 3.7r$

PRESS   **DISPLAY**   **COMMENTS**

⌈(2 ⌈×⌉ ⌈RCL 1⌉ ⌈+⌉ 3.1

⌈×⌉ ⌈RCL 2⌉ ⌈−⌉ 2.7 ⌈×⌉

⌈RCL 3⌉ ⌈)⌉ ⌈+/−⌉ ⌈+⌉   − 2.475   First parentheses of complex expression

⌈(4.1 ⌈×⌉ ⌈RCL 1⌉ ⌈+⌉

⌈RCL 3⌉ ⌈)⌉   10.425   Second parentheses of complex expression

⌈=⌉   7.950   Complex expression evaluated

⌈CLR⌉   0

2.1 ⌈×⌉ ⌈RCL 1⌉ ⌈−⌉ 3.1

⌈×⌉ ⌈RCL 2⌉ ⌈+⌉ 3.7

⌈×⌉ ⌈RCL 3⌉ ⌈=⌉   7.950   Simplified expression evaluated

1e: $-\left(2\frac{1}{5}g - 3\frac{1}{4}h\right) - \left(-1\frac{1}{2}g + 2\frac{1}{5}h\right) = -\frac{7}{10}g + \frac{21}{20}h$

PRESS   **DISPLAY**   **COMMENTS**

⌈(⌈(2 ⌈+⌉ 1 ⌈÷⌉ 5 ⌈)⌉

⌈×⌉ ⌈RCL 1⌉ ⌈−⌉ ⌈(

3 ⌈+⌉ 1 ⌈÷⌉ 4 ⌈)⌉ ⌈×⌉

⌈RCL 2⌉ ⌈)⌉ ⌈+ / −⌉   4.275   First parentheses evaluated with minus sign

PRESS	DISPLAY	COMMENTS
$\boxed{-}$	4.275	Display becomes minuend
$\boxed{(\!(}$ $\boxed{(\!(}$ 1 $\boxed{+}$ 1 $\boxed{\div}$ 2 $\boxed{)\!)}$		
$\boxed{+/-}$ $\boxed{\times}$ $\boxed{RCL\,1}$		
$\boxed{+}$ $\boxed{(\!(}$ 2 $\boxed{+}$ 1 $\boxed{\div}$		
5 $\boxed{)\!)}$ $\boxed{\times}$ $\boxed{RCL\,2}$ $\boxed{)\!)}$	2.875	Second parentheses evaluated
$\boxed{=}$	1.400	Complex expression evaluated
$\boxed{CLR}$	0	
$\boxed{(\!(}$ 7 $\boxed{\div}$ 10 $\boxed{\times}$ $\boxed{RCL\,1}$		
$\boxed{)\!)}$ $\boxed{+/-}$ $\boxed{+}$	-1.225	First term of simplified expression
$\boxed{(\!(}$ 21 $\boxed{\div}$ 20 $\boxed{\times}$		
$\boxed{RCL\,2}$ $\boxed{)\!)}$ $\boxed{=}$	1.400	Simplified expression evaluated

Using the same memory storage values, verify the nested parentheses problems in Example 2c and d.

2c: $\qquad x - (y + z - (2x + y) + 3z) = 3x - 4z$

PRESS	DISPLAY
$\boxed{RCL\,1}$ $\boxed{-}$ $\boxed{(\!(}$ $\boxed{RCL\,2}$ $\boxed{+}$ $\boxed{RCL\,3}$ $\boxed{-}$ $\boxed{(\!(}$ 2 $\boxed{\times}$	
$\boxed{RCL\,1}$ $\boxed{+}$ $\boxed{RCL\,2}$ $\boxed{)\!)}$ $\boxed{+}$ 3 $\boxed{\times}$ $\boxed{RCL\,3}$ $\boxed{)\!)}$ $\boxed{=}$	-7.750
$\boxed{CLR}$	0
3 $\boxed{\times}$ $\boxed{RCL\,1}$ $\boxed{-}$ 4 $\boxed{\times}$ $\boxed{RCL\,3}$ $\boxed{=}$	-7.750

2d: $\qquad u - (v - w - (2u - (2v - 2w))) = 3u - 3v + 3w$

PRESS	DISPLAY
$\boxed{RCL\,1}$ $\boxed{-}$ $\boxed{(\!(}$ $\boxed{RCL\,2}$ $\boxed{-}$ $\boxed{RCL\,3}$ $\boxed{-}$ $\boxed{(\!(}$ 2 $\boxed{\times}$	
$\boxed{RCL\,1}$ $\boxed{-}$ $\boxed{(\!(}$ 2 $\boxed{\times}$ $\boxed{RCL\,2}$ $\boxed{-}$ 2 $\boxed{\times}$ $\boxed{RCL\,3}$	
$\boxed{)\!)}$ $\boxed{)\!)}$ $\boxed{)\!)}$ $\boxed{=}$	7.500
$\boxed{CLR}$	0
3 $\boxed{\times}$ $\boxed{RCL\,1}$ $\boxed{-}$ 3 $\boxed{\times}$ $\boxed{RCL\,2}$ $\boxed{+}$	
3 $\boxed{\times}$ $\boxed{RCL\,3}$ $\boxed{=}$	7.500

PROBLEMS

Use your calculator ($\boxed{\text{Fix 3}}$) to verify Problems 1 to 8, choosing, 1.85 for the first variable $\boxed{\text{STO 1}}$, 2.35 for the second variable $\boxed{\text{STO 2}}$ and -2.55 for the third variable if used $\boxed{\text{STO 3}}$.

1. $(3.7b - 4.2c) + (1.6b + 5.7c) = 5.3b + 1.5c$

2. $(1.62x - 3.14y) + (-6.1x + 2.5y) = -4.48x - 0.64y$

3. $\left(2\frac{1}{4}c + 1\frac{1}{5}d\right) + \left(3\frac{1}{8}c - \frac{5}{6}d\right) = \frac{43}{8}c + \frac{11}{30}d$

4. $\left(\frac{2}{7}x - \frac{1}{6}y + \frac{1}{5}z\right) + \left(\frac{4}{21}x + \frac{5}{8}y - \frac{1}{6}z\right) = \frac{10}{21}x + \frac{11}{24}y + \frac{1}{30}z$

5. $(4.5s - 1.6) - (4.2s + 1.1t + 0.8) = 0.3s - 1.1t - 2.4$

6. $-\left(2\frac{1}{8}a + 3\frac{1}{2}b - \left(1\frac{1}{6}a - 1\frac{1}{5}b\right)\right) = -\frac{23}{24}a - \frac{47}{10}b$

7. $\left(3\frac{1}{5}u - \left(2\frac{1}{7}v - 3\frac{1}{2}w + 2\frac{5}{7}u - \left(v - w\right)\right)\right) = \frac{17}{35}u - 1\frac{1}{7}v + 2\frac{1}{2}w$

8. $-\left(3\frac{1}{2}a - \left(2\frac{1}{3}b + c\right) - \left(a - \left(2b - c\right) - 3\frac{1}{3}a\right) - 2\frac{1}{5}b\right)$

   $= -5\frac{5}{6}a + 2\frac{8}{15}b + 2c$

5.5 MULTIPLICATION OF ALGEBRAIC TERMS

Suppose that part of a problem involves the value of π times itself. Because this value is only approximately known, one answer can be 9.8696 (to $\boxed{\text{Fix 4}}$). However, the desired degree of accuracy may vary, so an exact way of expressing the result is π^2. If the variable X is to be multiplied by itself, the same reasoning applies. As a matter of fact, because the value of a variable at any given time may not always be known, under such circumstances the answer *must* be expressed as X^2, following the first law of exponents for pure numbers. Because variables are just numbers in a literal form, all four laws of exponents for numbers apply to variables of the *same* base.

Example 1

Perform the indicated multiplications.

a. $c \cdot c^2$

$c^{1+2} = c^3$

b. $d^{1/2} \cdot d^2$

$d^{(1/2)+2} = d^{5/2}$

c. $ab \cdot a^3 b^2$

$(a^{1+3})(b^{1+2}) = a^4 b^3$

d. $x^2 yz^3 \cdot xy^2 z^{-1} \cdot xy^{-3} z^2$

$(x^{2+1+1})(y^{1+2-3})(z^{3-1+2})$

$= x^4 y^0 z^4$

$= x^4 z^4 \qquad Note: y^0 = 1$

e. $\dfrac{tu}{s^2} \cdot \dfrac{s^3 u^2}{t^3}$

$s^{-2} tu \cdot s^3 t^{-3} u^2$

$= (s^{-2+3})(t^{1-3})(u^{1+2})$

$= st^{-2} u^3$

f. $(3.2ab)(1.7a^2 b^{-3})(2.5 \; bc^2)$

$(3.2 \times 1.7 \times 2.5)(a^{1+2})(b^{1-3+1})(c^2)$

$= 13.6a^3 b^{-1} c^2$

g. $\dfrac{2}{5} x^{-2} y^3 \cdot \dfrac{3}{4} xy^2 z \cdot \dfrac{5}{3} y^{-1} z^2$

$= \left(\dfrac{2}{\cancel{5}} \cdot \dfrac{\cancel{3}}{4} \cdot \dfrac{\cancel{5}}{\cancel{3}} \right)(x^{-2+1})(y^{3+2-1})(z^{1+2})$

$= \dfrac{1}{2} x^{-1} y^4 z^3$

These examples represent monomial terms times monomial terms. Sometimes either the multiplier or the multiplicand is a polynomial. (Cases where both terms are polynomials are discussed in the next chapter). Consider the following problem:

Multiplier is monomial $4 \cdot (21) = 84$
Multiplier is binomial $4 \cdot (10 + 11) = 4 \cdot 10 + 4 \cdot 11 = 40 + 44 = 84$
Multiplier is trinomial $4 \cdot (2 + 8 + 11) = 4 \cdot 2 + 4 \cdot 8 + 4 \cdot 11$
$= 8 + 32 + 44 = 84$

Notice in each case that the multiplicand is 4 and the multiplier totals 21. A general rule for multiplication is

> If the multiplier has two or more terms, you must multiply *each* and *every* term of the multiplier by the multiplicand to obtain the correct resultant.

Example 2

Perform the indicated multiplications.

a. $xy(x^2y + xyz)$
$x^3y^2 + x^2y^2z$

b. $ef^2(e^2 + f + ef)$
$e^3f^2 + ef^3 + e^2f^3$

c. $pq^{-1}(p^2q^3 - p^{-2}q - p^2q^2r)$
$p^3q^2 - p^{-1} - p^3qr$
$= p^3p^2 - \dfrac{1}{p} - p^3qr$

d. $3.4x^{-2}y^3(1.6xy - 2.5x^2y^{-1} + 4.8xy^3z^3)$
$5.44x^{-1}y^4 - 8.5y^2 + 16.32x^{-1}y^6z^3$
$= \dfrac{5.44y^4}{x} - 8.5y^2 + \dfrac{16.32y^6z^3}{x}$

e. $\dfrac{5}{6}ghj\left(\dfrac{3}{2}g^2hj^2 - \dfrac{4}{5}g^{-1}h^{-1}j^{-1}\right)$
$\dfrac{5}{4}g^3h^2j^3 - \dfrac{2}{3}$

PROBLEMS

Determine the resultants of the products in Problems 1 to 12.

1. $s^2 \cdot s^{-3} \cdot s^4$

2. $t^4 \cdot \dfrac{1}{t^3} \cdot t^2$

3. $c^{-1} \cdot c^2 \cdot c^3 \cdot d$

4. $ab \cdot a^2b$

5. $ef^2 \cdot e^{-2}f^3 \cdot e^{-1}f$

6. $x^2y^2 \cdot xy^{-1} \cdot xyz$

7. $tu^{-1} \cdot \dfrac{u}{v} \cdot u^{-2}v^2$

8. $\dfrac{g^{-1}h^{-1}}{j^{-2}} \cdot \dfrac{j}{g^{-2}h^{-3}} \cdot \dfrac{gh}{j^{-3}}$

9. $e^2f(e + e^2f + fg^2)$

10. $pr^{-2}(p^2r + pr^{-3} - p^{-2}r^{-4})$

11. $\dfrac{5}{7}dce^{-1}\left(\dfrac{2}{3}d^{-2} - \dfrac{1}{10}ce^2 + \dfrac{21}{15}d^{-2}c\right)$

12. $3\dfrac{1}{4}a^{-2}b^{-1}c\left(5\dfrac{1}{6}ab^2c^3 + 3\dfrac{5}{8}ab^2c - \dfrac{5}{8}a^4\right)$

5.6 DIVISION OF ALGEBRAIC TERMS

The addition, subtraction, and multiplication principles of working with individual terms one at a time extend equally to the division process. For example, a problem such as

$$\frac{4.2a^2b^3}{3.2ab^2}$$

can be separated into a series of both the arithmetic and the laws-of-exponents process when applied to the coefficients and variables, respectively. Thus the previous problem can be rewritten as

$$\underbrace{\left(\frac{4.2}{3.2}\right)}_{}\underbrace{(a^{2-1})}_{}\underbrace{(b^{3-2})}_{}$$
$$1.3125 \ \cdot \ a \ \cdot \ b$$

Remember that when dividing exponents of the same base, subtract the exponent of the denominator from the exponent of the numerator.

Example 1

Perform the indicated divisions.

a. $\dfrac{x^2 y^3}{xy}$

$(x^{2-1})(y^{3-1}) = xy^2$

b. $\dfrac{s^{-2} t^4}{s^{-3} t}$

$(s^{-2-(-3)})(t^{4-1}) = st^3$

c. $\dfrac{k^{-2} lm^4}{km^2}$

$(k^{-2-1})(l)(m^{4-2}) = k^{-3} lm^2$

d. $\dfrac{3.25xy^{-1}z^2}{1.6x^2 y^{-3} z}$

$\left(\dfrac{3.25}{1.6}\right)(x^{1-2})(y^{-1-(-3)})(z^{2-1}) = 2.03125x^{-1}y^2 z$

e. $\dfrac{\dfrac{4}{5} u^2 v^{1.5} w^{3.5}}{\dfrac{8}{25} u^{-3} vw^{1.5}}$

$\left(\dfrac{\overset{1}{\cancel{4}}}{\underset{1}{\cancel{5}}} \cdot \dfrac{\overset{5}{\cancel{25}}}{\underset{2}{\cancel{8}}}\right)(u^{2-(-3)})(v^{1.5-1})(w^{3.5-1.5})$

$\dfrac{5}{2} u^5 v^{0.5} w^2$

If the dividend of a division problem is a polynomial, you must follow rules similar to those for the multiplication process. Consider the following problem:

Dividend is monomial $\dfrac{48}{4} = 12$

Dividend is binomial $\dfrac{16 + 32}{4} = \dfrac{16}{4} + \dfrac{32}{4} = 4 + 8 = 12$

Dividend is trinomial $\dfrac{24 + 16 + 8}{4} = \dfrac{24}{4} + \dfrac{16}{4} + \dfrac{8}{4} = 6 + 4 + 2 = 12$

Notice in each case that the divisor is 4 and the dividend totals 48; therefore the quotient is 12. A general rule of division is:

If the dividend has two or more terms, you must divide *each* and *every* term of the dividend by the divisor to obtain the correct resultant.

Example 2

Perform the indicated division.

a. $\dfrac{xy^2 - x^2 y}{x}$

$x^{1-1}y^2 - x^{2-1}y = y^2 - xy$

b. $\dfrac{b^3 c + b^{-1}c^2}{bc}$

$b^{3-1}c^{1-1} + b^{-1-1}c^{2-1} = b^2 + b^{-2}c$

c. $\dfrac{p^2 r^{-3} - p^{-2}q + q^{-2}r^4}{pqr}$

$p^{2-1}q^{-1}r^{-3-1} - p^{-2-1}q^{1-1}r^{-1} + p^{-1}q^{-2-1}r^{4-1}$

$= pq^{-1}r^{-4} - p^{-3}r^{-1} + p^{-1}q^{-3}r^3$

$= \dfrac{p}{qr^4} - \dfrac{1}{p^3 r} + \dfrac{r^3}{pq^3}$

d. $\dfrac{3.8u^{-1.5}v^{3.2} - 6.1u^{3.4}w^2 + 2.8v^{1.6}w^{0.7}}{1.4u^2 v^{1.5}w}$

$\dfrac{3.8}{1.4}(u^{-1.5-2}v^{3.2-1.5}w^{-1}) - \dfrac{6.1}{1.4}(u^{3.4-2}v^{-1.5}w^{2-1}) + \dfrac{2.8}{1.4}(u^{-2}v^{1.6-1.5}w^{0.7-1})$

$= 2.7143u^{-3.5}v^{1.7}w^{-1} - 4.3571u^{1.4}v^{-1.5}w + 2u^{-2}v^{0.1}w^{-0.3}$

$= \dfrac{2.7143v^{1.7}}{u^{3.5}w} - \dfrac{4.3571u^{1.4}w}{v^{1.5}} + \dfrac{2v^{0.1}}{u^2 w^{0.3}}$

e. $\dfrac{\dfrac{3}{16}a^2 b^3 c^4 - \dfrac{5}{8}ac^{-1} + \dfrac{4}{5}b^{-2}c^4}{\dfrac{3}{4}abc}$

$\left(\dfrac{3}{16} \cdot \dfrac{4}{3}\right)(a^{2-1})(b^{3-1})(c^{4-1}) - \left(\dfrac{5}{8} \cdot \dfrac{4}{3}\right)(a^{1-1})(b^{-1})(c^{-1-1})$

$+ \left(\dfrac{4}{5} \cdot \dfrac{4}{3}\right)(a^{-1})(b^{-2-1})(c^{4-1})$

$= \dfrac{1}{4}ab^2 c^3 - \dfrac{5}{6}b^{-1}c^{-2} + \dfrac{16}{15}a^{-1}b^{-3}c^3$

$= \dfrac{ab^2 c^3}{4} - \dfrac{5}{6bc^2} + \dfrac{16c^3}{15ab^3}$

PROBLEMS

Determine the quotients in Problems 1 to 10.

1. $\dfrac{k^2 l}{k^{-1} l^2}$

2. $\dfrac{d^2 e^{-2} f}{de^{-1}}$

3. $\dfrac{g^2 k^{-1} l^3}{g^{-1} k^{-2}}$

4. $\dfrac{s^2 t + st^2}{s^{-1} t^3}$

5. $\dfrac{u^2 vw^{-1} - u^2 v^2 + w}{u^{-1} vw^{-2}}$

6. $\dfrac{3.6xy^2 + 2.1x^2 y}{2.6x^2 y^{-2}}$

7. $\dfrac{8.17a^{-2} b - 6.3bc^{-1} + 2.5abc}{2.62abc}$

8. $\dfrac{\frac{3}{5} a^{-2} b + \frac{1}{7} bc^{-1} - \frac{1}{6} ac^{-3}}{\frac{2}{10} ab^{-1} c^{-2}}$

9. $\dfrac{3\frac{1}{4} uv^{-2} w + 2\frac{1}{5} u^2 v}{3\frac{1}{2} u^{-1} vw^2}$

10. $\dfrac{1\frac{1}{5} st^{-2} v - 3\frac{1}{2} sv + 2st}{4\frac{1}{5} s^2 tv^3}$

5.7 CALCULATOR VERIFICATION OF PRODUCTS AND QUOTIENTS

You can use the calculator to verify the accuracy of reducing complex algebraic expressions involving products and quotients to simpler expressions, as you did in Section 5.4. Arbitrarily choose numeric values for each variable. Substitute these values into both the original and resultant expressions and perform the indicated operations. If both expressions yield the same numeric values, the algebraic operations were performed correctly. Note that the y^x function is frequently used for verifying such expressions.

Example 1

With your calculator at $\boxed{\text{Fix 2}}$, verify the accuracy of the following operations by using 1.5 for the first variable $\boxed{\text{STO 1}}$, 2.25 for the second variable $\boxed{\text{STO 2}}$, and 3.15 for the third variable if used $\boxed{\text{STO 3}}$. If your calculator has limited storage you will have to key in the second and/or third variable manually each time it is used.

a. $\dfrac{3.25xy^{-1}z^2}{1.6x^2y^{-3}z} = 2.03125x^{-1}y^2z$

PRESS   **DISPLAY**

$(\!(\ \ 3.25\ \boxed{\times}\ \boxed{\text{RCL 1}}\ \boxed{\times}\ \boxed{\text{RCL 2}}\ \boxed{1/x}\ \boxed{\times}\ \boxed{\text{RCL 3}}$

$\boxed{x^2}\ \boxed{)}\ \boxed{\div}\ (\!(\ 1.6\ \boxed{\times}\ \boxed{\text{RCL 1}}\ \boxed{x^2}\ \boxed{\times}\ \boxed{\text{RCL 2}}$

$\boxed{y^x}\ 3\ \boxed{+/-}\ \boxed{\times}\ \boxed{\text{RCL 3}}\ \boxed{)}\ \boxed{=}$ 21.59

$\boxed{\text{CLR}}$ 0

$2.03125\ \boxed{\times}\ \boxed{\text{RCL 1}}\ \boxed{1/x}\ \boxed{\times}\ \boxed{\text{RCL 2}}$

$\boxed{x^2}\ \boxed{\times}\ \boxed{\text{RCL 3}}\ \boxed{=}$ 21.59

b. $(3.2ab)(1.7a^2b^{-3})(2.5bc^2) = 13.6a^3b^{-1}c^2$

PRESS	DISPLAY	COMMENTS
$(\!(\ 3.2\ \boxed{\times}\ \boxed{\text{RCL 1}}$		
$\boxed{\times}\ \boxed{\text{RCL 2}}\ \boxed{)}\ \boxed{\times}$	10.80	Evaluates first parentheses
$(\!(\ 1.7\ \boxed{\times}\ \boxed{\text{RCL 1}}$		
$\boxed{x^2}\ \boxed{\times}\ \boxed{\text{RCL 2}}$		
$\boxed{y^x}\ 3\ \boxed{+/-}\ \boxed{)}$	0.34	Evaluates second parentheses
$\boxed{\times}\ (\!(\ 2.5\ \boxed{\times}\ \boxed{\text{RCL 2}}$		
$\boxed{\times}\ \boxed{\text{RCL 3}}\ \boxed{x^2}\ \boxed{)}\ \boxed{=}$	202.42	Evaluates complex expression
$\boxed{\text{CLR}}$	0	
$13.6\ \boxed{\times}\ \boxed{\text{RCL 1}}\ \boxed{y^x}$		
$3\ \boxed{\times}\ \boxed{\text{RCL 2}}\ \boxed{1/x}$		
$\boxed{\times}\ \boxed{\text{RCL 3}}\ \boxed{x^2}\ \boxed{=}$	202.42	Evaluates simplified expression

PROBLEMS

With your calculator at $\boxed{\text{Fix 3}}$, verify the operations in Problems 1 to 5 by using 2.75 for the first variable $\boxed{\text{STO 1}}$, 3.15 for the second variable $\boxed{\text{STO 2}}$, and 4.25 for the third variable if used $\boxed{\text{STO 3}}$.

1. $\dfrac{ab^2c^{-1}}{ab} = \dfrac{b}{c}$ **2.** $\dfrac{g^{-1}hi^2}{gh^2i} = \dfrac{i}{g^2h}$

3. $\dfrac{4.2x^2y + 1.67x^{3/2}y^{-1}}{2.73x^{1/2}y^2} = \dfrac{1.538x^{3/2}}{y} + \dfrac{0.612x}{y^3}$

4. $\dfrac{3\frac{1}{2}s^2t^{3/4} - 4\frac{1}{4}t^2v + 6sv^{-2}}{3\frac{1}{3}s^{1/2}tv^{1/2}}$

$= \dfrac{21}{20}s^{3/2}t^{-1/4}v^{-1/2} - \dfrac{51}{40}s^{-1/2}tv^{1/2} + \dfrac{9}{5}s^{1/2}t^{-1}v^{-5/2}$

5. $4\frac{1}{2}a^{-2}bc^3\left(3\frac{1}{4}ab^2c^{-1} - 2\frac{1}{5}a^{-1}bc\right) = \dfrac{117}{8}a^{-1}b^3c^2 - \dfrac{99}{10}a^{-3}b^2c^4$

SUMMARY

Algebraic expressions involve the use of variables; you can substitute numeric values for the variables and then evaluate the expressions. In the addition and subtraction processes, perform the indicated operations for the same variables on the coefficients. Each different variable is then associated with its resulting numeric coefficient.

In the multiplication and division processes, operate upon the numeric coefficients according to the operator sign, and apply the laws of exponents to the variables. When multiplying the same bases, add the exponents. When dividing the same bases, subtract the exponent of the denominator from the exponent of the numerator. To multiply a monomial by a polynomial or to divide a polynomial by a monomial, you have to multiply or divide each monomial by or into each term of the polynomial.

To use a calculator to verify the accuracy of algebraic manipulations, substitute any arbitrary numeric value for the variables and perform the indicated operations. If both the original and resultant expressions yield the same numeric evaluation, the algebraic operations have been correctly performed.

6 SPECIAL PRODUCTS, QUOTIENTS, AND FACTORING

6.1 ROOTS AND POWERS OF A MONOMIAL

When a group of variables comprise a monomial raised to a power, all variables and coefficients within the monomial are raised to that same power. For example,

$$(3xy^2)^2 = 3^2 \cdot x^2 \cdot (y^2)^2 = 9x^2y^4$$

Note that for the variable y the format is

$$(y^m)^n = y^{m \cdot n} \tag{6.1}$$

which is the third law of exponents as described in Equation (3.3). The solution is the base raised to a power that is the product of the two original powers. Technically, the coefficient 3 and the variable x follow the same format, but because their power (m) in Equation (6.1) is an implied 1, both merely assume the power to which the quantity is raised, namely, n.

Example 1

Expand the following monomials to the power indicated.

a. $(6x^2y^3)^2$
   $(6)^2(x^2)^2(y^3)^2 = 36x^4y^6$

b. $(2s^{-1}t^{-2})^3$
   $(2)^3(s^{-1})^3(t^{-2})^3$
   $= 8s^{-3}t^{-6}$
   $= \dfrac{8}{s^3t^6}$

c. $\left(\dfrac{3.1\mu^2}{v^{-1}}\right)^2$
   $\dfrac{(3.1)^2(\mu^2)^2}{(v^{-1})^2}$
   $= \dfrac{9.61\mu^4}{v^{-2}}$
   $= 9.61\mu^4v^2$

d. $\left(\dfrac{4.8w^{1.5}x^{2.5}}{z^{1.6}}\right)^3$

$\dfrac{(4.8)^3(w^{1.5})^3(x^{2.5})^3}{(z^{1.6})^3}$

$= \dfrac{110.592w^{4.5}x^{7.5}}{z^{4.8}}$

Sometimes, a power that is not an integer, such as $w^{4.5}$ in Example 1d, causes concern as to its meaning. Remember, this is just another way of expressing a variable when the exponent is a fraction. For example,

$$w^{4.5} = w^{9/2} = \sqrt[2]{w^9}$$

This expression states that when w is assigned a value, it is to be multiplied by itself eight times and then the square root of that resultant is extracted. The y^x function on a calculator, however, can perform this process directly. If w is 3.15, key in

$$3.15 \boxed{y^x}\ 4.5\ \boxed{=}$$

and the same result is obtained as with the slightly longer sequence of

$$3.15 \boxed{y^x}\ 9\ \boxed{=}\ \boxed{\sqrt{}}\ \boxed{=}$$

When powers are fractions, the denominator of the fraction defines the root to be extracted:

$$\sqrt{64a^4b^2} = (64a^4b^2)^{1/2} = (64)^{1/2}(a^4)^{1/2}(b^2)^{1/2} = 8a^2b$$

Both variables a and b assume the format of Equation (6.1); therefore you can apply the same rules.

Note that multiplying a number by a reciprocal is the same as dividing the number by the denominator of the reciprocal:

$$36\left(\frac{1}{5}\right) = \frac{36}{5}$$

Therefore,

To extract the root of a base number with an exponent, either multiply the exponent by the reciprocal or divide the exponent by the root designator (index).

Another aspect of roots is that all roots have as many answers as does the base number of the root. For example,

$$\sqrt[2]{16} = +4 \qquad \text{or} \qquad -4$$

because $(+4)(+4) = 16$ and $(-4)(-4) = 16$. The positive answer is called the *principal* root. If this problem is keyed into the calculator, however, only the plus (principal) answer is displayed. Thus this text restricts answers to the principal root unless another form is expressly needed for solutions, such as the solutions to the *quadratic equations* discussed in Chapter 10. If the root is preceded by a negative sign, the *explicit* negative sign is carried along in the solution:

$$-\sqrt{16} = -(+4) = -4$$

If this logic is further extended, then $\sqrt[3]{64}$ has three solutions, which is correct. One solution, 4, is obvious. The other two solutions are not obvious because they involve complex numbers, a topic covered in Chapter 16. To repeat, only the principal (positive) root is used in this book.

So far, any negative sign associated with a radical sign has *preceded* the radical (or parentheses, if in that format), which is mathematically permissible:

$$-\sqrt[3]{8} = -(8)^{1/3} = -(+2) = -2$$

A negative sign *under* the radical sign is a different matter. If the index is an *even* integer, the radicand cannot be negative because a negative sign multiplied by itself an even number of times is positive:

$$\sqrt{-4} \ne -2 \qquad \text{or} \qquad +2$$

because $(-2)(-2) = +4$ and $(+2)(+2) = +4$. If you key $\sqrt{-4}$ into the calculator, a flashing 2 (or, on some calculators, the word "ERROR") is displayed, indicating an illegal procedure. But if the index is an *odd* integer, a negative sign can be under the radical sign because a negative sign multiplied by itself an odd number of times is still negative:

$$\sqrt[3]{-64} = -4 \text{ because } (-4)(-4)(-4) = -64$$
$$\sqrt[5]{-32} = -2 \text{ because } (-2)(-2)(-2)(-2)(-2) = -32$$

We made this point to stress that our statement is *analytically* correct. Key $(-64)^{1/3}$ into the calculator and note the results.

PRESS   **DISPLAY**

64 $\boxed{+/-}$ $\boxed{y^x}$ $\boxed{(}$ 1 $\boxed{\div}$ 3 $\boxed{)}$ $\boxed{=}$ Flashing 4 (or ERROR)

When solving $\boxed{y^x}$ problems, the calculator does not permit entry of a negative radicand. However, you can get the correct solution for an odd-numbered root by changing the location of the $\boxed{+/-}$ entry. Key in

64 $\boxed{y^x}$ $\boxed{(}$ 1 $\boxed{\div}$ 3 $\boxed{)}$ $\boxed{=}$ $\boxed{+/-}$

The correct result of -4 is now obtained. In effect,

$$\sqrt[3]{-64} \qquad \text{is rewritten as} \qquad -\sqrt[3]{64}$$

Examples 2 and 3 illustrate these principals.

Example 2

Find the indicated roots for the following monomials.

a.　$\sqrt{9x^2}$

　　$(9)^{1/2}(x^2)^{1/2} = 3x$

b.　$-\sqrt{16b^4}$

　　$-(16)^{1/2}(b^4)^{1/2} = -4b^2$

c.　$\sqrt[3]{27x^6y^9}$

　　$(27)^{1/3}(x^6)^{1/3}(y^9)^{1/3} = 3x^2y^3$

d.　$\sqrt[3]{-64s^{12}z^{-6}}$

　　$-(64)^{1/3}(s^{12})^{1/3}(z^{-6})^{1/3} = -4s^4z^{-2} = \dfrac{-4s^4}{z^2}$

e.　$-\sqrt[5]{-32c^{10}d^{-15}}$

　　$-(-(32)^{1/5})(c^{10})^{1/5}(d^{-15})^{1/5} = 2c^2d^{-3} = \dfrac{2c^2}{d^3}$

Example 3

Use your calculator at $\boxed{\text{Fix 3}}$ to obtain the correct results for the following problems.

a.　$(\sqrt{17.5})(\sqrt[3]{21.3})$

PRESS　　　　　　　　　　　　　　　　　　　　　　　**DISPLAY**

17.5 $\boxed{\sqrt{}}$　$\boxed{\times}$　21.3 $\boxed{y^x}$　$\boxed{(}$　1　$\boxed{\div}$　3 $\boxed{)}$　$\boxed{=}$　　　11.596

b.　$\dfrac{(-\sqrt{38.2})(\sqrt[5]{117.6})}{\sqrt{10^3}}$

PRESS　　　　　　　　　　　　　　　　　　　　　　　**DISPLAY**

38.2 $\boxed{\sqrt{}}$　$\boxed{+/-}$　　$\boxed{\times}$　117.6 $\boxed{y^x}$

.2 $\boxed{\div}$　1 $\boxed{\text{EE}}$ 3 $\boxed{\sqrt{}}$　$\boxed{=}$　　　　　$-5.071 - 01$

Note that the answer to b is displayed in scientific notation. Also note that in this and the following problems, the root is frequently entered as a decimal when its reciprocal is obvious; for example: the fourth root = 0.25; the fifth root = 0.2; and so on.

c. $\sqrt[7]{\left(\dfrac{\sqrt[4]{121.6}}{\sqrt[5]{10^4}}\right)^3}$

PRESS	DISPLAY	COMMENT
⟨ 121.6 $\boxed{y^x}$ .25 $\boxed{\div}$ 1 $\boxed{EE}$ 4 $\boxed{y^x}$ .2 ⟩	5.263 − 01	Parentheses evaluated
$\boxed{y^x}$ ⟨ 3 $\boxed{\div}$ 7 ⟩ $\boxed{=}$	7.595 − 01	Parentheses raised to 3/7 power

d. $\sqrt[5]{(-11863.24)^{1/3}}$

PRESS	DISPLAY
11863.24 $\boxed{y^x}$ 3 $\boxed{1/x}$ $\boxed{=}$ $\boxed{y^x}$	
5 $\boxed{1/x}$ $\boxed{=}$ $\boxed{+/-}$	−1.869

or

11863.24 $\boxed{y^x}$ 15 $\boxed{1/x}$ $\boxed{=}$ $\boxed{+/-}$	−1.869

PROBLEMS

Perform the indicated operations for the monomials in Problems 1 to 14.

1. $(2.5xy^2z)^2$

2. $(1.7x^{-1}yz^{-2})^3$

3. $\left(\dfrac{3.4a}{bc^2}\right)^2$

4. $\left(\dfrac{1.5s^{-2}t}{v^{-3}}\right)^3$

5. $\left(\dfrac{5\mu^{1.2}t^{3/2}}{s^{2.3}}\right)^2$

6. $(1.6e^2f^4)^{1/2}$

7. $\left(\dfrac{8a^3b^6}{c^9}\right)^{1/3}$

8. $\sqrt[3]{-27x^{15}y^{12}}$

9. $-\sqrt{121\,x^2y^6}$

10. $-\sqrt[5]{32d^{-5}e^{-10}}$

11. $-\sqrt[3]{-64x^{-3}y^6}$

12. $\sqrt{\dfrac{17.2i^2j^4}{k^{-4}}}$

13. $\left(\dfrac{-32.5\mu^2}{t^4}\right)^{1/3}$

14. $-\left(\dfrac{62.5w^{-5}}{x^2y^{-2}}\right)^{1/4}$

Using your calculator, evaluate Problems 15 to 26.

15. $\sqrt{21.4} \cdot \sqrt[3]{16.2}$

16. $\sqrt{13.2^3} \cdot \sqrt[3]{13.2^2}$

17. $\sqrt{16.1^{2.4}} \cdot \sqrt[4]{105.9}$

18. $\sqrt[5]{10^4} \cdot \sqrt{\pi^3}$

19. $(\sqrt[4]{35.6^3})(16.2)^{1.75}$

20. $(\sqrt[3]{\pi})(\sqrt{1.75})^{0.6}$

21. $(93.62)^{1/7}(64.31)^{1/5}$

22. $\dfrac{\sqrt[3]{16.21^2}}{\sqrt{17.2^3}}$

23. $\dfrac{(\sqrt[4]{117.4^3})(32.5)^{2/3}}{\sqrt[3]{15.1}}$

24. $((\sqrt{62.4})(\sqrt[3]{13.1}))^{1/2}$

25. $\dfrac{\sqrt[5]{\pi^2}}{(161.7)^{1/3}(24.5)^{1/4}}$

26. $\left(\dfrac{(63.1)^{1/3}(\sqrt{26.5})}{17.2}\right)^{1/3}$

6.2 SIMPLIFYING THE REPRESENTATION OF POWERS AND RADICAL SIGNS

If two variables raised to the *same* power or under the same radical sign are to be multiplied or divided, you can combine the powers; the *same* power or radical sign is then associated with the combination. For example,

$$x^2 y^2 = (xy)^2$$

To verify this premise, assign the values 3 and 4 to x and y, respectively. Substituting these values,

$$3^2 \cdot 4^2 = 9 \cdot 16 = 144$$
$$(3 \cdot 4)^2 = (12)^2 = 144$$

Similarly, for radical signs,

$$\sqrt{25} \cdot \sqrt{16} = \sqrt{25 \cdot 16} = \sqrt{400} = 20$$

Also, $5 \cdot 4 = 20$

Because there is no restriction on what the power can be, the power may also indicate a root:

$$(\sqrt{3})(\sqrt{12}) = 3^{1/2} \cdot 12^{1/2} = (3 \cdot 12)^{1/2} = \sqrt{36} = 6$$

Note that just as the laws of exponents do not apply to the addition or subtraction of bases raised to powers, bases with the same radical sign cannot be combined for addition or subtraction. For example,

$$\sqrt{4} + \sqrt{9} \neq \sqrt{13}$$

rather, $\sqrt{4} + \sqrt{9} = 2 + 3 = 5$

Example 1

Simplify the following formats by combining powers where appropriate.

a. $s^3 \cdot t^3$
    $(st)^3$

b. $\dfrac{u^{1.5}}{v^{1.5}}$

    $\left(\dfrac{u}{v}\right)^{1.5}$

c. $x^2 y^4$
    $(x)^2(y^2)^2 = (xy^2)^2$

d. $\sqrt{x^2 y^3}$
    $\sqrt{x^2 y^2 y} = \sqrt{(xy)^2}\sqrt{y} = xy\sqrt{y}$

e. $\sqrt{45}$
    $\sqrt{9 \cdot 5} = \sqrt{9}\sqrt{5} = 3\sqrt{5}$

f. $\sqrt{3}(\sqrt{12} + \sqrt{27})$
    $\sqrt{3 \cdot 12} + \sqrt{3 \cdot 27} = \sqrt{36} + \sqrt{81} = 6 + 9 = 15$

g. $\sqrt{\dfrac{1}{27}}$

    $\dfrac{\sqrt{1}}{\sqrt{9 \cdot 3}} = \dfrac{\sqrt{1}}{\sqrt{9}\sqrt{3}} = \dfrac{1}{3\sqrt{3}}$

Example 1g illustrates a display form that is simpler than the original one but which for mathematical reasons is less than desirable. Radical expressions usually yield an irrational number, that is, a number not completely an integer or a quotient of an integer. Therefore, if you encounter a radical expression in the denominator of a fraction, *clear* the denominator of this condition, especially if the numerator is a polynomial, which would require that each term of the numerator be divided by the denominator. To clear the denominator, multiply both the numerator and the denominator by the radical expression. As explained in previous chapters, this effectively multiplies the fraction by 1:

$$\frac{1}{3\sqrt{3}} \cdot \frac{\sqrt{3}}{\sqrt{3}} = \frac{\sqrt{3}}{3(\sqrt{3})^2} = \frac{\sqrt{3}}{9}$$

Example 2

Simplify the format of the following expressions.

a. $\dfrac{25}{\sqrt{5}}$

$\dfrac{25}{\sqrt{5}} \cdot \dfrac{\sqrt{5}}{\sqrt{5}} = \dfrac{25\sqrt{5}}{5} = 5\sqrt{5}$

b. $\sqrt{\dfrac{8}{27}}$

$\dfrac{\sqrt{4 \cdot 2}}{\sqrt{9 \cdot 3}} = \dfrac{2\sqrt{2}}{3\sqrt{3}} \cdot \dfrac{\sqrt{3}}{\sqrt{3}} = \dfrac{2\sqrt{6}}{9}$

c. $\dfrac{1 + \sqrt{5}}{\sqrt{45}}$

$\dfrac{1 + \sqrt{5}}{\sqrt{9 \cdot 5}} = \dfrac{1 + \sqrt{5}}{3\sqrt{5}} \cdot \dfrac{\sqrt{5}}{\sqrt{5}} = \dfrac{\sqrt{5} + 5}{15}$

d. $\dfrac{x}{\sqrt{x^2 y}}$

$\dfrac{x}{x\sqrt{y}} \cdot \dfrac{\sqrt{y}}{\sqrt{y}} = \dfrac{\sqrt{y}}{y}$

PROBLEMS

Simplify Problems 1 to 15 by combining powers or removing radical signs. Do *not* use your calculator.

1. $x^4 y^4$

2. $\dfrac{p^{3/2}}{s^{3/2}}$

3. $u^3 v^6$

4. $\dfrac{x^{1.5} v^3}{w^{4.5}}$

5. $\sqrt{300}$

6. $\sqrt{12} + \sqrt{108}$

7. $\dfrac{\sqrt{28}}{\sqrt{63}}$

8. $(\sqrt{75a^2})(\sqrt{180b^2})$

9. $\sqrt{9c^4 d^6 \cdot 10^2}$

10. $\sqrt{\dfrac{10^3}{30}}$

11. $(1 + \sqrt{7})(3\sqrt{7})$

12. $5\sqrt{5}(\sqrt{20} - \sqrt{180})$

13. $\dfrac{1 + \sqrt{6}}{\sqrt{54}}$ **14.** $\dfrac{\sqrt{x^5}}{\sqrt{x^2 y}}$ **15.** $\dfrac{\sqrt{x^5 y}}{\sqrt{xy^5}}$

6.3 PRODUCTS OF POLYNOMIALS

The products of algebraic terms studied so far involved either a monomial multiplier or multiplicand. But frequently the multiplier and multiplicand are both polynomials. This section discusses methods for solving algebraic problems involving polynomials. We use integers written in the same format as the algebraic expressions, for example, 12×8, which yields the product 96. To restructure each number into a polynomial, first arbitrarily separate the multiplicand into $3 + 9$ and then arbitrarily break down the multiplier into $7 + 1$, as represented by Equation (6.2):

$$
\begin{aligned}
12 \cdot 8 & \\
= (3 + 9)(1 + 7) & \\
= 3 \cdot 1 + 3 \cdot 7 + 9 \cdot 1 + 9 \cdot 7 & \\
= 3 + 21 + 9 + 63 \qquad\qquad & = 96
\end{aligned}
\tag{6.2}
$$

Notice that to expand the product of the two binomials in Equation (6.2) and still obtain the correct answer, the first term of the multiplicand (3) has to be multiplied by both the 1 and the 7 of the multiplier; the second term of the multiplicand (9) then has to be multiplied by the same 1 and 7 of the multiplier. Also note that there are a total of four individual multiplications, which are then summed. Generally, there are as many individual multiplications as the number of terms in the multiplicand times the number of terms in the multiplier. Thus, from Equation (6.2), there are two terms in the multiplier times two terms in the multiplicand, for a total of four individual products. If a multiplier has two terms and the multiplicand three terms, then the total number of individual products summed is six.

Equation (6.3) expands this same concept, except literal terms are substituted for the numbers:

$$
(a + b)(c + d) = a \cdot c + a \cdot d + b \cdot c + b \cdot d
\tag{6.3}
$$

Examples 1 and 2 illustrate the principles involved.

Example 1

Perform the indicated multiplications.

a. $(x + 2y)(2x + 3y)$
   $(x)(2x) + (x)(3y) + (2y)(2x) + (2y)(3y)$
$= 2x^2 + 3xy + 4xy + 6y^2$
$= 2x^2 + 7xy + 6y^2$

b. $(2u + 3y)(u - 2y)$

$(2u)(u) + (2u)(-2y) + (3y)(u) + (3y)(-2y)$

$= 2u^2 - 4uy + 3uy - 6y^2$

$= 2u^2 - uy - 6y^2$

c. $(3a - 4.1b)(2c + 3.5d)$

$(3a)(2c) + (3a)(3.5d) + (-4.1b)(2c) + (-4.1b)(3.5d)$

$= 6ac + 10.5ad - 8.2bc - 14.35bd$

Notice that these three problems placed the multiplicand and multiplier side by side, and then the product was expanded according to Equation (6.3). Sometimes it is more convenient to place the multiplier below the multiplicand, and then during the expansion process align like terms for easier addition.

Example 2

Perform the indicated multiplications.

a.
$$3s + 2t$$
(times) $s - t$

$3s^2 + 2st$	product of $3s + 2t$ times s
$\underline{\quad - 3st - 2t^2}$	product of $3s + 2t$ times $(-t)$
$3s^2 - st - 2t^2$	product

b.
$$1.5d - 6.3e$$
(times) $2.6d + 1.7e$

$3.9d^2 - 16.38de$	product of $1.5d - 6.3e$ times $2.6d$
$\underline{\qquad\quad 2.55de - 10.71e^2}$	product of $1.5d - 6.3e$ times $1.7e$
$3.9d^2 - 13.83de - 10.71e^2$	product

c.
$$3.2xy^2 - 3.1z$$
(times) $1.6xy^2 - 5.2z$

$5.12x^2y^4 - 4.96xy^2z$	product of $3.2xy^2 - 3.1z$ times $1.6xy^2$
$\underline{\qquad\quad 16.64xy^2z + 16.12z^2}$	product of $3.2xy^2 - 3.1z$ times $(-5.2z)$
$5.12x^2y^4 - 21.6xy^2z + 16.12z^2$	product

Either method produces the correct result, but you are encouraged to become familiar with both formats. Several rules for simple polynomial expansion follow the format of Example 1 for rapid calculation; however, if the polynomial exceeds two or three terms, the simplified rules are impractical and so the format of Example 2 is then the most desired approach.

PROBLEMS

In Problems 1 to 15 perform the indicated multiplications, using the approach of either Example 1 or Example 2.

1. $(x + 2)(x + 3)$

2. $(2y + 4)(3y - 6)$

3. $(4.1s - 2.3)(3.6s - 4.7)$

4. $(a - b)(2a + 3b)$

5. $(2i + j)(3i - 2j)$

6. $(6.1k + 2.4m)(8.5k + 5.2m)$

7. $(x^2 + 2y^3)(-2x^2 + 3y^3)$

8. $(4.7p^4 - 2.1q)(5.2p^4 + 3.7q)$

9. $(4ab - c)(3ab + 2c)$

10. $(6a^2b + 2d)(7a^2b - 4d)$

11. $(3x + 2y)(4x - 3z)$

12. $(4.5w - 3.1x)(3.6w + 2.7y)$

13. $(8.7d^2 + 1.6f)(2.9d^2 - 3.7g)$

14. $(6.2a^2b - 1.4c^2)(-4.2a^2b + 5.1e)$

15. $(3.2x - 4.7y)(1.6w + 2.7z)$

6.4 SPECIAL POLYNOMIAL EXPANSIONS

Certain binomial products—squares, cubes, and trinomial squares—can be reduced to predictable forms so that they can be calculated mentally. A calculator may be necessary only when the coefficient is not a simple integer, such as 2.17^3. The following five cases and their examples explore such special cases.

Case 1: Product of the Sum and Difference of the Same Binomials

The product of two binomials that represent the sum and difference of the same two terms is a single binomial. For example,

$$\underbrace{(ax + by)}_{\substack{\text{Sum of} \\ \text{two terms}}}\underbrace{(ax - by)}_{\substack{\text{Difference of} \\ \text{same two terms}}} = (ax)^2 - (by)^2$$

When expanded, the following results:

$$
\begin{array}{l}
\phantom{(\text{times})\ } ax + by \\
(\text{times})\ \underline{ax - by} \\
\phantom{(\text{times})\ } (ax)^2 + (ax)(by) \\
\phantom{(\text{times})\ } \underline{\ - (ax)(by) - (by)^2} \\
\phantom{(\text{times})\ } (ax)^2 - (by)^2
\end{array}
$$

product of $ax + by$ times ax
product of $ax + by$ times $(-by)$
product

Notice that the middle term cancels out. A general rule is:

To form the product of the sum and difference of the same two terms of a binomial, square the first term and subtract the square of the second term.

Example 1

Mentally form the following products, using your calculator only when necessary to square the coefficients.

a. $(2x + 3y)(2x - 3y)$
$(2x)^2 - (3y)^2 = 4x^2 - 9y^2$

b. $(3x^2 + 4y^3)(3x^2 - 4y^3)$
$(3x^2)^2 - (4y^3)^2 = 9x^4 - 16y^6$

c. $(2ab + 5c^2)(2ab - 5c^2)$
$(2ab)^2 - (5c^2)^2 = 4a^2b^2 - 25c^4$

d. $\left(\dfrac{4.6s}{t} + 3.1v\right)\left(\dfrac{4.6s}{t} - 3.1v\right)$
$\left(\dfrac{4.6s}{t}\right)^2 - (3.1v)^2 = \dfrac{21.16s^2}{t^2} - 9.61v^2$

e. $(17.6i^2j^{-3} + 6.4)(17.6i^2j^{-3} - 6.4)$
$\left(\dfrac{17.6i^2}{j^3}\right)^2 - (6.4)^2 = \dfrac{309.76i^4}{j^6} - 40.96$

PROBLEMS

Mentally determine the products in Problems 1 to 10. Use your calculator only when necessary for coefficients.

1. $(a + 2b)(a - 2b)$

2. $(3x - y)(3x + y)$

3. $(2xy - z)(2xy + z)$

4. $(4s^2 - 2)(4s^2 + 2)$

5. $(6uv^2 - 5)(6uv^2 + 5)$

6. $\left(\dfrac{7p}{q} + r\right)\left(\dfrac{7p}{q} - r\right)$

7. $(1.5s^{-1} + t^3)(1.5s^{-1} - t^3)$

8. $(2.15v^2 + ut^2)(2.15v^2 - ut^2)$

9. $(8.62\pi + s^{-2}t^{-3})(8.62\pi - s^{-2}t^{-3})$

10. $\left(\dfrac{\sqrt{17}d^3}{6} - \sqrt{3}\right)\left(\dfrac{\sqrt{17}d^3}{6} + \sqrt{3}\right)$

Case 2: Product of Two Binomials with the Same Variables

When two binomials with the same variables are multiplied, and they do not fall into Case 1, the result is a trinomial. Consider the following generalized case, where a, b, c, d represent any coefficient and x and y are the variables:

$$ax + by$$
(times) $cx + dy$

$$(ac)x^2 + (bc)xy \qquad\qquad \text{product of } ax + by \text{ times } cx$$
$$(ad)xy + (bd)y^2 \qquad\qquad \text{product of } ax + by \text{ times } dy$$
$$(ac)x^2 + (bc + ad)xy + (bd)y^2 \qquad \text{product}$$

The following pattern generalizes the rule:

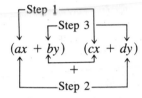

In other words,

Step 1.   Multiply the first terms of both binomials. Result $= (ac)x^2$
Step 2.   Add the product of the inside terms to the product of the outside terms. Result $= (bc + ad)xy$
Step 3.   Multiply the second terms of both binomials. Result $= (bd)y^2$
For the product, add the results of the three steps.

Example 2

Mentally form the following products, using your calculator only when necessary to form the coefficients.

a. $(3s + 2t)(4s + 5t)$
$(3 \cdot 4)s^2 + (2 \cdot 4 + 3 \cdot 5)st + (2 \cdot 5)t^2 = 12s^2 + 23st + 10t^2$

b. $(2x - 3y)(4x + 7y)$
$(2 \cdot 4)x^2 + ((-3) \cdot 4 + 2 \cdot 7)xy + ((-3) \cdot 7)y^2 = 8x^2 + 2xy - 21y^2$

c. $(6s^{-1} + 3t)(-4s^{-1} - 2t)$
$(6 \cdot (-4))s^{-2} + (3 \cdot (-4) + 6 \cdot (-2))s^{-1}t + (3 \cdot (-2))t^2$
$= -24s^{-2} - 24s^{-1}t - 6t^2$

d. $\left(\dfrac{7u}{v} + 2w\right)\left(\dfrac{3u}{v} - 4w\right)$

$(7 \cdot 3)\dfrac{u^2}{v^2} + (2 \cdot 3 + 7 \cdot (-4))\dfrac{uw}{v} + (2 \cdot (-4))w^2 = \dfrac{21u^2}{v^2} - \dfrac{22uw}{v} - 8w^2$

e. $\left(\dfrac{2.6s^{-2.5}}{u} + 4.3t^3\right)\left(\dfrac{1.7s^{-2.5}}{u} - 6.1t^3\right)$

$((2.6) \cdot (1.7))\dfrac{s^{-5}}{u^2} + ((4.3) \cdot (1.7) + (2.6) \cdot (-6.1))\dfrac{s^{-2.5}t^3}{u} + ((4.3) \cdot (-6.1))t^6$

$= \dfrac{4.42}{s^5 u^2} - \dfrac{8.55t^3}{s^{2.5}u} - 26.23t^6$

PROBLEMS

Mentally determine the products in Problems 1 to 10. Use your calculator only when necessary for coefficients.

1. $(x + y)(x - 2y)$

2. $(2x + y)(3x + 3y)$

3. $(x^2 - 4y^2)(x^2 + 5y^2)$

4. $\left(\dfrac{x}{y} + 2z\right)\left(\dfrac{x}{y} - 3z\right)$

5. $\left(\dfrac{2a}{b^2} + 6\right)\left(\dfrac{3a}{b^2} - 4\right)$

6. $(s^{-2} + t)(3s^{-2} - 4t)$

7. $(1.7xy - 3.4z)(5.6xy + 4.8z)$

8. $\left(\dfrac{2.14e^2}{f} - g\right)\left(\dfrac{3.19e^2}{f} + 3.2g\right)$

9. $(2\sqrt{5}ij^2 - \sqrt{7})(3.2\sqrt{5}ij^2 + \sqrt{6})$

10. $\left(\dfrac{\sqrt{\pi d}}{f^3} + 6.1e\right)\left(\dfrac{d}{f^3} - 5.2e\right)$

Case 3: Binomial Squared

A trinomial results from the squaring of a binomial. However, in this case the middle term is less complicated to calculate:

$$
\begin{array}{ll}
\quad ax + by & \\
(\text{times})\ \underline{ax + by} & \\
\quad (ax)^2 + \quad (ab)xy & \text{product of } ax + by \text{ times } ax \\
\qquad\qquad \underline{(ab)xy + (by)^2} & \text{product of } ax + by \text{ times } by \\
\quad (ax)^2 + 2(ab)xy + (by)^2 & \text{product}
\end{array}
$$

Thus the rule is:

A binomial squared is the first term squared plus twice the product of the two terms plus the last term squared.

Example 3

Mentally square the following binomials, using your calculator only when necessary to form the coefficients.

a. $(x + y)^2$
$x^2 + 2xy + y^2$

b. $(x - y)^2$
$x^2 + 2x(-y) + y^2 = x^2 - 2xy + y^2$

Note that the only difference between a and b is the sign of the middle term.

c. $(2x - 3y)^2$
$(2x)^2 + 2(2x)(-3y) + (-3y)^2 = 4x^2 - 12xy + 9y^2$

d. $(3ab + 4c^2)^2$
$(3ab)^2 + 2(3ab)(4c^2) + (4c^2)^2$
$= 9a^2b^2 + 24abc^2 + 16c^4$

e. $\left(\dfrac{4.65p}{s} + \dfrac{2.6}{q^2}\right)^2$

$\left(\dfrac{4.65p}{s}\right)^2 + 2\left(\dfrac{4.65p}{s}\right)\left(\dfrac{2.6}{q^2}\right) + \left(\dfrac{2.6}{q^2}\right)^2$

$= \dfrac{21.6225p^2}{s^2} + \dfrac{24.18p}{sq^2} + \dfrac{6.76}{q^4}$

PROBLEMS

Mentally calculate the products in Problems 1 to 10. Use your calculator only when necessary to determine coefficients.

1. $(x + 2y)^2$

2. $(3x - 4y)^2$

3. $(2xy - z)^2$

4. $\left(\dfrac{x}{y^2} + 2z\right)^2$

5. $\left(6\dfrac{s}{t} - 2v^2\right)^2$

6. $(3.1st^2 + 4.6uv)^2$

7. $(4.85df^{-2} - e^{-1})^2$

8. $(4.2\pi a + 3.2b^{-3})^2$

9. $\left(\dfrac{\sqrt{5x}}{y^2} - \sqrt{9.2}z^{-2}\right)^2$

10. $\left(\dfrac{4.95c^{-1}}{d^{-2}} - 3.8e^2\right)^2$

Case 4: Binomial Cubed

Expanding the square of a binomial into a cube requires one more multiplication of the binomial by itself. Beginning with the product of the square (case 3), the following expansion results:

$(ax)^2 + 2(ab)xy + (by)^2$	original binomial squared
(times) $ax + by$	original binomial
$(a^3)x^3 + 2(a^2b)x^2y +\ \ (ab^2)xy^2$	original binomial times ax
$\quad\quad\quad (a^2b)x^2y + 2(ab^2)xy^2 + (b^3)y^3$	original binomial times by
$(a^3)x^3 + 3(a^2b)x^2y + 3(ab^2)xy^2 + (b^3)y^3$	product

Simplifying the product, we get Equation (6.4):

$$(ax)^2 + 2(ab)(xy) + (by)^2 \times (ax + by)$$
$$= (ax)^3 + 3(ax)^2(by) + 3(ax)(by)^2 + (by)^3 \qquad (6.4)$$

The first observation is that the answer is a polynomial consisting of four terms. The second observation is the unusual pattern assumed by ax and by, the first and second terms, respectively, of the original binomial that is being cubed. If we use the symbol f to represent the *first* term (ax) of the binomial, meaning both the coefficient (a) and the *entire* variable (x), then f assumes the following progression of decreasing powers:

$$f^3 \cdot\ \cdot\ \cdot f^2 \cdot\ \cdot\ \cdot f^1 \cdot\ \cdot\ \cdot f^0$$

If we use the symbol s to represent the *second* term (by) of the binomial, meaning both the coefficient (b) and the *entire* variable (y), then s assumes the following progression of increasing powers:

$$s^0 \cdot\ \cdot\ \cdot s^1 \cdot\ \cdot\ \cdot s^2 \cdot\ \cdot\ \cdot s^3$$

Combining the definitions of f and s, we have

$$(f + s)^3 = f^3 + 3f^2s + 3fs^2 + s^3 \qquad (6.5)$$

Equation (6.5) represents mathematically the rule for the expansion of a binomial cubed:

> A binomial cubed is the first term cubed plus three times the product of the first term squared times the second term plus three times the product of the first term times the second term squared plus the second term cubed.

Example 4

Cube the following binomials, using your calculator for coefficients only when necessary.

a. $(x + y)^3$
    $x^3 + 3x^2y + 3xy^2 + y^3$

b. $(x - y)^3$
$$x^3 + 3(x^2)(-y) + 3(x)(-y)^2 + (-y)^3$$
$$= x^3 - 3x^2y + 3xy^2 - y^3$$

c. $(2x + 3y)^3$
$$(2x)^3 + 3(2x)^2(3y) + 3(2x)(3y)^2 + (3y)^3$$
$$= 8x^3 + 36x^2y + 54xy^2 + 27y^3$$

d. $(3x - 5y)^3$
$$(3x)^3 + 3(3x)^2(-5y) + 3(3x)(-5y)^2 + (-5y)^3$$
$$= 27x^3 - 135x^2y + 225xy^2 - 125y^3$$

e. $\left(2.1t^2 - \dfrac{3.6}{u}\right)^3$

$$(2.1t^2)^3 + 3(2.1t^2)^2\left(\dfrac{-3.6}{u}\right) + 3(2.1t^2)\left(\dfrac{-3.6}{u}\right)^2 + \left(\dfrac{-3.6}{u}\right)^3$$

$$= 9.261t^6 - \dfrac{47.628t^4}{u} + \dfrac{81.648t^2}{u^2} - \dfrac{46.656}{u^3}$$

Example 4e has coefficients that require calculator assistance to expand the cube. The two middle terms are not difficult because they need only the use of the square function on the calculator (which produces the same result whether the number is positive or negative). However, the first or last term of the cube expansion can present problems when using the $\boxed{y^x}$ calculator function if either term is negative. As explained for keying in a radicand with an odd-numbered root, if a number is negative, you must execute the $\boxed{+ / -}$ function after the $\boxed{y^x}$ function, or a flashing (or ERROR) indication occurs. Note the following key sequence to obtain each of the four coefficients for Example 4e.

Example 5

Use your calculator to work the following problems.

a. Determine the coefficients of Example 4e.

PRESS	DISPLAY	COMMENTS
2.1 $\boxed{y^x}$ 3 $\boxed{=}$	9.261	Coefficient of f^3 term
3 $\boxed{\times}$ 2.1 $\boxed{x^2}$ $\boxed{\times}$ 3.6 $\boxed{+/-}$ $\boxed{=}$	−47.628	Coefficient of $3f^2s$ term
3 $\boxed{\times}$ 2.1 $\boxed{\times}$ 3.6 $\boxed{+/-}$ $\boxed{x^2}$ $\boxed{=}$	81.648	Coefficient of $3fs^2$ term
3.6 $\boxed{y^x}$ 3 $\boxed{=}$ $\boxed{+/-}$	−46.656	Coefficient of s^3 term

b.   Determine ($\boxed{\text{Fix 3}}$) the solution of $(\sqrt{\pi}a - 3.29b)^3$. Note that to find the coefficient of $3f^2s$, it was *not* necessary to key in the sequence $\boxed{\sqrt{\ }}$ $\boxed{x^2}$ because the functions mutually canceled each other. This applies *only* here because the coefficient preceding a is a radical; otherwise the x^2 function is necessary.

PRESS	DISPLAY	COMMENTS
$\boxed{\pi}$ $\boxed{\sqrt{\ }}$ $\boxed{y^x}$ 3 $\boxed{=}$	5.568	Coefficient of f^3 term
3 $\boxed{\times}$ $\boxed{\pi}$ $\boxed{\sqrt{\ }}$ $\boxed{x^2}$ $\boxed{\times}$ 3.29 $\boxed{+/-}$ $\boxed{=}$	-31.008	Coefficient of $3f^2s$ term
3 $\boxed{\times}$ $\boxed{\pi}$ $\boxed{\sqrt{\ }}$ $\boxed{\times}$ 3.29 $\boxed{+/-}$ $\boxed{x^2}$ $\boxed{=}$	57.556	Coefficient of $3fs^2$ term
3.29 $\boxed{y^x}$ 3 $\boxed{=}$ $\boxed{+/-}$	-35.611	Coefficient of s^2 term

Answer is $5.568a^3 - 31.008a^2b + 57.556ab^2 - 35.611b^3$.

PROBLEMS

Mentally calculate the products in Problems 1 to 10, using your calculator only when necessary to determine coefficients.

1. $(s + 2t)^3$

2. $(2u + v)^3$

3. $(3a - b)^3$

4. $(3x - 2y)^3$

5. $(4x^2 - 5y)^3$

6. $(5xy^2 - 2z^2)^3$

7. $(-2.5d - 1.5e)^3$

8. $(4.8g^2 + 5.15hi^2)^3$

9. $(\sqrt{23}ab - \pi)^3$

10. $(3\sqrt{15}x^2y^2 + 2.4z)^3$

Case 5: Trinomial Squared

Keeping with earlier methods of analysis, a, b, and c represent the coefficients, and x, y, and z represent the variables. The square of the trinomial proceeds as follows:

$$
\begin{aligned}
&\quad ax + by + cz \\
(\text{times})\ &\quad ax + by + cz \\
\hline
&(a^2)x^2 + (ab)xy + (ac)xz \\
&\qquad\qquad (ab)xy \qquad\quad + (b^2)y^2 + (bc)yz \\
&\qquad\qquad\qquad\qquad (ac)xz \qquad\quad + (bc)yz + (c^2)z^2 \\
\hline
&(a^2)x^2 + 2(ab)xy + 2(ac)xz + (b^2)y^2 + 2(bc)yz + (c^2)z^2
\end{aligned}
$$

To begin the process of establishing a pattern that can be easily remembered, rewrite the product to combine ax, by, and cz. Remember that this product represents the first (f), second (s), and third (t) terms of the trinomial, respectively.

$$(ax)^2 + 2(ax)(by) + 2(ax)(cz) + (by)^2 + 2(by)(cz) + (cz)^2$$

or $\quad\quad f^2 \;+\;\; 2fs \;\;\;+\;\;\;\; 2ft \;\;+\; s^2 \;+\;\;\; 2st \;\;+\; t^2$

To analyze the pattern, rewrite the trinomial as the sum of its individual terms:

$$f + s + t$$

Beginning at the left, multiply the starting point f, by itself. Then add to this resultant twice the product of f times each remaining term to its right. Thus this initial resultant is

$$f^2 + 2fs + 2ft$$

Move right to the next term (s) and repeat the pattern, remembering that (s) is now the new starting point. This additional resultant is

$$s^2 + 2st$$

Now progress right to t. Repeating the process results only in t^2 because there is nothing to the right of t that still remains to be multiplied. When all the processes are summed, the resultant expansion of squaring the trinomial is obtained:

$$(f + s + t)^2 = f^2 + 2fs + 2ft + s^2 + 2st + t^2 \tag{6.6}$$

We can state the rule as:

A trinomial squared is the first term squared plus twice the product of the first and second terms plus twice the product of the first and third terms plus the second term squared plus twice the product of the second and third terms plus the third term squared.

Example 6

Mentally perform the indicated operations.

a. $(x + 2y + 3z)^2$
     $x^2 + 2(x)(2y) + 2(x)(3z) + (2y)^2 + 2(2y)(3z) + (3z)^2$
     $- x^2 + 4xy + 6xz + 4y^2 + 12yz + 9z^2$

b. $(2x + 3y + 4z)^2$
     $(2x)^2 + 2(2x)(3y) + 2(2x)(4z) + (3y)^2 + 2(3y)(4z) + (4z)^2$
     $= 4x^2 + 12xy + 16xz + 9y^2 + 24yz + 16z^2$

c. $(3s - 2t - 6v)^2$
     $(3s)^2 + 2(3s)(-2t) + 2(3s)(-6v) + (-2t)^2 + 2(-2t)(-6v) + (-6v)^2$
     $= 9s^2 - 12st - 36sv + 4t^2 + 24tv + 36v^2$

d. $(x^2 + 2xy + 3yz)^2$
$(x^2)^2 + 2(x^2)(2xy) + 2(x^2)(3yz) + (2xy)^2 + 2(2xy)(3yz) + (3yz)^2$
$= x^4 + 4x^3y + 6x^2yz + 4x^2y^2 + 12xy^2z + 9y^2z^2$

e. $(2i - 3j + 4kl)^2$
$(2i)^2 + 2(2i)(-3j) + 2(2i)(4kl) + (-3j)^2 + 2(-3j)(4kl) + (4kl)^2$
$= 4i^2 - 12ij + 16ikl + 9j^2 - 24jkl + 16k^2l^2$

PROBLEMS

Mentally calculate the products in Problems 1 to 10, using your calculator only when necessary to determine coefficients.

1. $(2a - b + 2c)^2$

2. $(e + 2f + 3g)^2$

3. $(3s - t + 4v)^2$

4. $(4w - 4x + 4y)^2$

5. $(-5i + 2k^2 - 3j^3)^2$

6. $(2ab - ac + bc^2)^2$

7. $(3.5x + 2.4y - 3.6z)^2$

8. $(3.7m^2 - 2.5n + \pi)^2$

9. $(\sqrt{5}x^2 - \sqrt{3}y + \sqrt{2}z)^2$

10. $\left(\dfrac{3.2s}{t} - 2.1uv + \dfrac{4.6}{w}\right)^2$

6.5 FACTORING

In Section 6.4, one form of polynomial expansion was the product of the sum and difference of the same terms of a binomial. For example,

$$(a + b)(a - b) = a^2 - b^2$$

Suppose that instead of being given the original two binomials, you are given $a^2 - b^2$ and asked, "What are the two original binomials that when multiplied together result in this product?" (This is similar to being given the answer and then asked "What is the question?") The process of breaking up a more complex term into its simpler components is known as *factoring*. Just as with expanding a polynomial, factoring can assume four different formats, as we see in this section.

Case 1: Polynomials with a Common Monomial Factor

The simplest form of a common factor in a polynomial is a common coefficient. If the variables x, y, and z are all multiplied by coefficient p, then the expression looks like

$$px + py + pz$$

Because p is common to all terms, it may be factored out; then it can function as a a monomial times the remaining polynomial:

$$p(x + y + z)$$

The following three rules apply when factoring expressions of this format:

1. For all the coefficients and all the variables common to each term within the expression, determine the highest multiple that will divide into each coefficient or like variable. This is called the highest common multiple (HCM).

2. Reduce the complexity of the expression by dividing each term by the HCM.

3. Group these reduced terms within parentheses, and then multiply it by the HCM.

Note that a variable not common to all terms of an expression *cannot* have a HCM.

Example 1

Factor the following expressions.

a. $6a + 3$

    HCM for 6, 3 = 3   step 1

    $\dfrac{6a}{3} + \dfrac{3}{3} = 2a + 1$ step 2

    Factors $= 3(2a + 1)$ step 3

b. $5x^3y^2 + 15x^2y^3$

    HCM for 5, 15 = 5   step 1

    for x^3, $x^2 = x^2$

    for y^2, $y^3 = y^2$

    $\therefore$ HCM $= 5x^2y^2$

    $\dfrac{5x^3y^2}{5x^2y^2} + \dfrac{15x^2y^3}{5x^2y^2} = x + y$ step 2

    Factors $= 5x^2y^2(x + y)$ step 3

c. $\dfrac{1}{6}xy^3 - \dfrac{1}{8}x^3y + \dfrac{1}{2}x^2y^2$

HCM for $\dfrac{1}{6}, \dfrac{1}{8}, \dfrac{1}{2} = \dfrac{1}{2}$ step 1

for x, x^3, $x^2 = x$

for y^3, y, $y^2 = y$

$\therefore$ HCM $= \dfrac{1}{2}xy$

$\dfrac{\frac{1}{6}xy^3}{\frac{1}{2}xy} - \dfrac{\frac{1}{8}x^3y}{\frac{1}{2}xy} + \dfrac{\frac{1}{2}x^2y^2}{\frac{1}{2}xy} = \dfrac{1}{3}y^2 - \dfrac{1}{4}x^2 + xy$ step 2

Factors $= \dfrac{1}{2}xy\left(\dfrac{1}{3}y^2 - \dfrac{1}{4}x^2 + xy\right)$ step 3

PROBLEMS

Factor the expressions in Problems 1 to 15.

1. $7R_1 + 21R_2$

2. $6ir + 24iz$

3. $\dfrac{I_1}{10} - \dfrac{3I_2}{40}$

4. $3a^2b + 21ab^2$

5. $5X_L^2 X_C^3 + 15X_L X_C - 75X_L^3 X_C^2$

6. $\dfrac{z}{4} + \dfrac{pz}{8}$

7. $3.6R_1R_2 + 1.8R_1^2 R_2 + 9R_1 R_2^2$

8. $\dfrac{\pi}{4} + \dfrac{\pi d}{16} - \dfrac{\pi d^2}{2}$

9. $4q^3 - 28q^2 + 16q$

10. $ab + a + cb + c$

11. $d_1\left(\dfrac{1}{a_1} + \dfrac{1}{b_1}\right) + d_2\left(\dfrac{1}{a_1} + \dfrac{1}{b_1}\right)$

12. $xy^2z^3 + x^2yz^2 + x^3y^3z$

13. $\dfrac{s}{b_1} + \dfrac{t}{b_1} + \dfrac{s}{b_2} + \dfrac{t}{b_2}$

14. $a^3 + a^2 + a + 1$

15. $u(v + 1) - w(v + 1)$

Case 2: Difference of Two Squares

Section 6.4 showed that the product of the sum and difference of the same two terms was the difference of their squares. This is probably the easiest format to recognize and reduce to its components: Merely take the square root of both terms, and then express the product of their sum and difference.

Example 2

Factor the following expressions.

a. $x^2 - y^2$
$(x + y)(x - y)$

b. $4a^2 - 9b^2$
$(2a)^2 - (3b)^2 = (2a + 3b)(2a - 3b)$

c. $16x^4 - 144y^8$
$(4x^2)^2 - (12y^4)^2 = (4x^2 + 12y^4)(4x^2 - 12y^4)$

d. $36s^2 - 15t^2$
$(6s)^2 - (\sqrt{15}t)^2 = (6s + \sqrt{15}t)(6s - \sqrt{15}t)$

e. $\dfrac{49u^2}{v^2} - 64w^2$

$\left(\dfrac{7u}{v}\right)^2 - (8w)^2 = \left(\dfrac{7u}{v} + 8w\right)\left(\dfrac{7u}{v} - 8w\right)$

PROBLEMS

Factor the expressions in Problems 1 to 10.

1. $a^2b^2 - 4$

2. $d^4e^6 - f^2$

3. $16 - 4x^2$

4. $25a^2 - 75b^2$

5. $\dfrac{f^2}{g^4} - \dfrac{h^6}{i^8}$

6. $(a - 1)^2 - (a + 1)^2$

7. $4(b + 2)^2 - 9b^2$

8. $(a + 1)^4 - (a + 1)^2$

9. $\dfrac{225s^{-2}}{t^4} - \dfrac{64}{v^6}$

10. $16(a^2 - b^2)^2 - 4(a^2 + b^2)^2$

Case 3: Factors of the Format $f^2 \pm 2fs + s^2$

Binomials squared may result in just two formats, only depending upon whether the binomial represents the sum or difference of two terms. Letting f represent both the coefficient and/or variable of the first term and s both the coefficient and/or variable of the second term, there are the following two possibilities:

$$(f + s)^2 = f^2 + 2fs + s^2$$
$$(f - s)^2 = f^2 - 2fs + s^2$$

or

Both conditions can be generalized as

$$(f \pm s)^2 = f^2 \pm 2fs + s^2 \qquad\qquad (6.7)$$

Working in reverse, any expression of the format of the trinomial in Equation (6.7) can be factored into a binomial squared. To test the trinomial for this condition,

1. Multiply the square root of the absolute value of the first term by the square root of the absolute value of the second term.
2. Double the product.

If the product of step 2 equals the absolute value of the middle term of the trinomial, the format of Equation (6.7) exists. Note that when referring to taking the square root of terms f or s, it is implied that either they are the two terms of the trinomial whose variable components represent the highest powers within the trinomial, or one term is a coefficient. The terms' relative positions within the trinomial are not important. Also note that if the middle term is positive, the binomial's sum of terms is squared; if the middle term is negative, the binomial's difference of terms is squared.

Example 3

Factor the following expressions.

a. $4x^2 + 4xy + y^2$ factor
$2\sqrt{4x^2}\sqrt{y^2} = 4xy$ test
$(2x + y)^2$ solution

b. $9a^2 - 24ab + 16b^2$ factor
$2\sqrt{9a^2}\sqrt{16b^2} = 24ab$ test
$(3a - 4b)^2$ solution

c. $36a^2b^2 + c^2 + 12abc$ factor
$2\sqrt{36a^2b^2}\sqrt{c^2} = 12abc$ test
$(6ab + c)^2$ solution

d. $\dfrac{4a^2}{b^2} - \dfrac{8a}{b} + 4$ factor

$2\sqrt{\dfrac{4a^2}{b^2}}\sqrt{4} = \dfrac{8a}{b}$ test

$\left(\dfrac{2a}{b} - 2\right)^2$ solution

e. $(xy)^8 - 2\sqrt{3}(xy)^4 + 3$ factor

$2\sqrt{(xy)^8}\sqrt{3} = 2\sqrt{3}(xy)^4$ test

$(x^4y^4 - \sqrt{3})^2$ solution

PROBLEMS

Factor the expressions in Problems 1 to 15.

1. $c^2 + 2c + 1$

2. $x^2 - 10xy + 25y^2$

3. $a^2 + 12ab + 36b^2$

4. $\dfrac{1}{4}s^2 + st + t^2$

5. $\dfrac{a^2}{9} - \dfrac{ab}{3} + \dfrac{b^2}{4}$

6. $9u^2v^2 + 54uvw + 81w^2$

7. $9R_1^2 - 36R_1R_2 + 36R_2^2$

8. $81d^2 - 90de + 25e^2$

9. $5g^2 + 2\sqrt{5}g + 1$

10. $10a^4 - 2\sqrt{10}a^2b + b^2$

11. $\dfrac{81u^4}{v^2} + \dfrac{72u^2}{v} + 16$

12. $\dfrac{36a^2}{25} + \dfrac{16b^2}{9} + \dfrac{48ab}{15}$

13. $\dfrac{3}{25}x^4y^2 + \dfrac{2\sqrt{3}}{5}x^2yz + z^2$

14. $(a - 1)^2 + 4b(a - 1) + 4b^2$

15. $(b^2 + 2)^4 - 18c(b^2 + 2)^2 + 81c^2$

Case 4: Factors of the Format $n_1x^2 + n_2xy + n_3y^2$

This format results from the product of two binomials with common variables but different coefficients associated with the variables. (This was Case 2 in Section 6.4.) The following product describes each n of the format:

$$(ax + by)(cx + dy) = \underbrace{ac}_{n_1}x^2 + \underbrace{(bc + ad)}_{n_2}xy + \underbrace{(bd)}_{n_3}y^2$$

The coefficients n_1 and n_3 are relatively easy to determine: They are the products of the coefficients of the first and second terms of each binomial, respectively. The value of n_2 is a little more difficult to determine; it is the sum of the products of the inside coefficients and the products of the outside coefficients.

This type of format does not always provide an immediately recognizable solution and thus may require several intuitive guesses before you obtain the correct factored solution. For example, to determine the factors of $x^2 + 20x + 36$, the solution must take the form $(x + ?)(x + ?)$.

The product of the first terms provides the required x^2 of the trinomial. The question now is what combinations of two coefficients (the question marks) result in a product of 36 (n_3) and at the same time result in 20 (n_2) when summing the inside and outside products. The following combinations produce an n_3 of 36:

n_3	Resulting n_2
1×36	37
2×18	20
3×12	15
4×9	13

Obviously, only the 2×18 combination produces the desired result, that is, an n_3 of 36 and simultaneously an n_2 of 20. Therefore,

$$x^2 + 20x + 36 = (x + 2)(x + 18)$$

Example 4

Factor the following expressions.

a. $x^2 + 5x + 6$
$(x + 3)(x + 2)$
$(x)(x) = x^2$, $(3)(x) + (2)(x) = 5x$, $(3)(2) = 6$ checks

b. $x^2 + x - 20$
$(x - 4)(x + 5)$
$(x)(x) = x^2$, $(-4)(x) + (5)(x) = x$, $(-4)(5) = -20$ checks

c. $2x^2 - 7x - 15$
$(2x + 3)(x - 5)$
$(2x)(x) = 2x^2$, $(3)(x) + (-5)(2x) = -7x$, $(3)(-5) = -15$ checks

d. $x^2 - 4xy + 3y^2$
$(x - y)(x - 3y)$
$(x)(x) = x^2$, $(-y)(x) + (-3y)(x) = -4xy$, $(-y)(-3y) = 3y^2$ checks

e. $6x^2 + 2xy - 8y^2$
    $(3x + 4y)(2x - 2y)$
    $(3x)(2x) = 6x^2$, $(4y)(2x) + (-2y)(3x) = 2xy$, $(4y)(-2y) = -8y^2$ checks

Note that the solution may be further factored to

$$2(x - y)(3x + 4y)$$

PROBLEMS

Factor the expressions in Problems 1 to 16.

1. $s^2 + 3s + 2$

2. $t^2 + 8t + 12$

3. $u^2 - 2u - 8$

4. $\pi^2 - 7\pi + 12$

5. $2R_1^2 - 7R_1 + 6$

6. $3w^2 - 25w + 28$

7. $6a^2 + 3a - 18$

8. $10x^2 - 5x - 15$

9. $2p^2 - 7pq + 6q^2$

10. $6v^2 + 19vw - 36w^2$

11. $v^4 - 12e^2 + 35$

12. $6f^4 + 7f^2g + 2g^2$

13. $3x^6 + 15x^3y^3 - 42y^6$

14. $\dfrac{1}{4}x^2 - \dfrac{1}{2}xy - 2y^2$

15. $r^2 - \dfrac{1}{4}r - \dfrac{1}{8}$

16. $\dfrac{i^2}{4} + \dfrac{ij}{12} - \dfrac{j^2}{18}$

6.6 DIVISION OF POLYNOMIALS

When both the dividend and the divisor of an expression are polynomials, before you can execute the division process you must arrange both so that the magnitude of the powers of each term form a decreasing progression. When aligned in this format, only the first term of the divisor is used to determine each term of the quotient. The process progresses in the same straightforward manner as used to divide pure numbers. The division algorithm continues to reduce the degree of complexity of the dividend until the dividend equals zero (an even division) or until it is less than the degree of complexity of the divisor. In the latter case, a remainder results because the division is not even. Example 1 illustrates both cases.

Example 1

Perform the following divisions.

a. Divide $X^3 - 5X^2 + 10X - 12$ by $X - 3$

$$
\begin{array}{r}
X^2 - 2X + 4 \\
X - 3 \overline{\smash{\big)}\ X^3 - 5X^2 + 10X - 12} \\
\underline{X^3 - 3X^2} \qquad\qquad\qquad \\
-2X^2 + 10X - 12 \\
\underline{-2X^2 + 6X} \qquad\quad \\
4X - 12 \\
\underline{4X - 12} \\
0
\end{array}
$$

	solution
	initialization
	step 1
	step 2
	step 3
	step 4
	step 5
	step 6

Initialization. Dividend and divisor are aligned in a decreasing progression of powers.

Step 1:   First term of dividend (X^3) is divided by first term of divisor (X). The resultant (X^2) is written over the divide symbol. The divisor is now multiplied by X^2, completing step 1.

Step 2:   Step 1 is subtracted from the original dividend. The resultant $(-2X^2 + 10X - 12)$ now becomes the new dividend. Note that the dividend's highest power of X has decreased from a cube to a square.

Step 3:   First term of the new dividend $(-2X^2)$ is divided by the first term of the divisor (X). The resultant $(-2X)$ is now added to the previous X^2 of the quotient, updating the solution to $X^2 - 2X$. The divisor is now multiplied by $(-2X)$, completing step 3.

Step 4:   Step 3 is subtracted from step 2; $4X - 12$ now becomes the new dividend.

Step 5:   First term of the new dividend $(4X)$ is divided by the first term of the divisor (X). The resultant (4) is now added to the previous quotient, updating the solution to $X^2 - 2X + 4$. The divisor is now multiplied by 4, completing step 5.

Step 6:   Step 5 is subtracted from step 4. Because the difference is now zero, the process is completed.

The answer was an even division, with no remainder. If the difference were not zero but of less complexity than the divisor, there would have been a remainder. Part d illustrates this case. Because the methodology is the same for all divisions of polynomials, parts b to d show only the results; the step-by-step process is analogous to that in part a.

b.   Divide $6X^4 - 4X^3 - 26X^2 + 40X - 16$ by $2X^2 - 4X + 2$

$$
\begin{array}{r}
3X^2 + 4X - 8 \\
\hline
2X^2 - 4X + 2 \,\big|\, 6X^4 - 4X^3 - 26X^2 + 40X - 16 \\
\underline{6X^4 - 12X^3 + 6X^2} \\
8X^3 - 32X^2 + 40X - 16 \\
\underline{8X^3 - 16X^2 + 8X} \\
-16X^2 + 32X - 16 \\
\underline{-16X^2 + 32X - 16} \\
0
\end{array}
$$

c.   Divide $X^3 - 8$ by $X - 2$. The dividend of this example is intentionally written with *gaps* in the decreasing power of X. Note that there are no X^2 or X terms. When solving problems with gaps in the dividend, it is frequently advantageous to write in the missing literal terms with coefficients of zero, to help align the decreasing powers:

$$
\begin{array}{r}
X^2 + 2X + 4 \\
\hline
X - 2 \,\big|\, X^3 + \cancel{0}X^2 + \cancel{0}X - 8 \\
\underline{X^3 - 2X^2} \\
2X^2 + \cancel{0}X - 8 \\
\underline{2X^2 - 4X} \\
4X - 8 \\
\underline{4X - 8} \\
0
\end{array}
$$

d.   Divide $X^3 - X^2Y - XY^2 + Y^3 + 1$ by $X + Y$. Note that in problems with two variables, usually only one can be aligned with a decreasing order of powers; the second variable usually progresses in an ascending order.

$$
\begin{array}{r}
X^2 - 2XY + Y^2 \\
\hline
X + Y \,\big|\, X^3 - X^2Y - XY^2 + Y^3 + 1 \\
\underline{X^3 + X^2Y} \\
-2X^2Y - XY^2 + Y^3 + 1 \\
\underline{-2X^2Y - 2XY^2} \\
XY^2 + Y^3 + 1 \\
\underline{XY^2 + Y^3} \\
1
\end{array}
$$

A coefficient of 1 remains. This is less complex than the divisor, so the solution is $X^2 - 2XY + Y^2$, with a remainder of $1/(X + Y)$.

PROBLEMS

Perform the indicated divisions in Problems 1 to 20.

1. $(a^3 + 4a^2 + a - 6) \div (a + 2)$

2. $(c^3 + 5c^2 + 2c - 12) \div (c + 3)$

3. $(i^4 + 2i^3 - 5i^2 + 17i + 20) \div (i + 4)$

4. $(2x^4 - 8x^3 - 29x^2 + 36x - 34) \div (x - 6)$

5. $(6y^4 + 13y^3 - 38y^2 - 31y + 7) \div (2y + 7)$

6. $(a^2 + 2ab + b^2 - 25) \div (a + b + 5)$

7. $(z^4 - 1) \div (z - 1)$

8. $(b^5 + 243) \div (b + 3)$

9. $\left(\dfrac{1}{8}a^3 + \dfrac{5}{12}a^2 - \dfrac{1}{6}a + \dfrac{2}{3}\right) \div \left(\dfrac{1}{2}a + 2\right)$

10. $\left(\dfrac{1}{15}g^4 - \dfrac{17}{30}g^3 - \dfrac{21}{20}g^2 + \dfrac{7}{10}g - 2\right) \div \left(\dfrac{1}{5}g - 2\right)$

11. $(6a^3 - a^2b - 20ab^2 + 6b^3) \div (2a - 3b)$

12. $(4c^4 - 6c^3d - 2c^2d^2 + 9cd^3 - 9d^4) \div (2c - 3d)$

13. $(6x^4 - x^3y - 16x^2y^2 + 23xy^3 - 10y^4) \div (2x^2 - 3xy + 2y^2)$

14. $(4c^3d + 4c^2d - 2c^2 - 6cd - 8d - 7) \div (2cd + 2d - 1)$

15. $\left(\dfrac{1}{8}a^2 - \dfrac{1}{60}ab - \dfrac{11}{15}b - \dfrac{1}{15}b^2 - 2\right) \div \left(\dfrac{1}{2}a + \dfrac{1}{3}b + 2\right)$

16. One factor of $12x^2 - 23xy + 10y^2$ is $3x - 2y$. What is the other factor?

17. One factor of $6y^3 + y^2 - 28y - 30$ is $2y - 5$. What is the other factor?

18. Is $x + 3$ a factor of $x^3 - 2x^2 + 6x - 3$? Why?

19.   Determine b so that $x - 3$ is a factor of $x^2 - 9x + b$.

20.   Determine k so that $3a - 5$ is a factor of $6a^2 + 11a - k$.

SUMMARY

The laws of exponents apply to algebraic terms exactly as they do to numeric terms. In a monomial containing both coefficients and literal variables raised to a power, all terms are raised to the same power. Because there is no restriction on the numeric value of the power, the laws of exponents apply equally to roots. Negative radicands can exist only for odd-numbered indexes. If the index is an even number, a complex number results (this topic is covered in Chapter 16).

You can verify the accuracy of algebraic manipulations by substituting arbitrarily chosen values into both the original expression and its solution. If the value calculated for the original problem equals the value calculated for the solution, the solution is correct.

When a group of terms is raised to the same power, you can group all terms by parentheses and raise the resulting monomial to that same power. It is mathematically preferable to manipulate fractions to remove radicals from the denominators.

There is a series of special cases for expanding polynomials or factoring certain expressions such that predictable formats can be constructed. The following chart summarizes these conditions; the coefficients are represented as a, b, and c, the variables as x, y, and z:

General Conditions	Factors
$ax + ay + az$	$a(x + y + z)$
$(ax)^2 - (by)^2$	$(ax + by)(ax - by)$
$(ax)x^2 + (bc + ad)xy + (bd)y^2$	$(ax + by)(cx + dy)$
$(ax)^2 + 2(ab)xy + (by)^2$	$(ax + by)^2$
$(ax)^2 - 2(ab)xy + (by)^2$	$(ax - by)^2$
$(ax)^3 + 3(ax)^2(by) + 3(ax)(by)^2 + (by)^3$	$(ax + by)^3$
$(ax)^3 - 3(ax)^2(by) + 3(ax)(by)^2 - (by)^3$	$(ax - by)^3$
$(ax)^2 + 2(ax)(by) + 2(ax)(cz) + (by)^2 + 2(by)(cz) + (cz)^2$	$(ax + by + cz)^2$

To divide one polynomial by another, arrange both the dividend and the divisor in a descending order of powers. Use only the first term of the divisor and the first term of the current dividend to determine each term of the quotient.

7 ALGEBRAIC FRACTIONS

7.1 REDUCTION OF FRACTIONS

Earlier chapters discussed performing the four basic arithmetic functions on purely numeric fractions and reducing the fractions to their lowest terms. These same rules equally apply to algebraic fractions; this section investigates methods to reduce fractions containing both numeric and literal terms.

Recall that the key to fraction reduction is finding common terms in both the numerator and denominator and then canceling these terms. For example, a numeric fraction can be reduced as follows:

$$\frac{25}{35} = \frac{5 \cdot \cancel{5}}{7 \cdot \cancel{5}} = \frac{5}{7}$$

If the fraction is algebraic, the same rules apply:

$$\frac{3\cancel{(a-b)}(a-c)}{9\cancel{(a-b)}} = \frac{3(a-c)}{9} = \frac{a-c}{3}$$

Frequently expressions can be factored, which quickly reveals their common terms:

$$\frac{a^2 - 2ab + b^2}{a - b} = \frac{(a-b)^2}{(a-b)} = \frac{\cancel{(a-b)}(a-b)}{\cancel{(a-b)}} = a - b$$

Example 1 presents several more illustrations of factoring expressions to find common terms for cancellation.

Example 1

Reduce the following fractions to their lowest terms.

a. $\dfrac{a^2 + 2ab + b^2}{a^2 - b^2}$

$$\frac{\cancel{(a+b)}(a+b)}{\cancel{(a+b)}(a-b)} = \frac{a+b}{a-b}$$

b. $\dfrac{x^2}{x^2 + xy^2}$

$$\frac{\cancel{x} \cdot x}{\cancel{x}(x+y^2)} = \frac{x}{x+y^2}$$

c. $\dfrac{a^2 + a - 12}{a^2 - 8a + 15}$

$\dfrac{\cancel{(a-3)}(a + 4)}{\cancel{(a-3)}(a - 5)} = \dfrac{a + 4}{a - 5}$

d. $\dfrac{2a^2 + 4a - 30}{8a^2 - 32a + 24}$

$\dfrac{2(a^2 + 2a - 15)}{8(a^2 - 4a + 3)} = \dfrac{2\cancel{(a-3)}(a + 5)}{8\cancel{(a-3)}(a - 1)} = \dfrac{a + 5}{4(a - 1)}$

PROBLEMS

Reduce the fractions in Problems 1 to 12 to their lowest common terms.

1. $\dfrac{9}{207}$

2. $\dfrac{a^2}{ab}$

3. $\dfrac{cd^2e}{c^2e^2}$

4. $\dfrac{ab + ac}{b + c}$

5. $\dfrac{0.25x + 0.75y}{1.5xy}$

6. $\dfrac{2a^2 + a - 3}{a - 1}$

7. $\dfrac{3s - 3t}{s^2 - t^2}$

8. $\dfrac{(ab)^2 - 1}{a^2b^2 - 2ab + 1}$

9. $\dfrac{3u^2 + uv - 2v^2}{4u^2 + 10uv + 6v^2}$

10. $\dfrac{9a^2 - 24ab + 16b^2}{6a^2 + 7ab - 20b^2}$

11. $\dfrac{x^3 + 3x^2y + 3xy^2 + y^3}{x^2 - y^2}$

12. $\dfrac{x^2 + 4xy + 2xz + 4y^2 + 4yz + z^2}{x + 2y + z}$

7.2 SIMPLIFICATION BY CHANGING SIGNS

When trying to combine or reduce fractional terms, it is frequently helpful to multiply both the numerator and denominator by some term that is the equivalent of 1. For example, when summing 1/3 and 1/5, multiply each fraction by the appropriate 1 to form a common denominator; this permits addition of the fractions:

$$\frac{1}{3} + \frac{1}{5} = \frac{1}{3} \cdot \left(\frac{5}{5}\right) + \frac{1}{5} \cdot \left(\frac{3}{3}\right) = \frac{5}{15} + \frac{3}{15} = \frac{8}{15}$$

Forming the optimum 1 multiplier may also involve two negatives, or, more appropriately, two negative 1s. For example:

$$(a - 2)(a - 3) = (-1)(a - 2)(-1)(a - 3) = (2 - a)(3 - a)$$

A negative numeric fraction can be expressed as

$$-\frac{2}{7} \quad \text{or} \quad \frac{-2}{7} \quad \text{or} \quad \frac{2}{-7}$$

Whether the negative sign is in the numerator or the denominator is of no consequence because the movement of the sign up to the numerator or down to the denominator is merely the result of multiplying by two negative 1s.

$$\frac{-a}{b} = \frac{(-1)(-a)}{(-1)(b)} = \frac{a}{-b}$$

Often, multiplying only one part of a fraction by -1 and retaining the other -1 can greatly simplify some fractions. For example:

$$\frac{a - b}{b - a} = \frac{(-1)(a - b)}{(-1)(b - a)} = \frac{(-1)\cancel{(a - b)}}{\cancel{(a - b)}} = -1$$

Example 1

Simplify the following fractions.

a. $-\dfrac{-5}{7}$

$\dfrac{5}{7}$

b. $\dfrac{3y - 3x}{x - y}$

$\dfrac{(-1)(3)(-1)(y - x)}{x - y} = \dfrac{(-3)\cancel{(x - y)}}{\cancel{(x - y)}} = -3$

c. $\dfrac{a^2 - 5a + 6}{12 - a - a^2}$

$\dfrac{(-1)(a^2 - 5a + 6)}{(-1)(12 - a - a^2)} = \dfrac{(-1)(a^2 - 5a + 6)}{(a^2 + a - 12)}$

$= \dfrac{(-1)\cancel{(a - 3)}(a - 2)}{\cancel{(a - 3)}(a + 4)} = \dfrac{2 - a}{a + 4}$

PROBLEMS

In Problems 1 to 8 simplify the fractions.

1. $\dfrac{x - y}{y - x}$

2. $\dfrac{c + d - e}{e - d - c}$

3. $\dfrac{s - t}{t^2 - s^2}$

4. $\dfrac{-(u - v)}{v^2 - u^2}$

5. $\dfrac{p^2 - q^2}{2q^2 - 2p^2}$

6. $\dfrac{i_1^2 - i_1 i_2 - 6i_2^2}{12i_2^2 + 11i_1 i_2 - 5i_1^2}$

7. $\dfrac{12m^2 - 17mn + 6n^2}{6n^2 - 17mn + 12m^2}$

8. $\dfrac{\pi^2 + 6\pi - 16}{12 - 10\pi + 2\pi^2}$

7.3 ADDITION AND SUBTRACTION OF ALGEBRAIC FRACTIONS

Finding the sum and the difference of algebraic fractions requires manipulating all terms so that each has the same denominator; then you can perform the plus and/or minus operations on each numerator. The resultant is associated with the common denominator.

In the following example, note how easily the same denominator is formed:

$$1 + \frac{1}{s} = \frac{s}{s} + \frac{1}{s} = \frac{s + 1}{s}$$

Because the denominator s is needed, merely replace the 1 with s/s and then combine terms. The problem could have been written as

$$\frac{1}{1} + \frac{1}{s}$$

and the LCD found. For this problem the LCD is s. Next, determine what each denominator must be multiplied by to become the LCD, and then multiply its associated numerator by the same quantity to retain the fractional equality. When you have performed this process on each term, all terms will have the same denominator (the LCD) and you can execute the indicated operations.

Example 1

Perform the indicated operations.

a. $\dfrac{a}{b} + \dfrac{c}{d}$ LCD $= bd$

To make the denominator b equal bd, multiply it by d. To make the denominator d also equal bd, multiply it by b. Therefore,

$$\frac{a}{b}\left(\frac{d}{d}\right) + \frac{c}{d}\left(\frac{b}{b}\right) = \frac{ad}{bd} + \frac{bc}{bd} = \frac{ad + bc}{bd}$$

b. $\dfrac{x}{y} + 1 + \dfrac{w}{z}$ LCD $= yz$

$$\frac{x}{y}\left(\frac{z}{z}\right) + \frac{yz}{yz} + \frac{w}{z}\left(\frac{y}{y}\right) = \frac{xz + yz + wy}{yz}$$

c. $\dfrac{a^2}{b + c} + \dfrac{b^2}{b - c}$ LCD $= (b + c)(b - c) = b^2 - c^2$

$$\frac{a^2}{(b + c)}\left(\frac{b - c}{b - c}\right) + \frac{b^2}{(b - c)}\left(\frac{b + c}{b + c}\right) = \frac{a^2b - a^2c + b^3 + b^2c}{b^2 - c^2}$$

PROBLEMS

Perform the addition or subtraction operations in Problems 1 to 15.

1. $\dfrac{1}{2} + \dfrac{1}{3} - \dfrac{1}{4}$

2. $\dfrac{3}{7} + \dfrac{4}{21} - \dfrac{5}{42}$

3. $\dfrac{2a}{3} + \dfrac{a}{12}$

4. $\dfrac{R_1}{R_2} + \dfrac{R_2}{R_1}$

5. $\dfrac{1}{a} + \dfrac{1}{b} - \dfrac{1}{c}$

6. $\dfrac{2a}{x} - \dfrac{2b}{y} + 1$

7. $\dfrac{s - 1}{t} + \dfrac{t - 1}{s}$

8. $\dfrac{3}{a - b} + \dfrac{4}{a + b}$

9. $\dfrac{5x}{x^2 - y^2} + \dfrac{4y}{x - y} - \dfrac{5}{x + y}$

10. $\dfrac{3a}{4a - 4b} + \dfrac{2b}{a - b} + 1$

11. $\dfrac{1}{a^2 + 2a - 15} - \dfrac{1}{a + 5}$

12. $\dfrac{x - 2}{x^2 - 8x + 12} + \dfrac{x - 6}{x - 2}$

13. $\dfrac{\pi}{2\pi^2 + 5\pi - 12} - \dfrac{\pi - 1}{2\pi - 3}$

14. $x + y - \dfrac{x^2 - y^2}{x + y} + x$

15. $\dfrac{s^2 - 9s + 14}{s - 2} - \dfrac{s^2 - 10s + 21}{s - 7}$

7.4 MULTIPLICATION AND DIVISION OF ALGEBRAIC FRACTIONS

The product of fractions is automatically a product of numerators divided by a product of denominators. For example,

$$\frac{2}{3} \cdot \frac{5}{7} \cdot \frac{3}{10} = \frac{2 \cdot 5 \cdot 3}{3 \cdot 7 \cdot 10} = \frac{30}{210} = \frac{1}{7}$$

You can cancel common terms either before you form the products or after, although the former method is preferable because the fractions are already in a more elementary format, and so terms mutually capable of being canceled are more easily discerned. Often, when algebraic fractions are multiplied, they can be factored, which reveals common terms that through cancellation simplify the products.

One fraction divided by another fraction looks like:

$$\frac{a}{b} \div \frac{c}{d}$$

or
$$\left.\begin{array}{c}\dfrac{a}{b}\end{array}\right\} \equiv N$$
$$\left.\begin{array}{c}\dfrac{c}{d}\end{array}\right\} \equiv D$$

The key to solving this type of problem is to multiply both N and D by the reciprocal of D, namely, d/c. In other words, invert the denominator fraction (D) and multiply it by the numerator fraction (N). This results in

$$\frac{\dfrac{a}{b}}{\dfrac{c}{d}} \cdot \frac{\dfrac{d}{c}}{\dfrac{d}{c}} = \frac{\dfrac{a}{b} \cdot \dfrac{d}{c}}{\dfrac{c}{d} \cdot \dfrac{d}{c}} = \frac{\dfrac{a}{b} \cdot \dfrac{d}{c}}{1} = \frac{a}{b} \cdot \frac{d}{c}$$

Example 1

Form the products of the following algebraic fractions.

a. $\dfrac{2}{3} \cdot \dfrac{4}{9}$

$\dfrac{8}{27}$

b. $\dfrac{ax}{b} \cdot \dfrac{cx}{d}$

$\dfrac{acx^2}{bd}$

c. $\dfrac{7w - 21y}{z} \cdot \dfrac{1}{w - 3y}$

$\dfrac{7(w - 3y)}{z(w - 3y)} = \dfrac{7}{z}$

d. $\dfrac{g^2 + 4gh - 12h^2}{g^2 - h^2} \cdot \dfrac{g - h}{2g^2 + 16gh + 24h^2}$

$\dfrac{(g - 2h)(g + 6h)}{(g - h)(g + h)} \cdot \dfrac{(g - h)}{2(g + 6h)(g + 2h)} = \dfrac{g - 2h}{2(g + h)(g + 2h)}$

Example 2

Form the quotients of the following algebraic fractions.

a. $\dfrac{2}{3} \div \dfrac{1}{4}$

$\dfrac{2}{3} \cdot \dfrac{4}{1} = \dfrac{8}{3}$

b. $2x \div \dfrac{x}{6}$

$\dfrac{2x}{1} \cdot \dfrac{6}{x} = \dfrac{12x}{x} = 12$

c. $\dfrac{c^2 + 2c}{c^2 - c - 12} \div \dfrac{c^2 + 4c + 4}{c^2 + 5c + 6}$

$\dfrac{c(c + 2)}{(c + 3)(c - 4)} \cdot \dfrac{(c + 2)(c + 3)}{(c + 2)(c + 2)} = \dfrac{c}{c - 4}$

d. $\dfrac{a^2 + 7a + 10}{a^2 - a - 6} \div \dfrac{a^2 - 2a - 35}{a^2 - 7a + 12}$

$\dfrac{(a + 2)(a + 5)}{(a - 3)(a + 2)} \cdot \dfrac{(a - 3)(a - 4)}{(a + 5)(a - 7)} = \dfrac{a - 4}{a - 7}$

PROBLEMS

In Problems 1 to 20 form the products and/or quotients of the algebraic fractions.

1. $\dfrac{3}{4} \cdot \dfrac{12}{21} \cdot \dfrac{14}{36}$

2. $\dfrac{7a}{6} \cdot \dfrac{18}{a^2} \cdot \dfrac{9}{b}$

3. $\dfrac{2a - 4b}{6} \cdot \dfrac{24}{a - b} \cdot \dfrac{c}{8}$

4. $\dfrac{a^2 - b^2}{c^2} \cdot \dfrac{cd}{3} \cdot \dfrac{33}{a + b}$

5. $\dfrac{a^2 - 4a - 21}{7c} \cdot \dfrac{21(cd)^2}{a - 7}$

6. $\dfrac{6x^2 - xy - 2y^2}{21d - 7c} \cdot \dfrac{7(c^2 - 9d^2)}{2x^2 + 21xy + 10y^2}$

7. $\dfrac{5x^2 + 22x + 8}{x^2 - 2xy + y^2} \cdot \dfrac{x - y}{10x^2 - 11x - 6}$

8. $\dfrac{x^3 + 3x^2y + 3xy^2 + y^3}{x^2 - y^2} \cdot \dfrac{x - y}{x^2 + 2xy + y^2}$

9. $\dfrac{16x^2 + 48xy + 36y^2}{4x^2 - 9y^2} \cdot \dfrac{6x - 9y}{4x + 6y}$

10. $\dfrac{2x^2 + 9x - 5}{12x^2 - 13x - 14} \cdot \dfrac{12x^2 + 5x - 2}{3x^2 + 3x - 30} \cdot \dfrac{3x + 12}{2x^2 + 11x - 6}$

11. $\dfrac{5}{9} \div \dfrac{15}{27}$

12. $\dfrac{\pi r^2 h}{8} \div 2\pi r$

13. $\dfrac{z - x}{R + x} \div (z - x)$

14. $\dfrac{R_1 R_2}{R_1 + R_2} \div R_1$

15. $x \div \dfrac{x}{y}$

16. $\dfrac{4a^2 - 1}{a^3 - 16a} \div \dfrac{2a - 1}{a - 4}$

17. $\dfrac{d^2 + 4d}{d} \div \dfrac{d^3 + 4d^2}{d^2}$

18. $\dfrac{4}{a - b} \div \dfrac{a^2 - b^2}{a^2 + 2ab + b^2}$

19. $\dfrac{x^2 - 4y^2}{xy + 2y^2} \div \dfrac{x - 2y}{2y}$

20. $\dfrac{u^2 - v^2}{u^2 + uv} \div \dfrac{u + v}{u^2 - uv}$

7.5 COMPLEX ALGEBRAIC FRACTIONS

Complex fractions contain several layers of fractions in the numerator, denominator, or both, and when they are algebraic, variables are usually intermixed with the numeric terms. Here is an example of a typical complex fraction:

$$\frac{1 + \dfrac{1}{a}}{a + \dfrac{a}{b}}$$

To simplify, reduce both the major numerator and the major denominator to individual fractions, so that the format above looks like

$$\frac{\dfrac{w}{x}}{\dfrac{y}{z}}$$

Once you reduce the problem to such a format, proceed by inverting the denominator and multiplying by the numerator:

$$\frac{w}{x} \cdot \frac{z}{y} = \frac{wz}{xy}$$

The terms wz and xy can often be factored, further simplifying the original complex fraction.

Example 1

Simplify the following complex fractions.

a. $$\frac{\dfrac{1}{2} - \dfrac{1}{3}}{3 - \dfrac{11}{18}}$$

Complex fractions involving only simple numeric terms can be most easily reduced by multiplying the numerator and denominator by the LCD of all the fractional parts. Because the integer 3 can be thought of as being divided by a denominator of 1, only the denominators 2, 3, and 18 are used to form an LCD of 18.

Step 1: Multiplying the numerator and denominator by 18 produces the following result.

$$\frac{18\left(\dfrac{1}{2} - \dfrac{1}{3}\right)}{18\left(3 - \dfrac{11}{18}\right)} = \frac{9 - 6}{54 - 11} = \frac{3}{43}$$

b. $\dfrac{1 + \dfrac{1}{a}}{1 + \dfrac{1}{b}}$

Step 1: Replace 1 of the numerator with a/a and 1 of the denominator with b/b, to get a result of

$$\frac{\dfrac{a}{a} + \dfrac{1}{a}}{\dfrac{b}{b} + \dfrac{1}{b}}$$

Step 2: Combine like terms in the denominators, to get a result of

$$\frac{\dfrac{a + 1}{a}}{\dfrac{b + 1}{b}}$$

Step 3: Invert the denominator and multiply by the numerator. The answer is

$$\frac{a + 1}{a} \cdot \frac{b}{b + 1} = \frac{b(a + 1)}{a(b + 1)} = \frac{ab + b}{ab + a}$$

c. $\dfrac{\left(\dfrac{3}{4}\right)^2 - \left(\dfrac{4}{5}\right)^2}{\dfrac{3}{4} - \dfrac{4}{5}}$

Step 1: Recognize the major numerator as of the format $a^2 - b^2$, which can be factored into $(a - b)(a + b)$. Cancel where appropriate. The result is

$$\frac{\left(\cancel{\dfrac{3}{4} - \dfrac{4}{5}}\right)\left(\dfrac{3}{4} + \dfrac{4}{5}\right)}{\left(\cancel{\dfrac{3}{4} - \dfrac{4}{5}}\right)} = \frac{3}{4} + \frac{4}{5}$$

Step 2: Convert the two remaining terms so they have equal denominators. The LCD = 20; the result is

$$\left(\frac{3}{4}\right)\left(\frac{5}{5}\right) + \left(\frac{4}{5}\right)\left(\frac{4}{4}\right) = \frac{15}{20} + \frac{16}{20}$$

Step 3: Combine terms. The answer is

$$\frac{15}{20} + \frac{16}{20} = \frac{15 + 16}{20} = \frac{31}{20}$$

d. $\dfrac{t + 1 - \dfrac{20}{t}}{t - 7 + \dfrac{12}{t}}$

Step 1: Clear out the fractional parts by multiplying the major numerator and the major denominator by t. The result is

$$\frac{t\left(t + 1 - \dfrac{20}{t}\right)}{t\left(t - 7 + \dfrac{12}{t}\right)} = \frac{t^2 + t - 20}{t^2 - 7t + 12}$$

Step 2: Factor both trinominals, for a result of

$$\frac{(t + 5)(t - 4)}{(t - 4)(t - 3)}$$

Step 3: Cancel where appropriate. The answer is

$$\frac{t + 5}{t - 3}$$

PROBLEMS

In Problems 1 to 15 simplify the complex fractions.

1. $\dfrac{1 + \dfrac{2}{3}}{\dfrac{3}{4} - \dfrac{5}{6}}$

3. $\dfrac{\dfrac{16}{25} - \dfrac{36}{81}}{\dfrac{6}{9} + \dfrac{4}{5}}$

5. $\dfrac{i + \dfrac{1}{i}}{I + \dfrac{1}{I}}$

2. $\dfrac{2\dfrac{1}{2} - \dfrac{1}{3}}{\dfrac{4}{5} + 3}$

4. $\dfrac{b + \dfrac{b}{c}}{c + \dfrac{c}{b}}$

6. $\dfrac{1}{1 + \dfrac{1}{1 + a}}$

7. $\dfrac{1 + \dfrac{1}{1 + b}}{1 - \dfrac{1}{1 - b}}$

10. $\dfrac{u - 2 - \dfrac{15}{u}}{u - 1 - \dfrac{12}{u}}$

13. $1 + \dfrac{1}{t} + \dfrac{1}{\dfrac{1}{t}}$

8. $\dfrac{\dfrac{1}{a}}{\dfrac{1}{\dfrac{1}{a}}}$

11. $\dfrac{2 - \dfrac{9}{v} - \dfrac{18}{v^2}}{2 + \dfrac{11}{v} + \dfrac{12}{v^2}}$

14. $\dfrac{\dfrac{1}{1 + \dfrac{1}{s}}}{1 + \dfrac{1}{s} + \dfrac{1}{1 + \dfrac{1}{s}}}$

9. $\dfrac{1 + \dfrac{1}{a + b}}{1 + \dfrac{1}{a + b}}$

12. $\dfrac{\dfrac{x + y}{x - y} - \dfrac{x - y}{x + y}}{\dfrac{x + y}{x - y} + \dfrac{x - y}{x + y}}$

13. $\dfrac{c}{c + \dfrac{c}{1 + c}} - \dfrac{c}{c - \dfrac{c}{1 - c}}$

7.6 VERIFICATION OF COMPLEX FRACTIONS

You can use your calculator to verify the solution of complex fractions exactly as you verified the simpler solutions in earlier chapters. Because problems containing multiple layers of fractions are more complex, be very careful when keying steps into the calculator. Parentheses become more critical. Sometimes it is even easier to start at the *bottom* of the fraction and work *up*, especially when there is a series of reciprocals. Examples 1 and 2 illustrate these two approaches. Note that in the first approach, which "attacks" the problem in a top-to-bottom, left-to-right fashion, several layers of nested parentheses are needed and little use can be made of the reciprocal function. This approach can sometimes overextend the capacities of calculators with limited "parentheses-pending operation" capacities and may result in a flashing or ERROR indication. The second approach, which starts at the *bottom* of the problem and progresses *up*, uses the reciprocal function and simultaneously avoids parentheses (which in essence is the RPN methodology). All scientific calculators should be capable of this latter approach.

Example 1

Verify the following solution using both of the previous approaches.

$$\text{Level 2} \left\{ \dfrac{1}{1 + \dfrac{1}{1 + \dfrac{1}{\pi}}} \right\} \text{Level 1} = \dfrac{\pi + 1}{2\pi + 1}$$

Approach 1 (18 steps)

PRESS	DISPLAY	COMMENT
1 $\div$ ((1 + ((1 $\div$ ((1 + 1 $\div$ π	3.142	Sets up problem for closing parentheses
))	1.318	Level 1
))	0.759	1/level 1
))	1.759	Level 2
=	0.569	Solution

Approach 2 (10 steps)

PRESS	DISPLAY	COMMENT
π $1/x$ + 1 =	1.318	Level 1
$1/x$ + 1 =	1.759	Level 2
$1/x$	0.569	Solution

Note that approach 2 is much less complicated. No parentheses are required, and significantly fewer steps are needed.

To verify:

PRESS	DISPLAY	COMMENT
π + 1 =	4.142	Numerator
$\div$ ((2 $\times$ π + 1))	7.283	Denominator
=	0.569	Solution verified

Example 2

Verify the following solutions of complex fractions. Let the first variable = 2.15 STO 0 and the second variable, if used, = 4.25 STO 1 . Note that STO 0 refers to storage location *zero*.

a. $\dfrac{1 + \dfrac{1}{1 + b}}{1 - \dfrac{1}{1 - b}} = \dfrac{b^2 + b - 2}{b^2 + b}$

PRESS	DISPLAY	COMMENT
$($ 1 $+$ $\boxed{\text{RCL 0}}$ $=$ $\boxed{1/x}$ $+$ 1 $)$	1.317	Major numerator
$\boxed{\div}$ $($		
1 $-$ $($ 1 $-$ $\boxed{\text{RCL 0}}$ $)$ $\boxed{1/x}$ $)$	1.870	Major denominator
$=$	0.705	Solution

To verify:

PRESS	DISPLAY	COMMENT
$\boxed{\text{RCL 0}}$ $\boxed{x^2}$ $+$ $\boxed{\text{RCL 0}}$		
$=$ $\boxed{\text{STO 2}}$ $-$ 2 $=$	4.773	Numerator
$\div$ $\boxed{\text{RCL 2}}$ $=$	0.705	Solution verified

Note that after $b^2 + b$ is calculated as part of the numerator, it is stored for later recall because it is also the value of the denominator. Whenever a group of terms are reused, it is usually advantageous to store that value after the initial calculation (for recall when needed again) rather than recalculate the value each time. The following problem emphasizes this point.

b. $\dfrac{\dfrac{u+v}{u-v} + \dfrac{u-v}{u+v}}{\dfrac{u+v}{u-v} - \dfrac{u-v}{u+v}} = \dfrac{u^2+v^2}{2uv}$

PRESS	DISPLAY	COMMENT
$\boxed{\text{RCL 0}}$ $+$ $\boxed{\text{RCL 1}}$ $=$ $\div$		
$($ $\boxed{\text{RCL 0}}$ $-$ $\boxed{\text{RCL 1}}$ $)$ $=$	-3.048	Calculates $\dfrac{u+v}{u-v}$
$\boxed{\text{STO 2}}$	-3.048	Stores displayed value in location 2
$\boxed{\text{CLR}}$ $\boxed{\text{RCL 0}}$ $-$ $\boxed{\text{RCL 1}}$		
$=$ $\div$ $($ $\boxed{\text{RCL 0}}$ $+$ $\boxed{\text{RCL 1}}$ $)$ $=$	-0.328	Calculates $\dfrac{u-v}{u+v}$
$\boxed{\text{STO 3}}$	-0.328	Stores displayed value in location 3

PRESS	DISPLAY	COMMENT
CLR RCL 2 + RCL 3		
= ÷ (RCL 2 −		
RCL 3) =	1.241	Solution

To verify:

PRESS	DISPLAY	COMMENT
CLR RCL 0 x^2 + RCL 1 x^2		
= ÷ (2 × RCL 0 ×		
RCL 1) =	1.241	Solution verified

PROBLEMS

In Problems 1 to 8, use your calculator to verify the solutions of the complex fractions. Let 2.78 = the first variable $\boxed{\text{STO 0}}$ and 3.62 = the second variable, if used $\boxed{\text{STO 1}}$. To help in the verification, the calculator displays for both the original problem and its solution are given when using these values.

1. $$\dfrac{a + \dfrac{a}{b}}{a - \dfrac{a}{b}} = \dfrac{a(b+1)}{a(b-1)} = 1.763$$

2. $$1 - \dfrac{1}{1 - \dfrac{1}{a}} = \dfrac{1}{1-a} = -0.562$$

3. $$\dfrac{1}{\dfrac{b}{b+c} - \dfrac{c}{b+c}} = \dfrac{b+c}{b-c} = -7.619$$

4. $$x - \dfrac{y}{x - \dfrac{y}{x - \dfrac{y}{x}}} = \dfrac{x^4 - 3x^2y + y^2}{x^3 - 2xy} = -8.174$$

5. $\dfrac{s}{s + \dfrac{t}{s - \dfrac{t}{s}}} - \dfrac{t}{s - \dfrac{t}{s + \dfrac{t}{s}}} = \dfrac{s^3 - s^2t - st - t^2}{s^3} = -1.380$

6. $\dfrac{\dfrac{1}{u + v} + \dfrac{1}{u - v}}{\dfrac{1}{u - v} - \dfrac{1}{u + v}} = \dfrac{u}{v} = 0.768$

7. $\dfrac{1 + x}{1 - \dfrac{x}{1 + \dfrac{x}{1 - x}}} = \dfrac{x + 1}{x^2 - x + 1} = 0.635$

8. $\dfrac{1}{\dfrac{1}{\dfrac{1}{c + d} + \dfrac{1}{c - d}} - \dfrac{1}{\dfrac{1}{c - d} - \dfrac{1}{c + d}}} = \dfrac{2cd}{cd(d + c) - (d^3 + c^3)} = -4.457$

SUMMARY

You can reduce algebraic fractions by cancellation when common terms, or groups of terms, are in both the numerators and denominators. Such grouping of terms usually occurs after factoring, and such terms are generally within parentheses. You can change the sign of each term within a group if the parentheses enclosing the group are preceded by a minus sign.

To add or subtract algebraic (or numeric) fractions, reduce each term to a common denominator, preferably the LCD, and perform the indicated operations on the numerators. The resultant is then associated with the common denominator. Multiplying fractions results in the product of the numerators being divided by the product of the denominators. Cancel common terms after forming the products. Division of a fraction by a fraction is the product of the numerator fraction and the reciprocal of the denominator fraction.

Complex fractions have multiple layers of fractions within them. To simplify, treat each sublayer of fraction individually when combining so that the overall complex fraction eventually reduces to a single numerator divided by a single denominator.

The calculator is a powerful and useful tool for verifying the accuracy of a complex fraction reduced to a simpler format. Frequently, when using the calculator for verification, beginning at the bottom of a complex fraction and progressing up eliminates the need for many nested parentheses.

8 EQUATIONS: LINEAR AND FRACTIONAL

8.1 THE EQUATION

It is very difficult when discussing mathematics to avoid the equation. As a matter of fact, this text has used equations ever since Chapter 1. For example, the first chapter stated that for two resistors in parallel,

The total resistance equals *the products of the two resistors divided by the sum of the two resistors.*

When this language is translated into mathematical symbols, the electrical problem takes the format

$$R_T = \frac{R_1 R_2}{R_1 + R_2} \tag{8.1}$$

An equation is a mathematical statement that one expression or quantity is equal to another expression or quantity.

Equations are the heart of mathematics and as such represent its most powerful tool. They are the media through which real-life problems, translated into mathematical symbols, are manipulated and reduced to a solution. Implicit in this statement are two separate operations:

1. The process of taking an abstract, real-life problem and reducing it to a mathematical expression.
2. Given a mathematical expression, mechanical rules that manipulate the equation so as to obtain the desired result must be applied.

The first step is referred to as a *word problem*. Performing this operation is mostly a matter of common sense, that is, writing symbolically the language of a grammatical statement, for example, studying a circuit diagram and expressing an unknown quantity in terms of known quantities to conform with an electronic law or principle.

Step 2 is more a mechanical function, one involving applying rules to manipulate equations to retain the *balance* of the equation while at the same time solving for the unknown variable.

This chapter mostly concentrates upon the rules for manipulating equations; only the last section considers general *word problems*. Chapter 9 is dedicated entirely to applying both operations to electrical problems.

154

8.2 LINEAR EQUATIONS

An equation is like a balance scale, with the equal sign being the pivot point. Figure 8.1 illustrates this concept for the simple equation

$$3x - 18 = 15$$

The problem is to solve for x. To do so first requires that all terms containing x be on one side of the equation and all terms not containing x be on the other side of the equation.

   Because $3x$ (the only term containing the unknown) is already on the left side, leave it there. However, the -18 is also on the left side, which by choice is being reserved for only x terms. Therefore, you must remove the -18. It is easy to cancel out -18, if $+18$ is added to it, because the resultant is then zero. However, to keep the balance, $+18$ must also be added to the right side (we know what happens to a real balance scale if equal amounts are not added to both sides). To help describe each step, we use the letters A, S, M, and D to imply the operations of add, subtract, multiply, and divide, respectively, to each side of the equation.

Step 1:   A; 18 $3x - 18 + 18 = 15 + 18$
$$3x = 33$$

At this point we now know what $3x$ equals. However, the problem is not what does $3x$ equal, but what does *one x* equal. To reduce $3x$ to $1x$ requires division by 3, again to *both* sides of the equation.

Step 2:   D; 3 $$\frac{3x}{3} = \frac{33}{3}$$
$$x = 11$$

We have now determined 11 as the unique solution of x that satisfies the original equation

$$3x - 18 = 15$$

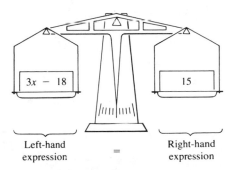

Left-hand   =   Right-hand
expression   expression

Figure 8.1   Symbolic representation of an equation.

To test the accuracy of this solution, substitute 11 for x and see if both sides of the equation do indeed equate:

$$3(11) - 18 = 15$$
$$33 - 18 = 15$$
$$15 = 15 \qquad \text{solution checks}$$

To summarize,

1. Isolate all terms containing the unknown on one side of the equation.
2. Isolate all terms not containing the unknown on the other side of the equation.
3. Perform the basic arithmetic functions necessary to both sides of the equation so that the unknown quantity is uniquely solved for a single numeric value or for other known variables.

Example 1

Solve the following equations for the indicated variable.

a. $4x - 7 = 3x + 5$

Step 1: S; $3x$
$$4x - 7 - 3x = 3x + 5 - 3x$$
$$x - 7 = 5$$

Step 2: A; 7
$$x - 7 + 7 = 5 + 7$$
$$x = 12$$

Check:
$$4(12) - 7 = 3(12) + 5$$
$$48 - 7 = 36 + 5$$
$$41 = 41$$

b. $5u + 6 = 8u - 3$

Step 1: A; 3
$$5u + 6 + 3 = 8u - 3 + 3$$
$$5u + 9 = 8u$$

Step 2: S; $5u$
$$5u + 9 - 5u = 8u - 5u$$
$$9 = 3u$$

Step 3: D; 3
$$\frac{9}{3} = \frac{3u}{3}$$
$$3 = u$$

Check:
$$5(3) + 6 = 8(3) - 3$$
$$15 + 6 = 24 - 3$$
$$21 = 21$$

c. $4(i + 7) + i = 3(i - 4)$

Step 1: Expand parentheses, combine like terms

$$4i + 28 + i = 3i - 12$$
$$5i + 28 = 3i - 12$$

Step 2: S; 28

$$5i + 28 - 28 = 3i - 12 - 28$$
$$5i = 3i - 40$$

Step 3: S; 3i

$$5i - 3i = 3i - 40 - 3i$$
$$2i = -40$$

Step 4: D; 2

$$\frac{2i}{2} = \frac{-40}{2}$$
$$i = -20$$

Check:

$$4(-20 + 7) + (-20) = 3(-20 - 4)$$
$$-80 + 28 - 20 = -60 - 12$$
$$-72 = -72$$

Note that when canceling a term preceded by a plus or minus sign, the net effect is to move the term across the equal sign and change its sign. For example, in the equation

$$7x - 20 = 29$$

you want to eliminate the 20 from the left side of the equation. Do this by adding 20 to both sides. The -20 is canceled on the left and now appears on the right as $+20$. Therefore, a *shortcut* for solving simple linear equations is to merely move the appropriate terms across the equal sign and change their signs. When this process is executed, the problem can be solved by the following simple sequence of steps:

$$7x - 20 = 29$$

is rewritten as

$$7x = 29 + 20 = 49$$
$$\therefore \quad x = 7$$

Example 2

Rewrite each original equation (after expanding, if appropriate) with all unknown terms on the left and all numeric terms on the right. Solve the equations.

a. $2x - 7 = -3x + 18$
    $2x + 3x = 18 + 7$
    $5x = 25$
    $x = 5$

b. $5s + 7 = 8s - 11$
    $5s - 8s = -11 - 7$
    $-3s = -18$
    $s = 6$

c. $4 - 2(t + 6) - 3(t + 4) = -6t - 10$
$4 - 2t - 12 - 3t - 12 = -6t - 10$
$-2t - 3t + 6t = -10 - 4 + 12 + 12$
$t = 10$

d. $8b - 7(2b - 3) \qquad = -7b + 3(b + 1)$
$8b - 14b + 21 \qquad = -7b + 3b + 3$
$8b - 14b + 7b - 3b = 3 - 21$
$-2b = -18$
$b = 9$

PROBLEMS

In Problems 1 to 20 solve for the unknowns in the equations.

1. $4x + 7 = 23$

2. $-19 - 5a = 6$

3. $3b - 4 = b + 4$

4. $6u + 1 = 3(u + 7) + 4$

5. $5(z - 3) = 3(z + 3)$

6. $6(v - 4) - 4(v + 2) = 2$

7. $13(-1 - 2c) + 7(c - 3) = 5(c + 2) + 4$

8. $13(d + 2(d - 1) + 3) = -65$

9. $3(x + 4(x - 1(x + 2) - 3) + 4) = -2(x - 1)$

10. $(i - 2)(i - 3) + i(6 - i) = 5$

11. $4(y + 2)(y - 5) = y(4y + 3) - 5$

12. $5(f-1)(f+3) = (5f+4)(f-6) - 27$

13. $6(2x + 3) - 2x = 28$

14. $2(t - 3) - 20 = 10 - 3(t + 2)$

15. $(y + 1)(y + 4) - 59 = y^2$

16. $(s + 3)(s + 2) = s(s + 2)$

17. $(3 - x)(2 + x) = 38 - (x^2 + 5x - 16)$

18. $6(2y + 1) - (6y + 10) - 3(4y - 3) = 3 - (4y - 3)$

19. $8x - (2 + x) = 2(3x - 1)$

20. $4 + y + y^2 = -y(2 - y) + 7$

8.3 SOLUTIONS OF VARIABLES

Often an electronic equation does not define the variable whose solution is sought. Instead, frequently the desired variable is intermixed with other variables. When this happens, the original equation has to be manipulated so it has a format that uniquely defines the unknown or desired quantity in terms of the other known quantities. For example, let us take Equation (8.1):

$$R_T = \frac{R_1 R_2}{R_1 + R_2} \tag{8.1}$$

We have two resistors (R_1 and R_2) and ask what equivalent resistance they represent if wired in parallel. But suppose we are given R_1 and want to know what other resistor (R_2) could be wired in parallel with R_1 to result in a given R_T. Equation (8.1) does not fit the bill anymore, so we have to manipulate it into the following format:

R_2 = An expression containing only terms of R_T and R_1

Examples 1 and 2 illustrate methods to accomplish these types of algebraic manipulations. Some steps are of necessity more complicated than those needed for simple linear equations, but the key is still the same:

Do the same thing to both sides of the equation to keep the balance.

One new factor to note. Up to this point variables have been designated by letters of the alphabet. However, many electronic symbols are represented by letters of the Greek alphabet. A classic example is omega (Ω), the symbol of resistance in ohms. Other electronic symbols and constants are represented by other Greek letters. Not all Greek letters have an electronic meaning; however, any Greek letter can be used to represent a variable, just as x or y have been used. Henceforth we use Greek symbols occasionally to represent variables in mathematical expressions.

Example 1

Solve the following expressions for the indicated variable.

a. $R_T = \dfrac{R_1 R_2}{R_1 + R_2}$ solve for R_2

Step 1:   M; $(R_1 + R_2)$ $R_T(R_1 + R_2) = \dfrac{R_1 R_2}{\cancel{R_1 + R_2}} \cancel{(R_1 + R_2)} = R_1 R_2$

Step 2: Expand binomial $R_T R_1 + R_T R_2 = R_1 R_2$

Step 3:   S; $R_T R_2$ $R_T R_1 = R_1 R_2 - R_T R_2$

Step 4:   Factor out R_2 $R_T R_1 = (R_1 - R_T)R_2$

Step 5:   D; $(R_1 - R_T)$ $\dfrac{R_T R_1}{R_1 - R_T} = R_2$

The key to solving for a variable when it exists more than once in an expression (such as the two R_2s in a) is to collect all terms of that variable on one side of the equation (step 3), factor it out (step 4), and divide by the entire parenthesis remaining after the factoring (step 5).

Interestingly (electronically speaking), given two resistors in parallel, the equivalent is the *product over the sum* of the two known values. However, given any one resistor and looking for the second resistor needed to equal a desired equivalent resistance results in the *product over the difference* of the two known values.

b. $\beta = \dfrac{\alpha}{1 - \alpha}$ solve for α

Step 1:   M; $(1 - \alpha)$ $\beta(1 - \alpha) = \dfrac{\alpha}{\cancel{1 - \alpha}} \cancel{(1 - \alpha)} = \alpha$

Step 2:   Expand binomial $\beta - \alpha\beta = \alpha$

Step 3:   A; $\alpha\beta$ $\beta = \alpha\beta + \alpha$

Step 4:   Factor out α $\beta = \alpha(\beta + 1)$

Step 5:   D; $(\beta + 1)$ $\dfrac{\beta}{\beta + 1} = \alpha$

This example is analogous to part a. Both the α and β expressions relate to transistor gain.

c. $M = k\sqrt{L_1 L_2}$ solve for L_2 (equation relates to mutual inductance between coils)

Step 1: Square the equation $M^2 = k^2 L_1 L_2$

Step 2: D; $k^2 L_1$ $\dfrac{M^2}{k^2 L_1} = L_2$

d. $W = \dfrac{Li^2}{2}$ solve for i (equation relates to energy stored in a coil)

Step 1: M; 2 $2W = Li^2$

Step 2: D; L $\dfrac{2W}{L} = i^2$

Step 3: $\sqrt{}$ $\sqrt{\dfrac{2W}{L}} = i$

e. $Z = \sqrt{R^2 + X_L^2}$ solve for X_L (equation relates to impedance of a reactance and resistance in series, or to opposition to current flow in AC circuits)

Step 1: Square the equation $Z^2 = R^2 + X_L^2$

Step 2: S; R^2 $Z^2 - R^2 = X_L^2$

Step 3: $\sqrt{}$ $\sqrt{Z^2 - R^2} = X_L$

When simplifying linear equations, you can move terms preceded by a plus or minus sign across the equal sign if the sign of the term being moved is changed. There are similar shortcuts for when either the left- or right-hand expression of an equation is a fraction. For example, consider

$$\frac{x}{2} = 3y$$

To solve for x, you must eliminate the denominator 2, otherwise, $x/2$ is expressed and not x. Multiplying the equation by 2 results in x as the left-hand expression and $2(3y)$ or $6y$ as the right-hand expression. Therefore, note that

$$\frac{x}{2} = 3y$$

may be rewritten as

$$x = 2(3y)$$

The net result is that

> You can move a fractional number and/or variable across the equal sign if you change its position from numerator to denominator or from denominator to numerator.

Example 2

Solve for x in each of the following.

a. $\dfrac{3x}{2} = y$

$x = \dfrac{2y}{3}$

b. $(\beta + 1)x = \alpha + 2$

$x = \dfrac{\alpha + 2}{\beta + 1} \cdot$

c. $\dfrac{3x(a + b)}{4d} = y + 7$

$x = \dfrac{4d(y + 7)}{3(a + b)}$

d. $\dfrac{(1 - \alpha^2)x}{2\sqrt{\beta}} = \dfrac{u}{v}$

$x = \dfrac{2u\sqrt{\beta}}{v(1 - \alpha^2)}$

PROBLEMS

Solve for the indicated variables in Problems 1 to 22.

1. $e = ir$ solve for i

2. $F = \dfrac{kQ_1Q_2}{r^2}$ solve for r

3. $\eta = \dfrac{DN_0}{M}$ solve for N_0

4. $R = \dfrac{\rho l}{A}$ solve for ρ

5. $R_2 = R_1 \left(\dfrac{A + T_2}{A + T_1} \right)$ solve for A

6. $R_2 = R_1 (1 + \alpha \, \Delta T)$ solve for α (Δ means "change in")

✗ **7.** $R_T = \dfrac{R_1 R_2 R_3}{R_1 R_2 + R_1 R_3 + R_2 R_3}$ solve for R_2

8. $I_1 = \dfrac{R_2}{R_1 + R_2} I_T$ solve for R_2

9. $P_{Rl} = \dfrac{V_s^2 R_L}{(R_s + R_L)^2}$ solve for R_s

10. $P_{Rl} = \dfrac{v_s^2}{4 R_L}$ solve for v_s

✗ **11.** $R_1 = \dfrac{R_a R_b}{R_a + R_b + R_c}$ solve for R_a

12. $F = \dfrac{M_1 M_2}{\mu d^2}$ solve for d

13. $F = \dfrac{B^2 A}{2\mu}$ solve for B

14. $L = \dfrac{N^2 \mu A}{l}$ solve for N

15. $L_T = \dfrac{1}{\dfrac{1}{L_1} + \dfrac{1}{L_2} + \dfrac{1}{L_3}}$ solve for L_2

✗ **16.** $V_1 = \dfrac{c_2 V}{c_1 + c_2}$ solve for c_2

17. $f_R = \dfrac{1}{2\pi \sqrt{LC}}$ solve for L

18. $Z_t = \dfrac{(Z_1 + Z_2) Z_3}{Z_1 + Z_2 + Z_3} + Z_4$ solve for Z_3

19. $Q = \dfrac{1}{R}\sqrt{\dfrac{L}{C}}$ solve for C

20. $f_0 = \dfrac{1}{2\pi\sqrt{LC}}\sqrt{1 - \dfrac{1}{Q^2}}$ solve for Q

21. $Z_{in} = Z_a + \dfrac{Z_b(Z_c + Z_L)}{Z_b + Z_c + Z_L}$ solve for Z_b

22. $Z_{in} = \dfrac{R}{\omega^2 R^2 C^2 + 1}$ solve for ω

8.4 FRACTIONAL EQUATIONS

The rules for solving linear equations equally apply to equations in which one or more terms are a fraction or decimal. You must retain the same balance on both sides as you apply different manipulative techniques for the equation to remain valid.

The simplest, single step to apply when one (or more) term(s) of an equation is a fraction is to multiply both sides of the equation by the LCD. This immediately clears all fractions and reduces the equation to a linear format.

Example 1

Solve the following fractional equations.

a. $\dfrac{x}{3} + \dfrac{2x}{9} = 5$

M; LCD $= 9$ $9\left(\dfrac{x}{3} + \dfrac{2x}{9}\right) = 9(5)$

$$3x + 2x = 45$$
$$5x = 45$$
$$x = 9$$

b. $\dfrac{6}{q} + 2 = \dfrac{3}{q} - 5$

M; LCD $= q$ $q\left(\dfrac{6}{q} + 2\right) = q\left(\dfrac{3}{q} - 5\right)$

$$6 + 2q = 3 - 5q$$
$$2q + 5q = 3 - 6$$
$$7q = -3$$
$$q = -\dfrac{3}{7}$$

c. $\dfrac{7}{z} - \dfrac{4}{5} = \dfrac{3 - z}{z}$

M; LCD = $5z$ $5z\left(\dfrac{7}{z} - \dfrac{4}{5}\right) = 5z\left(\dfrac{3 - z}{z}\right)$

$$35 - 4z = 15 - 5z$$
$$5z - 4z = 15 - 35$$
$$z = -20$$

d. $\dfrac{a - 5}{a + 3} + \dfrac{a + 6}{a - 2} = \dfrac{2a^2 - 3a + 4}{a^2 + a - 6}$

M; LCD = $(a + 3)(a - 2)$

$$(a + 3)(a - 2)\left(\dfrac{a - 5}{a + 3} + \dfrac{a + 6}{a - 2}\right) = (a + 3)(a - 2)\left(\dfrac{2a^2 - 3a + 4}{(a + 3)(a - 2)}\right)$$

$$\underbrace{(a - 2)(a - 5)} + \underbrace{(a + 3)(a + 6)} = 2a^2 - 3a + 4$$
$$a^2 - 7a + 10 + a^2 + 9a + 18 = 2a^2 - 3a + 4$$
$$2a^2 - 7a + 9a - 2a^2 + 3a = 4 - 10 - 18$$
$$5a = -24$$
$$a = -\dfrac{24}{5}$$

e. $\dfrac{0.6(y - 2)}{2.5} - \dfrac{0.4(y + 3)}{7.5} = \dfrac{y}{5}$

M; LCD = 15 $15\left(\dfrac{0.6(y - 2)}{2.5} - \dfrac{0.4(y + 3)}{7.5}\right) = 15\left(\dfrac{y}{5}\right)$

$$3.6(y - 2) - 0.8(y + 3) = 3y$$
$$3.6y - 7.2 - 0.8y - 2.4 = 3y$$
$$3.6y - 0.8y - 3y = 7.2 + 2.4$$
$$-0.2y = 9.6$$
$$y = -48$$

PROBLEMS

Solve the fractional equations in Problems 1 to 15 for the indicated variable.

1. $\dfrac{s}{7} - \dfrac{s}{21} = 3$

2. $\dfrac{t}{1.8} + 2 = \dfrac{t}{5.4}$

3. $\dfrac{3}{x} - 4 = \dfrac{5}{x}$

4. $\dfrac{1}{q} - 2 - \dfrac{2}{q} = \dfrac{3 - q}{q}$

5. $\dfrac{5}{3y} - \dfrac{2}{1.5y} = \dfrac{y + 1}{6y}$

6. $\dfrac{3\alpha + 2}{7} - \dfrac{2\alpha - 3}{28} = \alpha$

7. $\dfrac{1}{\beta - 1} - \dfrac{2}{\beta + 1} + \dfrac{3}{\beta - 1} = 0$

12. $\dfrac{R - 1}{R + 2} + \dfrac{R + 5}{R^2 - 5R - 14} = \dfrac{R + 3}{R - 7}$

8. $\dfrac{8 - \gamma}{\gamma + 2} + \gamma = \gamma + 2$

13. $\dfrac{2 - 3u}{u - 7} - \dfrac{5 + 2u}{u - 6} = \dfrac{8 - u - 5u^2}{u^2 - 13u + 42}$

9. $\dfrac{0.3(d - 1)}{2.1} - \dfrac{0.4(3 - 2d)}{6.3} - \dfrac{2.35}{3.15} = 0$

14. $\dfrac{3v^2 + 2v + 5}{2v^2 + 3v - 14} = 1 + \dfrac{v - 4}{2v + 7}$

10. $\dfrac{6}{4l - 8} - \dfrac{3}{16 - 8l} = 1$

15. $\dfrac{2p + 2}{p - 3} - \dfrac{p - 3}{3p + 5} - \dfrac{5p^2 - 1}{3p^2 - 4p - 15} = 0$

11. $\dfrac{4a^2 - 9}{2(2a + 3)} - \dfrac{6a^2 - 13a - 5}{3(3a + 1)} = a$

8.5 ANALYZING WORD PROBLEMS

Section 8.1 stressed that there are essentially two phases to the solution of a word problem: (1) the translation of the parameters of the problem into a mathematical equation, and (2) given an equation, the application of mechanical *balancing* rules to solve for the unknown variable. Sections 8.2 to 8.4 dealt with these balancing rules. Now that you are confident that once given an equation you know how to solve it, you can begin phase one: translating the grammar which defines a problem into a mathematical expression.

There are no steadfast rules for this phase, which is why most students initially approach word problems somewhat apprehensively. Electrical problems, like most real-life problems, are of this nature; they do not present themselves as an equation, where all that has to be done is to solve for x. They are in the form of failures in a given design that must be analyzed; or in the form of "given this input, what circuit is necessary to produce that output"; and so on. The key is to analyze what is given and then simply express it mathematically. This section illustrates phase one via simple analytical problems. They will lay the requisite foundation so as to better analyze the electronic problems throughout Chapter 9 and subsequent chapters.

Example 1

Solve the following problems by first writing the equation and then solving the equation.

a. The sum of two consecutive odd integers is 464. What are the integers?

Initially, name the first integer; let us arbitrarily say x. The next consecutive *odd* integer must be $x + 2$. The author has heard questions such as "How can it be $x + 2$ when everyone knows 2 is an even integer?" Such questions illustrate precisely why it is so important to *read* the problem. The problem does not state "two consecutive integers" but "two consecutive *odd* integers." Consecutive odd integers (or consecutive even integers) are always two digits apart; therefore our integers must be x and $x + 2$.

The next part of the problem states that first the integers are added and second, that the integers' sum is 464. Therefore

$$\underbrace{x}_{\substack{\text{First odd} \\ \text{integer}}} + \underbrace{x + 2}_{\substack{\text{Next consecutive} \\ \text{odd integer}}} = \underbrace{464}_{\text{Sum}}$$

$$x + (x + 2) = 464$$
$$2x = 464 - 2 = 462$$
$$x = 231$$
$$\therefore x + 2 = 231 + 2 = 233$$

Check: $\qquad\qquad 231 + 233 = 464$

b.   A rectangle is 12 m longer than its width; the perimeter of the rectangle is 96 m. What is the width of the rectangle?

From Figure 8.2, note that the perimeter consists of the sum of the four sides of the rectangle; therefore,

$$x + (x + 12) + x + (x + 12) = 96$$
$$4x + 24 = 96$$
$$4x = 96 - 24 = 72$$
$$x = 18 \text{ m}$$

Check: $\qquad 18 + (18 + 12) + 18 + (18 + 12) = 96$
$$96 = 96$$

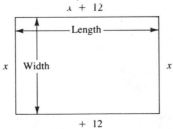

Figure 8.2

c. If a certain number plus 15 is the numerator of a fraction and the same number minus 15 is the denominator, the quotient is 4. What is the number? (Assign the following values):

x = number
$x + 15$ = numerator
$x - 15$ = denominator

$$\frac{x + 15}{x - 15} = 4$$
$$x + 15 = 4(x - 15) = 4x - 60$$
$$15 + 60 = 4x - x$$
$$75 = 3x$$
$$25 = x$$

PROBLEMS

Solve Problems 1 to 17.

1. Write the equation described by: The variable q equals the reciprocal of r times the square root of t.

2. Write the equation described by: The impedance z equals the square root of the sum of the reactance (x) squared plus the resistance (r) squared.

3. An integer plus 7 is twice as much as the same integer minus 8. What is the integer?

4. The sum of three consecutive numbers is 366. What are the numbers?

5. An integer plus 2/3 of the integer minus 1/2 the integer is 84. What is the integer?

6. When 2 1/2 times a certain number is then diminished by 55, the resultant is 290. What is the number?

7. Side two of a triangle is twice as big as side one and one-half as big as side three. The total perimeter of the triangle is 140 cm. What is the length of side three of the triangle?

8. Side two of a triangle is 11 cm longer than side one and 17 cm shorter than side three. If the perimeter is 111 cm, what is the length of side two?

9. The sum of the angles of a triangle is 180°. Angle two is three times larger than angle one and angle three is 55° larger than angle one. How many degrees are in angle one?

10. Angle three of a triangle is 2 3/4 times larger than angle one and angle one is 9° smaller than angle two. How many degrees are in angle two?

11. A student's math grades are 70, 84, and 88. What grade must be scored on the next test to result in an average of 85?

12. The average of three numbers is 45. The second number is three more than twice the first. The third is six less than three times the first. Find the numbers.

13. The numerator of a fraction is six less than the denominator. The value of the fraction is 5/8. Find the fraction.

14. When five times the reciprocal of a number is decreased by one-fourth, the result is 6. Find the number.

15. If one-fifth of a number is 11 less than three-fourths of the number, find the number.

16. The average of six consecutive even numbers is 203. Find the numbers.

17. What number must be subtracted from both the numerator and denominator of 21/29 so that the value of the resulting fraction is 3/5?

SUMMARY

Equations represent the most powerful tool of mathematics because they are the means by which one expression can be equated to another expression. In a practical sense, an unknown quantity can be expressed in terms of known quantities. Through the application of balancing rules, you can manipulate equations to find the unknown quantity, either as a numeric value or in terms of other measurable, known variables.

The rules for manipulating simple linear equations equally apply to fractional equations. Usually, you can reduce a fractional equation to a simple linear one by multiplying both sides of the equation by the LCD.

If a term on one side of an equation is preceded by a plus or minus sign, you can move the term to the other side of the equal sign if the term's sign is changed. If one side of an equation is a fraction, you can move the numerator or denominator of the fraction across the equal sign if it is transposed to the opposite position following the move.

If you perform any action on one side of an equation, you must perform the same action on the other side to retain the balance of the equation. You can solve word problems by applying two distinct steps: (1) Translate the grammar of the problem into a mathematical equation; (2) apply the routine balancing rules to the equation until you solve for the unknown.

APPLICATIONS 1: DC ELECTRONIC CIRCUITS

9.1 INTRODUCTION

Chapter 8 introduced the methodology for solving simple numeric word problems. The concepts introduced there extend to real-life electronic problems throughout this chapter. As stated in the preface, it is assumed that you have a basic understanding of DC electronic fundamentals. Therefore, this chapter concentrates upon analyzing the problems, reducing them to an equation(s), and calculating a numeric solution. When the formula is relatively simple, only the numeric solution is shown, but when the formula is more complex, the calculator steps and the solution are shown.

To facilitate solutions, we follow two dictums. First, we use electron flow, that is, current is emanating from the negative terminal of the source, through the external circuit and to the positive terminal of the source, to conform to usage in most electronic textbooks.

Second, only one subscript associated with voltage, such as V_x, implies the voltage magnitude and polarity of point x with respect to ground. Two subscripts associated with voltage, such as V_{xy}, implies that voltage magnitude and polarity are measured from point y to point x (reference). If point y is more positive than point x, the polarity is positive. If the converse is true, the polarity is negative. For example, in Figure 9.1,

$$V_A = 25 \text{ V}, \ V_B = 17.5 \text{ V}, \ V_C = 6.3 \text{ V}$$
$$V_{BA} = (25 - 17.5) \text{ V} = 7.5 \text{ V}$$
$$V_{CB} = (17.5 - 6.3) \text{ V} = 11.2 \text{ V}$$
$$V_{AC} = (6.3 - 25) \text{ V} = -18.7 \text{ V}$$

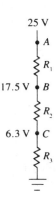

Figure 9.1

9.2 OHM'S LAW

Ohm's law can be stated as:

$$I = \frac{V}{R}$$

Current magnitude in a closed circuit is directly proportional to voltage and inversely proportional to resistance.

Example 1

See Figure 9.2. What is I? What is V_{R2}?

$$I = \frac{V}{R_T} = \frac{V}{R_1 + R_2 + R_3}$$

$$= \frac{12\ \text{V}}{(2.2 + 4.7 + 0.68)\ \text{k}\Omega} = \frac{12}{7.58}\ \text{mA} = 1.583\ \text{mA}$$

$$V_{R2} = IR_2 - (1.583\ \text{mA})(4.7\ \text{k}\Omega) = 7.441\ \text{V}$$

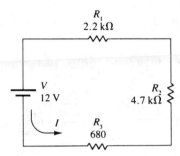

Figure 9.2

Example 2

See Figure 9.3. What is I?

$$I = \frac{V}{R_T} = \frac{V}{\dfrac{1}{\dfrac{1}{R_1} + \dfrac{1}{R_2} + \dfrac{1}{R_3}}} = V\left(\frac{1}{R_1} + \frac{1}{R_2} + \frac{1}{R_3}\right)$$

$$= 22\left(\frac{1}{4.7} + \frac{1}{5.6} + \frac{1}{6.8}\right)\ \text{mA}$$

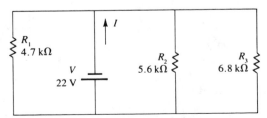

Figure 9.3

PRESS	DISPLAY	COMMENT

22 $\boxed{\times}$ $\boxed{(}$ 4.7 $\boxed{1/x}$ $\boxed{+}$ 5.6 $\boxed{1/x}$ $\boxed{+}$

6.8 $\boxed{1/x}$ $\boxed{)}$ $\boxed{=}$ 11.845 $I = 11.845$ mA

Example 3

See Figure 9.4. What value of R_1 results in I equal to 2 mA? Answer will be in kilohms (kΩ).

$$I = \frac{V}{R_T} = \frac{V}{R_1 + \dfrac{R_2 R_3}{R_2 + R_3}}$$

M both sides of equation by $R_1 + \dfrac{R_2 R_3}{R_2 + R_3}$: $I\left(R_1 + \dfrac{R_2 R_3}{R_2 + R_3}\right) = V$

D both sides of equation by I: $R_1 + \dfrac{R_2 R_3}{R_2 + R_3} = \dfrac{V}{I}$

Solve for R_1: $R_1 = \dfrac{V}{I} - \dfrac{R_2 R_3}{R_2 + R_3}$

PRESS	DISPLAY	COMMENT

27 $\boxed{\div}$ 2 $\boxed{-}$ 20 $\boxed{\times}$ 30 $\boxed{\div}$ $\boxed{(}$

20 $\boxed{+}$ 30 $\boxed{)}$ $\boxed{=}$ 1.5 $R_1 = 1.5$ kΩ

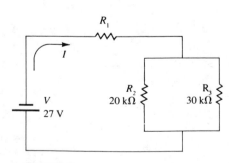

Figure 9.4

Example 4

See Figure 9.5. What must R_2 be so that $I = 1.25$ mA?

$$I = \frac{V}{R_T} = \frac{V}{R_1 + \dfrac{R_2 R_3}{R_2 + R_3}}$$

M numerator and denominator of right side of equation by $(R_2 + R_3)$:

$$I = \frac{V(R_2 + R_3)}{R_1(R_2 + R_3) + R_2 R_3} = \frac{V R_2 + V R_3}{R_1 R_2 + R_1 R_3 + R_2 R_3}$$

M; $(R_1 R_2 + R_1 R_3 + R_2 R_3)$:

$$I(R_1 R_2 + R_1 R_3 + R_2 R_3) = V R_2 + V R_3$$

Solve for R_2:

$$I R_1 R_2 + I R_1 R_3 + I R_2 R_3 = V R_2 + V R_3$$
$$I R_1 R_2 + I R_2 R_3 - V R_2 = V R_3 - I R_1 R_3$$
$$R_2(I R_1 + I R_3 - V) = R_3(V - I R_1)$$
$$R_2 = \frac{R_3(V - I R_1)}{I(R_1 + R_3) - V}$$

Note that the answer is automatically in kΩ because the current is in milliamperes (mA).

PRESS	DISPLAY	COMMENT
⟨ 33 × ⟨ 25 − 1.25 × 6.8 ⟩ ⟩	554.5	Numerator
÷ ⟨ 1.25 × ⟨ 6.8 + 33 ⟩ − 25 ⟩	24.75	Denominator
=	22	$R_2 = 22$ kΩ

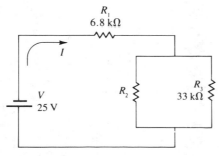

Figure 9.5

Example 5

See Figure 9.6. Calculate I. Answer will be in mA because all resistors are expressed in $k\Omega$.

$$I = \frac{V}{R_T} = \frac{V}{R_1 + R_7 + R_8} = \frac{V}{R_1 + \dfrac{R_2 R_3}{R_2 + R_3} + \dfrac{1}{\dfrac{1}{R_4} + \dfrac{1}{R_5} + \dfrac{1}{R_6}}}$$

$$= \frac{45}{3.3 + \dfrac{(2.2)(0.47)}{2.2 + 0.47} + \dfrac{1}{\dfrac{1}{1.5} + \dfrac{1}{4.7} + \dfrac{1}{10}}}$$

PRESS	DISPLAY	COMMENT
3.3 $\boxed{+}$ 2.2 $\boxed{\times}$ .47 $\boxed{\div}$ $\boxed{(}$		
2.2 $\boxed{+}$ .47 $\boxed{)}$ $\boxed{+}$	3.687	$R_1 + R_7$
$\boxed{(}$ 1.5 $\boxed{1/x}$ $\boxed{+}$ 4.7 $\boxed{1/x}$ $\boxed{+}$ 10 $\boxed{1/x}$ $\boxed{)}$ $\boxed{1/x}$ $\boxed{=}$	4.708	$R_1 + R_7 + R_8$
$\boxed{1/x}$ $\boxed{\times}$ 45 $\boxed{=}$	9.558	$I = 9.558$ mA

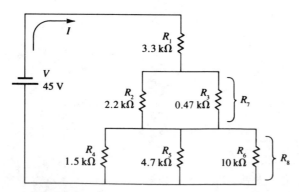

Figure 9.6

PROBLEMS

First, algebraically solve for the desired electronic quantity. Then, using your calculator, determine the following:

1. What are I and V_{R3} in Figure 9.7?

2. What are I and V_{R1} in Figure 9.8?

3. If R_2 in Figure 9.8 is unknown, what value is necessary to result in $I = 516.92$ μA?

4. For the three resistors in Figure 9.8, what voltage is required for I to equal 2.65 mA?

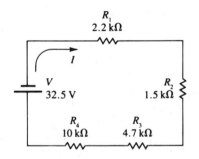

Figure 9.7

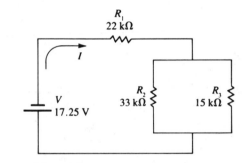

Figure 9.8

5. See Figure 9.9. What is the value of I?

6. What is the voltage across the parallel combination of R_4 and R_5 in Figure 9.9?

7. What value of source voltage is necessary to produce a current (I) of 3.245 mA in Figure 9.9?

8. See Figure 9.10. What is I?

9. What is the voltage across the parallel combination of R_3 and R_4 in Figure 9.10?

10. What value of V is necessary to produce a current of 360.75 μA in Figure 9.10?

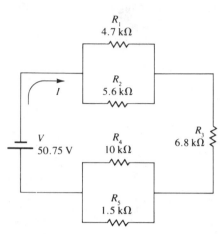

Figure 9.9

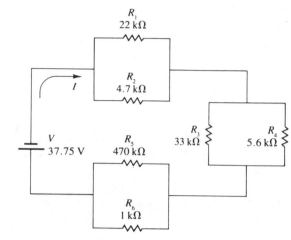

Figure 9.10

9.3 PROPORTIONAL VOLTAGE RATIO LAW

To more fully comprehend the derivation of the *proportional voltage ratio law*, refer to Figure 9.11 and follow its algebraic development.

By Ohm's law,

$$I = \frac{V}{R_1 + R_2}$$

The voltage across R_1 is the product of I times R_1, or

$$V_{R1} = IR_1 = \frac{V}{R_1 + R_2}(R_1) = \frac{R_1}{R_1 + R_2}(V) \qquad (9.2a)$$

Conversely, the voltage across R_2 is the product of I times R_2, or

$$V_{R2} = IR_2 = \frac{V}{R_1 + R_2}(R_2) = \frac{R_2}{R_1 + R_2}(V) \qquad (9.2b)$$

It appears that when resistors are in series, it is not at all necessary to know the value of the current in the circuit to determine the voltage across a resistor because nowhere in Equations (9.2a) or (9.2b) does I appear. All that is necessary is to multiply the applied voltage (V) by the ratio that the resistor in question (either R_1 or R_2) has to the total resistance. Now, because any closed loop can have any number of resistors, the more generalized expression for the voltage across any resistor (or any

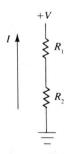

Figure 9.11

group of resistors within the loop, which are usually but not necessarily contiguous) is:

$$V_n = \frac{R_n}{R_T} \times V_T \qquad (9.2c)$$

The voltage across any resistor(s) in a closed loop equals the ratio of that resistance to the total resistance of the loop multiplied by the total applied voltage in the loop.

Example 1

See Figure 9.2. Determine the voltage across R_2 using Equation (9.2c). Compare with the answer of 7.441 V as determined in Example 1, Section 9.2.

$$V_{R2} = \frac{R_2(V)}{R_T} = \frac{R_2(V)}{R_1 + R_2 + R_3}$$

$$= \frac{(4.7)(12)}{2.2 + 4.7 + 0.68} = 7.441 \text{ V}$$

Determine the collective voltage across R_2 and R_3:

$$V_{R2 + R3} = \frac{(R_2 + R_3)\,V}{R_T} = \frac{(4.7 + 0.68)(12)}{(2.2 + 4.7 + 0.68)} = 8.517 \text{ V}$$

Determine the collective voltage across R_1 and R_3:

$$V_{R1 + R3} = \frac{(R_1 + R_3)\,V}{R_T} = \frac{(2.2 + 0.68)\,12}{(2.2 + 4.7 + 0.68)} = 4.559 \text{ V}$$

Example 2

If R_2 in Figure 9.5 is 15 kΩ, calculate the current through R_3 by direct application of the proportional voltage law. For convenience, refer to the voltage and resistance across the parallel combination as V_p and R_p, respectively.

By Ohm's law,

$$I_{R3} = \frac{V_p}{R_3} \qquad (9.3)$$

By the proportional voltage ratio:

$$V_p = \frac{R_p(V)}{R_T} = \frac{R_p(V)}{R_1 + R_p} = \frac{V}{\dfrac{R_1}{R_p} + 1} \qquad (9.4)$$

Using the product-over-sum formula for two parallel resistors results in

$$R_p = \frac{R_2 R_3}{R_2 + R_3} \tag{9.5}$$

Substituting R_p of Equation (9.5) into Equation (9.4) results in

$$V_p = \frac{V}{\dfrac{R_1}{\dfrac{R_2 R_3}{R_2 + R_3}} + 1} = \frac{V}{\dfrac{R_1(R_2 + R_3)}{R_2 R_3} + 1}$$

$$= \frac{V R_2 R_3}{R_1(R_2 + R_3) + R_2 R_3} \tag{9.6}$$

Substituting V_p of Equation (9.6) into Equation (9.3) results in an expression for the desired current:

$$I_{R3} = \frac{V_p}{R_3} = \frac{V R_2}{R_1 R_2 + R_1 R_3 + R_2 R_3} = \frac{V}{R_1 + R_3 + \dfrac{R_1 R_3}{R_2}}$$

$$= \frac{25}{6.8 + 33 + \dfrac{(6.8)(33)}{15}}$$

PRESS	DISPLAY	COMMENT
(| 6.8 [+] 33 [+] 6.8 [×] 33 [÷] 15 |)	54.76	Denominator
[1/x] [×] 25 [=]	0.457	$I_{R3} = 0.457$ mA

Example 3

Determine the current through R_5 in Figure 9.6 by direct application of the proportional voltage law.

By Ohm's law:

$$I_{R5} = \frac{V_{R8}}{R_5} \tag{9.7}$$

By the proportional voltage ratio:

$$V_{R8} = \frac{R_8(V)}{R_T} = \frac{R_8(V)}{R_1 + R_7 + R_8} \tag{9.8}$$

Substituting V_{R8} of Equation (9.8) into Equation (9.7) results in

$$I_{R5} = \frac{R_8(V)}{R_5(R_1 + R_7 + R_8)} \tag{9.9}$$

At about this point we are approaching a trade-off. Is it worth the additional time to substitute

$$R_7 = \frac{R_2 R_3}{R_2 + R_3}$$

and $\quad R_8 = \dfrac{1}{\dfrac{1}{R_4} + \dfrac{1}{R_5} + \dfrac{1}{R_6}}$

into Equation (9.9) and labor its algebraic reduction, or is it more advantageous to independently calculate R_7 and store it and then independently calculate R_8 and store it? If we use the latter approach, the stored values can then be recalled ($\boxed{\text{RCL}}$) into Equation (9.9) to determine I_{R5}. With more complex circuits this latter approach is more desirable, as the following display illustrates:

PRESS	DISPLAY	COMMENT
2.2 $\boxed{\times}$ .47 $\boxed{\div}$ $\boxed{(}$ 2.2 $\boxed{+}$ .47 $\boxed{)}$ $\boxed{=}$ $\boxed{\text{STO 0}}$	0.387	R_7
1.5 $\boxed{1/x}$ $\boxed{+}$ 4.7 $\boxed{1/x}$ $\boxed{+}$ 10 $\boxed{1/x}$ $\boxed{=}$ $\boxed{1/x}$ $\boxed{\text{STO 1}}$	1.021	R_8
$\boxed{(}$ 4.7 $\boxed{\times}$ $\boxed{(}$ 3.3 $\boxed{+}$ $\boxed{\text{RCL 0}}$		
$\boxed{+}$ $\boxed{\text{RCL 1}}$ $\boxed{)}$ $\boxed{)}$	22.129	Denominator of Equation (9.9)
$\boxed{1/x}$ $\boxed{\times}$ $\boxed{\text{RCL 1}}$ $\boxed{\times}$ 45 $\boxed{=}$	2.076	$I_{R5} = 2.076$ mA

9.4 INVERSE CURRENT RATIO LAW

In a manner analogous to the one used for the proportional voltage ratio law, let us algebraically develop the *inverse current ratio law* to better understand the mathematics described in the electronic theory. See Figure 9.12.

The equivalent parallel resistance is given as

$$R_T = \frac{R_1 R_2}{R_1 + R_2} \tag{9.10a}$$

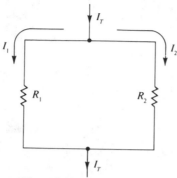

Figure 9.12

The voltage across the parallel combination is

$$V_T = I_T R_T = \frac{R_1 R_2}{R_1 + R_2}(I_T) \qquad (9.10b)$$

The current through R_1 is V_T divided by R_1, or

$$I_{R1} = \frac{V_T}{R_1} = \frac{R_1 R_2}{R_1(R_1 + R_2)}(I_T) = \frac{R_2}{R_1 + R_2}(I_T) \qquad (9.10c)$$

Conversely, the current through R_2 is V_T divided by R_2, or

$$I_{R2} = \frac{V_T}{R_2} = \frac{R_1 R_2}{R_2(R_1 + R_2)}(I_T) = \frac{R_1}{R_1 + R_2}(I_T) \qquad (9.10d)$$

It appears that when resistors are in parallel, it is not at all necessary to know the value of the voltage across the parallel to determine the current through either resistor because nowhere in Equations (9.10c) or (9.10d) does V appear. In words, given two parallel resistors with I_T entering the junction,

> The current through any one resistor equals the current entering the junction times the ratio of the other resistor's resistance to the sum of the two resistors' resistance.

Example 1

See Figure 9.13. Using the inverse current ratio, determine I_1 and I_2. Show that their sum equals the current entering the junction.

$$I_1 = \frac{R_2(I)}{R_1 + R_2} = \frac{(4.7)(95)}{4.7 + 3.3} = 55.813 \text{ mA}$$

$$I_2 = \frac{R_1(I)}{R_1 + R_2} = \frac{(3.3)(95)}{4.7 + 3.3} = 39.188 \text{ mA}$$

$$I_1 + I_2 = (55.813 + 39.188) \text{ mA} = 95.001 \text{ mA}$$

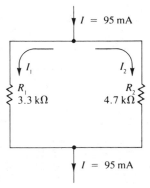

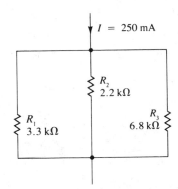

Figure 9.13   **Figure 9.14**

Note that the minute difference of 0.001 mA is due only to the round-off at the third decimal place. Example 1 illustrates one of Kirchhoff's laws, namely, the sum of the currents entering a junction equals the sum of the currents leaving the junction.

Example 2

See Figure 9.14. Using the inverse current law; determine the current through R_3 for three parallel resistors.

Combine R_1 and R_2 into one resistor called R_4:

$$R_4 = \frac{R_1 R_2}{R_1 + R_2}$$

Apply the inverse current ratio law to R_3 and R_4:

$$I_{R3} = \frac{R_4(I)}{R_3 + R_4} = \frac{\dfrac{R_1 R_2}{R_1 + R_2}(I)}{R_3 + \dfrac{R_1 R_2}{R_1 + R_2}}$$

M; $R_1 + R_2$: $$= \frac{R_1 R_2(I)}{R_3(R_1 + R_2) + R_1 R_2} = \frac{R_1 R_2(I)}{R_1 R_2 + R_1 R_3 + R_2 R_3}$$

In words,

   The current through any one resistor of three parallel resistors equals the product of the other two resistors divided by the sum of the products of all combinations of any two of the three resistors.

Note the striking similarity of this rule to the simple product-over-sum rule for three parallel resistors. Solving for I_{R3} results in

$$I_{R3} = \frac{(3.3)(2.2)(250)}{(3.3)(2.2) + (3.3)(6.8) + (2.2)(6.8)} \text{ mA} = 40.64 \text{ mA}$$

Example 3

If R_1 in Figure 9.4 is 6.8 kΩ, determine the current (I_2) through R_2 using the inverse current law.

Using Ohm's law,

$$I = \frac{V}{R_T} = \frac{V}{R_1 + \dfrac{R_2 R_3}{R_2 + R_3}} = \frac{V(R_2 + R_3)}{R_1 R_2 + R_1 R_3 + R_2 R_3} \tag{9.11a}$$

Using the inverse current ratio law,

$$I_2 = \frac{R_3}{R_2 + R_3}(I) \tag{9.11b}$$

Substituting the value of I in Equation (9.11a) into Equation (9.11b) results in

$$I_2 = \frac{V R_3}{R_1 R_2 + R_1 R_3 + R_2 R_3} = \frac{(27)(30)}{(6.8)(20) + (6.8)(30) + (20)(30)}$$
$$= 0.862 \text{ mA}$$

Example 4

Using the inverse current ratio law, determine the current (I_6) through R_6 in Figure 9.6.

In keeping with the principles established in Example 2 concerning the division of current among three parallel resistors,

$$I_6 = \frac{R_4 R_5 (I)}{R_4 R_5 + R_4 R_6 + R_5 R_6} \tag{9.12}$$

but $\quad I = \dfrac{V}{R_1 + R_7 + R_8} \tag{9.13}$

and $\quad R_7 = \dfrac{R_2 R_3}{R_2 + R_3} \qquad R_8 = \dfrac{1}{\dfrac{1}{R_4} + \dfrac{1}{R_5} + \dfrac{1}{R_6}}$

We again are faced with a somewhat tedious sequence of substitutions to solve for the final solution of I_6. As in earlier examples, an alternate approach is to use a series of subroutine-type calculations for temporary storage; these stored values can then be recalled as needed. Proceeding with the latter approach:

PRESS	DISPLAY	COMMENT
2.2 $\boxed{\times}$ .47 $\boxed{\div}$ $\boxed{(}$ 2.2 $\boxed{+}$ .47 $\boxed{)}$ $\boxed{=}$ $\boxed{\text{STO 0}}$	0.387	R_7 $\boxed{\text{STO 0}}$
1.5 $\boxed{1/x}$ $\boxed{+}$ 4.7 $\boxed{1/x}$ $\boxed{+}$ 10 $\boxed{1/x}$		
$\boxed{=}$ $\boxed{1/x}$ $\boxed{\text{STO 1}}$	1.021	R_8 $\boxed{\text{STO 1}}$
45 $\boxed{\div}$ $\boxed{(}$ 3.3 $\boxed{+}$ $\boxed{\text{RCL 0}}$ $\boxed{+}$		
$\boxed{\text{RCL 1}}$ $\boxed{)}$ $\boxed{=}$ $\boxed{\text{STO 2}}$	9.558	I (Equation (9.13)) $\boxed{\text{STO 2}}$
$\boxed{(}$ 1.5 $\boxed{\times}$ 4.7 $\boxed{+}$ 1.5 $\boxed{\times}$ 10 $\boxed{+}$ 4.7 $\boxed{\times}$ 10 $\boxed{)}$	69.05	Denominator of Equation (9.12)
$\boxed{1/x}$ $\boxed{\times}$ 1.5 $\boxed{\times}$ 4.7 $\boxed{\times}$ $\boxed{\text{RCL 2}}$ $\boxed{=}$	0.976	$I_6 = 0.976$ mA

PROBLEMS

Using combinations of the proportional voltage ratio law and the inverse current ratio law, calculate the quantities asked for.
Refer to Figure 9.15 for Problems 1 to 5.

1. Determine V_{R3}.

2. What is I?

3. What are I_1 and I_6?

4. What should V be for $I = 475$ μA?

5. What should V be for $I_5 = 225$ μA?

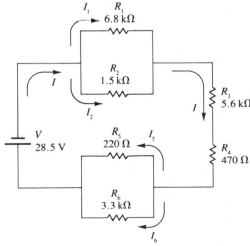

Figure 9.15

Refer to Figure 9.16 for Problems 6 to 8.

6. Determine I.

7. What is the value of I_3?

8. What value of V results in $I_2 = 173.5 \ \mu A$?

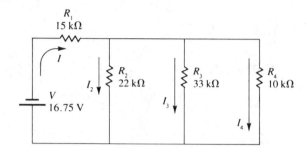

Figure 9.16

Refer to Figure 9.17 for Problems 9 and 10.

9. Calculate I_{R2} by algebraically extending the inverse current ratio law to four parallel resistors.

10. What should I be so $I_{R4} = 117.5$ mA?

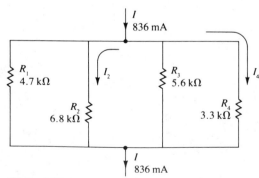

Figure 9.17

9.5 SERIES-PARALLEL CIRCUITS

Series-parallel circuits combine the application of many of the simpler electronic laws. You must reduce parallel resistors to a single equivalent resistance and then sum it with any series' resistors to form a total R_T. You then apply Ohm's law to determine the total current I_T. As I_T flows through the circuit, the voltages across individual series resistors or across parallel combinations of resistors divide according to the proportional voltage law. As current enters any parallel junction, it splits according to the inverse current ratio law. Thus series-parallel combinations require the application of several different electronic principles.

When using your calculator, storage of temporary results becomes more advantageous. In Examples 1 to 6—some are reasonably complex—it is not necessary to use more than four memory storage locations (numbers 0, 1, 2, and 3) because storage locations can be reused to store newer results after the "older" data is no longer needed. Naturally, if your calculator has only one storage location, you will have to manually record and key in additional data as needed, rather than recalling it. Examples 1 to 6 provide good insight into how to use calculators to solve electronic circuits.

Example 1

See Figure 9.18, which consists of a parallel circuit (R_2 in parallel with R_3) in series with R_1. Find I_1, I_2, I_3, and V_{R1}.

First, calculate the equivalent resistance of R_2 in parallel with R_3:

$$R = \frac{R_2 R_3}{R_2 + R_3}$$

PRESS	DISPLAY	COMMENT
2.2 $\boxed{\times}$ 5.6 $\boxed{\div}$ $\boxed{(}$ 2.2 $\boxed{+}$ 5.6 $\boxed{)}$ $\boxed{=}$ $\boxed{\text{STO 0}}$	1.579	Parallel resistance in kΩ; stored in location 0

Second, calculate I_1:

$$I_1 = \frac{V}{R_1 + R} = \frac{20}{1.5 + \boxed{\text{RCL 0}}}$$

PRESS	DISPLAY	COMMENT
20 $\boxed{\div}$ $\boxed{(}$ 1.5 $\boxed{+}$ $\boxed{\text{RCL 0}}$ $\boxed{)}$ $\boxed{=}$ $\boxed{\text{STO 1}}$	6.495	6.495 mA (I_1) stored in location 1

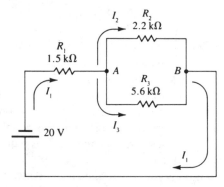

Figure 9.18

Third, calculate V_{R1}:

$$V_{R1} = I_1 R_1$$

PRESS	DISPLAY	COMMENT
1.5 ⊠ RCL 1 =	9.742	$V_{R1} = 9.742$ V

Fourth, calculate I_2 and I_3:

$$I_2 = \frac{R_3}{R_2 + R_3}(I_1) \qquad I_3 = I_1 - I_2$$

PRESS	DISPLAY	COMMENT
5.6 ÷ ((2.2 + 5.6)) ⊠		
RCL 1 = STO 2	4.663	$I_2 = 4.663$ mA STO 2
RCL 1 − RCL 2 =	1.832	$I_3 = 1.832$ mA

Example 2

See Figure 9.19. Example 2 is similar to Example 1 and Figure 9.18, with R_4 added into the circuit. Determine I, I_1, I_2, I_3, I_4, V_{R1}, V_{R2}, V_{R3}, and V_{R4}.

At first glance it might appear necessary to determine the R_T of all four resistors. However, from viewing the location of R_4, note that this resistor is connected completely across the source voltage; therefore Ohm's law is sufficient to calculate I_4. I_1 is 27.5 V divided by R_1 in series with the parallel combination of R_2 and R_3. Finally, $I_1 + I_4 = I$.

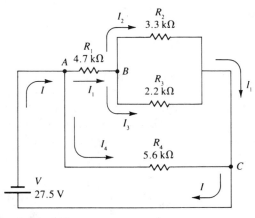

Figure 9.19

PRESS	DISPLAY	COMMENT
27.5 $\boxed{\div}$ 5.6 $\boxed{=}$ $\boxed{\text{STO 0}}$	4.911	$I_4 = \dfrac{V}{R_4}$: $\boxed{\text{STO 0}}$
27.5 $\boxed{\div}$ $\boxed{(}$ 4.7 $\boxed{+}$ 3.3 $\boxed{\times}$ 2.2 $\boxed{\div}$ $\boxed{(}$ 3.3 $\boxed{+}$ 2.2 $\boxed{)}$ $\boxed{)}$ $\boxed{=}$ $\boxed{\text{STO 1}}$	4.568	$I_1 = \dfrac{V}{R_1 + \dfrac{R_2 R_3}{R_2 + R_3}}$: $\boxed{\text{STO 1}}$
$\boxed{+}$ $\boxed{\text{RCL 0}}$ $\boxed{=}$	9.479	$I = I_1 + I_4$
2.2 $\boxed{\times}$ $\boxed{\text{RCL 1}}$ $\boxed{\div}$ $\boxed{(}$ 3.3 $\boxed{+}$ 2.2 $\boxed{)}$ $\boxed{=}$	1.827	$I_2 = \dfrac{R_3}{R_2 + R_3}(I_1)$
$\boxed{+/-}$ $\boxed{+}$ $\boxed{\text{RCL 1}}$ $\boxed{=}$	2.741	$I_3 = I_1 - I_2$
$\boxed{\times}$ 2.2 $\boxed{=}$	6.03	$V_{R3} = I_3 R_3$
$\boxed{+/-}$ $\boxed{+}$ 27.5 $\boxed{=}$	21.47	$V_{R1} = V - V_{R3}$
	6.03[1]	V_{R2}
	27.5[1]	V_{R4}

---

[1]These two voltages need no calculations because $V_{R2} = V_{R3}$ (voltages across parallel resistors are equal) and V_{R4} is across the source voltage. Note that few keystrokes and little memory are necessary to quickly and accurately calculate all the current and voltage parameters of the entire circuit.

Example 3

See Figure 9.20; determine I, I_1, and I_2. Note that values stored in this example are used in Example 4.

It is first necessary to calculate R_T before you can determine I_T. And it behooves us to temporarily store the sum of R_1 and R_2 in one location and the sum of R_3 and R_4 in another location because both sums will be needed when using the inverse current ratio law to determine how I_1 and I_2 split. Also note that only three memory locations are needed because they may be reused when their data is no longer needed.

PRESS	DISPLAY	COMMENT
4.7 $\boxed{+}$ 3.3 $\boxed{=}$ $\boxed{\text{STO 1}}$	8	$(R_1 + R_2)$ $\boxed{\text{STO 1}}$
2.2 $\boxed{+}$ 6.8 $\boxed{=}$ $\boxed{\text{STO 2}}$	9	$(R_3 + R_4)$ $\boxed{\text{STO 2}}$
$\boxed{\times}$ $\boxed{\text{RCL 1}}$ $\boxed{\div}$ $\boxed{(}$ $\boxed{\text{RCL 1}}$		
$\boxed{+}$ $\boxed{\text{RCL 2}}$ $\boxed{)}$ $\boxed{=}$	4.235	$(R_1 + R_2)\|(R_3 + R_4)$ $\|$ means "is parallel to"
$\boxed{+}$ 1.5 $\boxed{+}$ 1 $\boxed{=}$	6.735	R_T
$\boxed{1/x}$ $\boxed{\times}$ 30 $\boxed{=}$ $\boxed{\text{STO 0}}$	4.454	$I = \dfrac{V}{R_T}$ $\boxed{\text{STO 0}}$

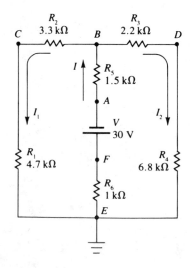

Figure 9.20

PRESS	DISPLAY	COMMENT

$\boxed{\times}$ $\boxed{\text{RCL 2}}$ $\boxed{\div}$ $\boxed{(\!(}$ $\boxed{\text{RCL 1}}$

$\boxed{+}$ $\boxed{\text{RCL 2}}$ $\boxed{)\!)}$ $\boxed{=}$ $\boxed{\text{STO 1}}$ 2.358* $I_1 = \dfrac{(R_3 + R_4)(I)}{(R_1 + R_2) + (R_3 + R_4)}$

$\boxed{\text{STO 1}}$

$\boxed{+/-}$ $\boxed{+}$ $\boxed{\text{RCL 0}}$ $\boxed{=}$ $\boxed{\text{STO 2}}$ 2.096* $I_2 = I - I_1$ $\boxed{\text{STO 2}}$

*Answer in mA units because resistors in kΩ units.

Example 4

Using the values stored from Example 3, show that V_{AD} is the same whether you progress by path *ABD* or path *AFECBD*.

Figures 9.21 and 9.22 show paths *ABD* and *AFECBD*, respectively, and also include the polarity across each resistor as distributed by electron current flow. Remember, going from a + to a − subtracts voltage because it represents a voltage *drop*; going from a − to a + adds voltage because it represents a voltage *rise*.

PRESS	DISPLAY	COMMENT

1.5 $\boxed{\times}$ $\boxed{\text{RCL 0}}$ $\boxed{+}$

   2.2 $\boxed{\times}$ $\boxed{\text{RCL 2}}$ $\boxed{=}$ 11.293 Path *ABD* (Figure 9.21)

30 $\boxed{-}$ $\boxed{\text{RCL 0}}$ $\boxed{-}$ $\boxed{(\!(}$ 4.7 $\boxed{+}$ 3.3 $\boxed{)\!)}$

   $\boxed{\times}$ $\boxed{\text{RCL 1}}$ $\boxed{+}$ 2.2 $\boxed{\times}$ $\boxed{\text{RCL 2}}$ $\boxed{=}$ 11.293 Path *AFECBD* (Figure 9.22)

Note from these calculations that any arbitrary or circuitous path can be taken between two points and that the net voltage is the same. This substantiates that voltages in parallel must be equal.

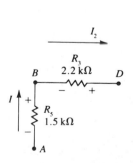

Figure 9.21 Voltage path *ABD* of Figure 9.20

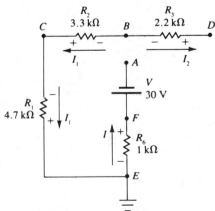

Figure 9.22 Voltage path *AFECBD* of Figure 9.20

Example 5

Determine all currents (in mA) in Figure 9.23. Store these values for Example 6.

PRESS	DISPLAY	COMMENT
5.6 $\boxed{+}$ 6.8 $\boxed{=}$ $\boxed{\text{STO 1}}$ $\boxed{\times}$ 10		
$\boxed{\div}$ $\boxed{((}$ 10 $\boxed{+}$ $\boxed{\text{RCL 1}}$ $\boxed{)}$ $\boxed{=}$		
$\boxed{+}$ 3.3 $\boxed{=}$ $\boxed{\text{STO 1}}$	8.836	$R_2 + (R_3\|(R_4 + R_5))$ $\boxed{\text{STO 1}}$; define as R_8
$\boxed{\text{CLR}}$ 1.5 $\boxed{+}$ 1 $\boxed{=}$ $\boxed{\text{STO 2}}$	2.5	$R_6 + R_7$ $\boxed{\text{STO 2}}$; define as R_9
$\boxed{\times}$ $\boxed{\text{RCL 1}}$ $\boxed{\div}$ $\boxed{((}$ $\boxed{\text{RCL 1}}$		
$\boxed{+}$ $\boxed{\text{RCL 2}}$ $\boxed{)}$ $\boxed{=}$	1.949	$R_8\|R_9$; define as R_{10}
$\boxed{+}$ 4.7 $\boxed{=}$ $\boxed{1/x}$ $\boxed{\times}$ 22.5 $\boxed{=}$ $\boxed{\text{STO 0}}$	3.384	$I = \dfrac{V}{R_1 + R_{10}}$ $\boxed{\text{STO 0}}$
$\boxed{\times}$ $\boxed{\text{RCL 2}}$ $\boxed{\div}$ $\boxed{((}$ $\boxed{\text{RCL 1}}$ $\boxed{+}$		
$\boxed{\text{RCL 2}}$ $\boxed{)}$ $\boxed{=}$ $\boxed{\text{STO 1}}$	0.746	$I_1 = \dfrac{R_9(I_0)}{R_8 + R_9}$ $\boxed{\text{STO 1}}$
$\boxed{\times}$ 10 $\boxed{\div}$ $\boxed{((}$ 10 $\boxed{+}$ 5.6		
$\boxed{+}$ 6.8 $\boxed{)}$ $\boxed{=}$ $\boxed{\text{STO 3}}$	0.333	$I_3 = \dfrac{R_3(I_1)}{R_3 + R_4 + R_5}$ $\boxed{\text{STO 3}}$
$\boxed{+/-}$ $\boxed{+}$ $\boxed{\text{RCL 1}}$ $\boxed{=}$ $\boxed{\text{STO 2}}$	0.413	$I_2 = I_1 - I_3$ $\boxed{\text{STO 2}}$
$\boxed{\text{RCL 0}}$ $\boxed{-}$ $\boxed{\text{RCL 1}}$ $\boxed{=}$	2.638	$I_4 = I - I_1$

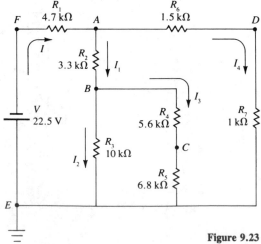

Figure 9.23

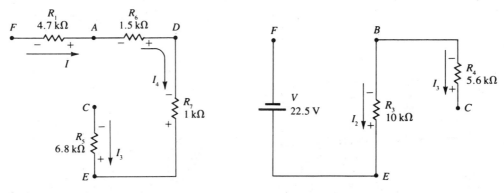

Figure 9.24 Voltage path *CEDAF*.   **Figure 9.25** Voltage path *CBEF*.

Example 6

Using the values stored in Example 5, show that V_{CF} is the same whether taking path *CEDAF* or path *CBEF*. Figures 9.24 and 9.25 show these two paths, with the polarities produced by electron current flow.

PRESS	DISPLAY	COMMENT
6.8 $\boxed{\times}$ $\boxed{\text{RCL 3}}$ $\boxed{-}$ $\boxed{(}$ $\boxed{\text{RCL 0}}$ $\boxed{-}$		
$\boxed{\text{RCL 1}}$ $\boxed{)}$ $\boxed{\times}$ $\boxed{(}$ 1 $\boxed{+}$ 1.5 $\boxed{)}$		
$\boxed{-}$ 4.7 $\boxed{\times}$ $\boxed{\text{RCL 0}}$ $\boxed{=}$	-20.234	Path *CEDAF* (Figure 9.24)
5.6 $\boxed{+/-}$ $\boxed{\times}$ $\boxed{\text{RCL 3}}$ $\boxed{+}$ 10		
$\boxed{\times}$ $\boxed{\text{RCL 2}}$ $\boxed{-}$ 22.5 $\boxed{=}$	-20.234	Path *CBEF* (Figure 9.25)

In a manner similar to that used in Example 4, this illustrates that the voltage between any two points must be the same regardless of the path over which the voltage drops (or rises) are summed.

PROBLEMS

Using your calculator and any appropriate electronic principle or law, determine the solution of the following:

Refer to Figure 9.26 for Problems 1 to 4.

1. What is the value of I?

2. What is the value of I_2?

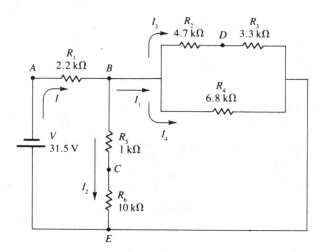

Figure 9.26

3. Determine the polarity and magnitude of V_{BD}.

4. Determine the polarity and magnitude of V_{CD}.

Refer to Figure 9.27 for Problems 5 to 8.

5. What is the value of I?

6. Determine the magnitude of I_{R5}.

7. Determine the polarity and magnitude of V_{CD}.

8. Determine the polarity and magnitude of V_{DB}.

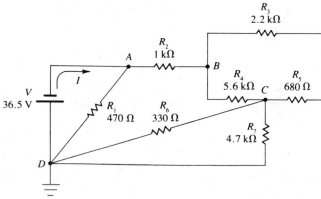

Figure 9.27

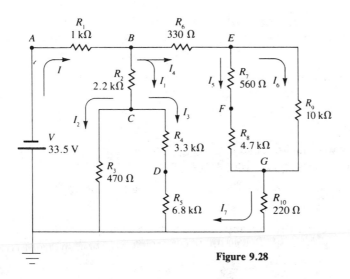

Figure 9.28

Refer to Figure 9.28 for Problems 9 to 12.

9. Does I_4 equal I_7?

10. Determine the magnitude of all currents.

11. Determine the voltage of all referenced points with respect to ground.

12. What are the polarity and magnitude of V_{AC}, V_{DE}, and V_{CG}?

SUMMARY

A calculator, because of its storage feature, is ideally suited for solving electronic problems. For complex problems—such as many series-parallel circuits—temporary storage of intermediate results for later recall greatly facilitates accurate solutions. In most cases series-parallel circuits involve the application of several electronic principles.

   For simple circuits it is often more convenient to algebraically solve directly for the desired quantity. Then, directly key in the known quantities and obtain the desired solution without intermediate storage. Practice and commonsense analysis quickly determine the trade-off between these two approaches.

   Although there are many electronic principles, six of the more important laws and/or principles are:

   1.   Ohm's law—the current in a closed loop is directly proportional to the applied voltage and inversely proportional to the resistance.

2. Proportional voltage law—the voltage across any resistor is the ratio of that resistor's resistance times the net voltage divided by the total resistance of the circuit.

3. Inverse current law—the current through any single resistor of two parallel resistors is the ratio of the other (parallel) resistor's resistance times the current entering the junction divided by the sum total of both resistors' resistance.

4. Kirchhoff's voltage law—the sum of all the voltages (including source voltages and voltage drops across any resistors) in a closed loop is zero.

5. Kirchhoff's current law—the sum of all currents entering a junction equals the sum of all currents leaving the junction.

6. Parallel voltages—the voltage between any two points in a circuit must be the same regardless of the path taken between those two points.

10 EQUATIONS: QUADRATIC

10.1 INTRODUCTION

Up to this point equations have been of the *first degree*, which means that when solving for an unknown, the power of the variable has been 1 (implied), or namely, x. If the highest power of the unknown had been 2 (called degree 2) the equation could have been broadly classified as quadratic. In its most general form, a quadratic equation looks like

$$ax^2 + bx + c = 0 \qquad (10.1)$$

This means that there must exist at least the term ax^2 (even if the coefficient a equals 1, which then reduces to simply x^2) and that at least b *or* c not both equal zero. The reason b *and* c cannot both equal zero is that the expression would then reduce to

$$ax^2 = 0$$

in which case the variable x is zero (assuming that a is finite) and no problem exists. Given the stipulation that to have a quadratic equation the variable x must be of degree 2, there are only three possible formats. These are generalized in an ascending order of difficulty:

1. $ax^2 + c \qquad = 0$
2. $ax^2 + bx \qquad = 0$
3. $ax^2 + bx + c = 0$

This chapter investigates the solutions of these three formats.

10.2 FORMS OF QUADRATICS

Preceding chapters demonstrated that there are two possible solutions when taking the square root of a number. For example,

$$\sqrt{49} = \pm 7$$

because $(+7)^2 = 49$ and $(-7)^2 = 49$. Two solutions always exist for quadratic equations, although in some cases both solutions might be the same number.

Case 1: $b = 0$

If, within the generalized form $ax^2 + bx + c = 0$ the coefficient b equals zero, the expression reduces to

$$ax^2 + c = 0 \qquad\qquad (10.2)$$

We learned that for a real-number solution, the number under the radical sign must be positive. Therefore, Equation (10.2) can be further refined

to $\qquad\qquad ax^2 + (-c) = 0$

or $\qquad\qquad ax^2 - c = 0 \qquad\qquad (10.3)$

(In Chapter 16, "Complex Numbers," this restriction will no longer be necessary.) To solve Equation (10.3), proceed as follows:

$$\text{A; } c: ax^2 = c$$

$$\text{D; } a: x^2 = \frac{c}{a}$$

$$\sqrt{}\,; \qquad x = \pm\sqrt{\frac{c}{a}}$$

Example 1

Solve for both solutions of the following quadratic equations.

a. $y^2 - 36 = 0$

$\quad y^2 = 36$

$\quad y = \sqrt{36} = \pm 6$

b. $3\beta^2 - 432 = 0$

$\quad 3\beta^2 = 432$

$\quad \beta^2 = 144$

$\quad \beta = \sqrt{144} = \pm 12$

c. $\dfrac{\alpha^2}{3} - 25 = 0$

$\quad \dfrac{\alpha^2}{3} = 25$

$\quad \alpha^2 = 75$

$\quad \alpha = \sqrt{75} = \sqrt{25 \cdot 3} = \pm 5\sqrt{3}$

d. $\dfrac{a + 2}{a - 2} + \dfrac{a - 2}{a + 2} = \dfrac{13}{a^2 - 4}$

M; LCD is $(a - 2)(a + 2)$:

$$(a + 2)^2 + (a - 2)^2 = 13$$
$$a^2 + 4a + 4 + a^2 - 4a + 4 = 13$$
$$2a^2 + 8 = 13$$
$$2a^2 = 5$$
$$a^2 = \dfrac{5}{2}$$
$$a = \pm \sqrt{\dfrac{5}{2}} = \pm \dfrac{\sqrt{10}}{2}$$

Case 2: $c = 0$

When c in Equation (10.1) is zero, the equation reduces to

$$ax^2 + bx = 0 \tag{10.4}$$

Factoring out x results in

$$x(ax + b) = 0 \tag{10.5}$$

Notice that Equation (10.5) states that a monomial (x) times a binomial $(ax + b)$ equals zero. This is possible only if one of the two expressions is zero; therefore the first possible solution is always

$$x = 0$$

and the second possible solution is

$$ax + b = 0$$
$$ax = -b$$
$$x = -\dfrac{b}{a}$$

Example 2

Solve for both solutions of the following quadratic equations.

a. $x^2 + 7x = 0$
   $x = 0, -7$

b. $3x^2 - \sqrt{7}x = 0$
   $x = 0, -\dfrac{-\sqrt{7}}{3}$ or $\dfrac{\sqrt{7}}{3}$

c. $\dfrac{2x^2}{3} + \dfrac{x}{4} = 0$

$$x = 0, \quad \dfrac{-\dfrac{1}{4}}{\dfrac{2}{3}} \quad \text{or} \quad -\dfrac{3}{8}$$

d. $(\alpha + 2)^2 x^2 - \beta^2 x = 0$

$$x = 0, \quad -\dfrac{-\beta^2}{(\alpha + 2)^2} \quad \text{or} \quad \left(\dfrac{\beta}{\alpha + 2}\right)^2$$

Case 3: $(b$ and $c) \neq 0$

When neither b nor c in Equation (10.1) is zero, the full form of the quadratic exists. This section deals with only those specific problems in which the coefficients a, b, and c are recognized as being factorable; all others are discussed in the next section. Suppose we want to solve the quadratic

$$x^2 + 5x + 6 = 0$$

for both possible solutions of x. Note that the coefficients are such that the equation can be rewritten in factored form:

$$(x + 2)(x + 3) = 0$$

If the product of two binomials is to be zero, then at least one binomial *must* be zero; therefore, either

$$x + 2 = 0 \qquad \text{first possible solution}$$
$$x = -2$$

or

$$x + 3 = 0 \qquad \text{second possible solution}$$
$$x = -3$$

Then write the solutions as

$$x = -2, -3$$

To test this premise, substitute both solutions (one at a time) into the original equation to see if it does indeed equal to zero:

$$(-2)^2 + 5(-2) + 6 = 0$$
$$4 - 10 + 6 = 0$$
$$0 = 0$$

and
$$(-3)^2 + 5(-3) + 6 = 0$$
$$9 - 15 + 6 = 0$$
$$0 = 0$$

Both solutions check.

We stated that sometimes both solutions can be the same number. Consider the following quadratic:

$$s^2 + 6s + 9 = 0$$
$$(s + 3)(s + 3) = 0$$

One binomial must be zero, but because the binomials are identical, equating to zero and solving either one produces the same answer: -3. Therefore,

$$s = -3, -3$$

Thus, when the factors of the quadratic are a binomial squared, both answers are the same number.

Example 3

Solve for both solutions of the following quadratic equations.

a. $x^2 - 2x - 35 = 0$
   $(x - 7)(x + 5) = 0$
   $\therefore x - 7 = 0$ $x + 5 = 0$
                and
   $x = 7$ $x = -5$

b. $8a^2 - 2a - 15 = 0$
   $(2a - 3)(4a + 5) = 0$
   $\therefore 2a - 3 = 0$ $4a + 5 = 0$
   $2a = 3$ and $4a = -5$
   $a = \dfrac{3}{2}$ $a = -\dfrac{5}{4}$

c. $\dfrac{y^2}{6} - \dfrac{5y}{6} - 6 = 0$

   $\left(\dfrac{y}{2} + 2\right)\left(\dfrac{y}{3} - 3\right) = 0$

   $\therefore \dfrac{y}{2} + 2 = 0$ $\dfrac{y}{3} - 3 = 0$

   $\dfrac{y}{2} = -2$ and $\dfrac{y}{3} = 3$

   $y = -4$ $y = 9$

Sometimes, as in Example 3, the factors of a fractional quadratic are not easy to recognize. If you use the earlier trick of multiplying both sides of the equation by the LCD, a quadratic still results, but often in a form easier to factor. For example, multiplying Example 3c by 6 (the LCD) results in

$$y^2 - 5y - 36 = 0$$
$$(y + 4)(y - 9) = 0$$
$$y = -4, 9$$

PROBLEMS

In Problem 1 to 15, solve for both solutions of the variable.

1. $2c^2 - 25 = 0$

2. $\dfrac{4e^2}{9} - 1 = 0$

3. $5\phi^2 - 180 = 0$

4. $\dfrac{\beta + 5}{\beta - 5} + \dfrac{\beta - 5}{\beta + 5} - \dfrac{150}{\beta^2 - 25} = 0$

5. $x^2 - x = 0$

6. $\gamma^2 - 5\gamma = 0$

7. $4a^2 - 9a = 0$

8. $6f^2 + \sqrt{25f^2} + f = 0$

9. $\dfrac{3}{4}f^2 - \dfrac{f}{2} = 0$

10. $u^2 - 3u - 18 = 0$

11. $v^2 - 3v = 40$

12. $3s^2 + 5s - 2 = 0$

13. $7\lambda^2 - 29\lambda - 30 = 0$

14. $\dfrac{\theta^2}{3} - 6 = \dfrac{7\theta}{3}$

15. $1 + \dfrac{8}{m} = \dfrac{33}{m^2}$

10.3 THE QUADRATIC EQUATION

One benefit of applying the rules of algebra to equations is that it permits solutions in a generalized format. Then, all that is necessary to solve any given problem of that format is to plug in the coefficients and "crank out" the answer. Because of a quadratic's generalized nature, the coefficients can have any value; they do not have to be readily factorable, and b or c can be zero. In short, once it is generalized, the quadratic equation works every time, with no exceptions. This section develops and applies the quadratic equation.

Before starting the development, it is important to review a factoring principle studied in Chapter 6. Recall that if two terms of a trinomial are perfect squares and the third term is twice the square root of each of the perfect squares, the trinomial can be factored into a binomial squared. For example,

$$t^2 + 8t + 16 = (t + 4)^2$$

Note that the first and third terms of the trinomial are perfect squares: $(t)^2$ and $(4)^2$. The middle term $(8t)$ is twice the product of the square root of the perfect squares, or $(2)(t)(4)$. This is an important fact because during the development of the quadratic equation, a key maneuver is to manipulate terms into this format. This process is mathematically referred to as "completing the square." To proceed:

$$ax^2 + bx + c = 0$$

D; a:
$$x^2 + \frac{bx}{a} + \frac{c}{a} = 0$$

S; $\frac{c}{a}$:
$$x^2 + \frac{bx}{a} = -\frac{c}{a}$$

To complete the square, take one-half the coefficient of x, which is $b/2a$, then square it, to get $b^2/4a^2$.

A; $\frac{b^2}{4a^2}$:
$$x^2 + \frac{b}{a}x + \frac{b^2}{4a^2} = -\frac{c}{a} + \frac{b^2}{4a^2}$$

Now that we have manipulated the trinomial into the desired format, factor the left side of the equation and combine terms on the right side:

$$\left(x + \frac{b}{2a}\right)^2 = \frac{b^2 - 4ac}{4a^2}$$

$\sqrt{}$:
$$x + \frac{b}{2a} = \frac{\pm\sqrt{b^2 - 4ac}}{2a}$$

S; $\frac{b}{2a}$:
$$x = -\frac{b}{2a} \pm \frac{\sqrt{b^2 - 4ac}}{2a}$$

Combine terms:
$$x = \frac{-b \pm \sqrt{b^2 - 4ac}}{2a} \tag{10.6}$$

Equation (10.6) defines the two solutions of *any* quadratic equation:

$$x_1 = \frac{-b + \sqrt{b^2 - 4ac}}{2a} \tag{10.7a}$$

and
$$x_2 = \frac{-b - \sqrt{b^2 - 4ac}}{2a} \tag{10.7b}$$

All that is necessary is to substitute the coefficients of the quadratic into Equations (10.7a) and (10.7b) and crank out the solutions x_1 and x_2.

One precaution is again stressed. Note that if $4ac > b^2$ in Equation (10.6), a negative number exists under the radical sign. The complete expression $\sqrt{b^2 - 4ac}$ is called the *discriminant*. As long as $b^2 \geq 4ac$, then the discriminant results in a real number. For now, examples are restricted to this condition; in chapter 16, which covers complex numbers, we investigate the negative number condition.

Example 1

Using Equation (10.6), solve for both solutions of the following quadratics.

a. $2x^2 + 7x - 15 = 0$: $a = 2, b = 7, c = -15$

$$x_1, x_2 = \frac{-7 \pm \sqrt{49 - 4(2)(-15)}}{2(2)}$$

$$= \frac{-7 \pm \sqrt{169}}{4}$$

$$= \frac{-7 + 13}{4}, \frac{-7 - 13}{4}$$

$$= \frac{3}{2}, -5$$

b. $6x^2 + 11x - 35 = 0$: $a = 6, b = 11, c = -35$

$$x_1, x_2 = \frac{-11 \pm \sqrt{121 - 4(6)(-35)}}{2(6)}$$

$$= \frac{-11 \pm \sqrt{961}}{12}$$

$$= \frac{-11 + 31}{12}, \frac{-11 - 31}{12}$$

$$= \frac{5}{3}, -\frac{7}{2}$$

c. $7x^2 - 36 = 0$: $a = 7, b = 0, c = -36$

$$x_1, x_2 = \frac{-0 \pm \sqrt{0 - 4(7)(-36)}}{2(7)}$$

$$= \frac{\pm \sqrt{1008}}{14}$$

$$= \frac{6\sqrt{7}}{7}, \frac{-6\sqrt{7}}{7}$$

When coefficients are not so convenient, a calculator is indispensable. Because the discriminant is used in both solutions, it is convenient to calculate and store this value for recall. Reasonable shortcuts can be executed. For example, if the b term is negative, then $-b$ is keyed in as positive. The same logic is valid for the b^2 term in the discriminant.

Example 2

Using your calculator at $\boxed{\text{Fix 3}}$, determine the solutions of the following quadratic.
Verify the solutions by substituting the answers into the original equation.

$$3.75t^2 - 14.2t + 6.1 = 0$$

PRESS	DISPLAY	COMMENT
14.2 $\boxed{x^2}$ $\boxed{-}$ 4 $\boxed{\times}$ 3.75		
$\boxed{\times}$ 6.1 $\boxed{=}$ $\boxed{\sqrt{}}$ $\boxed{\text{STO 0}}$	10.495	Discriminant $\boxed{\text{STO 0}}$
$\boxed{\text{CLR}}$ $\boxed{(}$ 14.2 $\boxed{+}$ $\boxed{\text{RCL 0}}$ $\boxed{)}$ $\boxed{\div}$		
$\boxed{(}$ 2 $\boxed{\times}$ 3.75 $\boxed{)}$ $\boxed{=}$ $\boxed{\text{STO 1}}$	3.293	x_1
$\boxed{\text{CLR}}$ $\boxed{(}$ 14.2 $\boxed{}$ $\boxed{\text{RCL 0}}$ $\boxed{)}$ $\boxed{\div}$		
$\boxed{(}$ 2 $\boxed{\times}$ 3.75 $\boxed{)}$ $\boxed{=}$ $\boxed{\text{STO 2}}$	0.494	x_2

To verify the answer:

$\boxed{\text{CLR}}$ 3.75 $\boxed{\times}$ $\boxed{\text{RCL 1}}$ $\boxed{x^2}$ $\boxed{-}$		
14.2 $\boxed{\times}$ $\boxed{\text{RCL 1}}$ $\boxed{+}$ 6.1 $\boxed{=}$	0.000	x_1 value satisfies equation
$\boxed{\text{CLR}}$ 3.75 $\boxed{\times}$ $\boxed{\text{RCL 2}}$ $\boxed{x^2}$ $\boxed{-}$		
14.2 $\boxed{\times}$ $\boxed{\text{RCL 2}}$ $\boxed{+}$ 6.1 $\boxed{=}$	0.000	x_2 value satisfies equation

PROBLEMS

Using the quadratic equation (and your calculator whenever necessary), determine
the solutions to Problems 1 to 12.

1. $x^2 - 121 = 0$

2. $4x^2 = 225$

3. $3x^2 + 5x = 0$

4. $x^2 + x - 30 = 0$

5. $2x^2 + 5x - 63 = 0$

6. $16y^2 - 9 = 0$

7. $9x^2 + 4x - 5 = 0$

8. $\beta^2 - \beta = 1$

9. $2\alpha^2 + 4\alpha - 2 = 0$

10. $13t^2 - 14.5t + 1 = 0$

11. $-2.75u^2 - 3.6u + 1.7 = 0$

12. $1.823p^2 = p + 6.8$

SUMMARY

Quadratic equations are different from other types of equations in that they involve a term of the second degree. In its most complete form, a quadratic equation can also have a term of degree 1 and a coefficient. The entire expression is equated to zero.

If the expression can be easily factored, you can determine both solutions by equating each factor to zero and solving for the unknown, which at this point is now of degree 1. Under any condition, you can use the quadratic equation to solve for both values of the unknown.

The expression $\sqrt{b^2 - 4ac}$ is called the discriminant. Although this chapter discussed only positive values, the expression $b^2 - 4ac$ can be positive, zero, or negative. Negative values are discussed in subsequent chapters, although all three values can be significant when electronic circuits are being analyzed, for example, determining whether electronic oscillations are under-, critically, or overdamped.

EQUATIONS: SIMULTANEOUS

11.1 INTRODUCTION

In the preceding chapters there was only one unknown in any given equation that required a solution. This was even true for quadratic equations because even though the unknown was of degree 2, it was still only one unknown. Now consider this very simple problem.

A digital multimeter and hand-held calculator together cost $258. Suppose we ask "How much did each device cost?" What conclusions can we make?

First, there are obviously now two unknowns: the cost of the multimeter and the cost of the calculator. If we arbitrarily assign the variables m to the multimeter and c to the calculator, we can write the following equation:

$$m + c = 258 \tag{11.1}$$

Second, we cannot work the problem because there are an infinite number of combinations of two numbers whose sum is 258. More information is needed.

If the multimeter costs $216 more than the calculator, we can write another equation:

$$m = c + 216 \tag{11.2}$$

We now seem ready for a solution. Knowing the value of m in terms of c, if we substitute Equation (11.2) into Equation (11.1) we get the following results:

$$\underbrace{(c + 216)}_{m} + c = 258 \tag{11.3}$$

Eureka! Equation (11.3) has only one unknown, and we know how to solve for it:

$$2c + 216 = 258$$
$$2c = 42$$
$$c = 21$$

If the calculator costs $21, Equation (11.2) states that the multimeter costs $216 more, or

$$m = 21 + 216 = \$237$$

To check, the sum is $21 plus $237 for a total of $258. These two conclusions, which eventually resulted in a unique solution, lead to a mathematical law:

There must be at least as many unique equations as there are unknowns before a unique solution can be determined for each unknown.

Note the frequent use of the word "unique." A unique equation is an equation that cannot be derived from another equation(s) constituting the problem. For example, if Equation (11.2) is $2m + 2c = 516$, no solution can be determined because this is merely multiplying Equation (11.1) by 2, which does not really provide any *new* information.

As you may have guessed, the term for this method of solution is called *substitution*. You can take several different approaches when solving simultaneous equations. This chapter investigates three general methods; Chapter 12 expands this topic by dealing with *determinants* and *matrices*, methods broadly described as manipulating number arrays. When dealing with only *two* unknowns, any of the three methods described here will probably provide the most rapid solutions, but when dealing with *three or more* unknowns, determinants or matrices usually provide the most useful method of solution.

11.2 METHODS OF SOLUTION—TWO UNKNOWNS

Case 1: Substitution

The introduction demonstrated the methodology of this approach. Solve one of the two equations for any one variable in terms of the other variable and substitute this value into the other equation. The resultant is now one equation with only one unknown. You can then apply the standard laws of balancing equations until you reach a solution.

Example 1

Solve the following simultaneous equations by substitution.

a. $3s - 6t = -24$ (1)
    $s + 6t = 32$ (2)

Solve Equation (2) for s in terms of t:

$$s = 32 - 6t \qquad (3)$$

Substitute $s = 32 - 6t$ into Equation (1):

$$\underbrace{3(32 - 6t)}_{s} - 6t = -24$$

$$96 - 18t - 6t = -24$$
$$-24t = -120$$
$$t = 5$$

Substitute $t = 5$ into Equation (3):
$$s = 32 - 6(5) = 2$$

b. $-4x + 5y = 11.5$ (1)
     $2x - y = -0.5$ (2)

Solve Equation (2) for y:

$$y = 2x + 0.5 \qquad (3)$$

Substitute y into Equation (1):

$$-4x + \underbrace{5(2x + 0.5)}_{y} = 11.5$$

$$-4x + 10x + 2.5 = 11.5$$
$$6x = 9$$
$$x = 1.5$$

Substitute $x = 1.5$ into Equation (3):

$$y = 2(1.5) + 0.5 = 3.5$$

c. $\dfrac{x}{4} + \dfrac{3y}{8} = 5.875$ (1)

    $-\dfrac{x}{5} - \dfrac{y}{2} = -6.7$ (2)

When you encounter fractional equations, it is recommended that you multiply each equation through by its LCD.

M; Equation (1) by 8: $2x + 3y = 47$ (1′)
M; Equation (2) by 10: $-2x - 5y = -67$ (2′)

Solve Equation (1′) for $2x$ (why $2x$ instead of x?):

$$2x = 47 - 3y \qquad (3)$$

Substitute $2x$ into Equation (2′):

$$-\underbrace{(47 - 3y)}_{2x} - 5y = -67$$

$$-47 + 3y - 5y = -67$$
$$-2y = -20$$
$$y = 10$$

Substitute $y = 10$ into Equation (3):

$$2x = 47 - 3(10) = 17$$
$$x = \frac{17}{2} = 8.5$$

Case 2: Addition or Subtraction of Equations

If two sets of equations are added or subtracted, a valid equation still exists. For example,

$$1 \text{ dozen} = 12 \qquad (11.4)$$
$$\underline{+ \ 2 \text{ dozen} = 24} \qquad (11.5)$$
$$3 \text{ dozen} = 36$$

If Equation (11.4) is already balanced, adding or subtracting either side of Equation (11.5) to either side of Equation (11.4) still retains the balance.

The selection of whether to add or subtract equations to cancel one of the variables depends upon the sign of that variable in each of the two equations. If their magnitudes are equal but of opposite sign, adding the equations cancels the variable. If the magnitudes are equal and of the same sign, subtracting the equations cancels the variable. Finally, if the magnitude of the coefficients preceding the variable that you want to cancel are not identical, multiply one or both equations by the appropriate quantity to make them identical and then either add or subtract the equations, as appropriate, to cancel the variable. Example 2 illustrates each preceding condition.

Example 2

Solve the following simultaneous equations by the addition or subtraction of equations.

a. $3x - 7y = 16$ (1)
 $x + 7y = -4$ (2)

Add Equations (1) and (2) to cancel the y variable:

$$4x + 0 = 12$$
$$x = 3$$

Substitute $x = 3$ into either Equation (1) or (2); let us arbitrarily pick Equation (2):

$$3 + 7y = -4 \qquad (2)$$
$$7y = -7$$
$$y = -1$$

b. $x + y = 5$ (1)
 $3x + y = 0$ (2)

Subtract Equation (1) from Equation (2) to cancel the y variable:

$$2x = -5$$
$$x = -2.5$$

Substitute $x = -2.5$ into Equation (1) and solve for y:

$$y = 5 - x \qquad (1)$$
$$y = 5 - (-2.5) = 7.5$$

c. $3x - 4y = 18$ (1)
  $4x + y = 5$ (2)

Multiply Equation (2) by 4 to obtain the same coefficient (absolute value) for y:

  $16x + 4y = 20$ (3)

Add Equations (1) and (3) to eliminate y:

  $19x = 38$
  $x = 2$

Substitute $x = 2$ into Equation (2) and solve for y:

  $y = 5 - 4x$
  $y = 5 - 4(2) = -3$

Case 3: Comparison

Comparison is the last process easily adaptable to the solution of simultaneous equations. Comparison involves solving each equation for the same variable; each equation is then equated to the other. This is valid and based upon the mathematical law that:

  Things equal to the same thing are equal to each other.

Example 3

Solve the following simultaneous equations by the method of comparison.

a. $2u - 3v = -14.7$ (1)
  $3u + v = -8.3$ (2)

Multiply Equation (2) by -3:

  $-9u - 3v = 24.9$ (2′)

Solve Equations (1) and (2′) for $-3v$ (the same thing).

From Equation (1),
  $-3v = -14.7 - 2u$

From Equation (2′),
  $-3v = 9u + 24.9$

Because Equations (1) and (2′) are both equated to the same thing ($-3v$), they can be equated to each other:

  $-14.7 - 2u = 9u + 24.9$

Solve for u:

$$-11u = 39.6$$
$$u = -3.6$$

Mathematically, $u = -3.6$ can be substituted into either Equation (1) or (2). Because it is easier to substitute into Equation (2), this equation is used to solve for v:

$$v = -8.3 - 3u$$
$$= -8.3 - 3(-3.6)$$
$$= -8.3 + 10.8$$
$$= 2.5$$

b. $3.1f - 2.6g = 4.015$ (1)
    $-1.5f + g = -0.975$ (2)

Multiply Equation (2) by -2.6 to get an expression of $-2.6g$ for both equations:

$3.9f - 2.6g = 2.535$ (2')

Solve Equations (1) and (2') each for $-2.6g$:

$-2.6g = 4.015 - 3.1f$ (1)
$-2.6g = 2.535 - 3.9f$ (2')

Equating Equation (1) to Equation (2'),

$$4.015 - 3.1f = 2.535 - 3.9f$$

Solve for f:

$$0.8f = -1.48$$
$$f = -1.85$$

Substitute $f = -1.85$ into Equation (2) to solve for g:

$$g = -0.975 + 1.5f$$
$$= -0.975 + 1.5(-1.85)$$
$$= -3.75$$

Verify this problem by substituting the calculated values of f and g into the original equations:

Using algebra:

$$3.1f - 2.6g = 4.015 \qquad \text{solution 1}$$
$$3.1(-1.85) - 2.6(-3.75) = 4.015$$
$$4.015 = 4.015$$

$$-1.5f + g = -0.975 \qquad \text{solution 2}$$
$$-1.5(-1.85) + (-3.75) = -0.975$$
$$-0.975 = -0.975$$

Using the calculator:

PRESS	DISPLAY	COMMENT

$3.1\ \boxed{\times}\ 1.85\ \boxed{+/-}\quad\boxed{-}\ 2.6\ \boxed{\times}$

$\quad 3.75\ \boxed{+/-}\quad\boxed{=}$ 4.015   Equation (1) verified

$1.5\ \boxed{+/-}\quad\boxed{\times}\ 1.85\ \boxed{+/-}\quad\boxed{+}$

$\quad 3.75\ \boxed{+/-}\quad\boxed{=}$ −0.975   Equation (2) verified

PROBLEMS

Solve the simultaneous equations in Problems 1 to 5 by substitution.

1. $3h - 2i = 23$
      $-\ h +\ i = -10$

4. $3.1x + 2.4y = 0.37$
      $-\ x + 1.7y = -5.81$

2. $4\alpha - 5\beta = 1.25$
      $-\ \alpha -\ \beta = -4.25$

5. $3.6c - 1.8d = 2.286$
      $4.5c + 0.3d = 9.105$

3. $5a +\ b = 12.9$
      $-2a - 3b = 1.6$

Solve the simultaneous equations in Problems 6 to 10 by adding or subtracting equations.

6. $3\theta + 2\varphi = -4$
      $-4\theta - 2\varphi = 2$

9. $-1.6x - 2.1y = 1.86$
      $3x +\ \ y = 1.8$

7. $4u + 5v = 4$
      $3u - 4v = 34$

10. $2.4x = 3.7y + 6.45$
      $1.8y = 1.6x - 3.7$

8. $3.5a -\ \ b = 3.25$
      $7a + 2.5b = 15.5$

Solve the simultaneous equations in Problems 11 to 15 by comparison.

11. $c + 2f = -1$
      $c - 5f = 13$

14. $1.5x - 3.9y = 18.585$
      $4.5x + 1.6y = 3.885$

12. $2i + 3j = 9.7$
      $i -\ j = 0.6$

15. $m =\ n - 3.28$
      $-8.52 = 2m - 3n$

13. $2k + 4j = -0.8$
      $4k + 2j = 8.9$

11.3 METHODS OF SOLUTION—THREE UNKNOWNS

If there are three unknown variables, there must be at least three unique equations for a solution. As expected, even though the algebra is methodical, it does become more involved. Either the methods of substitution or addition (or subtraction) of equations is usually employed to solve simultaneous equations of three unknowns.

To use the method of substitution, match any two equations against the third to eliminate one variable. The resulting two new equations then have only two unknowns, and are now at a point where we can use the methods described in Section 11.2. After applying the rules of substitution to one of the two unknowns, substitute the resultant into the second equation to determine the second unknown. Having now determined two of the original three unknowns, return to any one of the original three equations and, after substituting, determine the third unknown.

To use the method of addition or subtraction of variables involves a somewhat similar approach: Manipulate one equation in such a manner that the other two equations can be added (or subtracted, if appropriate) to eliminate one variable. After this initial phase, there is a similar condition as in the method of substitution: two equations with only two unknowns. From this point on, the steps to the final solution of all variables are virtually the same for either method. Example 1 uses both methods for the same problem to better clarify the subtle differences.

Example 1

Solve the following simultaneous equations of three unknowns by both the methods of substitution and addition (or subtraction).

$$3a - 2b + 4c = -33.5 \tag{1}$$
$$-a + 4.5b - c = 24 \tag{2}$$
$$1.5a + 3b - 5c = 51.75 \tag{3}$$

Method 1: Substitution

Arbitrarily choose any equation and solve for one variable in terms of the other two variables. Let us choose Equation (2) and solve for a:

$$-a = 24 + c - 4.5b$$
$$a = 4.5b - 24 - c \tag{4}$$

Substitute the value of a into Equations (1) and (3):

$$3(4.5b - 24 - c) - 2b + 4c = -33.5$$
$$13.5b - 72 - 3c - 2b + 4c = -33.5$$
$$11.5b + c = 38.5 \tag{1'}$$

$$1.5(4.5b - 24 - c) + 3b - 5c = 51.75$$
$$6.75b - 36 - 1.5c + 3b - 5c = 51.75$$
$$9.75b - 6.5c = 87.75 \tag{3'}$$

Multiply Equation (1′) by 6.5 to make the coefficients of c between Equations (1′) and (2′) identical:

$$74.75b + 6.5c = 250.25 \tag{1′}$$
$$9.75b - 6.5c = 87.75 \tag{3′}$$

Add Equations (1′) and (3′), thereby eliminating c:

$$84.5b = 338$$
$$b = 4$$

Having determined the value of b, substitute this into the original Equation (1′) and solve for c:

$$c = 38.5 - 11.5(b)$$
$$= 38.5 - 11.5(4)$$
$$= -7.5$$

Now, knowing the value of b and c, substitute both into Equation (4) to solve for a:

$$a = 4.5(b) - 24 - c$$
$$= 4.5(4) - 24 - (-7.5)$$
$$= 1.5$$

To summarize: $a = 1.5$, $b = 4$, $c = -7.5$.

If verification is necessary, substitute all three values into the original three equations.

Method 2: Addition or Subtraction of Equations

Repeating the original three equations:

$$3a - 2b + 4c = -33.5 \tag{1}$$
$$-a + 4.5b - c = 24 \tag{2}$$
$$1.5a + 3b - 5c = 51.75 \tag{3}$$

Multiply Equation (2) by 3:

$$-3a + 13.5b - 3c = 72 \tag{2′}$$

Add Equations (2′) and (1), thereby eliminating a:

$$11.5b + c = 38.5 \tag{4}$$

Multiply Equation (3) by -2:

$$-3a - 6b + 10c = -103.5 \tag{3′}$$

Add Equations (3′) and (1), thereby eliminating a:

$$-8b + 14c = -137 \tag{5}$$

Equations (4) and (5) are now two equations with only two unknowns. They are repeated for convenience:

$$11.5b + \quad c = 38.5 \tag{4}$$
$$-8b + 14c = -137 \tag{5}$$

Multiply Equation (4) by -14:

$$-161b - 14c = -539 \tag{6}$$

Add Equations (6) and (5) to eliminate c:

$$-169b = -676$$
$$b = 4$$

Substitute $b = 4$ into Equation (4):

$$c = 38.5 - 11.5b$$
$$= 38.5 - 11.5(4)$$
$$= -7.5$$

Substitute $b = 4$ and $c = -7.5$ into Equation (2):

$$-a = 24 + c - 4.5b$$
$$a = -24 - c + 4.5b$$
$$= -24 - (-7.5) + 4.5(4)$$
$$= 1.5$$

To summarize: $a = 1.5$, $b = 4$, $c = -7.5$.

Either approaches 1 or 2 gives the same answer. Because each method requires about the same amount of calculation, you are encouraged to use both when solving the following problems, to develop familiarity and expertise with both approaches.

PROBLEMS

Use both methods in Section 11.3 to solve the simultaneous equations of three unknowns in Problems 1 to 8.

1.
$$2a + b - 3c = -42$$
$$-a - b - c = -8$$
$$4a - 5b + c = 56$$

3.
$$2\alpha - \beta + 3\gamma = 18.2$$
$$-3\alpha + 2\beta - 7\gamma = -38.9$$
$$6\alpha - 2\beta - 7\gamma = -12.6$$

2.
$$6x - y - 2z = 34$$
$$-3x - y + 5z = -98$$
$$x + y - 3z = 62$$

4.
$$3u + 2v - w = 19.2$$
$$-4u - v + 6w = -64.3$$
$$u + 7v + 3w = -12.5$$

5. $2.6s - 4.1t + v = 17.02$

$-3.8s + t - 1.6v = -11.66$

$-1.1s - t + 2.4v = -4.89$

7. $3.6f = 1.7g - 4.1h + 3.95$

$0.95g + 1.62h = 3.9f - 14.44$

$-11.43 = 1.2g + h - 1.7f$

6. $2.1i - 3.6j + 4.7k = -11.82$

$-1.1i + 2.6j - 3.7k = 9.02$

$3.6i + j - k = -5.02$

8. $3.6q = 1.7p + 2.1r + 27.135$

$-3.15r + 2.6p = 1.6q + 8.7175$

$-1.45p + 2.1r = -2.65q - 4.225$

SUMMARY

Simultaneous equations are unlike linear equations in that they involve the solutions of more than one unknown. There are three different ways to solve two simultaneous equations: (1) substitution, (2) addition or subtraction of equations, and (3) comparison.

Substitution requires that one equation be solved for one unknown in terms of the other unknown. When this value is then substituted into the other equation, only one unknown remains. You then solve for this unknown by standard algebraic methods. Further substitution of the solution into the second equation solves for the second unknown.

Addition or subtraction of equations involves manipulating one or both of the original equations so that the absolute value of the coefficients of at least one unknown are identical. The addition (or subtraction) process then proceeds according to the net sign of the coefficients. The resultant, as in the method of substitution, is the elimination of one variable. Then proceed as in the method of substitution.

Comparison involves equating a common term within each equation. Then, based upon the mathematical law that "things equal to the same thing are equal to each other," equate the two resultants. Again, this results in one equation with one unknown.

When the number of simultaneous equations reaches three (implying three unknowns), you must apply a successive series of substitutions and/or addition or subtraction of equations for a solution.

Alternate methods of algebraically manipulating number arrays to easily solve for three or more unknowns are called *determinants* and *matrices*, both of which are discussed in the next chapter. The applications of word problems as they apply to both multiple unknowns and electronic circuits are deferred until Chapter 13. If you thoroughly understand the principles for solving simultaneous equations, you can readily transfer these methodologies to electronic circuits.

12 DETERMINANTS AND MATRIX ANALYSIS

12.1 INTRODUCTION

Chapter 11 was a strictly mathematical approach to solving simultaneous equations. There was no attempt to relate this theory to electronic circuits in particular or to word problems in general. This chapter continues the mathematical discussion of solving simultaneous equations by introducing two additional and most powerful methods of solution, powerful because they are so easily adapted to calculators, especially those with built-in algorithms to solve matrices. Because these approaches are so readily applicable to circuit analysis, electronic theory is intermixed with the mathematical theory to better clarify the examples.

12.2 DETERMINANTS

In electric circuits with clearly defined series and parallel elements, it is relatively easy to determine current magnitude and direction through each element and the voltage across each element. However, in many practical circuits elements are neither in series with nor parallel to the other elements. Such a circuit is shown in Figure 12.1; this circuit is frequently called a Wheatstone bridge.

The current I_T, provided by the voltage source, enters junction A, where it divides between R_1 and R_3. Note that R_1 and R_3 are not parallel to each other. Whether any current flows through the current meter depends solely upon the ratio of the four resistors in the bridge. Suppose resistor R_1 equals R_2 and resistor R_3 equals R_4. No current flows through the current meter because by the proportionate voltage law, point B equals $V_T/2$ and point C also equals $V_T/2$. Because points B and C are at the same potential, the bridge is considered balanced.

This circuit is practical because it can very accurately measure the resistance of an unknown resistor (R_3) by varying R_4 until it equals the value of the unknown resistor. After the bridge has been balanced (no current through the meter), the value of R_3 can be determined by observing the calibrated dials of R_4. If R_3 and R_4 are not equal, the current through the meter will be in one direction or the other, depending upon whether R_4 is less than or greater than R_3.

In this example, the meter provided a visual reading of current magnitude and direction. But if the meter is replaced by its internal resistance (or any resistance, for that matter) all visual indicators are removed. With the resistor's values known and an unbalanced state assumed, an analytical calculation is required to determine the current between points B and C. At this point the difficulty of the solution may now

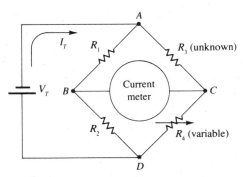

Figure 12.1   Simplified Wheatstone bridge.

be apparent. If resistors are not quite in series or quite parallel, how do you analyze such a circuit? First, let us discuss loop equations and determinants.

Consider the following equations, with two unknown variables (a and b):

$$2a + 4b = 14 \tag{12.1}$$
$$3a + 3b = 15 \tag{12.2}$$

A way to solve these simultaneous equations is to eliminate one variable. If we multiply Equation (12.1) by 3—the coefficient of b in Equation (12.2)—and Equation (12.2) by 4—the coefficient of b in Equation (12.1)—we get the following two equations:

$$6a + 12b = 42 \tag{12.3}$$
$$12a + 12b = 60 \tag{12.4}$$

Subtract Equation (12.3) from Equation (12.4):

$$6a + 0b = 18 \tag{12.5}$$

Solve for a:

$$a = \frac{18}{6} = 3$$

Substitute this value of a into Equation (12.2) and solve for b:

$$3(3) + 3b = 15$$
$$3b = 6$$
$$b = 2$$

The example used specific coefficients for the two variables, but it is preferable to adopt a generalized solution to the problem. Equations (12.6) and (12.7) are solved in a similar fashion, but the solution of the unknown variables x and y is expressed in the general terminology of their algebraic coefficients a, b, c, d, e, and f:

$$ax + by = c \tag{12.6}$$
$$dx + ey = f \tag{12.7}$$

Multiply Equation (12.6) by e and Equation (12.7) by b:

$$aex + bey = ce \tag{12.8}$$
$$bdx + bey = bf \tag{12.9}$$

Subtract Equation (12.9) from Equation (12.8):

$$(ae - bd)x = ce - bf$$

Solve for x:

$$x = \frac{ce - bf}{ae - bd} \tag{12.10}$$

Similarly,

$$y = \frac{af - cd}{ae - bd} \tag{12.11}$$

Notice that in both solutions the denominator is the same.

We have now established the mechanics of determinants. To write a determinant,

1. Write the equations with like variables aligned, as in Equations (12.6) and (12.7).

2. Using Equations (12.6) and (12.7) as examples, write an array of coefficients in the following determinant format:

$$x = \frac{\begin{vmatrix} c & b \\ f & e \end{vmatrix}}{\begin{vmatrix} a & b \\ d & e \end{vmatrix}} \tag{12.12}$$

$$y = \frac{\begin{vmatrix} a & c \\ d & f \end{vmatrix}}{\begin{vmatrix} a & b \\ d & e \end{vmatrix}} \tag{12.13}$$

Observe that both denominators are the same and the coefficients are in the same order as if reading Equations (12.6) and (12.7) from left to right. Also note that the numerators are arranged so that the coefficients c and f are substituted for the coefficients of the unknow variable being solved. This array is still in a left-to-right direction. To complete the solutions of x and y,

3. Form the cross product of the upper left and lower right coefficients within each vertical.

4. Subtract the cross product of the lower left and upper right coefficients within each vertical. This results in

$$x = \frac{ce - bf}{ae - bd}$$

$$y = \frac{af - cd}{ae - bd}$$

Thus, a determinant is just an organized array of numbers. The numbers are related to the coefficients of the variables and the coefficients to which each equation is equated. The solution of the determinant, and hence the solution of the variables, is purely a mechanical operation.

Multiply in this direction:

Upper left ↘ Lower right

Subtract in this direction:

Lower left ↗ Upper right

Example 1

Solve the following determinant.

$$q = \frac{\begin{vmatrix} 3 & 2 \\ 6 & 1 \end{vmatrix}}{\begin{vmatrix} 1 & 4 \\ 3 & 2 \end{vmatrix}}$$

$$= \frac{(3)(1) - (6)(2)}{(1)(2) - (3)(4)} = \frac{3 - 12}{2 - 12} = \frac{-9}{-10} = 0.9$$

Example 2

Solve the following determinant.

$$t = \frac{\begin{vmatrix} -3 & 2 \\ -1 & 6 \end{vmatrix}}{\begin{vmatrix} 2 & -5 \\ -1 & -7 \end{vmatrix}}$$

$$= \frac{(-3)(6) - (-1)(2)}{(2)(-7) - (-1)(-5)}$$

$$= \frac{-16}{-19} = 0.842$$

Example 3

Using determinants, solve the following simultaneous equations.

$$4r - 5s = -32$$
$$-3r + 2s = 17$$

$$r = \frac{\begin{vmatrix} -32 & -5 \\ 17 & 2 \end{vmatrix}}{\begin{vmatrix} 4 & -5 \\ -3 & 2 \end{vmatrix}} = \frac{(-64) - (-85)}{(8) - (15)} = \frac{21}{-7} = -3$$

$$s = \frac{\begin{vmatrix} 4 & -32 \\ -3 & 17 \end{vmatrix}}{-7} = \frac{(68) - (96)}{-7} = \frac{-28}{-7} = 4$$

Notice that the denominator of the determinant for s was already solved when we found the solution of r. Generally, whatever the order of the determinant, once the denominator (D) of the determinant of the first variable has been solved, the value of the denominator of all other variables is established because it is the same for all the variable quantitites.

Example 4

Solve the following simultaneous equations using determinants.

$$y = 21 - 4x$$
$$-2x - 3 = 5y$$

First, arrange each equation into an aligned array as represented in Equations (12.6) and (12.7):

$$4x + y = 21$$
$$-2x - 5y = 3$$

Now solve determinants for the unknown quantities:

$$x = \frac{\begin{vmatrix} 21 & 1 \\ 3 & -5 \end{vmatrix}}{\begin{vmatrix} 4 & 1 \\ -2 & -5 \end{vmatrix}} = \frac{(-105) - (3)}{(-20) - (-2)} = \frac{-108}{-18} = 6$$

$$y = \frac{\begin{vmatrix} 4 & 21 \\ -2 & 3 \end{vmatrix}}{-18} = \frac{(12) - (-42)}{-18} = \frac{54}{-18} = -3$$

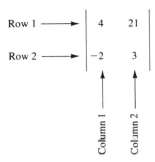

Figure 12.2 Determinant row and
column designation.

You can also use simplification rules to solve determinants. Figure 12.2 illustrates
the numerator of y in Example 4 and indicates both the column and row numbers of
the determinant. If any column or row can be evenly divided by a coefficient, each
term of that column or row can be reduced to the quotient of that division if you later
multiply the determinant by the divisor. (This is somewhat analogous to reducing
36/72 to 1/2 in that both fractions are the same, but the numbers in 1/2 are smaller
and therefore easier to manipulate.) Column 1 is divisible by 2 and column 2 is divisible by 3. Reducing both columns results in:

$$(2)(3) \begin{vmatrix} 2 & 7 \\ -1 & 1 \end{vmatrix} = 6((2) - (-7)) = 54$$

This is the same value as determined in the original answer. The simplification process is best utilized when there are large, factorable numbers within the determinant;
calculating subsequent products then is much easier.

Example 5

Solve the following determinant, using simplification processes wherever possible.

$$x = \frac{\begin{vmatrix} 25 & 225 \\ 15 & 15 \end{vmatrix}}{\begin{vmatrix} 16 & 225 \\ 4 & 15 \end{vmatrix}}$$

In the numerator, column 1 can be factored by 5 and column 2 by 15. In the denominator, column 1 can be factored by 4 and column 2 by 15:

$$x = \frac{(5)(15) \begin{vmatrix} 5 & 15 \\ 3 & 1 \end{vmatrix}}{(4)(15) \begin{vmatrix} 4 & 15 \\ 1 & 1 \end{vmatrix}}$$

The 15s cancel, and row 1 in the numerator is further factorable by 5:

$$x = \frac{(5)(5)\begin{vmatrix} 1 & 3 \\ 3 & 1 \end{vmatrix}}{(4)\begin{vmatrix} 4 & 15 \\ 1 & 1 \end{vmatrix}} = \frac{25((1) - (9))}{4((4) - (15))} = \frac{-200}{-44}$$

$$= \frac{50}{11} = 4.545$$

Examples 1 to 5 were for second-order determinants, that is, determinants solving for two unknowns. However, the principles involved apply to determinants of any order. For example, in three equations with three unknowns, you can solve each unknown in a like manner:

$$ax + by + cz = d \tag{12.14}$$
$$ex + fy + gz = h \tag{12.15}$$
$$ix + jy + kz = \ell \tag{12.16}$$

$$x = \frac{\begin{vmatrix} d & b & c \\ h & f & g \\ \ell & j & k \end{vmatrix}}{\begin{vmatrix} a & b & c \\ e & f & g \\ i & j & k \end{vmatrix} \equiv D} \tag{12.17}$$

$$y = \frac{\begin{vmatrix} a & d & c \\ e & h & g \\ i & \ell & k \end{vmatrix}}{D} \tag{12.18}$$

$$z = \frac{\begin{vmatrix} a & b & d \\ e & f & h \\ i & j & \ell \end{vmatrix}}{D} \tag{12.19}$$

Notice that when writing *all* determinants, you should follow these mechanics:

1. Align like variables.
2. In the numerator, from left to right, substitute the coefficients to which each equation is equated for the coefficients of the variable being solved.
3. In the denominator, D is the same for all unknowns and, reading from left to right, represents the coefficients of all the variables.

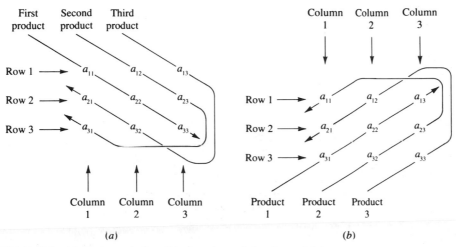

Figure 12.3 (*a*) Products of all combinations (upper left to lower right).
(*b*) Products of all combinations (lower left to upper right).

When solving third-order determinants, you must do the following mechanics:

1. Sum all combinations of the products of three terms, from upper left to lower right.
2. From step 1, subtract all combinations of the products of three terms, from lower left to upper right.

Figure 12.3*a* illustrates all combinations of three terms from upper left to lower right; Figure 12.3*b* illustrates all combinations from lower left to upper right. Observe that the first subscript of each term indicates the row; the second subscript indicates the column. This designation is very useful when we discuss matrices in Sections 12.5 and 12.6.

There is an alternate to the "loop-around" format just described to develop all the "products of three terms" necessary to solve a 3 × 3 determinant. If the first two columns are recopied to the right of the original determinant, as illustrated in Figure 12.4, then products 1, 2, and 3 represent all combinations of three terms from upper left to lower right, and products 4, 5, and 6 represent all combinations of three terms from lower left to upper right. Note that this method exhibits all the necessary products as the multiplication of numbers whose relative position within the expanded array forms a straight line. The loop-around method of Figure 12.3*a* and *b* or the straight-line product method of Figure 12.4 produces the same desired result. Whichever method you use is your personal decision. The loop-around method requres a little getting acquainted with, and the straight-line product method requires writing two additional columns.

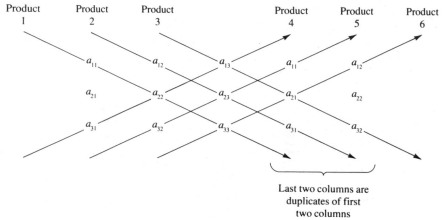

Product 1 Product 2 Product 3 Product 4 Product 5 Product 6

a_{11} a_{12} a_{13} a_{11} a_{12}

a_{21} a_{22} a_{23} a_{21} a_{22}

a_{31} a_{32} a_{33} a_{31} a_{32}

Last two columns are
duplicates of first
two columns

Figure 12.4 Straight-line products method.

Example 6

Solve the following determinant.

$$x = \begin{vmatrix} 1 & 4 & 7 \\ 2 & 5 & 8 \\ 3 & 6 & 9 \end{vmatrix}$$

$$= (1)(5)(9) + (4)(8)(3) + (7)(6)(2)$$
$$\quad -(3)(5)(7) - (6)(8)(1) - (9)(4)(2)$$

$$= 45 + 96 + 84 - 105 - 48 - 72$$

$$= 225 - 225 = 0$$

Example 7

Solve the following determinant, simplifying where possible.

$$y = \begin{vmatrix} 15 & 25 & 15 \\ 6 & 36 & 36 \\ 9 & 36 & 81 \end{vmatrix}$$

First, simplify the rows. Row 1 is divisible by 5, row 2 is divisible by 6, and row 3 is divisible by 9:

$$y = \underbrace{(5)(6)(9)}_{270} \begin{vmatrix} 3 & 5 & 3 \\ 1 & 6 & 6 \\ 1 & 4 & 9 \end{vmatrix}$$

Next, column 3 is divisble by 3:

$$y = \underbrace{(270)(3)}_{810} \begin{vmatrix} 3 & 5 & 1 \\ 1 & 6 & 2 \\ 1 & 4 & 3 \end{vmatrix}$$

$$= 810(54 + 10 + 4 - 6 - 24 - 15)$$

$$= 810(23) = 18,630$$

Example 8

Solve the following simultaneous equations with three unknowns.

$$2a - 3b - c = 9 \tag{1}$$
$$-a + 4b - 3c = -26 \tag{2}$$
$$-3a - 4b + 6c = 30 \tag{3}$$

$$D \equiv \begin{vmatrix} 2 & -3 & -1 \\ -1 & 4 & -3 \\ -3 & -4 & 6 \end{vmatrix} = (48 - 27 - 4) - (12 + 24 + 18) = -37$$

$$a = \frac{\begin{vmatrix} 9 & -3 & -1 \\ -26 & 4 & -3 \\ 30 & -4 & 6 \end{vmatrix}}{D} = \frac{(216 - 104 + 270) - (-120 + 468 + 108)}{-37}$$

$$= \frac{-74}{-37} = 2$$

$$b = \frac{\begin{vmatrix} 2 & 9 & -1 \\ -1 & -26 & -3 \\ -3 & 30 & 6 \end{vmatrix}}{D} = \frac{(-312 + 30 + 81) - (-78 - 54 - 180)}{-37}$$

$$= \frac{111}{-37} = -3$$

After solving for any two unknowns, it is usually easier to substitute these values into any of the three original equations to solve for the remaining unknown. Choosing the first equation,

$$c = 2a - 3b - 9$$
$$= 2(2) - 3(-3) - 9 = 4$$

Check by substituting the values of a, b, and c into the remaining two equations:

$$-(2) + 4(-3) - 3(4) = -26 \tag{2}$$
$$-2 - 12 - 12 = -26$$
$$-26 = -26 \quad \text{checks}$$

$$-3(2) - 4(-3) + 6(4) = 30 \tag{3}$$
$$-6 + 12 + 24 = 30$$
$$30 = 30 \quad \text{checks}$$

Example 9

Solve the following simultaneous equations, simplifying where appropriate.

$$6 - 4c = 8a + 2b$$
$$-2b + 16a = -6 - 8c$$
$$6c = 16 + 4a$$

First, rearrange the equations so like variables are aligned:

$$8a + 2b + 4c = 6 \tag{12.20}$$
$$16a - 2b + 8c = -6 \tag{12.21}$$
$$-4a + 6c = 16 \tag{12.22}$$

Note the absence of a b variable in Equation (12.22). This can be advantageous because if the a variable is solved, then only one unknown remains in Equation (12.22). This one unknown is then best solved by substitution rather than by solving another determinant. In addition, the value of zero for the b variable in Equation (12.22) greatly simplifies the determinant.

$$a = \frac{\begin{vmatrix} 6 & 2 & 4 \\ -6 & -2 & 8 \\ 16 & 0 & 6 \end{vmatrix}}{\begin{vmatrix} 8 & 2 & 4 \\ 16 & -2 & 8 \\ -4 & 0 & 6 \end{vmatrix}} = \frac{(2)(2)(2)\begin{vmatrix} 3 & 1 & 2 \\ -3 & -1 & 4 \\ 8 & 0 & 3 \end{vmatrix}}{(4)(2)(2)\begin{vmatrix} 2 & 1 & 2 \\ 4 & -1 & 4 \\ -1 & 0 & 3 \end{vmatrix}} \begin{pmatrix} reduce \\ rows \end{pmatrix} \begin{pmatrix} reduce \\ columns \end{pmatrix}$$

$$= \frac{1}{2}\left(\frac{(-9 + 32) - (-16 - 9)}{(-6 - 4) - (2 + 12)} \right)$$

$$= \frac{1}{2}\left(\frac{48}{-24} \right) = -1$$

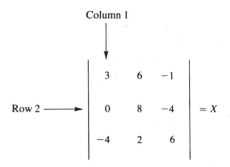

Figure 12.5 Selection of "zero" term for cofactor.

From Equation (12.22):

$$6c = 16 + 4a = 16 + 4(-1) = 12$$
$$c = 2$$

From Equation (12.20):

$$2b = 6 - 4c - 8a = 6 - 4(2) - 8(-1) = 6 - 8 + 8 = 6$$
$$b = 3$$

Another major simplification process for solving determinants involves *cofactors*. Cofactors are best used in four-order or higher determinants or when a third-order determinant has one or more zeros. In keeping with the general theme of determinants, cofactors are merely a mechanical manipulation of the numbers of the array. Two rules are employed in this method.

Rule 1: Select a given column or row, usually the one with the most zeros. Using Figure 12.5 as an example, this can be either column 1 or row 2. Multiply cofactors of each term of that column or row by the term for which the cofactor was created.

Suppose we select column 1. The first term (a_{11}) is 3, the second term (a_{21}) is 0, and the third term (a_{31}) is -4. Draw imaginary lines, horizontally and vertically, through each coefficient in the same column or row of the term for which the cofactor is being written. Write a two-order determinant from the remaining four numbers. For example, in Figure 12.6a the resultant cofactor of 3 (a_{11}) is

$$\begin{vmatrix} 8 & -4 \\ 2 & 6 \end{vmatrix}$$

In a like manner, from Figure 12.6b, the cofactor of 0 (a_{21}) is

$$\begin{vmatrix} 6 & -1 \\ 2 & 6 \end{vmatrix}$$

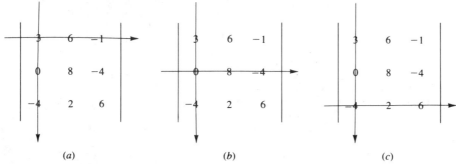

Figure 12.6 (*a*) Cofactor of a_{11} (3); (*b*) cofactor of a_{21} (0); (*c*) cofactor of a_{31} (−4).

and from Figure 12.6*c*, the cofactor of −4 (a_{31}) is

$$\begin{vmatrix} 6 & -1 \\ 8 & -4 \end{vmatrix}$$

We can now sum the solution of the determinant for *x* as

$$x = (\text{sign})\underbrace{(3)}_{a_{11}}\begin{vmatrix} 8 & -4 \\ 2 & 6 \end{vmatrix} + (\text{sign})\underbrace{(0)}_{a_{21}}\begin{vmatrix} 6 & -1 \\ 2 & 6 \end{vmatrix} + (\text{sign})\underbrace{(-4)}_{a_{31}}\begin{vmatrix} 6 & -1 \\ 8 & -4 \end{vmatrix}$$

$$\underbrace{}_{\substack{\text{Cofactor} \\ \text{of } a_{11}}} \qquad \underbrace{}_{\substack{\text{Cofactor} \\ \text{of } a_{21}}} \qquad \underbrace{}_{\substack{\text{Cofactor} \\ \text{of } a_{31}}}$$

We can now state three observations:

1. The original third-order determinant is reduced to the sum of a series of second-order determinants.

2. We need not evaluate the determinant cofactor preceded by zero because any quantity multiplied by zero is still zero. This is the advantage of picking the column or row with the most zeros.

3. We must incorporate a new sign quantity into the series.

Observation 3 is essentially the second rule, namely:
   Rule 2: All terms of the reduced determinant must be preceded by a sign according to the following, alternating pattern:

$$\begin{vmatrix} +(a_{11}) & -(a_{12}) & +(a_{13}) \\ -(a_{21}) & +(a_{22}) & -(a_{23}) \\ +(a_{31}) & -(a_{32}) & +(a_{33}) \end{vmatrix}$$

Applying rule 2 to the solution for x:

$$x = +(3)\begin{vmatrix} 8 & -4 \\ 2 & 6 \end{vmatrix} - (0)\begin{vmatrix} 6 & -1 \\ 2 & 6 \end{vmatrix} - (4)\begin{vmatrix} 6 & -1 \\ 8 & -4 \end{vmatrix}$$

$$= 3((48) - (-8)) - 4((-24) - (-8))$$

$$= 232$$

Example 10

Solve the determinant for a in Example 9 using cofactors. Verify that the solution is still -1. From Example 9,

$$a = \left(\frac{1}{2}\right)\frac{\begin{vmatrix} 3 & 1 & 2 \\ -3 & -1 & 4 \\ 8 & 0 & 3 \end{vmatrix}}{\begin{vmatrix} 2 & 1 & 2 \\ 4 & -1 & 4 \\ -1 & 0 & 3 \end{vmatrix}}$$

Utilizing row 3 to form cofactor terms:

$$2a = \frac{(8)\begin{vmatrix} 1 & 2 \\ -1 & 4 \end{vmatrix} - (0) + (3)\begin{vmatrix} 3 & 1 \\ -3 & -1 \end{vmatrix}}{(-1)\begin{vmatrix} 1 & 2 \\ -1 & 4 \end{vmatrix} - (0) + (3)\begin{vmatrix} 2 & 1 \\ 4 & -1 \end{vmatrix}}$$

$$= \frac{8((4) - (-2)) + 3((-3) - (-3))}{-1((4) - (-2)) + 3((-2) - (4))}$$

$$= \frac{48}{-24} = -2$$

$$a = -1$$

The answer agrees with the calculation for a in Example 9.

Example 11

Solve the following simultaneous equations by using the most advantageous cofactors of the determinant. Complete the solution by using the most advantageous substitutions.

$$3a - 4b - c = 1 \qquad\qquad (12.23)$$
$$3b + 2c = 1 \qquad\qquad (12.24)$$
$$-a + 2b - c = -9 \qquad\qquad (12.25)$$

We use column 1 of both N and D when writing the cofactors:

$$a = \frac{\begin{vmatrix} 1 & -4 & -1 \\ 1 & 3 & 2 \\ -9 & 2 & -1 \end{vmatrix} \equiv N}{\begin{vmatrix} 3 & -4 & -1 \\ 0 & 3 & 2 \\ -1 & 2 & -1 \end{vmatrix} \equiv D}$$

$$= \frac{1\begin{vmatrix} 3 & 2 \\ 2 & -1 \end{vmatrix} - 1\begin{vmatrix} -4 & -1 \\ 2 & -1 \end{vmatrix} - 9\begin{vmatrix} -4 & -1 \\ 3 & 2 \end{vmatrix}}{3\begin{vmatrix} 3 & 2 \\ 2 & -1 \end{vmatrix} - 1\begin{vmatrix} -4 & -1 \\ 3 & 2 \end{vmatrix}}$$

$$= \frac{(-3 - 4) - (4 + 2) - 9(-8 + 3)}{3(-3 - 4) - (-8 + 3)}$$

$$= \frac{-7 - 6 + 45}{-21 + 5} = \frac{32}{-16} = -2$$

$$b = \frac{\begin{vmatrix} 3 & 1 & -1 \\ 0 & 1 & 2 \\ -1 & -9 & -1 \end{vmatrix}}{D} = \frac{3\begin{vmatrix} 1 & 2 \\ -9 & -1 \end{vmatrix} - \begin{vmatrix} 1 & -1 \\ 1 & 2 \end{vmatrix}}{-16}$$

$$= \frac{3(-1 + 18) - (2 + 1)}{16} = \frac{48}{-16} = -3$$

From Equation (12.24):

$$2c = 1 - 3b = 1 - (3)(-3) = 10$$
$$c = 5$$

Example 12

Solve the following fourth-order determinant.

$$x = \begin{vmatrix} 2 & 6 & 2 & -1 \\ 1 & 0 & -7 & 0 \\ -5 & -1 & 1 & 4 \\ 3 & 4 & 9 & 5 \end{vmatrix}$$

Make cofactors from row 2, to use two zeros:

$$x = (\text{sign})(1)\begin{vmatrix} 6 & 2 & -1 \\ -1 & 1 & 4 \\ 4 & 9 & 5 \end{vmatrix} + 0 + (\text{sign})(-7)\begin{vmatrix} 2 & 6 & -1 \\ -5 & -1 & 4 \\ 3 & 4 & 5 \end{vmatrix} + 0$$

$$= (-)1\begin{vmatrix} 6 & 2 & -1 \\ -1 & 1 & 4 \\ 4 & 9 & 5 \end{vmatrix}(-) - 7\begin{vmatrix} 2 & 6 & -1 \\ -5 & -1 & 4 \\ 3 & 4 & 5 \end{vmatrix}$$

$$= -((30 + 9 + 32) - (-4 - 10 + 216))$$
$$\quad + 7((-10 + 20 + 72) - (3 - 150 + 32))$$
$$= -(71 - 202) + 7(82 - (-115))$$
$$= 131 - 1379 = 1510$$

PROBLEMS

1. Solve the following determinants.

   a. $\begin{vmatrix} 3 & 6 \\ -2 & 4 \end{vmatrix}$

   b. $\begin{vmatrix} -2 & 5 \\ -4 & 12 \end{vmatrix}$

   c. $\begin{vmatrix} -3 & 5 & 4 \\ 2 & -5 & -3 \\ 1 & -10 & 8 \end{vmatrix}$

   d. $\begin{vmatrix} 3 & 6 & -9 \\ 2 & 5 & 18 \\ -4 & 1 & -21 \end{vmatrix}$

   e. $\begin{vmatrix} 4 & 2 & 0 \\ 0 & -3 & -5 \\ -6 & 1 & -7 \end{vmatrix}$

   f. $\begin{vmatrix} 4 & 2 & 0 & 2 \\ -1 & -6 & 9 & -3 \\ 0 & 8 & -5 & 0 \\ 3 & -4 & -1 & 7 \end{vmatrix}$

2. Solve the following equations, using determinants.

a.
$$6a - 5c = -14 - 4b$$
$$-3b - 4c = 11 + 2a$$
$$a - 2b + 4c = -20$$

b.
$$2a - 3b = 2c + 22$$
$$-3a + 7 = 6b - 2c$$
$$-3b = a + c + 8$$

c.
$$3x + y = 2z + 1$$
$$-3y + 4z = -5x - 11$$
$$-2z = -2x - y + 2$$

d.
$$10 = -3\alpha + \beta - \gamma$$
$$\beta - 3\gamma = 2\alpha + 2$$
$$5\gamma = 3\alpha + 2\beta + 7$$

e.
$$2s + 3t - v = -23$$
$$-2t + 3v = 5s + 38$$
$$43 - 5v = -2s - 3t$$

3. Write an expression for the solution of the following determinant, using cofactors of row 2.

$$x = \begin{vmatrix} a & d & g \\ b & e & h \\ c & f & i \end{vmatrix}$$

4. Solve the following equations, using determinants.

$$-w + 3x - z = 3$$
$$2x - 3y + 4z = 29$$
$$w + x + y + z = 2$$
$$2w + 3y - 4z = -27$$

12.3 USING CALCULATORS FOR DETERMINANT SOLUTIONS

Both the examples and problems in Section 12.2 used "nice" numbers, so it was not too difficult when applying the rules of determinants to mentally calculate the solutions. But as you are by now well aware, many resistor values are "not so nice": 470 Ω, 5.6 kΩ, 22 kΩ, and so on. The very nature of these numbers renders mental calculations of determinants difficult if not impractical. This section covers applications of determinants to electronic circuits, so you can expect the not so nice numbers. Thus you should study the following calculator applications for solving determinants and note that because the denominator is the same for all determinants of a problem, once calculated it should be stored for later usage.

Example 1

Using your calculator, solve the following determinant.

$$\begin{vmatrix} 3.4 & 2.5 & 9.8 \\ 5.6 & -1.3 & -1.1 \\ -1.7 & 7.6 & -0.9 \end{vmatrix}$$

$3.4 \times (-1.3) \times (-.9) + 2.5 \times (-1.1) \times (-1.7)$
$+9.8 \times 5.6 \times 7.6 - (-1.7) \times (-1.3) \times 9.8$
$-5.6 \times 2.5 \times (-.9) - 3.4 \times 7.6 \times (-1.1)$
$= 445.107$

PRESS	DISPLAY
3.4 $\boxed{\times}$ 1.3 $\boxed{+/-}$ $\boxed{\times}$ .9 $\boxed{+/-}$ $\boxed{+}$	3.978
2.5 $\boxed{\times}$ 1.1 $\boxed{+/-}$ $\boxed{\times}$ 1.7 $\boxed{+/-}$ $\boxed{+}$	8.653
9.8 $\boxed{\times}$ 5.6 $\boxed{\times}$ 7.6 $\boxed{-}$	425.741
1.7 $\boxed{+/-}$ $\boxed{\times}$ 1.3 $\boxed{+/-}$ $\boxed{\times}$ 9.8 $\boxed{-}$	404.083
5.6 $\boxed{\times}$ 2.5 $\boxed{\times}$ .9 $\boxed{+/-}$ $\boxed{-}$	416.683
3.4 $\boxed{\times}$ 7.6 $\boxed{\times}$ 1.1 $\boxed{+/-}$ $\boxed{=}$	445.107

Example 2

Using your calculator at $\boxed{\text{Fix 3}}$, solve the following simultaneous equations for all unknowns.

$$\begin{aligned} 2.3x - 1.5y - 3.6z &= 17.1 & (1) \\ -1.4x + 3.2y + 7.1z &= -3.6 & (2) \\ 3.4x + 1.5y - 5.7z &= 6.8 & (3) \end{aligned}$$

First solve for D:

$$D = \begin{vmatrix} 2.3 & -1.5 & -3.6 \\ -1.4 & 3.2 & 7.1 \\ 3.4 & 1.5 & -5.7 \end{vmatrix}$$

PRESS	DISPLAY
2.3 $\boxed{\times}$ 3.2 $\boxed{\times}$ 5.7 $\boxed{+/-}$ $\boxed{+}$	-41.952
1.5 $\boxed{+/-}$ $\boxed{\times}$ 7.1 $\boxed{\times}$ 3.4 $\boxed{+}$	-78.162
3.6 $\boxed{+/-}$ $\boxed{\times}$ 1.4 $\boxed{+/-}$ $\boxed{\times}$ 1.5 $\boxed{-}$	-70.602

PRESS	DISPLAY	COMMENT
3.4 $\boxed{\times}$ 3.2 $\boxed{\times}$ 3.6 $\boxed{+/-}$ $\boxed{-}$	-31.434	
1.4 $\boxed{+/-}$ $\boxed{\times}$ 1.5 $\boxed{+/-}$ $\boxed{\times}$ 5.7 $\boxed{+/-}$ $\boxed{-}$	-19.464	
2.3 $\boxed{\times}$ 1.5 $\boxed{\times}$ 7.1 $\boxed{=}$ $\boxed{\text{STO 0}}$	-43.959	D $\boxed{\text{STO 0}}$

Second, solve for x:

$$x = \begin{vmatrix} 17.1 & -1.5 & -3.6 \\ -3.6 & 3.2 & 7.1 \\ 6.8 & 1.5 & -5.7 \end{vmatrix} \div D$$

PRESS	DISPLAY	COMMENT
17.1 $\boxed{\times}$ 3.2 $\boxed{\times}$ 5.7 $\boxed{+/-}$ $\boxed{+}$	-311.904	
1.5 $\boxed{+/-}$ $\boxed{\times}$ 7.1 $\boxed{\times}$ 6.8 $\boxed{+}$	-384.324	
3.6 $\boxed{+/-}$ $\boxed{\times}$ 3.6 $\boxed{+/-}$ $\boxed{\times}$ 1.5 $\boxed{-}$	-364.884	
6.8 $\boxed{\times}$ 3.2 $\boxed{\times}$ 3.6 $\boxed{+/-}$ $\boxed{-}$	-286.548	
3.6 $\boxed{+/-}$ $\boxed{\times}$ 1.5 $\boxed{+/-}$ $\boxed{\times}$ 5.7 $\boxed{+/-}$ $\boxed{-}$	-255.768	
17.1 $\boxed{\times}$ 1.5 $\boxed{\times}$ 7.1 $\boxed{=}$	-437.883	
$\boxed{\div}$ $\boxed{\text{RCL 0}}$ $\boxed{=}$ $\boxed{\text{STO 1}}$ [1]	9.961	x

Third, solve for y:

$$y = \begin{vmatrix} 2.3 & 17.1 & -3.6 \\ -1.4 & -3.6 & 7.1 \\ 3.4 & 6.8 & -5.7 \end{vmatrix} \div D$$

PRESS	DISPLAY	COMMENT
2.3 $\boxed{\times}$ 3.6 $\boxed{+/-}$ $\boxed{\times}$ 5.7 $\boxed{+/-}$ $\boxed{+}$	47.196	
17.1 $\boxed{\times}$ 7.1 $\boxed{\times}$ 3.4 $\boxed{+}$	459.99	
3.6 $\boxed{+/-}$ $\boxed{\times}$ 1.4 $\boxed{+/-}$ $\boxed{\times}$ 6.8 $\boxed{-}$	494.262	
3.4 $\boxed{\times}$ 3.6 $\boxed{+/-}$ $\boxed{\times}$ 3.6 $\boxed{+/-}$ $\boxed{-}$	450.198	
1.4 $\boxed{+/-}$ $\boxed{\times}$ 17.1 $\boxed{\times}$ 5.7 $\boxed{+/-}$ $\boxed{-}$	313.74	
2.3 $\boxed{\times}$ 6.8 $\boxed{\times}$ 7.1 $\boxed{=}$	202.696	
$\boxed{\div}$ $\boxed{\text{RCL 0}}$ $\boxed{=}$ $\boxed{\text{STO 2}}$	-4.611	y

[1]Assumes a calculator with more than one (ultimately, at least four) storage locations. In this example, these are locations zero through three, inclusive.

Fourth, solve for z:

$$z = \begin{vmatrix} 2.3 & -1.5 & 17.1 \\ -1.4 & 3.2 & -3.6 \\ 3.4 & 1.5 & 6.8 \end{vmatrix} \div D$$

PRESS	DISPLAY	COMMENT
2.3 ☒ 3.2 ☒ 6.8 ☐+	50.048	
1.5 ☐+/− ☒ 3.6 ☐+/− ☒ 3.4 ☐+	68.408	
17.1 ☒ 1.4 ☐+/− ☒ 1.5 ☐−	32.498	
3.4 ☒ 3.2 ☒ 17.1 ☐−	−153.55	
1.4 ☐+/− ☒ 1.5 ☐+/− ☒ 6.8 ☐−	−167.83	
2.3 ☒ 1.5 ☒ 3.6 ☐+/− ☐=	−155.41	
☐÷ RCL 0 ☐= STO 3	3.535	z

To verify the solutions, substitute the values of x, y, and z into the original three equations. Again note that if your calculator has only one storage location, you must manually record excess stored values in the example. Use these values in lieu of the following RCL instructions.

PRESS	DISPLAY	COMMENT
2.3 ☒ RCL 1 ☐− 1.5 ☒ RCL 2		
☐− 3.6 ☒ RCL 3 ☐=	17.1	Equation (1) verified
1.4 ☐+/− ☒ RCL 1 ☐+ 3.2 ☒ RCL 2		
☐+ 7.1 ☒ RCL 3 ☐=	−3.6	Equation (2) verified
3.4 ☒ RCL 1 ☐+ 1.5 ☒ RCL 2		
☐− 5.7 ☒ RCL 3 ☐=	6.8	Equation (3) verified

We can thus conclude that all solutions are correct.

PROBLEMS

After constructing determinants, use your calculator at ☐Fix 3 to evaluate each unknown in the simultaneous equations in Problems 1 to 6.

1. $2.4c + 3.1d = -6.5$
$-1.6c - 2.5d = 3.4$

2. $1.25a - 6.4b = 7.2$
$3.17a + 9.6 = 2.14b$

3. $1.2x + 3.4y - 5.6z = -3.5$
$-2.6x - 1.7y + 3.9z = 6.2$
$-5.5x + 3.9y - 1.1z = 4.85$

4. $3.56\alpha - 2.52\beta \qquad = 3.7\gamma + 7.2$
$1.41\gamma + 5.6 - 3.71\beta = 2.62\alpha$
$-2.16\beta - 9.52 + 0.9\alpha = 3.19\gamma$

5. $\qquad 6.2u + 5.16v = 3.5w + 2.41$
$-6.7w - 7.62 + 2.34v = 3.92u$
$\qquad 6.14v - 1.97u = -2.16w + 4.32$

6. $1.3r - 2.69s - 6.1t + 3.2v = 1.52$
$-2.61r + 7.16s - 2.13t - 9.5v = 3.16$
$-0.96r + 1.17s - 4.21t + 7.6v = -7.05$
$8.17r - 3.42s - 1.09t - 1.1v = -3.27$

12.4 ELECTRONIC APPLICATIONS OF DETERMINANTS

Sections 12.2 and 12.3 extensively discussed a method for mechanically cranking out solutions to simultaneous equations. But where do the equations come from so that this practical process can be applied? You now have all the background knowledge necessary to solve the Wheatstone bridge problem in Section 12.2; Kirchhoff's voltage law permits writing the necessary equations and the methodology of determinants facilitates their solution.

The Wheatstone bridge configuration in Figure 12.1 is redrawn in Figure 12.7. There are three uniquely enclosed areas, so three equations must be constructed.

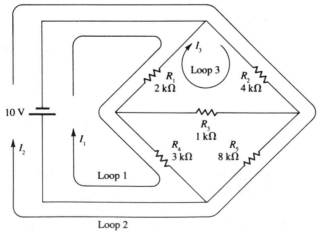

Figure 12.7   Selecting loops for a determinant.

Currents I_1, and I_2, and I_3 are arbitrarily drawn in a clockwise direction. Any direction could have been chosen because the solution of the determinant gives the correct current magnitude. If the solution is negative, it only means that the assumed current direction must be reversed but the magnitude is correct.

You should carefully select the three loops. Each choice must close back unto itself, and every element of the circuit must have at least one current flowing through it. One of the many other possible selections is shown in Figure 12.8. Notice again that each element has at least one current flowing through it and all loops close.

From Figure 12.7,

$$I_{R1} = I_1 - I_3 \qquad \text{current choices in opposite directions} \qquad (12.26)$$
$$I_{R2} = I_2 + I_3 \qquad \text{current choices in same direction} \qquad (12.27)$$
$$I_{R3} = I_3 \qquad (12.28)$$
$$I_{R4} = I_1 \qquad (12.29)$$
$$I_{R5} = I_2 \qquad (12.30)$$

From Kirchhoff's voltage law, the sum of the voltages in a closed loop is zero. Because all resistors are in units of kilohms, current is in units of milliamperes, so we drop the kilohm designation to keep the equations simpler.

From loop 1:

$$I_1R_1 + I_1R_4 - I_3R_1 = 10$$
$$I_1(R_1 + R_4) - I_3R_1 = 10$$
$$5I_1 + 0I_2 - 2I_3 = 10 \qquad (12.31)$$

From loop 2:

$$I_2(R_2 + R_5) + I_3R_2 = 10$$
$$0I_1 + 12I_2 + 4I_3 = 10 \qquad (12.32)$$

From loop 3:

$$-I_1R_1 + I_2R_2 + I_3(R_1 + R_2 + R_3) = 0$$
$$-2I_1 + 4I_2 + 7I_3 = 0 \qquad (12.33)$$

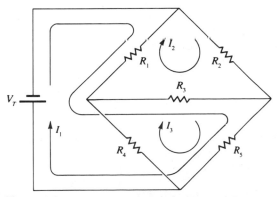

Figure 12.8   Alternate loop choices for Figure 12.7.

Collect the three simultaneous equations:

$$5I_1 + 0I_2 - 2I_3 = 10 \qquad (12.31)$$
$$0I_1 + 12I_2 + 4I_3 = 10 \qquad (12.32)$$
$$-2I_1 + 4I_2 + 7I_3 = 0 \qquad (12.33)$$

Example 1

Calculate the current magnitude and direction through each resistor in Figure 12.7, using the three equations just determined from the application of Kirchhoff's voltage law. Equations (12.26) to (12.30) specify the currents through each resistor according to the choice of loops and assumed direction in Figure 12.7. Using determinants, solve Equations (12.31) to (12.33) for the currents I_1, I_2, and I_3.

$$I_1 = \frac{\begin{vmatrix} 10 & 0 & -2 \\ 10 & 12 & 4 \\ 0 & 4 & 7 \end{vmatrix}}{\begin{vmatrix} 5 & 0 & -2 \\ 0 & 12 & 4 \\ -2 & 4 & 7 \end{vmatrix}}$$

$$= \frac{(840 - 80) - (160)}{(420) - (48 + 80)} = \frac{600}{292} = 2.05 \text{ mA}$$

From Equation (12.31):

$$2I_3 = 5I_1 - 10 = 5(2.05) - 10 = 0.25 \text{ mA}$$
$$I_3 = \frac{0.25}{2} \text{mA} = 0.125 \text{ mA}$$

From Equation (12.32):

$$12I_2 = 10 - 4I_3 = 10 - 4(0.125) = 9.5 \text{ mA}$$
$$I_2 = \frac{9.5}{12} \text{ mA} = 0.79 \text{ mA}$$

The current directions were assumed correctly because all answers are positive. Figure 12.9 depicts the current distribution and directions for the bridge circuit in Figure 12.7.

Example 2

Solve for the currents in Figure 12.10. Current flow is assumed in the directions indicated.

From loop 1:

$$9.3I_1 - 4.7I_2 + 2.4I_3 = 10$$

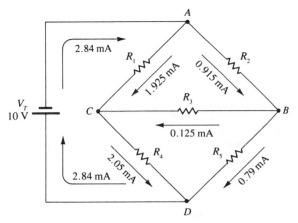

Figure 12.9   Current distribution for Figure 12.7.

From loop 2:

$$-4.7I_1 + 14.8I_2 + 6.8I_3 = 12$$

From loop 3:

$$2.4I_1 + 6.8I_2 + 13.4I_3 = 20$$

$$I_1 = \frac{\begin{vmatrix} 10 & -4.7 & 2.4 \\ 12 & 14.8 & 6.8 \\ 20 & 6.8 & 13.4 \end{vmatrix}}{\begin{vmatrix} 9.3 & -4.7 & 2.4 \\ -4.7 & 14.8 & 6.8 \\ 2.4 & 6.8 & 13.4 \end{vmatrix}}$$

$$= \frac{(1983.2 + 195.8 - 639.2) - (710.4 - 755.8 + 462.4)}{(1844.4 - 76.7 - 76.7) - (85.3 + 296 + 430)}$$

$$= \frac{1122.8}{879.7} = 1.28 \text{ mA}$$

$$I_2 = \frac{\begin{vmatrix} 9.3 & 10 & 2.4 \\ -4.7 & 12 & 6.8 \\ 2.4 & 20 & 13.4 \end{vmatrix}}{D}$$

$$= \frac{(1495.4 - 225.6 + 163.2) - (69.1 + 1264.8 - 629.8)}{879.7}$$

$$= \frac{728.9}{879.7} = 0.83 \text{ mA}$$

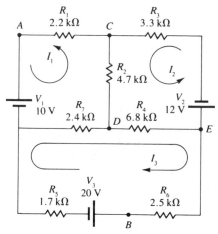

Figure 12.10

From loop 2:

$$6.8I_3 = 12 + 4.7I_1 - 14.8I_2$$
$$6.8I_3 = 12 + 6.02 - 12.28 = 5.74$$
$$I_3 = \frac{5.74}{6.8} = 0.84 \text{ mA}$$

Example 3

Using the current magnitudes and directions determined in Example 2, find V_{AB} in Figure 12.10. Observe subscript notation. One path between points A and B could be

$$V_{AB} = V_1 + V_{R5} + V_3$$
$$= (-10 - (1.7)(0.84) + 20) \text{ V}$$
$$= 8.57 \text{ V}$$

Example 4

Using the current magnitudes and directions determined in Example 2, find V_{AB} in Figure 12.10 using path $V_{AC} + V_{CD} + V_{DE} + V_{EB}$. Verify that the answer is the same for V_{AB} as determined in Example 3. Figure 12.11 illustrates the desired path, accounting for the specified current direction and voltage polarities.

$$V_{AC} = -I_1R_1 = -(1.28)(2.2) \text{ V} = -2.82 \text{ V}$$
$$V_{CD} = -(I_1 - I_2)R_2 = -(0.45)(4.7) \text{ V} = -2.12 \text{ V}$$
$$V_{DE} = (I_2 + I_3)R_4 = (1.67)(6.8) \text{ V} = 11.36 \text{ V}$$
$$V_{EB} = I_3R_6 = (0.84)(2.5) \text{ V} = 2.1 \text{ V}$$
$$\therefore V_{AB} = (-2.82 - 2.12 + 11.36 + 2.1) \text{ V} = 8.52 \text{ V}$$

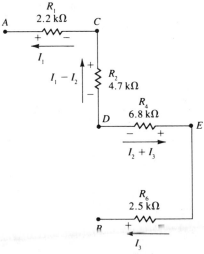

Figure 12.11 V_{AB} along prescribed path.

This is the same answer as determined in Example 3; the minor difference in voltage is caused by the round-off of numbers.

Example 5

Determine the current through R in Figure 12.12. Again note that in dropping the kilohm term, current is automatically in milliamperes. Choosing the current directions shown, the three loop equations are

$$18.3I_1 - 1.5I_2 - 6.8I_3 = 30$$
$$-1.5I_1 + 8.4I_2 - 4.7I_3 = 0$$
$$-6.8I_1 - 4.7I_2 + 13I_3 = 0$$

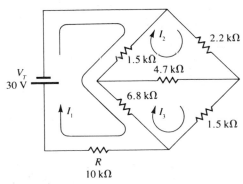

Figure 12.12

Because I_1 is the only current through R, we need solve only one determinant:

$$I_R = I_1 = \frac{\begin{vmatrix} 30 & -1.5 & -6.8 \\ 0 & 8.4 & -4.7 \\ 0 & -4.7 & 13 \end{vmatrix} \equiv N}{\begin{vmatrix} 18.3 & -1.5 & -6.8 \\ -1.5 & 8.4 & -4.7 \\ -6.8 & -4.7 & 13 \end{vmatrix} \equiv D}$$

Column 1 of the numerator has two zeros, so it is ideally suited for solution by co-factors:

$$N = 30 \begin{vmatrix} 8.4 & -4.7 \\ -4.7 & 13 \end{vmatrix} = 30(109.2 - 22.09)$$

$$= 2613.3$$
$$D = (1998.4 - 47.94 - 47.94) - (388.42 + 29.25 + 404.25)$$
$$= 1080.6$$

$$I_1 = \frac{N}{D} = \frac{2613.3}{1080.6} = 2.42 \text{ mA}$$

PROBLEMS

1. Using determinants, find the current through each resistor in the circuit in Figure 12.13.

2. Using determinants, find the current in each resistor in Figure 12.14.

3. Find V_{BG} in Figure 12.14 by the following path:

$$V_{BE} + V_{EF} + V_{FC} + V_{CH} + V_{HG}$$

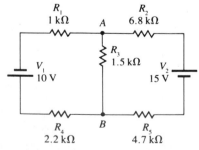

Figure 12.13

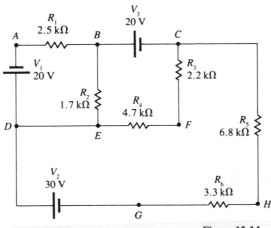

Figure 12.14

4. Find V_{BG} in Figure 12.14 by the following path, and verify that this is the same answer as in Problem 3:

$$V_{BA} + V_{AD} + V_{DG}$$

5. Using determinants, find the current through each resistor in Figure 12.15.

6. Find V_B with reference to ground in Figure 12.15. Find V_F with reference to ground. What is V_{BF}?

7. Determine V_{BF} in Figure 12.15 by the following path, and verify that this is the same answer determined for V_{BF} in Problem 6:

$$V_{BC} + V_{CE} + V_{ED} + V_{DF}$$

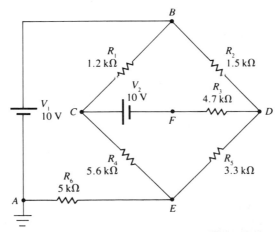

Figure 12.15

12.5 THEORY OF THE MATRIX

Determinants—arrays of numbers—facilitate the solution of complex circuits. A set of rules for manipulating these numbers produces a solution for each unknown, one at a time. This section studies another, more advanced form of number manipulation: solution by matrix, which has a distinct advantage:

Unlike a determinant, a matrix solves *all* unknown quantities simultaneously. One manipulation of the number array gives the answer to all unknown quantities in the simultaneous equations.

Figure 12.16 is the general configuration of a matrix array. Again, remember that each subscripted a merely represents a coefficient as determined by an original set of simultaneous equations. The first subscript indicates the row in which a particular a is located; the second subscript indicates the column in which the a is located. Together, the two-digit subscript defines a unique position in the number array for each coefficient.

The following theory sets forth rules for how to manipulate each coefficient so that the final answer is a simultaneous solution of all variables. Suppose an equation is written as

$$AB = C \qquad (12.34)$$

and we want to solve for B. If we multiply both sides of the equation by A^{-1}, we get the following equation:

$$(A^{-1})(AB) = (A^{-1})(C)$$
$$A^{0}B = A^{-1}C$$
$$B = A^{-1}C \qquad (12.35)$$

The term on the right-hand side (A^{-1}) is the entire key to the matrix. All the rules that follow deal with the meaning of this term, which is called the *inverse matrix*.

To illustrate matrix solution, let us solve the following three equations with three unknowns:

$$2A - B + C = 3 \qquad (12.36)$$
$$-A - B + 2C = 3 \qquad (12.37)$$
$$2A + B - 3C = -5 \qquad (12.38)$$

$$\begin{bmatrix} a_{11} & a_{12} & a_{13} \cdots a_{1j} \\ a_{21} & a_{22} & a_{23} \cdots a_{2j} \\ a_{31} & a_{32} & a_{33} \cdots a_{3j} \\ \vdots & \vdots & \vdots \\ a_{i1} & a_{i2} & a_{i3} \cdots a_{ij} \end{bmatrix}$$

Figure 12.16 General configuration of a matrix.

Write the initial matrix in the following fashion:

$$\underbrace{\begin{bmatrix} 2 & -1 & 1 \\ -1 & -1 & 2 \\ 2 & 1 & -3 \end{bmatrix}}_{X} \underbrace{\begin{bmatrix} A \\ B \\ C \end{bmatrix}}_{Y} = \underbrace{\begin{bmatrix} 3 \\ 3 \\ -5 \end{bmatrix}}_{Z} \qquad (12.39)$$

Notice that the coefficients of the unknown variables are written (reading from left to right) within the X matrix. This assumes that the equations have been aligned. The Y matrix contains the three variables. The Z matrix contains the coefficients to which each equation is equated. Multiplying both sides of Equation (12.39) by the inverse of matrix X produces

$$\underbrace{\begin{bmatrix} A \\ B \\ C \end{bmatrix}}_{Y} = \underbrace{\begin{bmatrix} 2 & -1 & 1 \\ -1 & -1 & 2 \\ 2 & 1 & -3 \end{bmatrix}^{-1}}_{X} \underbrace{\begin{bmatrix} 3 \\ 3 \\ -5 \end{bmatrix}}_{Z} \qquad (12.40)$$

First, to solve X^{-1} replace each term by its cofactor. For example, the cofactor of a_{11} is

$$\begin{bmatrix} -1 & 2 \\ 1 & -3 \end{bmatrix} = (-1)(-3) - (1)(2) = 1$$

If we similarly solve all terms up to and including a_{33} for their respective cofactors, we get the following new matrix when each cofactor is substituted for its respective a term:

$$\begin{bmatrix} 1 & -1 & 1 \\ 2 & -8 & 4 \\ -1 & 5 & -3 \end{bmatrix}$$

Second, in the solution of X^{-1}, precede each a_{ij} cofactor with a sign that a determinant cofactor would have according to its location in the array. This sign designation is redrawn as Figure 12.17. Following the application of rule 2, X^{-1} now takes the form

$$\begin{bmatrix} +(1) & -(-1) & +(1) \\ -(2) & +(-8) & -(4) \\ +(-1) & (5) & +(-3) \end{bmatrix} = \begin{bmatrix} 1 & 1 & 1 \\ -2 & -8 & -4 \\ -1 & -5 & -3 \end{bmatrix}$$

$$\begin{bmatrix} +(a_{11}) & -(a_{12}) & +(a_{13}) \\ -(a_{21}) & +(a_{22}) & -(a_{23}) \\ +(a_{31}) & -(a_{32}) & +(a_{33}) \end{bmatrix}$$

Figure 12.17   Cofactor sign designation.

Third, to solve the inverse matrix, transpose each row for each column. Row 1, which is now 1, 1, 1, becomes column 1 of the same coefficients. Similarly, rewrite row 2 as column 2 and row 3 as column 3. At this point X^{-1} now appears as

$$
\begin{array}{ccc}
\text{Formerly} & \text{Formerly} & \text{Formerly} \\
\text{row 1} & \text{row 2} & \text{row 3} \\
\downarrow & \downarrow & \downarrow
\end{array}
$$

$$
\begin{bmatrix}
1 & -2 & -1 \\
1 & -8 & -5 \\
1 & -4 & -3
\end{bmatrix}
\tag{12.41}
$$

Finally, multiply the matrix of Equation (12.41) by $1/D$. D is the solution of the determinant of the original matrix X: (See Eq. 12.39)

$$
D = (6 - 1 - 4) - (-2 + 4 - 3) = 2
$$

The final solution of X^{-1} is

$$
X^{-1} = \frac{1}{2}
\begin{bmatrix}
1 & -2 & -1 \\
1 & -8 & -5 \\
1 & -4 & -3
\end{bmatrix}
\tag{12.42}
$$

Substituting Equation (12.42) into Equation (12.40), the matrix array of the three unknown variables is

$$
\underbrace{\begin{bmatrix} A \\ B \\ C \end{bmatrix}}_{Y}
= \left(\frac{1}{2}\right)
\underbrace{\begin{bmatrix}
1 & -2 & -1 \\
1 & -8 & -5 \\
1 & -4 & -3
\end{bmatrix}}_{X^{-1}}
\underbrace{\begin{bmatrix} 3 \\ 3 \\ -5 \end{bmatrix}}_{Z}
$$

The evaluation of the first unknown (A) is

$A = 1/D$ times the sum of the products formed by multiplying each term of row 1 of the matrix portion of X^{-1} by each term of matrix Z.

Figure 12.18 illustrates the mechanics involved.

$$
A = \frac{1}{2}\Big((1)(3) + (-2)(3) + (-1)(-5) \Big) = 1
$$

In a similar fashion,

$$
B = \frac{1}{2}\Big((1)(3) + (-8)(3) + (-5)(-5) \Big) = 2
$$

$$
C = \frac{1}{2}\Big((1)(3) + (-4)(3) + (-3)(-5) \Big) = 3
$$

Again, this process is merely mechanically applying a set of rules to an array of numbers. The final resultant is the simultaneous solution of all unknowns.

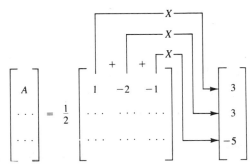

Figure 12.18   Determination of first matrix term.

12.6 EXAMPLES OF MATRIX SOLUTIONS

Examples 1 to 5 help clarify how to write and solve a matrix for simultaneous solutions of unknown quantities.

Example 1

Determine the inverse of the following matrix.

$$X = \begin{bmatrix} 3 & -1 & 7 \\ -2 & 6 & -3 \\ 4 & -5 & 1 \end{bmatrix}$$

Rule 1: Replace each a_{ij} term by its cofactor:

$$\begin{bmatrix} -9 & 10 & -14 \\ 34 & -25 & -11 \\ -39 & 5 & 16 \end{bmatrix}$$

Rule 2: Precede each new u_{ij} term by the sign configuration in Figure 12.17:

$$\begin{bmatrix} -9 & -10 & -14 \\ -34 & -25 & 11 \\ -39 & -5 & 16 \end{bmatrix}$$

Rule 3: Transpose the rows and columns:

$$\begin{bmatrix} -9 & -34 & -39 \\ -10 & -25 & -5 \\ -14 & 11 & 16 \end{bmatrix}$$

Rule 4: Multiply the new matrix by $1/D$:

$$D = \left((18 + 70 + 12) - (168 + 2 + 45)\right) = -115$$

$$\therefore X^{-1} = \frac{1}{-115} \begin{bmatrix} -9 & -34 & -39 \\ -10 & -25 & -5 \\ -14 & 11 & 16 \end{bmatrix}$$

Example 2

Solve the following simultaneous equations using matrix solutions.

$$3b - c = -(4 + 2a)$$
$$-3a - 13 = 2(b - c)$$
$$a - 3b - 3 = -3c$$

Simplify by expanding and aligning like unknowns:

$$2a + 3b - c = -4$$
$$-3a - 2b + 2c = 13$$
$$a - 3b + 3c = 3$$

$$\begin{bmatrix} a \\ b \\ c \end{bmatrix} = \begin{bmatrix} 2 & 3 & -1 \\ -3 & -2 & 2 \\ 1 & -3 & 3 \end{bmatrix}^{-1} \begin{bmatrix} -4 \\ 13 \\ 3 \end{bmatrix}$$

Rule 1:

$$\begin{bmatrix} a \\ b \\ c \end{bmatrix} = \begin{bmatrix} 0 & -11 & 11 \\ 6 & 7 & -9 \\ 4 & 1 & 5 \end{bmatrix} \begin{bmatrix} -4 \\ 13 \\ 3 \end{bmatrix}$$

Rule 2:

$$\begin{bmatrix} a \\ b \\ c \end{bmatrix} = \begin{bmatrix} 0 & 11 & 11 \\ -6 & 7 & 9 \\ 4 & -1 & 5 \end{bmatrix} \begin{bmatrix} -4 \\ 13 \\ 3 \end{bmatrix}$$

Rule 3:

$$\begin{bmatrix} a \\ b \\ c \end{bmatrix} = \begin{bmatrix} 0 & -6 & 4 \\ 11 & 7 & -1 \\ 11 & 9 & 5 \end{bmatrix} \begin{bmatrix} -4 \\ 13 \\ 3 \end{bmatrix}$$

Rule 4:

$$
\begin{bmatrix} a \\ b \\ c \end{bmatrix} = \frac{1}{22} \begin{bmatrix} 0 & -6 & 4 \\ 11 & 7 & -1 \\ 11 & 9 & 5 \end{bmatrix} \begin{bmatrix} -4 \\ 13 \\ 3 \end{bmatrix}
$$

The solution is

$$
\begin{bmatrix} a \\ b \\ c \end{bmatrix} = \frac{1}{22} \begin{bmatrix} (-78 & +12) \\ (-44 & +91 & -3) \\ (-44 & +117 & +15) \end{bmatrix} = \begin{bmatrix} -3 \\ 2 \\ 4 \end{bmatrix}
$$

$$
\therefore a = -3, \qquad b = 2, \qquad c = 4
$$

Example 3

Find the currents through each resistor in Figure 12.19 by using matrix analysis.

$$
7.18I_1 - I_2 + 2.2I_3 = 10
$$
$$
-I_1 + 5.8I_2 + 1.5I_3 = 15
$$
$$
2.2I_1 + 1.5I_2 + 8.4I_3 = 20
$$

$$
\begin{bmatrix} I_1 \\ I_2 \\ I_3 \end{bmatrix} = \begin{bmatrix} 7.18 & -1 & 2.2 \\ -1 & 5.8 & 1.5 \\ 2.2 & 1.5 & 8.4 \end{bmatrix}^{-1} \begin{bmatrix} 10 \\ 15 \\ 20 \end{bmatrix}
$$

Rules 1 and 2 combined:

$$
\begin{bmatrix} I_1 \\ I_2 \\ I_3 \end{bmatrix} = \begin{bmatrix} 46.5 & 11.7 & -14.3 \\ 11.7 & 55.5 & -13.0 \\ -14.3 & -13.0 & 40.6 \end{bmatrix} \begin{bmatrix} 10 \\ 15 \\ 20 \end{bmatrix}
$$

Rules 3 and 4 combined:

$$
\begin{bmatrix} I_1 \\ I_2 \\ I_3 \end{bmatrix} = \frac{1}{290.5} \begin{bmatrix} 46.5 & 11.7 & -14.3 \\ 11.7 & 55.5 & -13 \\ -14.3 & -13 & 40.6 \end{bmatrix} \begin{bmatrix} 10 \\ 15 \\ 20 \end{bmatrix}
$$

The solutions are

$$
\begin{bmatrix} I_1 \\ I_2 \\ I_3 \end{bmatrix} = \frac{5}{290.5} \begin{bmatrix} 46.5 & 11.7 & -14.3 \\ 11.7 & 55.5 & -13 \\ -14.3 & -13 & 40.6 \end{bmatrix} \begin{bmatrix} 2 \\ 3 \\ 4 \end{bmatrix} = \begin{bmatrix} 1.22 \\ 2.37 \\ 1.63 \end{bmatrix} \text{mA}
$$

$$
\therefore I_1 = 1.22 \text{ mA}, \qquad I_2 = 2.37 \text{ mA}, \qquad I_3 = 1.63 \text{ mA}
$$

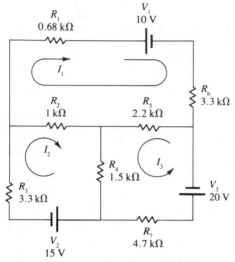

Figure 12.19

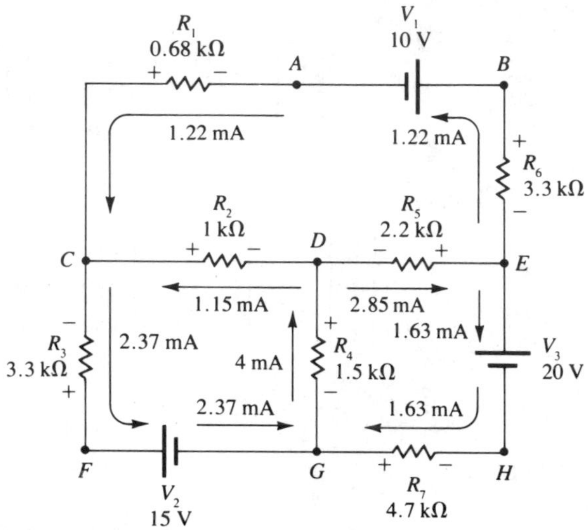

Figure 12.20   Current and voltage polarity distribution of Figure 12.19.

The assumptions for the current directions were correct because all currents resulted in positive values:

$$I_{R1} = I_1 = 1.22 \text{ mA}$$
$$I_{R2} = I_2 - I_1 = (2.37 - 1.22) \text{ mA} = 1.15 \text{ mA}$$
$$I_{R3} = I_2 = 2.37 \text{ mA}$$
$$I_{R4} = I_2 + I_3 = (2.37 + 1.63) \text{ mA} = 4 \text{ mA}$$
$$I_{R5} = I_1 + I_3 = (1.22 + 1.63) \text{ mA} = 2.85 \text{ mA}$$
$$I_{R6} = I_1 = 1.22 \text{ mA}$$
$$I_{R7} = I_3 = 1.63 \text{ mA}$$

The currents and voltage polarities in Figure 12.19 are summarized in Figure 12.20.

Example 4

Using the currents found in Example 3, find V_{AG} in Figure 12.20 by the following two paths:

$$V_{AC} + V_{CD} + V_{DG} \qquad \text{path 1}$$
$$V_{AB} + V_{BE} + V_{EB} + V_{BG} \qquad \text{path 2}$$

Verify that both paths (or any other path between these two points) result in the same answer. For path 1:

$$V_{AG} = (1.22 \text{ mA})(0.68 \text{ k}\Omega) - (1.15 \text{ mA})(1 \text{ k}\Omega) - (4 \text{ mA})(1.5 \text{ k}\Omega)$$
$$= 0.83 \text{ V} - 1.15 \text{ V} - 6 \text{ V} = -6.32 \text{ V}$$

For path 2:

$$V_{AG} = 10 - (1.22 \text{ mA})(3.3 \text{ k}\Omega) - 20 + (1.63 \text{ mA})(4.7 \text{ k}\Omega)$$
$$= 10 \text{ V} - 4.0 \text{ V} - 20 \text{ V} + 7.66 \text{ V} = -6.34 \text{ V}$$

Example 5

The current meter in Figure 12.21 has 100 Ω resistance. Using matrix analysis, determine the magnitude and direction of current flow through the meter and all resistors. The currents are assumed in the directions shown:

$$1.33I_1 - 0.33I_2 - 1.0I_3 = 3$$
$$-0.33I_1 + 1.11I_2 - 0.1I_3 = 0$$
$$-1.0I_1 - 0.1I_2 + 1.57I_3 = 0$$

$$\begin{bmatrix} I_1 \\ I_2 \\ I_3 \end{bmatrix} = \begin{bmatrix} 1.33 & -0.33 & -1.0 \\ -0.33 & 1.11 & -0.1 \\ -1.0 & -0.1 & 1.57 \end{bmatrix}^{-1} \begin{bmatrix} 3 \\ 0 \\ 0 \end{bmatrix}$$

Rules 1 and 2 combined:

$$\begin{bmatrix} I_1 \\ I_2 \\ I_3 \end{bmatrix} = \begin{bmatrix} 1.733 & 0.618 & 1.143 \\ 0.618 & 1.088 & 0.463 \\ 1.143 & 0.463 & 1.367 \end{bmatrix} \begin{bmatrix} 3 \\ 0 \\ 0 \end{bmatrix}$$

Rules 3 and 4 combined:

$$\begin{bmatrix} I_1 \\ I_2 \\ I_3 \end{bmatrix} = \frac{1}{0.958} \begin{bmatrix} 1.733 & 0.618 & 1.143 \\ 0.618 & 1.088 & 0.463 \\ 1.143 & 0.463 & 1.367 \end{bmatrix} \begin{bmatrix} 3 \\ 0 \\ 0 \end{bmatrix}$$

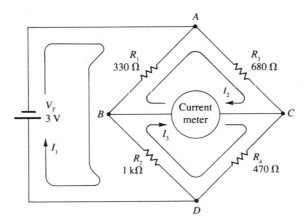

Figure 12.21

The solutions are

$$I_1 = \frac{(1.733)(3)}{0.958} = 5.43 \text{ mA}$$

$$I_2 = \frac{(0.618)(3)}{0.958} = 1.935 \text{ mA}$$

$$I_3 = \frac{(1.143)(3)}{0.958} = 3.579 \text{ mA}$$

From Figure 12.21:

$$I_{R1} = I_1 - I_2 = (5.43 - 1.935) \text{ mA} = 3.495 \text{ mA}$$
$$I_{R2} = I_1 - I_3 = (5.43 - 3.579) \text{ mA} = 1.851 \text{ mA}$$
$$I_{meter} = I_3 - I_2 = (3.579 - 1.935) \text{ mA} = 1.644 \text{ mA}$$
$$I_{R3} = I_2 = 1.935 \text{ mA}$$
$$I_{R4} = I_3 = 3.579 \text{ mA}$$

Figure 12.22 depicts the current distribution in Figure 12.21.

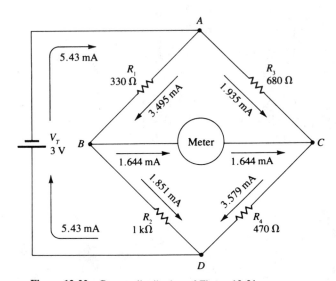

Figure 12.22   Current distribution of Figure 12.21.

PROBLEMS

Find the inverse matrix of the matrices in Problems 1 to 5. Use your calculator wherever appropriate.

1. $\begin{bmatrix} 2 & 4 & 1 \\ -1 & -6 & 5 \\ 3 & 7 & -9 \end{bmatrix}$ 3. $\begin{bmatrix} 1.2 & -5.6 & 7.4 \\ -6.7 & 1.8 & 3.9 \\ 3.4 & -9.2 & -2.7 \end{bmatrix}$ 5. $\begin{bmatrix} 3 & 2 & 11 & -8 \\ -7 & 8 & 19 & -4 \\ 14 & -5 & 7 & 3 \\ 6 & 9 & -1 & 17 \end{bmatrix}$

2. $\begin{bmatrix} 6 & 1 & 2 \\ -4 & 7 & -5 \\ 3 & -8 & 9 \end{bmatrix}$ 4. $\begin{bmatrix} 8.3 & -7.41 & 9.26 \\ -1.72 & 5.29 & -4.44 \\ -3.19 & 3.17 & 1.09 \end{bmatrix}$

Using the methods of matrix analysis and your calculator at $\boxed{\text{Fix 3}}$, solve for the unknowns in Problems 6 to 10.

6. $2a - 3b - 4c = -22$
    $-a + 7b + c = -20$
    $3a - b - 4c = -29$

7. $x + y + z = 0$
    $-2x + 3y + 4z = -40$
    $3x + 5y - 7z = 46$

8. $4s - t + v = -23$
    $-3s + 4t - 5v = 45$
    $2s - 2t + 4v = -32$

9. $3.1u + 2.6v - 4.7w = 6.8$
    $-1.8u + 2.9v + 1.6w = -7.3$
    $-5.1u - 3.7v - 6.8w = 14.1$

10. $2e + 4f - 5g + 4h = 16$
    $-3e - 5f + g - h = -2$
    $e + 2f - 3g + 4h = -17$
    $-4e - f + 7g - 5h = 10$

11. Using solutions by matrix, determine the currents in Figure 12.23.

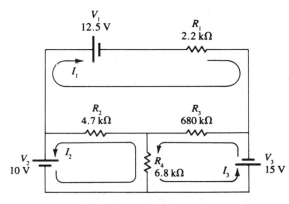

Figure 12.23

12. Using solutions by matrix, determine the currents in Figure 12.24.

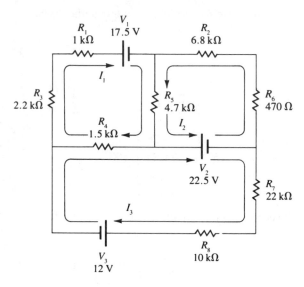

Figure 12.24

SUMMARY

The application of determinants and matrices involves mechanically manipulating number arrays, resulting in solutions for the variables of simultaneous equations. Determinants solve for one solution at a time; matrices solve all unknowns simultaneously. If you want only one variable solution, such as the current through one specific resistor, determinants are easier to apply than matrices. If you want all (or most) variable solutions, matrices represent the easier application, especially if computer assistance is available.

13 APPLICATIONS 2: NETWORK THEOREMS

13.1 INTRODUCTION

One of the best ways to study the mathematics relevant to electronics (and to gain an appreciation of its usefulness) is to periodically apply the concepts studied to practical circuit analysis. This chapter is the second of three dedicated to enhancing the learning process by practical application.

By now you have learned the basic algebraic principals, how to solve simultaneous equations, and the mechanical process for solving both determinant and matrix structures. This chapter applies the rules for these concepts to *network theorems*. There are many such theorems, most of which are more appropriately left to a textbook about electronics. Here we thoroughly analyze only three such theorems—*Millman's* theorem, the *ladder* theorem, and Δ–Y transformations. The three reasons for these specific selections are:

1. The resulting equations for each of the three theorems involve more than simple ratios, so they are ideal for calculator application.

2. Most textbooks about basic electronics cover more traditional theorems, such as Thevenin's, Norton's, maximum power, and so on. The theorems in this chapter are more apt to present new theory for the electronics student.

3. The derivation of the ladder theorem and the Δ–Y transformations are excellent illustrations of how to find solutions by cascading algebraic expressions to arrive at the desired results. As such, the application of previously learned mathematical concepts is vividly illustrated.

13.2 MILLMAN'S THEOREM

Millman's theorem, which determines voltages at a common point, is useful when applied to analog summing circuits. In a simple voltage divider, such as in Figure 13.1, the voltage at point A can be easily determined by simply applying the proportional voltage ratio law studied in Chapter 9 (Applications 1: DC Electronic Circuits). The specific numbers used to solve the problem depend upon which end of the network we choose as reference. Regardless of the choice, the correct answer is always obtained when the mathematics are correctly applied.

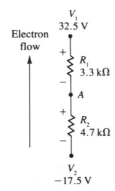

Figure 13.1   Simple voltage divider.

Mathematically, first list the reference point. Then add or subtract the proportionate resistor ratio of the voltage difference, depending upon whether the voltage potential is in a rising or falling direction.

First, choosing V_1, the 32.5-V source as reference,

$$V_A = V_1 - \frac{R_1}{R_1 + R_2}(V_1 - V_2)$$

$$= 32.5 - \frac{3.3}{3.3 + 4.7}(32.5 - (-17.5))$$

PRESS	DISPLAY	COMMENT
32.5 $\boxed{-}$ 3.3 $\boxed{\div}$ $\boxed{(\!(}$ 3.3 $\boxed{+}$ 4.7 $\boxed{)\!)}$ $\boxed{\times}$		
$\boxed{(\!(}$ 32.5 $\boxed{-}$ $\boxed{(\!(}$ 17.5 $\boxed{+/-}$ $\boxed{)\!)}$ $\boxed{)\!)}$ $\boxed{=}$	11.875	V_A using V_1 as reference

Second, choosing V_2, the −17.5-V source as reference,

$$V_A = V_2 + \frac{R_2}{R_1 + R_2}(V_1 - V_2)$$

$$= -17.5 + \frac{4.7}{3.3 + 4.7}(32.5 - (-17.5))$$

PRESS	DISPLAY	COMMENT
17.5 $\boxed{+/-}$ $\boxed{+}$ 4.7 $\boxed{\div}$ $\boxed{(\!(}$ 3.3 $\boxed{+}$ 4.7 $\boxed{)\!)}$		
$\boxed{\times}$ $\boxed{(\!(}$ 32.5 $\boxed{-}$ $\boxed{(\!(}$ 17.5 $\boxed{+/-}$ $\boxed{)\!)}$ $\boxed{)\!)}$ $\boxed{=}$	11.875	V_A using V_2 as reference

Note that as stated, the same answer is obtained regardless of which source voltage is selected as reference.

These solutions are not difficult because current can flow through only one path between the two sources of voltage. But suppose there are several paths for current, such as in Figure 13.2. It is in such types of applications that Millman's theorem is extremely useful:

The voltge at the common point where several branches meet is the ratio of the sum of the currents through each branch, calculated as though that were the only branch present, to the sum of the conductances of each branch, again calculated as though that were the only branch present.

Expressed mathematically,

$$V_A = \frac{\Sigma I}{\Sigma G} \tag{13.1}$$

Example 1

Determine the voltage at point A in Figure 13.2.
    If point A is temporarily considered zero volts (ground potential), then:

$$I_1 = \frac{V_1}{R_1} = \frac{12.75}{0.47} \text{ mA}$$

$$I_2 = \frac{V_2}{R_2} = \frac{-6.85}{0.68} \text{ mA}$$

$$I_3 = \frac{V_3}{R_3} = \frac{2.9}{0.33} \text{ mA}$$

Note that resistance is expressed in kΩ units, so each I is in mA. The numerator in Equation (13.1), when applied to Figure 13.2, is algebraically expressed as

$$\Sigma I = I_1 + I_2 + I_3 = \frac{V_1}{R_1} + \frac{V_2}{R_2} + \frac{V_3}{R_3}$$

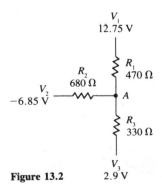

Figure 13.2

Each term of the denominator in Equation (13.1) can be expressed as

$$G_1 = \frac{1}{R_1} = \frac{1}{0.47} \text{ mS}$$

S stands for siemens, which is the preferred term for mho

$$G_2 = \frac{1}{R_2} = \frac{1}{0.68} \text{ mS}$$

$$G_3 = \frac{1}{R_3} = \frac{1}{0.33} \text{ mS}$$

Algebraically, when applied to Figure 13.2, the denominator in Equation (13.1) is expressed as follows:

$$\Sigma G = \frac{1}{R_1} + \frac{1}{R_2} + \frac{1}{R_3}$$

When both the numerator and denominator in Equation (13.1) are combined, we have the ratio of a *sum of ratios* to a *sum of reciprocals*. This is a breeze for a calculator, especially when milliunits in the numerator cancel milliunits in the denominator, leaving only the relative numbers to be entered into the calculator:

$$V_A = \frac{\dfrac{V_1}{R_1} + \dfrac{V_2}{R_2} + \dfrac{V_3}{R_3}}{\dfrac{1}{R_1} + \dfrac{1}{R_2} + \dfrac{1}{R_3}}$$

$$= \frac{\dfrac{12.75}{0.47} + \dfrac{-6.85}{0.68} + \dfrac{2.9}{0.33}}{\dfrac{1}{0.47} + \dfrac{1}{0.68} + \dfrac{1}{0.33}}$$

PRESS	DISPLAY	COMMENT
((12.75 [÷] .47 [+] 6.85 [+/−]		
[÷] .68 [+] 2.9 [÷] .33))	25.842	Numerator
[÷] ((.47 [1/x] [+] .68 [1/x] [+] .33 [1/x]))	6.629	Denominator
[=]	3.899	V_A

As a variation of Example 1, turn the problem around so that you have to determine what value of resistor is required to produce a given voltage at the common point. This requires algebraically manipulating the basic format of Millman's theorem to express the required resistance in terms of all the remaining variables.

Example 2

See Figure 13.3. What value of R_3 is necessary to produce 1.82 V at point A?
    Start with the generalized version of Millman's theorem:

$$V_A = \frac{\dfrac{V_1}{R_1} + \dfrac{V_2}{R_2} + \dfrac{V_3}{R_3}}{\dfrac{1}{R_1} + \dfrac{1}{R_2} + \dfrac{1}{R_3}} \qquad (13.2)$$

Solve R_3 by applying the basic rules for manipulating equations.

$$V_A\left(\frac{1}{R_1} + \frac{1}{R_2} + \frac{1}{R_3}\right) = \frac{V_1}{R_1} + \frac{V_2}{R_2} + \frac{V_3}{R_3}$$

Eliminate parentheses:

$$\frac{V_A}{R_1} + \frac{V_A}{R_2} + \frac{V_A}{R_3} = \frac{V_1}{R_1} + \frac{V_2}{R_2} + \frac{V_3}{R_3}$$

Multiply by R_3:

$$\frac{V_A R_3}{R_1} + \frac{V_A R_3}{R_2} + V_A = \frac{V_1 R_3}{R_1} + \frac{V_2 R_3}{R_2} + V_3$$

Transfer all R_3 terms to the left side of the equation, all non-R_3 terms to the right side of the equation:

$$\frac{V_A}{R_1}R_3 + \frac{V_A}{R_2}R_3 - \frac{V_1}{R_1}R_3 - \frac{V_2}{R_2}R_3 = V_3 - V_A$$

Factor out R_3 from the left-hand expression:

$$R_3\left(\frac{V_A}{R_1} + \frac{V_A}{R_2} - \frac{V_1}{R_1} - \frac{V_2}{R_2}\right) = V_3 - V_A$$

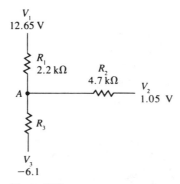

Figure 13.3

Divide the quantity in parentheses into the right-hand expression to uniquely solve for R_3:

$$R_3 = \frac{V_3 - V_A}{V_A\left(\dfrac{1}{R_1} + \dfrac{1}{R_2}\right) - \dfrac{V_1}{R_1} - \dfrac{V_2}{R_2}} \tag{13.3}$$

Substitute the appropriate numeric values for each variable:

$$R_3 = \frac{-6.1 - 1.82}{1.82\left(\dfrac{1}{2.2} + \dfrac{1}{4.7}\right) - \dfrac{12.65}{2.2} - \dfrac{1.05}{4.7}} \tag{13.3'}$$

Note that the answer is in kΩ.

PRESS	DISPLAY	COMMENT
$($ 6.1 $\boxed{+/-}$ $\boxed{-}$ 1.82 $)$	-7.920	Numerator of Equation (13.3')
$\boxed{\div}$ $($ 1.82 $\boxed{\times}$ $($ 2.2 $\boxed{1/x}$ $\boxed{+}$ 4.7 $\boxed{1/x}$		
$)$ $\boxed{-}$ 12.65 $\boxed{\div}$ 2.2 $\boxed{-}$ 1.05 $\boxed{\div}$ 4.7 $)$	-4.759	Denominator of Equation (13.3')
$\boxed{=}$ $\boxed{STO\ 1}$	1.664	Value of R_3 in kΩ; $\boxed{STO\ 1}$ [1]

Check the answer in Example 2 by substituting R_3 into Equation (13.2) and then verify 1.82 V at point A:

$$V_A = \frac{\dfrac{12.65}{2.2} + \dfrac{1.05}{4.7} + \dfrac{-6.1}{\boxed{RCL\ 1}}}{\dfrac{1}{2.2} + \dfrac{1}{4.7} + \dfrac{1}{\boxed{RCL\ 1}}} \tag{13.4}$$

---

[1] R_3 is accurately expressed to three decimal places. If you want to substitute R_3 into Equation (13.2) to see if the calculated solution really results in 1.82 V, store the entire value of R_3 (hence the optional $\boxed{STO}$ command).

PRESS	DISPLAY	COMMENT
[(12.65 [÷] 2.2 [+] 1.05 [÷] 4.7		
[+] 6.1 [+ / −] [÷] [RCL 1] [)]	2.308	Numerator of Equation (13.4)
[÷] [(2.2 [1/x] [+] 4.7 [1/x]		
[+] [RCL 1] [1/x] [)]	1.268	Denominator of Equation (13.4)
[=]	1.820	Voltage at point A verified for calculated value of R_3

Example 3

Use Millman's theorem to determine V_x in Figure 13.4. Then, knowing the total voltage across each resistor, calculate all currents. Notice how this method can eliminate simultaneous equations and reduce each current solution to a simple Ohm's law approach. Show that the sum of all currents at point X is zero.

$$V_x = \frac{\dfrac{11.75}{1.5} + \dfrac{-5.6}{4.7} + \dfrac{0}{6.8} + \dfrac{7.45}{0.33}}{\dfrac{1}{1.5} + \dfrac{1}{4.7} + \dfrac{1}{6.8} + \dfrac{1}{0.33}} = 7.202 \text{ V}$$

$$I_{R1} = \frac{V_1 - V_x}{R_1} = \frac{(11.75 - 7.202) \text{ V}}{1.5 \text{ k}\Omega} = 3.032 \text{ mA} \qquad \text{from point } x$$

$$I_{R2} = \frac{V_2 - V_x}{R_3} = \frac{(-5.6 - 7.202) \text{ V}}{4.7 \text{ k}\Omega} = -2.724 \text{ mA} \qquad \text{toward point } x$$

$$I_{R3} = \frac{\text{ground} - V_x}{R_3} = \frac{(0 - 7.202) \text{ V}}{6.8 \text{ k}\Omega} = -1.059 \text{ mA} \qquad \text{toward point } x$$

$$I_{R4} = \frac{V_4 - V_x}{R_4} = \frac{(7.45 - 7.202) \text{ V}}{0.33 \text{ k}\Omega} = 0.752 \text{ mA} \qquad \text{from point } x$$

$\Sigma I = (3.032 - 2.724 - 1.059 + 0.752) \text{ mA} = 0.001 \text{ mA} \approx 0 \text{ mA}$

Note that the minor answer of 0.001 mA is caused by the round-off of numbers. When carried to more decimal places, the final answer is truly zero.

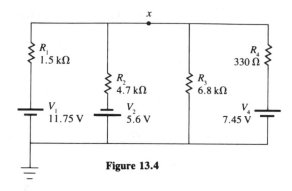

Figure 13.4

Example 4

In the circuit in Figure 13.5a find V_0 for the input shown. The source V varies between the two DC voltage levels $+5$ V and $+10$ V.

When the input is at $+5$ V:

$$V_0 = \frac{\dfrac{5}{3.3} + \dfrac{-5}{2.2} + \dfrac{0}{1.5}}{\dfrac{1}{3.3} + \dfrac{1}{2.2} + \dfrac{1}{1.5}}$$

$$= \frac{1.52 - 2.27}{0.30 + 0.45 + 0.67} = \frac{-0.75}{1.42} = -0.53 \text{ V}$$

When the input is at $+10$ V:

$$V_0 = \frac{\dfrac{10}{3.3} + \dfrac{-5}{2.2}}{1.42} = \frac{3.30 - 2.27}{1.42} = 0.53 \text{ V}$$

Figure 13.5b illustrates V_0.

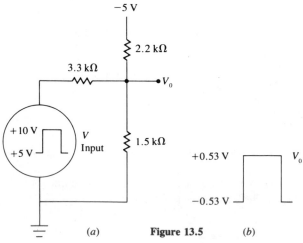

(a)   **Figure 13.5**   (b)

PROBLEMS

1. Find the voltage of point x in Figure 13.6 by the method of proportionate voltage. Apply Millman's theorem to verify that the same voltage is obtained.

2. Using Millman's theorem, find the voltage at point x in Figure 13.7.

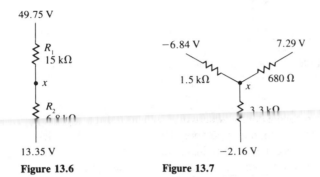

Figure 13.6   Figure 13.7

3. What should V be in Figure 13.8 so that point x has 0–V potential?

4. Repeat Problem 3, but solve for point x at -1.15 V.

5. Determine the value of R in Figure 13.9 so that point x has a 7.223-V potential.

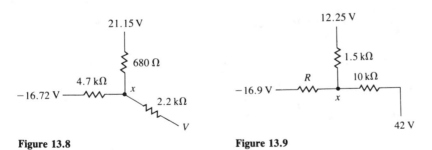

Figure 13.8   Figure 13.9

13.3 Δ–Y AND Y–Δ TRANSFORMATIONS

Recall that one method for solving simultaneous equations was to add or subtract equations to eliminate the unknown variables. This principle has excellent application in the development of the following electronic equations. Observe Figure 13.10, which is the classical bridge circuit. Suppose we want to determine the total generator current I_g. In a circuit of this nature it is not possible to define explicitly those elements in series or parallel, so we must resort to a series of loop equations. This approach can be tedious and easily subject to error.

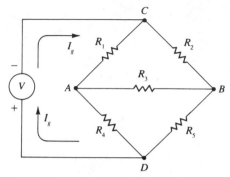

Figure 13.10 Classical bridge configuration.

This section describes how to reduce a complex network, such as in Figure 13.10, to a simple series-parallel network and then how to apply Ohm's law to find the desired solution. If we reduce the section between points ABD or ABC (Δ configurations) to an equivalent Y configuration, we can easily solve the resulting series-parallel network, assuming that the values of the substitute resistors can be so calculated that they are equivalent to the original circuit.

To approach the methodology of such conversions, consider the Δ and Y configurations in Figure 13.11. For the two circuits to be equivalent, the resistances measured between AD, AB, and BD of the Δ configuration should equal the resistances measured between AD, AB, and BD, respectively, of the Y configuration.

$$R_{\Delta AD} = \frac{R_A(R_B + R_C)}{R_A + R_B + R_C} = R_{YAD} = R_1 + R_3 \tag{13.5}$$

$$R_{\Delta AB} = \frac{R_B(R_A + R_C)}{R_A + R_B + R_C} = R_{YAB} = R_1 + R_2 \tag{13.6}$$

$$R_{\Delta AB} = \frac{R_C(R_A + R_B)}{R_A + R_B + R_C} = R_{YBD} = R_2 + R_3 \tag{13.7}$$

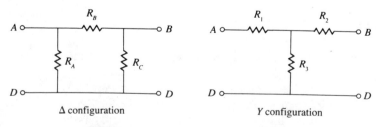

Figure 13.11 Isolated Δ and Y configurations.

Adding Equations (13.5) and (13.6) and subtracting Equation (13.7) produces the following interesting result:

$$\frac{R_A R_B + R_A R_C + R_A R_B + R_B R_C - R_A R_C - R_B R_C}{R_A + R_B + R_C} = R_1 + R_3 + R_1 + R_2 - R_2 - R_3$$

After we cancel the sum and difference of like terms, we get the following equations:

$$R_1 = \frac{R_A R_B}{R_A + R_B + R_C} \tag{13.8}$$

Similarly,
$$R_2 = \frac{R_B R_C}{R_A + R_B + R_C} \tag{13.9}$$

$$R_3 = \frac{R_A R_C}{R_A + R_B + R_C} \tag{13.10}$$

Defining $\Delta R = R_A + R_B + R_C$, then

$$R_1 = \frac{R_A R_B}{\Delta R} \tag{13.11}$$

$$R_2 = \frac{R_B R_C}{\Delta R} \tag{13.12}$$

$$R_3 = \frac{R_A R_C}{\Delta R} \tag{13.13}$$

If we use Figure 13.12 to relate the resistors of a Δ configuration to the resistors of a Y configuration, the following verbal description describes these transformations:

Each resistor of the Y configuration that is replacing the Δ configuration equals the product of the Δ resistors on each side of the Y resistor, divided by ΔR, which is the sum of the three Δ resistors.

To convert a Y configuration into a Δ configuration, use the following quotients:

$$\frac{\text{Equation (13.8)}}{\text{Equation (13.9)}} - \frac{R_1}{R_2} = \frac{R_A}{R_C}$$

In a similar manner,

$$\frac{\text{Equation (13.8)}}{\text{Equation (13.10)}} = \frac{R_1}{R_3} = \frac{R_B}{R_C}$$

and

$$\frac{\text{Equation (13.9)}}{\text{Equation (13.10)}} = \frac{R_2}{R_3} = \frac{R_B}{R_A}$$

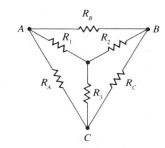

Figure 13.12 Δ-Y resistors connected between common points.

Dividing both numerator and denominator of Equation (13.8) by R_B results in

$$R_1 = \frac{R_A}{\dfrac{R_A}{R_B} + 1 + \dfrac{R_C}{R_B}}$$

Solving for R_A:

$$R_A = R_1\left(\frac{R_3}{R_2} + 1 + \frac{R_3}{R_1}\right)$$

$$= \frac{R_1R_2 + R_1R_3 + R_2R_3}{R_2} \qquad (13.14)$$

If we define $YR = R_1R_2 + R_1R_3 + R_2R_3$, then we can simplify Equation (13.14):

$$R_A = \frac{YR}{R_2} \qquad (13.15)$$

Similarly,

$$R_B = \frac{YR}{R_3} \qquad (13.16)$$

and

$$R_C = \frac{YR}{R_1} \qquad (13.17)$$

If we use Figure 13.12 to relate the elements of a Y configuration to a Δ configuration, the following verbal description describes these transformations:

Each Δ resistor is equal to YR divided by the opposite Y resistor, where YR equals the sum of all the products of two Y resistors.

Example 1

Convert the Δ configuration in Figure 13.13 to a Y configuration:

$$\Delta R = 2 + 3 + 5 = 10 \ \Omega$$

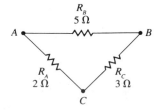

Figure 13.13

Use Figure 13.12 as a reference. After you apply Equations (13.11) to (13.13), the following results:

$$R_1 = \frac{(2)(5)}{10} = 1\ \Omega$$

$$R_2 = \frac{(5)(3)}{10} = 1.5\ \Omega$$

$$R_3 = \frac{(2)(3)}{10} = 0.6\ \Omega$$

Example 2

Convert the Y configuration in Figure 13.14 to a Δ configuration:

$$YR = (5)(10) + (10)(20) + (20)(5)$$
$$= 50 + 200 + 100$$
$$= 350$$

Use Figure 13.12 as a reference. After you apply Equations (13.15) to (13.17), the following results:

$$R_A = \frac{350}{5} = 70\ \Omega$$

$$R_B = \frac{350}{20} = 17.5\ \Omega$$

$$R_C = \frac{350}{10} = 35\ \Omega$$

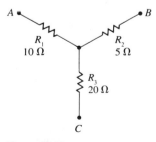

Figure 13.14

Example 3

Determine the total generator current in Figure 13.15 by using transformations. $\triangle ABC$ in Figure 13.15, when converted into a Y configuration, results in Figure 13.16:

$$\Delta R = (6.8 + 15 + 10) \text{ k}\Omega = 31.8 \text{ k}\Omega$$

$$R_1 = \frac{(6.8)(15)}{31.8} = 3.21 \text{ k}\Omega$$

$$R_2 = \frac{(6.8)(10)}{31.8} = 2.14 \text{ k}\Omega$$

$$R_3 = \frac{(10)(15)}{31.8} = 4.72 \text{ k}\Omega$$

$$R_2 + R_4 = (2.14 + 3.3) \text{ k}\Omega = 5.44 \text{ k}\Omega \equiv R_6$$

$$R_3 + R_5 = (4.72 + 47) \text{ k}\Omega = 51.72 \text{ k}\Omega \equiv R_7$$

$$R_6 \| R_7 = \frac{(5.44)(51.72)}{57.16} = 4.92 \text{ k}\Omega \equiv R_8$$

$$R_1 + R_8 = (3.21 + 4.92) \text{ k}\Omega = 8.13 \text{ k}\Omega \equiv R_9$$

$$I_g = \frac{V}{R_9} = \frac{27.5 \text{ V}}{8.13 \text{ k}\Omega} = 3.38 \text{ mA}$$

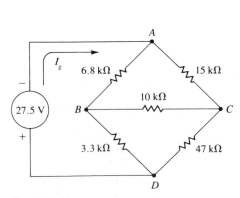

Figure 13.15

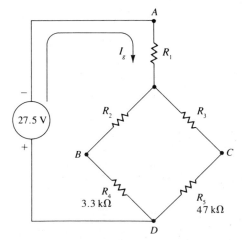

Figure 13.16 $\triangle ABC$ of Figure 13.7 transformed into a Y configuration.

PROBLEMS

1. Convert the Δ configuration in Figure 13.17 into a Y configuration.

2. Assume each resistor of a Δ configuration has the value of R. What is the equivalent Y configuration?

3. Convert the Y configuration in Figure 13.18 into a Δ configuration.

4. Assume each resistor of a Y configuration has the value of R. What is the equivalent Δ configuration?

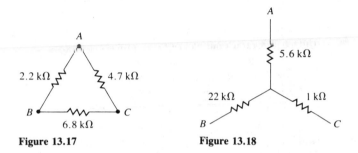

Figure 13.17   Figure 13.18

5. See Figure 13.19. Assume each edge of a cube is a resistor R. Using Δ–Y transformations where appropriate, calculate the total resistance between points A and B (across the major diagonal).

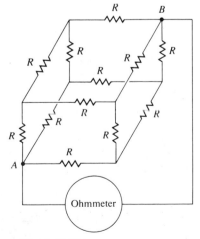

Figure 13.19

13.4 LADDER THEOREM

Earlier examples of solving simultaneous equations used the principle of successive substitution. For example, consider three simultaneous equations with the three unknown variables x, y, and z. After manipulation, we are ultimately left with one equation (let us call it Equation I) containing x, y, and z; one equation (call it Equation II) containing y and z; and one equation (Equation III) containing only z. We can now easily solve Equation III for the variable z, and then substitute this value of z into Equation II, permitting y to be solved. We can then substitute both calculated values of y and z into Equation I, permitting the calculation of x. This principle of successively substituting previously calculated algebraic values, when working toward the completion of a problem, is aptly illustrated by developing the mathematics of the ladder theorem.

Some circuitry in communication electronics (such as television circuits) or in digital electronics (such as computer delay lines and analog-to-digital converters) exhibit a geometric structure resembling a ladder. For example, Figure 13.20 represents part of the vertical integration section of a television circuit and Figure 13.21 represents a delay line used in some computer memory circuits. Both circuits contain reactive components, a mathematical analysis of which requires knowledge of the j operator. As stated several times in earlier chapters, complex numbers are covered in Chapter 16. For now we study only a resistive ladder network (such as in Figure 13.22). Following Chapter 16, the principles established here can be transferred to circuits such as those illustrated in Figures 13.20 and 13.21.

Consider Figure 13.23, which has four sections; in general, the value of each resistor is $R\,\Omega$. The approach is to label the current through each resistor, the voltage across each resistor, and the voltage at the junction point between each resistor. Computation of all labeled unknowns proceeds as follows.

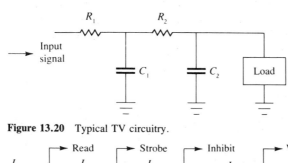

Figure 13.20 Typical TV circuitry.

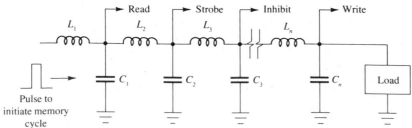

Figure 13.21 Typical computer delay line.

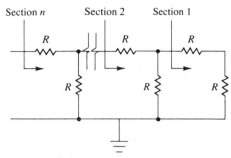

Figure 13.22 Resistive ladder network.

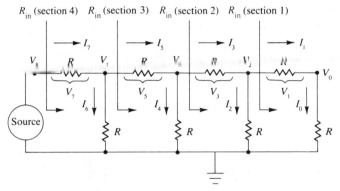

Figure 13.23 Resistive ladder network labeled for analysis.

Assume $V_0 = 1$ V. This may seem a strange approach because in all probability the odds are highly against V_0 being exactly 1 V in any given circuit. This really does not matter; what we want to find is the ratio of V_{in} and I_{in} that produces this 1 V. Because voltage, current, and resistance are proportionate in DC resistive circuits, any difference between the actual value of V_0 and the assumed 1 V is accompanied by a corresponding change in all currents and voltages. This ratio remains a constant; hence R_{in} (as seen by any section) can be evaluated by starting with this simple assumption. From Figure 13.23:

$$V_0 = 1 \text{ V}$$

$$I_0 = \frac{V_0}{R} = \frac{1}{R}$$

$$\text{sequence } 1 \left\{ \begin{array}{l} I_1 = I_0 = \dfrac{1}{R} \\[2ex] V_1 = (I_1)(R) = \left(\dfrac{1}{R}\right)(R) = 1 \\[2ex] V_2 = V_1 + V_0 = 2 \\[2ex] I_2 = \dfrac{V_2}{R} = \dfrac{2}{R} \end{array} \right\} \quad R_{in1} = \frac{V_2}{I_1} = \frac{2}{\dfrac{1}{R}} = \frac{2}{1}R$$

Kirchhoff's current law states that $I_3 = I_2 + I_1$. We calculated I_1 and I_2 in generalized terms in sequence 1, so now we can substitute these values into sequence 2 to begin the determination of R_{in2}:

sequence 2
$$\begin{cases} I_3 = I_2 + I_1 = \dfrac{2}{R} + \dfrac{1}{R} = \dfrac{3}{R} \\[2mm] V_3 = (I_3)(R) = \left(\dfrac{3}{R}\right)(R) = 3 \\[2mm] V_4 = V_3 + V_2 = 3 + 2 = 5 \\[2mm] I_4 = \dfrac{V_4}{R} = \dfrac{5}{R} \end{cases} \quad R_{in2} = \dfrac{V_4}{I_3} = \dfrac{5}{\dfrac{3}{R}} = \dfrac{5}{3}R$$

Kirchhoff's current law states that $I_5 = I_4 + I_3$. We calculated I_3 and I_4 in generalized terms in sequence 2, so now we can substitute these values into sequence 3 to begin the determination of R_{in3}:

sequence 3
$$\begin{cases} I_5 = I_4 + I_3 = \dfrac{5}{R} + \dfrac{3}{R} = \dfrac{8}{R} \\[2mm] V_5 = (I_5)(R) = \left(\dfrac{8}{R}\right)(R) = 8 \\[2mm] V_6 = V_5 + V_4 = 8 + 5 = 13 \\[2mm] I_6 = \dfrac{V_6}{R} = \dfrac{13}{R} \end{cases} \quad R_{in3} = \dfrac{V_6}{I_5} = \dfrac{13}{\dfrac{8}{R}} = \dfrac{13}{8}R$$

Kirchhoff's current law states that $I_7 = I_6 + I_5$. We calculated I_5 and I_6 in generalized terms in sequence 3, so now we can substitute these values into sequence 4 to begin the determination of R_{in4}:

sequence 4
$$\begin{cases} I_7 = I_5 + I_6 = \dfrac{8}{R} + \dfrac{13}{R} = \dfrac{21}{R} \\[2mm] V_7 = (I_7)(R) = \left(\dfrac{21}{R}\right)(R) = 21 \\[2mm] V_8 = V_7 + V_6 = 21 + 13 = 34 \\[2mm] (I_8 \text{ if there are additional sections}) \end{cases} \quad R_{in4} = \dfrac{V_8}{I_7} = \dfrac{34}{\dfrac{21}{R}} = \dfrac{34}{21}R$$

Again notice how previously calculated values are substituted into pending equations to continue the mathematical sequence.

You may have already observed an interesting and somewhat unique geometric progression to the total input resistance that is seen by a source, if looking into any given number of sections of a ladder-type resistive network. Extracting the multiplying factor of R, the coefficients for any given number of sections can be extrapolated as follows. Write the first two whole integers:

$$1, 2 \qquad \text{pair 1}$$

Now continue to expand these integers by successively adding digits in pairs:

$$1 + 2 = 3$$
$$2 + 3 = 5$$

therefore, the next pair is

$$3, 5 \qquad \text{pair 2}$$

Continuing,

$$3 + 5 = 8$$
$$5 + 8 = 13$$

therefore, the next pair is

$$8, 13 \qquad \text{pair 3}$$

Continuing,

$$8 + 13 = 21$$
$$13 + 21 = 34$$

therefore, the next pair is

$$21, 34 \qquad \text{pair 4}$$

If we consider the first digit of each pair the denominator of a fraction and the second digit the numerator, the following fractions result for the four pairs:

$$\frac{2}{1} \qquad \frac{5}{3} \qquad \frac{13}{8} \qquad \frac{34}{21}$$

Note that the four fractions just happen to be the multiplying factor of R for each of the four sections of the resistive ladder network.

It is not surprising that such a unique geometric progression exists, because a ladder network possesses symmetry. In the following example, the uniqueness of any symmetrical circuit is further expanded.

Example 1

Suppose the resistive network of Figure 13.23 has five sections. If each resistor is 47 kΩ, what is the total input resistance of the ladder network?

Section 4 resulted in a number pair of

$$21, 34 \qquad \text{pair 4}$$

Continuing the pairing,

$$21 + 34 = 55$$
$$34 + 55 = 89$$

therefore, R_{in} for five sections is

$$\frac{89}{55} \times (R) = \frac{89}{55}(47 \text{ k}\Omega)$$
$$= 76.05 \text{ k}\Omega$$

PROBLEMS

1. If R in Figure 13.24 is 22 kΩ, what is R_{in} for three sections of a ladder-type resistive network? What is R_{in} for seven sections?

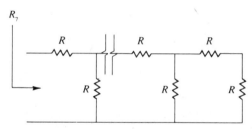

Figure 13.24

2. Determine the multiplying coefficients of R_{in} for each section in Figure 13.25.

3. Using the methods described in the text, determine the input resistance of the bridged T network in Figure 13.26 if $R = 47$ kΩ.

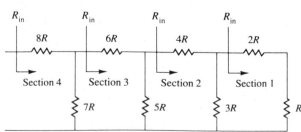

Figure 13.25

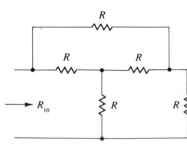

Figure 13.26

SUMMARY

Basic mathematics are not restricted to simply solving analytical problems; the principles studied for manipulating equations are also used in the derivation of electronic network theorems. Although there are many such theorems, a complete analysis is more appropriate for an electronics textbook. The three theorems analyzed in this chapter exemplify the application of the mathematical principles studied to date.

Apply Millman's theorem when determining the voltage at a common point where several branches connect. The voltage is easy to calculate because the theorem uses a summation of both the reciprocal functions and simple Ohm's law functions.

$\Delta-Y$ conversions are derived from the principles of adding and subtracting equations. These conversions share equations that make respective substitutions simple. Frequently these substitutions render a circuit more applicable for analysis.

The ladder theorem uses the concepts of successive algebraic substitutions to analyze a circuit. Basically, you can determine the input impedance of n number of sections of a symmetrical electronic pattern by assuming an output voltage and then working backward toward the source to determine the values that would produce the assumed result.

REVIEW PROBLEMS, PART II

The following problems are a review of Part II of the text and directly relate to the numbered sections.

Section 5.1

Evaluate the following expressions.

1. $18.5 \, \epsilon^{-t/0.35 \times 10^{-4}}$ where $\epsilon = 2.718$

$t = 5.2 \times 10^{-5}$

2. $\dfrac{1}{2\pi RC}$ where $R = 1.5 \text{ k}\Omega$

$C = 0.02 \, \mu\text{F}$

3. $\dfrac{R_1 R_2}{R_1 + R_2}$ where $R_1 = 250 \, \Omega$

$R_2 = 500 \, \Omega$

4. $\left(\dfrac{R_2}{R_1 + R_2}\right) i$ where $R_1 = 150 \, \Omega$

$R_2 = 50 \, \Omega$

$i = 20 \text{ mA}$

5. $12(1 - \epsilon^{-t/RC})$ where $\epsilon = 2.718$

$R = 3.2 \text{ k}\Omega$

$C = 2 \, \mu\text{F}$

$t = 4.5 \times 10^{-3}$

Section 5.2

Add.

6. $(12a - 15b) + (-2a - 6b)$

7. $(-3.6r + 3.9s) + (5.9r - 10.6s)$

8. $\left(2\dfrac{1}{3}x + 5\dfrac{1}{4}y\right) + \left(3\dfrac{1}{2}x - 6\dfrac{1}{3}y\right)$

9. $\left(\dfrac{1}{5}x + \dfrac{1}{4}y - \dfrac{1}{3}z\right) + \left(\dfrac{4}{15}x - \dfrac{5}{12}y - z\right)$

10. $4.5c + 15d - 4.1e$
$- \ 7d + 6.7e$
$\underline{-9.6c + \ \ 5d - 3.8e}$

11. $-7e + 15d + 16f - 5e + 7f - 4d - 17g$

Section 5.3

Simplify.

12. $(3.2x - 4.1y) - (3.5x + 5.9y)$

13. $3a + (2.5a - (-3a + 4.1a) - 5a) + a$

14. $-(2.1s - 3.4t) - (-3.1s + 9.2t) - t + 5.1s$

15. $-3.1u - (8.2v + 4.7w - (4u - 2.1v + 3w))$

16. $-(-x + y - (-x + y)) - (x - (-x + y))$

17. $\dfrac{3}{5}r + \left(5\dfrac{1}{4}s - \dfrac{3}{5}t\right) - (3t - (4.1s - 5r))$

Section 5.4

Simplify, and then use your calculator at $\boxed{\text{Fix 3}}$ to verify Review Problems 18 to 23; let 2.64 be the first variable $\boxed{\text{STO 1}}$, 1.29 the second variable $\boxed{\text{STO 2}}$, and -1.89 the third variable if used $\boxed{\text{STO 3}}$.

18. $5.2a - 2.1b - 1.6c + 2.5a$

19. $(3.21x - 2.98y) - (-5.8x + 2.3y)$

20. $\left(\dfrac{1}{5}x - \dfrac{2}{3}y + \dfrac{7}{30}z\right) + \left(\dfrac{2}{15}x - \dfrac{1}{6}y - \dfrac{3}{5}z\right)$

21. $\left(2\dfrac{1}{4}r + 5\dfrac{1}{9}s\right) + \left(4\dfrac{1}{6}r - 4\dfrac{1}{3}s\right)$

22. $-(4.1u + 7.1v - (3.2u - 5.3v))$

23. $4\dfrac{1}{2}x - \left(3\dfrac{1}{4}y - 6\dfrac{1}{3}z + 5\dfrac{1}{4}x - (y - 2.1z)\right)$

Section 5.6

Perform the indicated divisions.

24. $\dfrac{24x^3y - 16x^2y^2 - 64xy^3}{-8xy}$

25. $\dfrac{18a^4 + 36a^3b - 24a^2b^2}{6a^{-2}}$

26. $\dfrac{r^2s - r^2s^2 - 6rs^2}{-rs}$

27. $\dfrac{3u^2v - uv^3}{6u^{-2}v^{-3}}$

28. $\dfrac{5.43x^{-4}y^1 - 4.2xy^{-2} + 3.1xy^5}{2.4x^5y^{-3}}$

29. $\dfrac{k^3l^{-3}m^{-1}}{k^{-2}l^2}$

30. $\dfrac{3\frac{1}{2}a^{-2}bc - 2\frac{1}{2}bc^2 + ac}{3\frac{1}{2}a^2bc^{-1}}$

Section 5.7

First simplify Review Problems 31 to 36, and then check with your calculator at $\boxed{\text{Fix 3}}$. Let 2.31 be the first variable $\boxed{\text{STO 1}}$, 1.98 the second variable $\boxed{\text{STO 2}}$, and 5.21 the third variable if used $\boxed{\text{STO 3}}$.

31. $\dfrac{xy^3z}{x^{-1}z^2}$

32. $\dfrac{r^{-2}st^4}{rs^{-3}t^0}$

33. $\dfrac{3.1a^2b + 3.4a^{1/2}b^{-1}}{6.5a^{3/2}b^3}$

34. $\dfrac{xy^2z^{-5} - x^4y^{-1}z^0 + 4xy^2z}{xy^{-1}z^2}$

35. $\dfrac{2\frac{1}{2}u^2v^{1/2} - 3\frac{1}{5}u^4v^{1/4} + 5uv}{2\frac{1}{2}u^2v^{-1/2}}$

36. $\dfrac{rs^{-2}t^3}{4.5}(2.5r^2st^{-1} - 4.1r^{-2}s^3t)$

Section 6.1

Perform the indicated operations.

37. $(3.1\ ab^2c^{-3})^2$

38. $\left(\dfrac{2.8x}{yz^3}\right)^2$

39. $\left(\dfrac{3.2\ l^2m^{-2}}{m^2n^{-3}}\right)^3$

40. $(2.4\ r^{3/2}s^{-5/4})^2$

41. $\left(\dfrac{64a^6b^{-9}}{c^{-3}}\right)^{1/3}$

42. $-\sqrt{144a^4b^{-2}}$

43. $\sqrt[3]{\dfrac{25.2\ i^7j^7}{ij^{-2}}}$

44. $\left(-\dfrac{16x^5}{y^5}\right)^{1/3}$

Using your calculator, evaluate the following.

45. $\sqrt{164} \cdot \sqrt[3]{984}$

46. $\sqrt{9.8^{2.3}} + \sqrt[3]{109}$

47. $\pi^{2/3} + (\sqrt{152})^{0.3}$

48. $\dfrac{\sqrt[3]{15.2^2}}{(\sqrt{25.2})^{1/3}}$

49. $\left(\dfrac{(3.96)^{1/5}(\sqrt{210})}{29^2}\right)^2$

Section 6.2

Simplify by combining powers or removing radicals. Do not use the root key on your calculator.

50. $a^4 b^0$

51. $\dfrac{x^{-1/2}}{y^{-1/2}}$

52. $\dfrac{r^7 s^{14}}{t^{-21}}$

53. $\sqrt{48}$

54. $\sqrt[3]{-27}$

55. $\sqrt{20} + \sqrt{45}$

56. $2\sqrt{175} + \sqrt{63}$

57. $\sqrt{8} + \sqrt{18}$

58. $\dfrac{\sqrt{60}}{\sqrt{45}}$

59. $(3 + \sqrt{2})(-2\sqrt{2})$

60. $(\sqrt{72x^2})(\sqrt{200y^4z}$

61. $r^2\sqrt{21r^6 s^5 t^8}$

62. $2\sqrt{3}(\sqrt{48} - 3\sqrt{75})$

63. $\dfrac{2 + \sqrt{5}}{\sqrt{5}}$

64. $\dfrac{2\sqrt{3} - 1}{\sqrt{3}}$

65. $\dfrac{\sqrt{a^5}}{\sqrt{a^3 b^2 c}}$

Section 6.3

Perform the indicated multiplications.

66. $3x - 4y$
  $\underline{\times\ 5x - 2y}$

67. $2.1a^2 b - 3.1a$
  $\underline{\times\ -5.3ab\ \ + 4.3b}$

68. $(3r - 2s)(5r - 4s)$

69. $(x^2 + 5y^3)(2x^2 - 3y^3)$

70. $(2.1u^3 - 5v)(3u^3 - 6.2v)$

71. $(5x^2 y - 9z)(3x^2 y + 17z)$

72. $(5.9a^2 b - 3.1c^3)(-2.1a^2 b + 6.8c^3)$

73. $(7x - 9y)(4z - 2y)$

Section 6.4

Find the products.

74. $(x - 2)(x + 2)$

75. $(2a - 4b)(2a + 4b)$

76. $(x^2 + 4)(x^2 + 1)$

77. $\left(\frac{1}{2}y + 3\right)(y - 2)$

78. $(a - 5)^2$

79. $(x^3 - 4)^3$

80. $(5x - y)(-4x + 3y)$

81. $(6r - 7s)^2$

82. $\left(2.9e - \frac{1.6}{f}\right)^3$

83. $\left(\frac{3a^2}{b} + 5c^{-1}\right)\left(\frac{2a^2}{b} - c^{-1}\right)$

84. $\left(\frac{\sqrt{2}a}{b} - \sqrt{6.4b}\right)^2$

85. $(0.3x + 1)(0.3x - 1)$

86. $(5 - 2x)^3$

87. $(3a - 2b)(4b - 2a)$

88. $(2x^3 - 3)(x^3 - 3x - 8)$

89. $(3x + 2y - z)^2$

90. $(a - b)(a^2 + 2ab + b^2)$

91. $(6xy^2 - 2z^3)^3$

92. $\left(\frac{5r}{s} + t\right)\left(\frac{5r}{s} - t\right)$

93. $(20x - 9)^2$

94. $(3.9c^2 + 4.6d^3)^2$

95. $(3ab^2 - ac + bc^2)^2$

96. $(3.2x^2 - y^{-3}z^2)(3.2x^2 + y^{-3}z^2)$

97. $(3t^2 - 2)(4t^2 + 7)$

98. $\left(a + \frac{1}{2}\right)\left(a - \frac{1}{2}\right)$

99. $3x(2x - 5)(3x + 2)$

100. $(2y + 1)(y + 6)$

101. $(3r - 2s + 6t)^2$

102. $(m - 15)(m + 2)$

103. $a(3a + b)(4a - 2b)$

104. $(4a - 3b)(3a + b)$

105. $(a + 5)(a - 5)(a^2 + 25)$

Section 6.5

Factor Problems 106 to 145.

106. $3a + 3b$

107. $2my - 8m^2y$

108. $ax^2 + 3ax^3$

109. $32x + x^2$

110. $16.2x^2 + 16.2y^2$

111. $x^2 - 49$

112. $a^2 - 8a + 16$

113. $\pi r^2 + \pi R^2$

114. $x^3 - 4x$

115. $2x^2 + 8x + 2$

116. $9z^2 + 42z + 49$

117. $3x^2 - 8x + 4$

118. $\dfrac{1}{2}hb + \dfrac{1}{2}hc$

119. $a^2 + 10a + 25$

120. $(x - 2)^2 - (x + 1)^2$

121. $4c^2 - 12c + 9$

122. $15x^3y^3z^3 - 5xyz$

123. $a^4 - 625$

124. $2y^3 - 54$

125. $y^2 + 10y + 9$

126. $5y^2 - 30y + 45$

127. $8a^4b^2c^3 + 12a^2b^2c^2$

128. $16 + 17d + d^2$

129. $s^2 - \dfrac{1}{100}$

130. $-x^2 + 36x - 324$

131. $a^4 - 16$

132. $r^2 + 12r + 11$

133. $18t^2 - 23t - 6$

134. $28m^4n^3 - 70m^2n^4$

135. $3x^2 + 10x + 3$

136. $9x^2 + 30x + 25$

137. $r^2s^2 - 144$

138. $9xy^2 - x$

139. $2a^2 + 11a + 12$

140. $a(x + y) + b(x + y)$

141. $a^3 + 7a^2 + 10a$

142. $av - aw + bv - bw$

143. $x^2 - 0.64$

144. $2ax^2 - 2ax - 12a$

145. $3u^2 - 7uv + 2v^2$

Section 6.6

Divide.

146. $(6x^3 + 11x^2 - 4x - 4) \div (3x - 2)$

147. $\dfrac{-15a^3 + 30a^2 + 5}{5a^3}$

148. $(y^2 + 7y + 52) \div (y + 4)$

149. $(8z^3 - 22z^2 - 5z + 12) \div (4z + 3)$

150. $(x^3 + y^3) \div (x + y)$

151. $(3x^3 + 9x - 4) \div (3x + 3)$

152. $(3c^3 + 14c^2 + 4c - 4) \div (c + 4)$

153. $(10x^2 - 5y^2 + 38xy) \div (2x + 8y)$

154. $(2a^4 - 9a^3 + 6a + 6 - 14a^2) \div (2 - 3a - a^2)$

Section 7.1

Reduce the following fractions to their lowest common terms.

155. $\dfrac{36x^4y^2}{48xy^3}$

156. $\dfrac{9y - 18}{3y^2 - 12}$

157. $\dfrac{m^2 - 3m}{m^2 - 4m + 3}$

158. $\dfrac{-15a^2b}{90ab^2}$

159. $\dfrac{a^2 - 4a}{a^2 - 6a + 8}$

160. $\dfrac{s^2 - 4s - 5}{s^2 - 2s - 15}$

161. $\dfrac{6x + 6y}{9x + 9y}$

162. $\dfrac{3x - 9}{x^2 - 9}$

163. $\dfrac{7x^2 - 31xy + 12y^2}{x^2 + xy - 20y^2}$

164. $\dfrac{s^2(s + t)}{s(s + t) + 3(s + t)}$

Section 7.2

Simplify.

165. $\dfrac{4 - a}{a^2 - 16}$

166. $\dfrac{2a - 2b}{2b^2 - ab - a^2}$

167. $\dfrac{2n - 32n^3}{20n^3 - 3n^2 - 2n}$

168. $\dfrac{32d^2 - 18c^2}{6c^2 - cd - 12d^2}$

169. $\dfrac{6 - 6y^2}{18y + 18}$

170. $\dfrac{48 + 8r - r^2}{r^2 + r - 12}$

Section 7.3

Perform the indicated operations.

171. $\dfrac{2}{s^3} - \dfrac{4}{s^2} + \dfrac{3}{s}$

172. $\dfrac{x - 1}{4} - \dfrac{x - 2}{2}$

173. $\dfrac{2a + 3}{12a} - \dfrac{3a - 6}{8a} - \dfrac{5}{a}$

174. $\dfrac{a}{a - b} - \dfrac{ab}{a^2 - b^2}$

175. $\dfrac{x + 4}{9 - x^2} - \dfrac{x + 2}{x + 3} - \dfrac{x - 2}{3 - x}$

176. $\dfrac{1}{r - 9} + \dfrac{1}{r + 9}$

177. $\dfrac{5}{x + 7} - \dfrac{3}{x} + 4$

178. $\dfrac{3b}{b^2 - 4} - \dfrac{4}{2b - 4}$

179. $\dfrac{x}{8x - 4y} - \dfrac{y}{12x - 6y}$

Section 7.4

Perform the indicated operations.

180. $\dfrac{x^2 - 9}{2y^4} \cdot \dfrac{10y^3}{x + 3}$

185. $\dfrac{y^2 - 16}{27} \div \dfrac{y - 4}{9}$

181. $\dfrac{5c^4}{a^2 - b^2} \cdot \dfrac{(a - b)^2}{10c^3}$

186. $\dfrac{3a - 3b}{ab^2} \div \dfrac{a^2 - b^2}{a^2b}$

182. $\dfrac{a^2 - a - 20}{3a - 15} \cdot \dfrac{3}{a + 4}$

187. $\dfrac{3x + 6}{2x + 6} \div \dfrac{x^2 - 4}{7x + 21}$

183. $\dfrac{y^2 - 4y - 5}{y^2 \quad 11y + 30} \cdot \dfrac{y^2 - 4y - 12}{y^2 - y - 2}$

188. $\dfrac{2x - 8}{x^2 - 4} \div \dfrac{x - 4}{x^2 + 6x + 8}$

184. $\dfrac{(s + 4)(s - 6)}{s^2 - 16} \cdot (s - 6)^{-1}$

Section 7.5

Simplify the following complex fractions.

189. $\dfrac{4 + \dfrac{3}{5}}{\dfrac{1}{3} - \dfrac{3}{6}}$

191. $\dfrac{2}{2 + \dfrac{7}{x + 1}}$

193. $\dfrac{2 - \dfrac{1}{r^2}}{\dfrac{1}{r^2}}$

190. $\dfrac{\dfrac{a}{b} - \dfrac{b}{c}}{\dfrac{b}{a} - \dfrac{c}{b}}$

192. $\dfrac{x^2 - \dfrac{x^2 - 1}{x}}{1 - \dfrac{x - 1}{x}}$

194. $\dfrac{\dfrac{2x}{1 - x^2} + \dfrac{1}{1 - x}}{\dfrac{1}{1 + x}}$

Section 7.6

Using your calculator, verify the following. Let 3.25 be the first variable $\boxed{\text{STO } 1}$ and 2.39 the second variable $\boxed{\text{STO } 2}$.

195. $\dfrac{1 + \dfrac{1}{B}}{\dfrac{A}{B} + B} = \dfrac{B + 1}{A + B^2}$

196. $\dfrac{x + \dfrac{1}{x}}{\dfrac{1}{y} + x} = \dfrac{x^2y + y}{x + x^2y}$

197. $\dfrac{\dfrac{1}{a-b} - \dfrac{1}{a+b}}{\dfrac{1}{a-b} + \dfrac{1}{a+b}} = \dfrac{b}{a}$

198. $\dfrac{w}{w - \dfrac{v^2}{w}} + \dfrac{\dfrac{1}{w}}{1 - \dfrac{v}{w}} = \dfrac{w^2 + w + v}{w^2 - v^2}$

Section 8.2

Solve for the unknowns.

199. $6y = 2y + 20$

200. $5b = 28 + b$

201. $7v - 3 = 5v - v + 36$

202. $7(y + 2) = 5(y + 4)$

203. $(13 - 3d) + 7d = 93$

204. $7y - (4y - 42) = 0$

205. $2(x + 1) - 3x = 3(3 + 2x)$

206. $15d - 4(2d + 2) = 13 + 4d$

207. $4(3a + 1) - 3(2a - 5) = 29 + 4a$

208. $x(x + 1) = (x + 4)(x - 6)$

209. $(B - 1)(B - 5) - B^2 = -13$

Section 8.3

Solve for the indicated variable.

210. $p = \dfrac{v^2}{R}$ solve for v

211. $v_1 = \dfrac{R_1}{R_1 + R_2} v$ solve for R_2

212. $f = \dfrac{1}{2\pi RC}$ solve for C

213. $f_\beta = f_\alpha(1 - \alpha)$ solve for α

214. $I_E = \dfrac{V_B - V_{BE}}{R_E}$ solve for V_{BE}

215. $V_i = \left(I_i + \dfrac{V_o - V_i}{R_F}\right) h_{ie}$ solve for V_o

216. $\alpha = \dfrac{\beta}{\beta + 1}$ solve for β

Section 8.4

Solve the following fractional equations.

217. $\dfrac{3}{y} = \dfrac{2}{5 - y}$

218. $\dfrac{4}{a^2 - 1} - \dfrac{1}{a + 1} = \dfrac{1}{a - 1}$

219. $\dfrac{y-3}{y-1} - \dfrac{y-1}{y} = \dfrac{5}{y^2 - y}$

221. $\dfrac{z^2+1}{z-1} - \dfrac{3z+2}{2z-2} - z = 0$

220. $\dfrac{x}{x+3} + \dfrac{2}{x-3} = \dfrac{x^2+9}{x^2-9} + \dfrac{5}{x+3}$

222. $\dfrac{x+3}{x^2+x-2} + \dfrac{x-7}{x+2} = \dfrac{x-2}{x-1}$

Section 8.5

223. One number is 12 more than another. When the larger number is divided by the smaller number, the quotient is 8/5. Find the numbers.

224. When twice the reciprocal of a number is increased by one-third, the result is 5. Find the number.

225. John drove 250 mi one day and 310 mi the second day. How many miles must he drive the third day so he will average 325 mi/day for his whole trip?

226. The denominator of a fraction is 15 more than the numerator. If 5 is subtracted from the numerator but the denominator is unchanged, the value of the resulting fraction is 2/3. Find the original fraction.

227. The larger of two numbers is 16 more than three times the smaller. If the smaller number equals one-fifth the larger number, find the numbers.

Section 9.2

See Figure II.1 for Review Problems 228 to 230.

228. Determine the value of I.

229. Determine V_{R1}, V_{R2}, and V_{R3}.

230. If R_1 and R_2 remain as shown, what should R_3 be for I to be 0.5 mA?

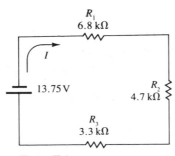

Figure II.1

See Figure II.2 for Review Problems 231 to 233.

231. What is the value of I?

232. What is the value of I_2?

233. If R_2 and R_3 are as shown, what should the new value of R_1 be to restrict I to 1 mA?

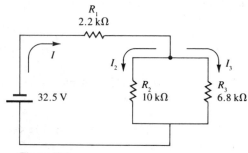

Figure II.2

See Figure II.3 for Review Problems 234 and 235.

234. Determine the value of I.

235. What should R be to limit I to 1 mA?

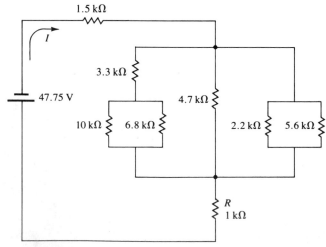

Figure II.3

Sections 9.3, 9.4

236. In Figure II.4, use the proportional voltage ratio law to determine V_{R2} and $(V_{R2} + V_{R3})$.

See Figure II.5 for Review Problems 237 and 238.

237. What should R be so that $V_{BA} = 6.275$ V?

238. Using the inverse current ratio law, what is I_R for the value of R found in Review Problem 237?

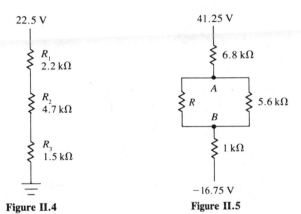

Figure II.4   Figure II.5

See Figure II.6 for Review Problems 239 and 240.

239. Using the inverse current ratio law, what is the current through the 3.3-kΩ resistor?

240. If I through the 6.8-kΩ resistor in Figure II.6 is 200 mA, what is the value of I_T?

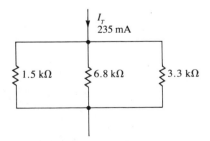

Figure II.6

Section 9.5

See Figure II.7 for Review Problems 241 and 242.

241. Determine all the currents in Figure II.7.

242. What is V_{BC}?

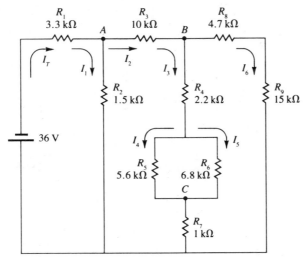

Figure II.7

Section 10.2

Solve by factoring.

243. $x^2 - 8x + 16 = 0$

244. $a^2 + 8a + 7 = 0$

245. $x(x - 3) = 4$

246. $3b^2 - 12 = 0$

247. $x^2 - 2x = 0$

248. $z^2 = 4$

249. $\frac{1}{3}a^2 + \frac{4}{3}a + 1 = 0$

Section 10.3

Solve by using the quadratic formula.

250. $c^2 + 8c = 0$

251. $x^2 - 9x + 20 = 0$

252. $5x^2 - 3x - 2 = 0$

253. $x^2 - 2x - 2 = 0$

254. $\frac{1}{4}x^2 - \frac{3}{2}x = 1$

255. $2b^2 - 10b - 9 = 0$

Section 11.2

Solve the simultaneous equations by *substitution*.

256. $2x - 4y = -6$
         $-x + y = 1$

257. $3m + n = 13$
         $m + 6n = -7$

258. $-x + 4y = -3$
         $3y - 2x = -11$

259. $3r - 2s = 11$
         $r + 2s = 9$

260. $y - 3x = -1$
         $7x + 2y = 37$

261. $2a - b = 1$
         $2a + b = 8$

Solve the simultaneous equations by adding or subtracting equations.

262. $2x - y = 4$
         $x + y = 5$

263. $5u - 3b = -2$
         $4a + 2b = 5$

264. $3x + 2y = 7$
         $4x + 2y = -14$

265. $7m - 3n = 14$
         $3m + 8n = 6$

266. $4a + 5b = 23$
         $4a - b = 5$

Solve the simultaneous equations by comparison.

267. $2a - b = 1$
         $2a + b = 8$

268. $5x + 2y = 3$
         $2x + 3y = -1$

269. $2m - 3n = 7$
         $3m + n = 5$

270. $x - y = 7$
         $3x - 2y = 18$

271. $x + y = 24$
         $y = 2x$

Section 11.3

Solve the following equations of three unknowns.

272. $2x - y + 2z = -8$
         $3x - 2y - z = -9$
         $4x + y - 3z = -3$

273. $r + s + t = 2$
         $2r + s + 2t = 3$
         $r - s + 3t = 4$

274. $3l - 5m + 4n = 7$
         $l + 2m + 3n = 9$
         $3l - m + 2n = -1$

275. $a - b - 2c = 2$
         $4a + b + c = 1$
         $3a - 2b + 3c = 5$

276. $3x - 4y + 2z = -1$
         $2x - y + z = 3$
         $x + 2y - z = 3$

Section 12.2

277. Solve the following determinants.

a. $\begin{vmatrix} 5 & -3 \\ 6 & 2 \end{vmatrix}$

b. $\begin{vmatrix} -1 & 2 & -3 \\ 5 & -3 & 2 \\ 1 & -1 & -3 \end{vmatrix}$

c. $\begin{vmatrix} 0 & -2 & 1 \\ 2 & 3 & 2 \\ -1 & 4 & 0 \end{vmatrix}$

d. $\begin{vmatrix} 6 & 5 & 4 & -2 \\ 12 & 2 & 8 & 7 \\ 3 & 2 & 2 & 1 \\ 9 & -3 & 6 & -5 \end{vmatrix}$

278. Solve the following unknowns by using determinants.

a. $2x + y - z = 5$
$3x - 2y + 2z = -3$
$x - 3y - 3z = -2$

b. $a + 2b - 3c = -7$
$3a - b + 2c = 8$
$2a - b + c = 5$

c. $2u - 2v = 6 - 3w$
$v - w = 4 - 2u$
$3u + 2w = v - 1$

Section 12.3

After constructing determinants, use your calculator to evaluate each unknown in the simultaneous equations.

279. $3.5x + 4.2y = -3.1$
$-1.6x + 6.7y = 9.5$

281. $4.2a + 3.1b + c = -3$
$-5.4a - 2.9b + 1.2c = -4$
$1.6a + 4.8b - 7.5c = 2$

280. $2.1x - 9.5y = 4$
$3.8y - 5.1 = 3x$

282. $3.6x - 8.5y + 4.2z = 1.2$
$4.7x - 5.1y - 2.8z = 3.8$
$1.5x + 4.3y + 2.0z = 2.4$

Section 12.4

See Figure II.8 for Review Problems 283 and 284.

283. Solve for the three currents shown in Figure II.8. If any current is a negative number, remember that its actual direction is opposite that shown.

284. Using the currents as determined in Review Problem 283, find V_{AB}.

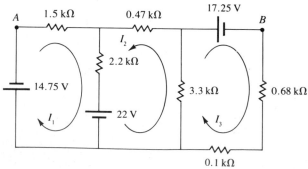

Figure II.8

285. In Figure II.9, determine the current through each resistor.

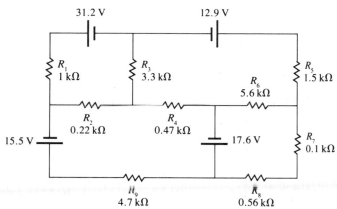

Figure II.9

Section 12.6

Find the inverse matrix of the following. Use your calculator when appropriate.

286. $\begin{bmatrix} 2 & 5 & -3 \\ 3 & 4 & 2 \\ 1 & -2 & 2 \end{bmatrix}$ **287.** $\begin{bmatrix} 3 & -2 & 4 & 0 \\ 2 & 1 & 5 & -3 \\ 1 & -2 & -2 & 2 \\ -3 & 2 & 1 & 1 \end{bmatrix}$ **288.** $\begin{bmatrix} 3.2 & 4.1 & -2.1 \\ 6.4 & -3.1 & -0.1 \\ -3.0 & -6.7 & 2.9 \end{bmatrix}$

Using the methods of matrix analysis, solve for the following unknowns.

289. **a.** $\begin{aligned} 3x - 4y - z &= 4 \\ 2x + y + 2z &= -1 \\ x - 2y + z &= 5 \end{aligned}$ **b.** $\begin{aligned} x - 2y + z - 3w &= 4 \\ 2x + 3y - z - 2w &= -4 \\ 2x - y - 3z + 2w &= -2 \\ 3x - 4y + 2z - 4w &= 12 \end{aligned}$

290. The currents i_1, i_2, i_3, i_4 (in amperes) can be found from the following equations. Find i_4.

$$\begin{aligned} i_1 + 3i_4 - i_3 &= 4 \\ 2i_2 - i_1 - 4i_4 &= 0 \\ 3i_3 + 4i_2 + i_1 &= 5 \\ 2i_4 + i_3 - 3i_2 &= -5 \end{aligned}$$

Section 13.2

See Figure II.10 for Review Problems 291 and 292.

291. Using Millman's theorem, determine the voltage at point X.

292. What should V_1 be (all other quantities remaining the same) so that point X is 0 V? What is the polarity of the terminal of V_1 closest to R_1?

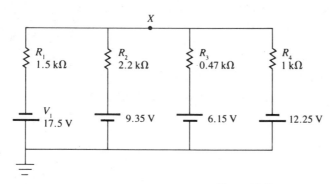

Figure II.10

Section 13.3

See Figure II.11 for Review Problems 293 and 294.

293. Determine I_T by converting triangle ABC into a Y configuration and then applying series-parallel combinations.

294. Determine I_T by converting triangle BCD into a Y configuration and then applying series-parallel combinations. Compare with the answer found for Review Problem 293.

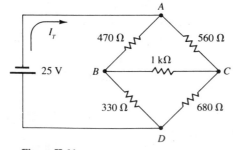

Figure II.11

295. In Figure II.12, convert the back-to-back Y configurations into a Wheatstone bridge configuration.

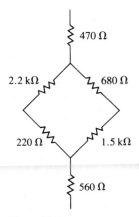

470 Ω

2.2 kΩ   680 Ω

220 Ω   1.5 kΩ

560 Ω

Figure II.12

Section 13.4

296. In Figure II.13, determine the multiplying coefficients of R_{in} for each section.

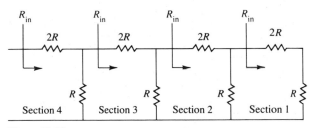

R_{in} R_{in} R_{in} R_{in}

2R   2R   2R   2R

R   R   R   R

Section 4   Section 3   Section 2   Section 1

Figure II.13

TRIGONOMETRY AND COMPLEX ARITHMETIC

14/ ANGLES AND TRIANGLES

14.1 GENERATION AND DEFINITION OF AN ANGLE

A major area of study germane to electronics is trigonometry. To master this topic, you must thoroughly understand angles and triangles. This chapter lays the necessary foundation for the study of trigonometry by analyzing the properties associated with angles and triangles.

An angle is the figure created from the intersection of two straight lines. Because any angle is actually a measurement of the degree of divergence between the two lines, it is really more appropriate to think of an angle as being created by two lines emanating from a common point called the *vertex*. It is also preferable to have the vertex coincide with an origin point, so we create such an origin by first intersecting two lines perpendicular to one another. Besides creating a point of reference, this intersection sets the stage for the two-dimensional-plane grid system (commonly called the Cartesian coordinate system) on which to locate points. (The benefits of plotting points become more apparent in Chapter 16 when we discuss phasors.)

Figure 14.1 illustrates the Cartesian coordinate system (the reference line OP, is explained shortly). Notice that the two intersecting perpendicular lines automatically create four equal directions (and quadrants), somewhat analogous to the north, south, east, and west directions on a map. By convention, the X and $-X$ directions correspond to east and west, respectively, and the Y and $-Y$ correspond to north and south, respectively. If you are looking at a map, the origin (O) of the directions is relative to where you are standing; in this case the origin is the intersection of the lines. Mathematics defines the four quadrants of a circle as shown in Figure 14.1. Conventionally the starting point for generating an angle is line OX, which is called the *initial* side of the angle. It is fixed in position. Line OP coincides with OX. Line OP is the rotating, *terminal* side of the angle. In the unique position shown in Figure 14.1, the angle is evaluated as zero because no rotation has yet begun. When OP is rotated (again, by convention, this is always in a counterclockwise direction), an angle is generated. The magnitude is directly proportional to the amount of rotation. Figure 14.2 shows that *terminal* line OP has rotated through an angle whose magnitude is defined as α. To simplify writing angles, we use the symbol $\angle$ for angle:

$$\text{Angle } XOP \equiv \angle \alpha$$

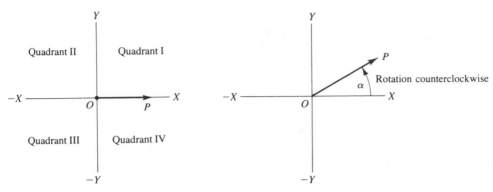

Figure 14.1   Cartesian coordinate system for generating an angle.

Figure 14.2   Components constituting an angle.

Figure 14.2 now contains all the requisite parts to qualify α as an angle: two straight lines (OX and OP) emanating from a common vertex (which coincides with a reference origin) and diverging from each other. Although OX and OP in Figure 14.1 exhibit no divergence, it still qualifies as an angle of magnitude zero.

14.2 ANGULAR MEASUREMENT METHODS

There are essentially three ways to measure angles: degrees, radians, and grads. We first discuss each method and then show you a table of conversions to help you easily change measurements from one form to another.

Degrees

If line OP of Figure 14.2 is rotated one complete revolution, it will once again coincide with OX, and point P will have generated a circle. A *degree* (°) is a measurement of an angle and equal to 1/360th of a circle; therefore there are 360° in a circle. Because the X and Y axes form four equal quadrants, the angle represented by OP rotating from OX to OY must be one-fourth of the circle of 360°, or 90°. The sides of the angle that formed 90° are perpendicular to each other. This particular angle is so important, especially in the construction field, that it has its own name: *right angle*.

$$\underline{/90°} \equiv \text{a right angle}$$

Other angles have special names also, but these angles are not generally defined in magnitudes as exact as that of the right angle. Consider Figure 14.3; $\underline{/\alpha}$ is less than (<) 90°. Any angle less than 90° is called an *acute* angle. Note how general this definition is—it does not contain the precise measurement of 90° that defines the right angle. An acute angle is of any magnitude more than (>) 0° and less than

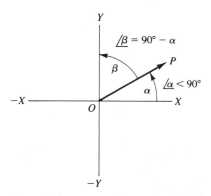

Figure 14.3   Acute angle α: $\alpha + \beta = 90°$.

90°. For further illustration, let us arbitrarily assume α in Figure 14.3 equals 39.1°. If $\beta = 90° - 39.1°$ (50.9°), then β is defined as the *complementary* angle of α. Generally, any two angles whose sum is 90° are complementary to each other. Figure 14.4 depicts α as an angle of more than 90° and less than 180°; such an angle is called *obtuse*. Furthermore, if $\alpha + \beta$ in Figure 14.4 is exactly 180°, angles α and β are *supplementary* to each other; that is, either $\underline{/\alpha}$ or $\underline{/\beta}$ is the supplement of the other angle.

Figure 14.5 depicts an angle greater than 360°. *OP* has rotated counterclockwise more than one complete revolution (let us arbitrarily say 401°). Because 360° returns us to the origin, the final terminal position of *OP* is really 401° − 360°, or 41° beyond the *OX* reference. If an angle exceeds 360°, the magnitude of the angle between *OP* (the terminal line) and *OX* (the initial line) is

Final angle = total ° of rotation − greatest number of multiples
of 360° not exceeding total
rotation

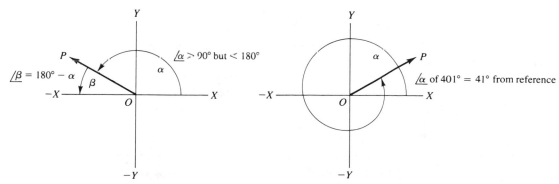

Figure 14.4   Obtuse angle α: $\alpha + \beta = 180°$.   **Figure 14.5**   Angles > 360°.

Example 1

Determine the final angle of the following angles whose rotation exceeds 360°.

a.   412.8°

$\underline{/\alpha}$ = 412.8° − 360° = 52.8°   quadrant I
one multiple

b.   621.7°

$\underline{/\alpha}$ = 621.7° − 360° = 261.7°   quadrant III
one multiple

c.   1053.6°

$\underline{/\alpha}$ = 1053.6° − 2(360°) = 333.6°   quadrant IV
two multiples

What is the meaning of negative angles? Figure 14.6 depicts $\underline{/\alpha}$ in quadrant IV. Let us arbitrarily specify $\underline{/\alpha}$ = 305°; doing so implies that $\underline{/\alpha}$ is 305° counterclockwise from the origin. However, if we rotate 360° − 305° (or 55°) clockwise from the origin, we are at the same point; therefore

$$\underline{/\alpha} = \underline{/-\beta} \quad \text{or} \quad \underline{/305°} = \underline{/-55°}$$

Generally, negative angles rotate clockwise from their reference point.

"Degree" notation has a unique peculiarity (not found in the subsequent two types of notation) that requires a special calculator function. Degrees equal 1/360th of a circle, but for greater accuracy, each degree is further subdivided into 1/60th parts called minutes (denoted by the symbol ′). Note that this is *not* equivalent to a decimal notation. For example, each minute equates as follows on the calculator:

$$1' = (1/60)° = .01667° \quad \boxed{\text{Fix 5}}$$

To complicate the picture, for further accuracy when defining angles, each minute is subdivided into 1/60th parts called seconds (symbolized ″). Each second therefore equates as follows on the calculator:

$$1'' = (1/60)' = (1/3600)° = .0002778° \quad \boxed{\text{Fix 7}}$$

Complications arise when you try to perform the basic arithmetic operations on two or more angles expressed in the degree-minute-second (°, ′, ″) format because calculators operate in a strictly decimal format. Angles, if they are expressed in a format such as 72° 13′ 22″, are not equivalent in decimal format to 72.1322° but to (72 + 13/60 + 22/3600)°. As you can see, to do something as simple as adding with the calculator two angles expressed in the °, ′, ″ format, you must convert each into an equivalent decimal number, add the forms, and then convert back again into °, ′, ″ format. By the way, before you throw up your hands in exasperation over what seem to be unwarranted complications, consider the analogous situation of adding time spans, such as 3 hours, 17 minutes, and 24 seconds to 7 hours, 54 minutes, and 59 seconds. Time is also segregated into unusual combinations: 24 hours constitute

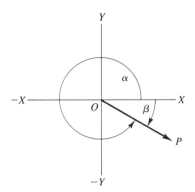

Figure 14.6   Relationship between α and $-\beta$.

1 day, and hours are divided into minutes and seconds, which are each in turn respectively subdivided into 1/60th parts the same way degrees are subdivided into minutes and seconds. Are you starting to see the parallelism? How does a calculator solve these types of problems?

The answer is the $\boxed{\text{D.MS}}$ key [DMS represents degree (or hour) minute second]. This key converts an angle from the °, ′, ″ format into degrees, plus any fractional equivalent of the degree, and expresses it in decimal format. Note that the $\boxed{\text{D.MS}}$ key requires a decimal point between the value of degrees and the value of minutes seconds. Because the number of degrees can be one, two, or three digits in magnitude, the decimal point indicates to the calculator where the last digit indicating degrees ends and the first digit of minutes begins. For a calculator with an eight-digit display capacity, the entire format is

DDD.MMSSs

1.   Degrees (or hours) can be as large as three digits (DDD).

2.   Then the decimal point.

3.   Minutes must be expressed as two digits (MM).

4.   Seconds must be expressed as two digits (SS).

5.   If accuracy is so great that seconds are expressed in tenths, like 12.8″, s is keyed in.

Condition 5 rarely occurs in angular measurements because 1″ of a circle is already an extremely fine measurement. To illustrate,

$1'' = 1/(360 \times 60 \times 60)$th part of circle   or $1'' = 0.0000007716$ part of circle

However, it is not uncommon to find "time" referred to in tenths of a second, for example, 1 min, 12.7 s. As a matter of fact, many major sporting events, such as in the Olympics, are measured in hundredths of a second, as Examples 2 to 6 illustrate.

Example 2

Convert the following angular or time measurements into the degree/minute/second format for use of the calculator D.MS key.

a. $6°22'13'' = 6.2213$

b. $145°6'7'' = 145.0607$

c. $17°3.2'' = 17.00032$

d.   1 h 4 min 21 s $= 1.0421$

e.   13.2 s $= 0.00132$

Example 3

Convert $71°23'51''$ into a decimal equivalent of degrees Fix 4 .

PRESS	DISPLAY	COMMENT
71.2351 D.MS	71.3975	Decimal equivalent of $71°23'51''$ is $71.3975°$

Example 4

Convert $123.896°$ into a °, ', " format Fix 4 .

PRESS	DISPLAY	COMMENT
123.896 INV D.MS [1]	123.5346	$123°53'46''$

Example 5

Add $21°7'42''$ to $35°21'29''$, giving your answer in the °, ', " format Fix 4 .

PRESS	DISPLAY	COMMENT
21.0742 D.MS	21.1283	Converts first angle into decimal format
+ 35.2129 D.MS	35.3581	Converts second angle into decimal format
=	56.4864	Sums the decimal angles
INV D.MS	56.2911	Answer is $56°29'11''$

---

[1]Note that as in all previous examples, the INV key performs the inverse function indicated. In this example we converted from an angle expressed in pure decimal format into an angle expressed in the °, ', " format.

Example 6

You are buying computer time. Three lengthy, scientific problems each run 1 min, 23.6 s; 2 min, 8.9 s; and 1 min, 17.8 s. If time is $17.85/min, what is the total bill? [Fix 4]

PRESS	DISPLAY	COMMENT
.01236 [D.MS]	0.0232	Decimal fractional equivalent of first time
[+] .02089 [D.MS]	0.0358	Decimal fractional equivalent of second time
[+] .01178 [D.MS]	0.0216	Decimal fractional equivalent of third time
[=]	0.0806	Total decimal fractional part of an hour
[×] 60 [=]	4.8383	Total decimal number of minutes
[×] 17.85 [=]	86.3643	$86.36 rounded to nearest whole cent

Radians

See Figure 14.7. OP begins to sweep out an angle counterclockwise and stops when arc $PP' = OP$. The magnitude of $\angle \alpha$ is defined as 1 radian (rad). $\angle \alpha$ must always be some fixed number of degrees because PP' always equals 1 radius of a circle regardless of how big the circle is. To calculate this value, determine how many times 1 radius divides into its circumference. Because the circumference of a circle is $2\pi r$, this magnitude divided by r is

$$\frac{2\pi \cancel{r}}{\cancel{r}} = 2\pi \approx 6.28$$

Figure 14.8 illustrates this relationship. Because one circle is also 360°, then

$$1 \text{ rad} = \frac{360°}{2\pi} \approx 57.296°$$

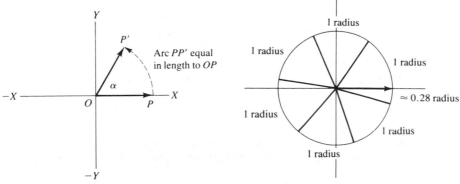

Figure 14.7 $\alpha = 1$ rad when arc $PP' = OP$.   **Figure 14.8**   Circumference length ≈ 6.28 rad.

This may seem like an awkward magnitude to deal with, but many electronic formulas use the radian in their calculation. In the next chapter ("Fundamentals of Trigonometry") you will see that magnitudes not explicitly expressed in degrees are automatically implied in radians.

To establish an exact relationship between radians and degrees, consider that

$$1 \text{ rad} = \left(\frac{360}{2\pi}\right)^\circ = \left(\frac{180}{\pi}\right)^\circ \qquad (14.1)$$

Using the values established in Equation (14.1), we can write the converse:

$$1^\circ = \left(\frac{\pi}{180}\right) \text{ rad} \qquad (14.2)$$

Example 7 illustrates the conversions between radians and degrees.

Example 7

a. Convert 1.62 rad into degrees.
    From Equation (14.1):

$$\frac{1.62 \times 180}{\pi} = 92.82^\circ$$

b. Convert 21.63° into radians.
    From Equation (14.2):

$$\frac{21.63 \times \pi}{180} = 0.377 \text{ rad}$$

Grade Units

We illustrated that dividing a circle into the °, ', " format does not directly translate into decimal notation; for example, 23°50' does *not* equal 23.5°. Furthermore, radian measurements, which often involve the π symbol, are also somewhat difficult to convert into exact decimal equivalents. Thus *grade* units (abbreviated grad) developed. This method divides each quadrant of a circle into 100 equal parts. Each grade is therefore slightly smaller than a degree, and both a grade and a degree are significantly smaller than a radian. For greater accuracy, each grade is further subdivided into 100 equal parts, called *centigrades* (cgrad). This latter term is not frequently encountered because an expression such as

$$32 \text{ grad } 56 \text{ cgrad} \equiv 32.56 \text{ grad}$$

In short, because a centigrade is a direct decimal subdivision of a grade, the original expression can be written in decimal notation rather than being expressed as a separate unit.

You can add or subtract two angles expressed in grades directly; for example,

$$\angle 14.3 \text{ grad} + \angle 21.5 \text{ grad} = \angle 35.8 \text{ grad}$$

To convert from grades into degrees, proceed as follows:

$$1 \text{ quadrant} = 100 \text{ grad} = 90°$$
$$\therefore 1 \text{ grad} = 0.9° \qquad (14.3)$$

Conversely,
$$1° = \frac{10}{9} \text{ grad} \approx 1.111 \text{ grad} \qquad (14.4)$$

Equations (14.1) and (14.2) relate radians to degrees. Because Equations (14.3) and (14.4) relate grades to degrees, we now have a common link between radians and grades:

$$1° = \frac{\pi}{180} \text{ rad} = \frac{10}{9} \text{ grad}$$

$$\therefore 1 \text{ rad} = \frac{10}{\cancel{9}} \cdot \frac{\cancel{180}^{20}}{\pi} = \frac{200}{\pi} \text{ grad} \qquad (14.5)$$

Conversely,
$$1 \text{ grad} = \frac{\pi}{\cancel{180}_{20}} \cdot \frac{\cancel{9}}{10} = \frac{\pi}{200} \text{ rad} \qquad (14.6)$$

Table 14.1 combines the conversion ratios of all three methods. Note that these ratios are based upon degrees being expressed as a decimal number. If degrees are expressed in the °, ', " format, you must first convert them into decimal notation. Using these ratios, it is not difficult to convert from one form into another, but when you attempt to add or subtract angles expressed in mixed formats, the process is a little more difficult. You can add directly some forms, like grades, but for other systems, like degrees (if expressed in the °, ', " format), you must first convert using the $\boxed{\text{D.MS}}$ function before you add them.

TABLE 14.1 CONVERSION RATIOS
FOR °, RAD, AND GRAD

Multiplying Conversion Ratios	To		
	°	rad	grad
From:			
°	. . .	$\frac{\pi}{180}$	$\frac{10}{9}$
rad	$\frac{180}{\pi}$	. . .	$\frac{200}{\pi}$
grad	0.9	$\frac{\pi}{200}$	. . .

Example 8

Using Table 14.1, perform the following conversions [Fix 4] .

a.   87°23′ into radian and grade formats.

PRESS	DISPLAY	COMMENT
87.23 [D.MS]	87.3833	Decimal equivalent
[×] [π] [÷] 180 [=]	1.5251	Radian equivalent
[×] 200 [÷] [π] [=]	97.0926	Grade equivalent

b.   2.641 rad into grade and °,′,″ formats.

PRESS	DISPLAY	COMMENT
2.641 [×] 200 [÷] [π] [=]	168.1313	Grade equivalent
[×] .9 [=] [INV] [D.MS]	151.1906	151°19′06″

Example 9

Add the following angles, expressing your answer in the °, ′, ″ format.

$$36.21° + 1.17 \text{ rad} + 51.08 \text{ grad}$$

PRESS	DISPLAY	COMMENT
36.21 [+]	36.2100	36.21°
[(] 1.17 [×] 180 [÷] [π] [)]	67.0361	rad→°
[+]	103.2461	$\angle 1 + \angle 2$
[(] 51.08 [×] .9 [)]	45.9720	grad→°
[=]	149.2181	$\angle 1 + \angle 2 + \angle 3$
[INV] [D.MS]	149.1305	149°13′05″

PROBLEMS

1.   What is the complementary angle of the following?

a.   24°

c.   12°13′

e.   89°0′17″

b.   51.6°

d.   37°21′35″

2. What is the supplementary angle of the following?

    a. 91°

    b. 124.64°

    c. 151°24′

    d. 170°6′

    e. 179°32′18″

3. Reduce the following angles to a magnitude of less than 1 revolution:

    a. 419°

    b. 776°12′

    c. 1054°6′21″

    d. 2001°0′14″

4. What negative angle describes the following?

    a. 312.4°

    b. 101°22′

    c. 217°7′29″

    d. 14°19″

5. Convert the following into decimal format:

    a. 21°18′

    b. 75°21′24″

    c. 105°32″

    d. 191°1′1″

    e. 51°7″

    f. 24″

6. Convert the following into °, ′, ″ format:

    a. 17.216°

    b. 105.206°

    c. 201.772°

    d. 305.011°

7. Convert the following formats into the other two formats:

    a. 14.6°

    b. 71°17′21″

    c. 2.152 rad

    d. 0.08 rad

    e. 41.26 grad

    f. 0.889 grad

8. Determine the magnitude of $\angle\alpha$ by performing the following operations on the angles indicated. Give your answer in the format requested.

    a. $21.2° + 27°34′ = \alpha\ °\ ′\ ″$

    b. $106°21′54″ + 55°54′7″ = \alpha$ rad

    c. $34.2° + 1.67$ rad $= \alpha\ °\ ′\ ″$

    d. $2.89° - 39.92° = \alpha$ rad

    e. $17.6° + 55.19$ grad $= \alpha$ rad

    f. $33°18′17″ + 85.61$ grad $= \alpha$ grad

    g. 0.986 rad $- 13.62$ grad $= \alpha°\ ′\ ″$

14.3 SIMILAR AND RIGHT TRIANGLES

One law governing the mathematics of geometry is

A single line intersecting two parallel lines creates an equal angle with each parallel.

Figure 14.9 illustrates this concept. Lines PQ and $P'Q'$ are parallel; line OO' intersects both the parallels. Angle OPQ (γ) equals angle $OP'Q'$ (γ'). If a second line (OO'') is drawn to intersect the same parallels (Figure 14.10), two triangles are formed. By the same logic, $\underline{/\phi} = \underline{/\phi'}$. If $\underline{/\gamma}$ and $\underline{/\phi}$ of triangle OPQ and $\underline{/\gamma'}$ and $\underline{/\phi'}$ of triangle $OP'Q'$ are equal, these triangles are similar to each other.

Regardless of their respective magnitudes, when any two angles of one triangle equal the same two angles of another triangle, the triangles are similar.

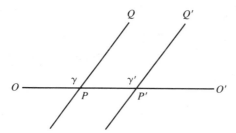

Figure 14.9 Constructing $\alpha = \alpha'$.

Because both triangles share OO' and OO'' as common lines, $\underline{/\theta}$ is also shared by both triangles; therefore all three angles of each triangle are equal. However, even if the triangles do not coincide as shown, when any two angles of one triangle equal the same two angles of another triangle, the third angles of each triangle are automatically equal because the sum of the three interior angles of any triangle is always 180°. Therefore, $180° - (\gamma + \phi)°$ must equal $180° - (\gamma' + \phi')°$ whenever

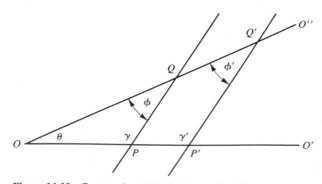

Figure 14.10 Constructing OPQ similar to $OP'Q'$.

$\underline{/\gamma} = \underline{/\gamma'}$ and $\underline{/\phi} = \underline{/\phi'}$. In conclusion, for both triangles to be similar, two angles of one triangle must equal the same two angles of the other triangle. This point is stressed because similar triangles possess proportionality properties that help facilitate their solutions.

With reference to Figure 14.10, if OP' is 40 percent greater than OP, $P'Q'$ is also 40 percent greater than PQ and OQ' is 40 percent greater than OQ. Similar triangles possess proportions such that their similar ratios are always equal.

Example 1

See Figure 14.10. If $OP' = 10$ cm and $OP = 7.5$ cm, what is $P'Q'$ if $PQ = 6.4$ cm? By similar ratios:

$$\frac{P'Q'}{OP'} = \frac{PQ}{OP}$$

$$P'Q' = \frac{PQ}{OP}(OP') = \frac{6.4}{7.5}(10 \text{ cm}) = 8.53 \text{ cm}$$

If lines PQ and $P'Q'$ were perpendicular to OO', then $\underline{/\gamma}$ would still equal $\underline{/\gamma'}$, but then both angles would be 90°. From the discussion in Section 14.2, these triangles would be right triangles; Figure 14.11 depicts this condition. Similar right triangles still retain the properties of proportionality, but right triangles in general also possess properties not common to other triangles.

Figure 14.12 is a right triangle; the three sides are a, b, and h. The three interior angles are γ, ϕ, and θ, with $\underline{/\gamma}$ being 90°. Side b is called the *base* of the triangle, side a is the *altitude* of the triangle, and side h is the *hypotenuse* of the triangle. The hypotenuse is always the longest of the three sides and always the side opposite the 90° angle. A unique property of right triangles is the relationship known as the Pythagorean theorem:

$$a^2 + b^2 = h^2$$

In any right triangle, the altitude squared plus the base squared equals the hypotenuse squared.

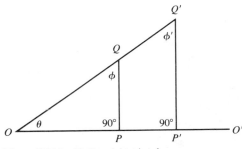

Figure 14.11 Similar right triangles.

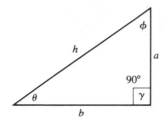

Figure 14.12 Right triangle with standard designations.

Example 2

With reference to Figure 14.12, calculate the missing side for the following.

a. $a = 3, b = 4$

$$a^2 + b^2 = h^2$$
$$\therefore h = \sqrt{a^2 + b^2} = \sqrt{3^2 + 4^2} = \sqrt{25} = 5$$

b. $a = 3.5, h = 12.7$

$$b^2 = h^2 - a^2$$
$$b = \sqrt{h^2 - a^2} = \sqrt{12.7^2 - 3.5^2} = \sqrt{149.04} = 12.208$$

PROBLEMS

1. See Figure 14.11. If $OP = 12.2$ cm, $OP' = 26.1$ cm, and $OQ = 17.2$ cm, what is OQ'?

2. Given the following dimensions for the right triangle in Figure 14.12, determine the values requested.

   a. $\theta = 23.1°$; find ϕ.

   b. $\phi = 53.71°$; find θ.

   c. $a = 1.23$ m and $b = 3.16$ m. Find h.

   d. $a = 1.59$ m and $h = 8.16$ m. Find b.

   e. $b = 82.11$ cm and $h = 3.14$ m. Find a.

SUMMARY

An angle is the degree of divergence between two intersecting straight lines. The lines emanate from a common vertex that is the origin of a grid system.

The three methods for measuring the magnitude of an angle are

1. The *degree*, which is 1/360th of a circle. The degree (°) is subdivided into 60 equal parts called minutes ('), and each minute is further subdivided into 60 equal parts called seconds ("). Arithmetic operations on the calculator in the °, ', " format require changing the notation into a decimal form by using the $\boxed{\text{D.MS}}$ key.

2. The *radian* unit is used in electronic formulas. There are 2π rad in a circle; therefore, 1 rad ≈ 57.296°.

3. The *grade* provides an easy way to convert angular measurements into decimal notation.

By using conversion ratios you can easily change from one form into the other form(s).

Similar triangles (two angles of one triangle equal the same two angles of another triangle) possess properties of proportionality. If you know two sides of triangle A, you have to know only one corresponding side of triangle B to calculate the remaining side of triangle B.

Right triangles possess properties that readily facilitate their solution. One of the most powerful properties is the Pythagorean theorem, which states that the base squared plus the altitude squared equals the hypotenuse squared.

15 FUNDAMENTALS OF TRIGONOMETRY

15.1 THE BASIC TRIGONOMETRIC FUNCTIONS

The basic trigonometric functions refer exclusively to *right* triangles. This may seem restrictive, but consider that most practical architectural structures contain the right angle. Certainly, every attempt is made to build structures with at least some intersections perpendicular to each other.

Chapter 14 demonstrated that similar right triangles were merely special cases of the "general" category of similar triangles. When you attempted to apply the properties of proportionality to similar triangles, you had to know at least three of the four lengths that relate ratios before you could determine the fourth (unknown) length. But because right triangles do possess certain unique properties, you applied the Pythagorean theorem rather than the ratio method to find an unknown length. However, this too required the knowledge of at least two lengths to determine the third (unknown) length. In both cases it was always a "length" that had to be known; at no time was any angle of the triangle involved. Can we be freed of such dependence upon length and instead use angles to determine elements of a right triangle? Well obviously, with a buildup like this, the answer is "yes." The essence of the trigonometric functions therefore is the utilization of angles to determine the unknown variables of a right triangle.

Consider Figure 15.1. QP and $Q'P'$ are parallel lines. OO' is a perpendicular intersect of the parallels, so γ and γ' are 90° angles. OO'' also intersects the parallels, thereby making $\angle\phi$ equal to $\angle\phi'$. Because $\angle\gamma$ equals $\angle\gamma'$ and $\angle\phi$ equals $\angle\phi'$ and both triangles OPQ and $OP'Q'$ share θ as a common angle, both triangles are certainly similar. However, at this point our attention now centers on θ, the common angle. For the purpose of hanging a lable on $\angle\theta$, we arbitrarily specify its magnitude as 30°. Now if we accurately draw Figure 15.1 and arbitrarily structure its dimensions such that h is 10 cm and h' is 16 cm, we get the following measurements:

$$a = 5 \text{ cm} \qquad a' = 8 \text{ cm}$$
$$b = 5\sqrt{3} \text{ cm} \qquad b' = 8\sqrt{3} \text{ cm}$$

We emphasize the word "arbitrary" when constructing the lengths of h and h' because the length of the construction is of no consequence. The conclusions that follow are the only important things; these conclusions are reached regardless of the lengths of h and h'.

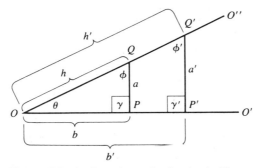

Figure 15.1   Configuration to develop the significance of ratios.

For $\underline{/\theta} = 30°$, the ratio of

$$\frac{a}{b} = \frac{5 \text{ cm}}{5\sqrt{3} \text{ cm}} = \frac{1}{\sqrt{3}}$$

and the ratio of

$$\frac{a'}{b'} = \frac{8 \text{ cm}}{8\sqrt{3} \text{ cm}} = \frac{1}{\sqrt{3}}$$

Even though triangle OPQ is smaller than triangle $OP'Q'$, the ratio of the side opposite $\underline{/\theta}$ (the altitude) to the right side adjacent to $\underline{/\theta}$ (the base) is the same. As a matter of fact, it does not matter how big the triangles are; for a 30° angle, the ratio of the side opposite the angle to the side adjacent to the angle is always $1/\sqrt{3}$. This simple but most important fact occurs because whenever one side changes dimensions, the other side changes by the same proportion; therefore the ratios remain a constant. For example, the ratio of

$$\frac{20}{5} = 4$$

Now, if we structure a new numerator whose value is 60 percent of 20 (12) and a new denominator whose value is 60 percent of 5 (3), the ratio of

$$\frac{(60\%)(20)}{(60\%)(5)} = \frac{12}{3} \text{ still equals 4}$$

In effect, whenever the numerator and denominator of a fraction change by the same proportion, the ratio of the fraction is unchanged. Although this example was for a 30° angle, any angle has a unique value for the ratio of the side opposite the angle to the side adjacent to the angle, and that number remains a constant for that angle, regardless of the size of the triangle. This predictability of a ratio of lengths for a given angle adds a new dimension to the solutions of right triangles and frees us from the limiting dependency upon lengths.

A total of six ratio combinations can be structured from the three sides of a right triangle. The first three ratios are unique definitions; the last three ratios are merely the reciprocals of the first three. Figure 15.2 lists the key parts used to define the trigonometric functions. The following definitions in Table 15.1 all refer to $\angle\theta$, although the verbal descriptions can also apply to $\angle\phi$ if the sides of Figure 15.2 are redefined:

TABLE 15.1   TRIGONOMETRIC FUNCTIONS

Name of Function		Abbreviation	Ratios Involved
sine θ		sin θ	$\dfrac{\text{Side opposite } \theta}{\text{Hypotenuse}}$
cosine θ	Unique function definitions	cos θ	$\dfrac{\text{Side adjacent } \theta}{\text{Hypotenuse}}$
tangent θ		tan θ	$\dfrac{\text{Side opposite } \theta}{\text{Side adjacent } \theta}$
cosecant θ		csc $\theta\left(\dfrac{1}{\sin\theta}\right)$	$\dfrac{\text{Hypotenuse}}{\text{Side opposite } \theta}$
secant θ	Reciprocal function definitions	sec $\theta\left(\dfrac{1}{\cos\theta}\right)$	$\dfrac{\text{Hypotenuse}}{\text{Side adjacent } \theta}$
cotangent θ		cot $\theta\left(\dfrac{1}{\tan\theta}\right)$	$\dfrac{\text{Side adjacent } \theta}{\text{Side opposite } \theta}$

Calculators do not have keying-in functions for determining the values of the reciprocal trigonometric functions. If you need these values, determine the appropriate basic function (sine, cosine, or tangent) on your calculator and then use the $\boxed{1/x}$ (reciprocal) key.

Observe that the previous definitions of the functions involved a description of the positions of the sides of a triangle in relation to the location of θ. For example, in Figure 15.2, tan θ was specifically defined as

$$\frac{\text{Side opposite } \theta}{\text{Side adjacent } \theta}$$

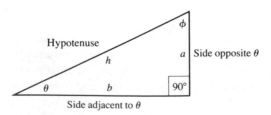

Figure 15.2   Configuration to define the trigonometric functions relative to θ.

It is also true that for Figure 15.2

$$\tan\theta = \frac{a}{b}$$

but this is only for Figure 15.2 because of the labeling selected. The definition of

$$\frac{\text{Opposite}}{\text{Adjacent}}$$

for tangent is more general and is therefore preferable to a/b because it is less restrictive. Figure 15.2 is redrawn as Figure 15.3 with the same labels for a, b, h, θ, and ϕ; however, the sides are now listed with reference to ϕ instead of θ. Notice that

$$\tan\phi \equiv \frac{\text{side opposite }\phi}{\text{side adjacent }\phi} = \frac{b}{a}$$

When $\tan\theta = a/b$ and $\tan\phi = b/a$, you can see that specific lables refer to specific figures, whereas the definition of

$$\frac{\text{Opposite}}{\text{Adjacent}}$$

is more general and therefore not so restricted to a specific figure.

The following sections deal with calculator methodologies and electronic applications of the trigonometric functions.

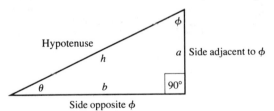

Figure 15.3   Configuration to define the trigonometric functions relative to ϕ.

15.2 TRIGONOMETRIC FUNCTIONS AND THE CALCULATOR

Mode of Operation

Once most calculators are "powered up," they are automatically in the degree mode. To change to the radian or grade mode, you have to indicate this to the calculator by keying in the mode desired. Additionally, you can change modes while in the middle of a problem if you execute the change for the function desired before you complete the calculation. Examples 1 to 5, interspersed with clarifying narrative, illustrate the mechanics of using the calculator with trigonometric functions.

Example 1

Select a convenient number of degrees, such as 60°. Determine the sine, cosine, and tangent of the angle. Translate 60° into the equivalent radian and grade units, and then verify the same answers. Power up your calculator (use $\boxed{\text{Fix 4}}$) and/or key in $\boxed{\text{Deg}}$.

PRESS	DISPLAY	COMMENT
60 $\boxed{\text{sin}}$ [1]	0.8660 (record this value)	sin 60°
60 $\boxed{\text{cos}}$	0.5000 (record this value)	cos 60°
60 $\boxed{\text{tan}}$	1.7321 (record this value)	tan 60°
60 $\boxed{\times}$ 10 $\boxed{\div}$ 9 $\boxed{=}$ $\boxed{\text{STO 1}}$	66.6667	Convert 60° into grads
$\boxed{\text{Grad}}$	66.6667	Shifts into grad mode
$\boxed{\text{sin}}$	0.8660 (compare with previous value)	Same as sin 60°
$\boxed{\text{RCL 1}}$ $\boxed{\text{cos}}$	0.5000 (compare with previous value)	Same as cos 60°
$\boxed{\text{RCL 1}}$ $\boxed{\text{tan}}$	1.7321 (compare with previous value)	Same as tan 60°
60 $\boxed{\times}$ $\boxed{\pi}$ $\boxed{\div}$ 180 $\boxed{=}$ $\boxed{\text{STO 1}}$	1.0472	Convert 60° into rads
$\boxed{\text{Rad}}$	1.0472	Shifts into rad mode
$\boxed{\text{sin}}$	0.8660 (compare with previous value)	Same as sin 60°
$\boxed{\text{RCL 1}}$ $\boxed{\text{cos}}$	0.5000 (compare with previous value)	Same as cos 60°
$\boxed{\text{RCL 1}}$ $\boxed{\text{tan}}$	1.7321 (compare with previous value)	Same as tan 60°

---

[1]Note that because the trigonometric functions, just like all the special functions, act immediately upon the value displayed, they must *follow* the value of the angle. For example, suppose you are evaluating the value of 12 sin 30°. Assuming that the calculator is in the degree mode, the correct sequence is

$$12 \boxed{\times} 30 \boxed{\text{sin}} \boxed{=}$$

If you attempt to key in

$$12 \boxed{\times} \boxed{\text{sin}} \cdots$$

the calculator immediately evaluates sin 12°, which is erroneous because the number 12 is in the display register when the sin function is keyed in. Remember, any time you depress a *special-function* key, the key acts upon the value displayed. *Conclusion*: When evaluating a special function, key in the operand first and then the operator.

This example illustrates that the trigonometric functions of equivalent degrees, radians, and grades produce the same result and the mode of the calculator can be randomly changed during an operation.

Interpolation and Extrapolation

Prior to the universal availability of the calculator, math books included (and many still do) extensive tables of the trigonometric functions. Usually these tables are a compilation of the 90 equal degrees of the first quadrant (in more refined tables they are broken down into tenths of a degree), with the numerical evaluation of the sine, cosine, and tangent functions. If you needed values of the trigonometric functions during the solution of a problem, you had to refer to the tables. However, if you needed the value of a function that was not one of the *exact* listings, you had to perform the process(es) of interpolation and/or extrapolation to arrive at a value. Interpolation is the finding of terms between any two consecutive terms of a series that conforms to the law of the series. Extrapolation is the estimating or inferring of a value beyond the known range of the series based upon certain values within the known range of the series. To oversimplify the process of interpolation for clarification: If $2.5x = 14$ and $3.5x = 16$, then $3x$ (one-half the difference between 2.5 and 3.5) = 15 (one-half the difference between 14 and 16).

If the function is *linear*, that is, if the amount of change of the dependent variable is consistently equal to an equal amount of change of the independent variable, the above methodology is accurate. Figure 15.4 illustrates a linear function, a graph of the current in the series circuit in Figure 15.5. Each time V changes by 1 V, the current I changes by 1 mA. Because this represents a simple one-to-one relationship, we can *extrapolate* and predict that the current will be 4.5 mA when V equals 4.5 V. Similarly, given that $I = 2.1$ mA when V is 2.1 V and $I = 2.3$ mA when V is 2.3 V, we can *interpolate* and state that I will equal 2.17 mA when V equals 2.17 V. Unfortunately, trigonometric functions are not linear, so these described processes have inherent built-in errors.

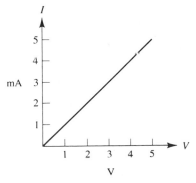

Figure 15.4   Linear function of Figure 15.5.

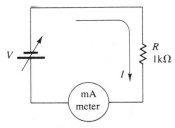

Figure 15.5   Simple series circuit producing the linear graph in Figure 15.4.

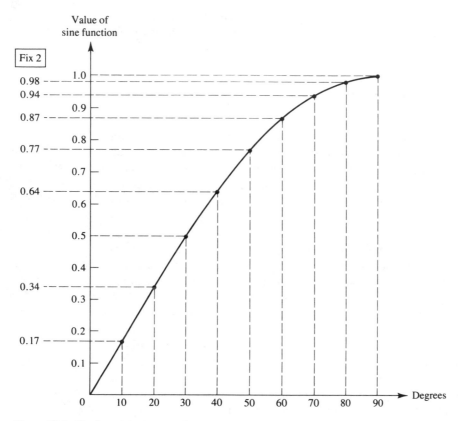

Figure 15.6   Nonlinear sine function.

Note the sine function graph in Figure 15.6. Even though the initial portion of the graph (from 0° to approximately 30°) looks linear, it is not. To magnify this concept, Figure 15.7 is a graph of the sine function from 80° to 90°, the region where the non-linearity is the most pronounced. We stress this point to reinforce the applicability of the calculator to *accurately* solve problems because the calculator can

1.   Automatically convert between degrees, radians, and grades during the calculation of the trigonometric functions.

2.   Provide the value of the trigonometric functions more rapidly than can a look-up table.

3.   Significantly increase the accuracy of the values of the trigonometric function by the number of significant digits displayed and by possessing algorithms. These algorithms more precisely calculate any trigonometric value without relying upon the interpolation and/or extrapolation methods. When a high degree of accuracy is required, this is most important.

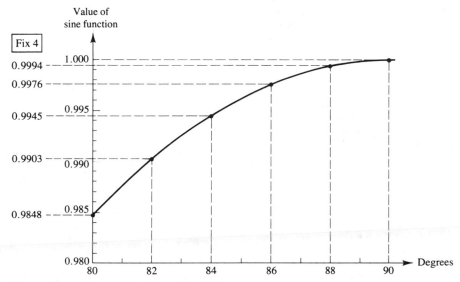

Figure 15.7   Expanded view of the sine function.

Example 2

In a linear function, the dependent variable (y) equals 13.257 when the independent variable (x) equals 3.62. When x increases to 5.02, y increases to 19.356. What is the value of y when x equals 4.52?

The key to solutions of this type is to set up ratios analogous to those used to solve similar triangles. Each dimension can be constructed as the side of a triangle (as in Table 15.2); see Figure 15.8.

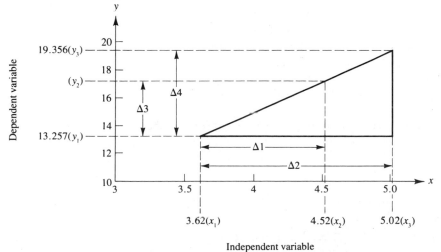

Figure 15.8   Construction of similar triangles.

TABLE 15.2   TABULAR SUMMARY OF FIGURE 15.8

Independent Variable x	Dependent Variable y
(x_1) 3.62 $\left.\vphantom{\begin{matrix}a\\b\end{matrix}}\right\} \triangle 1$ $\left.\vphantom{\begin{matrix}a\\b\\c\end{matrix}}\right\} \triangle 2$ (x_2) 4.52 (x_3) 5.02	(y_1) 13.257 $\left.\vphantom{\begin{matrix}a\\b\end{matrix}}\right\} \triangle 3$ $\left.\vphantom{\begin{matrix}a\\b\\c\end{matrix}}\right\} \triangle 4$ (y_2) ? (y_3) 19.356

$$x_2 - x_1 \equiv \triangle 1 = 4.52 - 3.62 = 0.9$$
$$x_3 - x_1 \equiv \triangle 2 = 5.02 - 3.62 = 1.4$$
$$y_2 - y_1 \equiv \triangle 3 = y_2 - 13.257$$

Note that $\triangle 3$ is the only term containing the unknown.

$$y_3 - y_1 \equiv \triangle 4 = 19.352 - 13.257 = 6.095$$

By similar triangular ratios:

$$\frac{\triangle 3}{\triangle 4} = \frac{\triangle 1}{\triangle 2}$$
$$\frac{y_2 - 13.257}{6.095} = \frac{0.9}{1.4}$$
$$y_2 = \frac{(0.9)(6.095)}{1.4} + 13.257 = 17.1752$$

Example 3

This is a linear interpolation problem similar to Example 2 but introducing a transition in the polarity of the independent variable x. Calculate y_2.

$$(x_1) -7.612 \qquad (y_1) \ \ 3.215$$
$$(x_2) -1.009 \qquad (y_2) \quad ?$$
$$(x_3) \ \ \ 3.62 \qquad (y_3) \ 17.869$$
$$x_2 - x_1 \equiv \triangle 1 = -1.009 - (-7.612) = 6.603$$
$$x_3 - x_1 \equiv \triangle 2 = 3.62 - (-7.612) = 11.232$$
$$y_2 - y_1 \equiv \triangle 3 = y_2 - 3.215$$
$$y_3 - y_1 \equiv \triangle 4 = 17.869 - 3.215 = 14.654$$

Constructing similar triangles analogous to Figure 15.8 but substituting the values of Example 3, we get the following results:

$$\frac{\triangle 3}{\triangle 4} = \frac{\triangle 1}{\triangle 2}$$
$$\frac{y_2 - 3.215}{14.654} = \frac{6.603}{11.232}$$
$$y_2 = \frac{(6.603)(14.654)}{11.232} + 3.215 = 11.8297$$

Example 4

This is linear interpolation, but the dependent variable y has a negative slope (that is, y decreases in value as x increases in value). Calculate y_2.

$$
\begin{array}{ll}
(x_1)\ -61.328 & (y_1)\ \ 13.613 \\
(x_2)\ \ \ \ 1.096 & (y_2)\ \ \ ? \\
(x_3)\ \ 75.083 & (y_3)\ -7.148
\end{array}
$$

Note that both x and y are changing polarities: The independent variable x is changing with a positive slope, and the dependent variable y is changing with a negative slope. Using the methodologies of the previous examples, the following results:

$$x_2 - x_1 \equiv \triangle 1 = 1.096 - (-61.328) = 62.424$$
$$x_3 - x_1 \equiv \triangle 2 = 75.083 - (-61.328) = 136.411$$
$$y_2 - y_1 \equiv \triangle 3 = y_2 - 13.613$$
$$y_3 - y_1 \equiv \triangle 4 = -7.148 - 13.613 = -20.761$$

$$\frac{\triangle 3}{\triangle 4} = \frac{\triangle 1}{\triangle 2}$$

$$\frac{y_2 - 13.613}{-20.761} = \frac{62.424}{136.411}$$

$$y_2 = \frac{(62.424)(-20.761)}{136.411} + 13.613 = 4.1124$$

The calculator's ability to more precisely express trigonometric function values of any angle magnitude is illustrated in Example 5. The calculator's accuracy over manual interpolation methods varies from slight to significant, depending upon the function and the magnitude of the angle. For example, the error for manual interpolation methods for the tangent function at small angles is slight, but at large angles it can be significant.

Example 5

Interpolate the following trigonometric functions using linear manual methods. Then use your calculator to determine the requested value and compare the results.

a.   Determine the tangent of $8.16°$ given the following:

$$\tan 8.1° = 0.14232$$
$$\tan 8.2° = 0.14410$$

Using definitions of the previous examples, linear interpolation results in

$$\triangle 1 = 8.16 - 8.1 = 0.06$$
$$\triangle 2 = 8.2 - 8.1 = 0.1$$
$$\triangle 3 = x - 0.14232$$
$$\triangle 4 = 0.14410 - 0.14232 = 0.00178$$

$$\frac{\triangle 3}{\triangle 4} = \frac{\triangle 1}{\triangle 2}$$

$$\frac{x - 0.14232}{0.00178} = \frac{0.06}{0.1}$$

$$x = \frac{(0.06)(0.00178)}{0.1} - 0.14232 = 0.143388$$

From the calculator $\boxed{\text{Fix 6}}$: tan 8.16° = 0.143390. Linear interpolation conclusion: a slight difference.

b. Determine the tangent of 88.86° given the following:

$$\tan 88.8° = 47.740$$
$$\tan 88.9° = 52.081$$

Using definitions of the previous examples, linear interpolation results in

$$\triangle 1 = 88.86 - 88.8 = 0.06 \qquad \text{same as in part a}$$
$$\triangle 2 = 88.9 - 88.8 = 0.1 \qquad \text{same as in part a}$$
$$\triangle 3 = x - 47.740$$
$$\triangle 4 = 52.081 - 47.740 = 4.341$$

$$\frac{\triangle 3}{\triangle 4} = \frac{\triangle 1}{\triangle 2}$$

$$\frac{x - 47.740}{4.341} = \frac{0.06}{0.1}$$

$$x = \frac{(0.06)(4.341)}{0.1} + 47.740 = 50.3446$$

From the calculator $\boxed{\text{Fix 4}}$: tan 88.86° = 50.2528. Linear interpolation conclusion: a significant difference.

PROBLEMS

1. Calculate the value of the dependent variable (y_2) for the following *linear* functions.

    a. $x_1 = 3.216$ $y_1 = 1.055$
         $x_2 = 5.128$ $y_2 = \ ?$
         $x_3 = 7.189$ $y_3 = 14.862$

b. $x_1 = -12.234$ $y_1 = 3.887$
   $x_2 = -1.655$ $y_2 = \ ?$
   $x_3 = 2.815$ $y_3 = 13.894$

c. $x_1 = -5.111$ $y_1 = -8.621$
   $x_2 = 0.86$ $y_2 = \ ?$
   $x_3 = 5.283$ $y_3 = -1.854$

d. $x_1 = 0.152$ $y_1 = -1.053$
   $x_2 = 3.816$ $y_2 = \ ?$
   $x_3 = 10.555$ $y_3 = -14.862$

2. Using extrapolation methods determine the value of the variable indicated for the following *linear* functions.

a. $x_1 = 4.931$ $y_1 = 4.586$
   $x_2 = 6.705$ $y_2 = 3.158$
   $x_3 = 8.511$ $y_3 = \ ?$

b. $x_1 = -15.113$ $y_1 = \ ?$
   $x_2 = -5.125$ $y_2 = 5.186$
   $x_3 = -1.117$ $y_3 = 8.997$

c. $x_1 = -1.055$ $y_1 = 2.651$
   $x_2 = 4.157$ $y_2 = 7.191$
   $x_3 = \ ?$ $y_3 = 15.828$

d. $x_1 = \ ?$ $y_1 = -2.593$
   $x_2 = -3.562$ $y_2 = -7.118$
   $x_3 = 2.050$ $y_3 = -15.289$

3. To develop more proficiency when using your calculator, determine the value of the following exercise problems (Fix 4) .

a. $(\sin 32.1°)(\tan 16.1°)$

b. $\dfrac{\csc 21.3°}{\sec 47.8°}$

c. $(\cos 1.3 \text{ rad})(\tan 24.8 \text{ grad})$

d. $\dfrac{(\cot 32.9°)\left(\sin \dfrac{\pi}{6} \text{ rad} \right)}{\sec 72.1 \text{ grad}}$

e. $\dfrac{\cos \sqrt{\dfrac{\pi}{2}} \text{ rad}}{\cot 21.5°}$

f. $\dfrac{\tan \, (\cos 39.2°) \text{ rad}}{\csc 61.9 \text{ grad}}$

g. $\dfrac{\cos \dfrac{\pi}{3} \text{ rad} - \tan 11.2 \text{ grad}}{\cot \, (95 \sin 23°)°}$

h. $\dfrac{\tan \dfrac{\pi}{4} \text{ rad} + \tan 21\pi°}{\sin 81.6 \text{ grad} + \cos 0.3 \text{ rad}}$

i. $\cos \, (\csc 21.8 \text{ grad} + 50 \cot 0.41\pi \text{ rad})°$

j. $\dfrac{\sin \, (17 \cos 0.1\pi \text{ rad} + 25 \tan 20.6 \text{ grad})°}{\cos \, (41 \cot 41.2° - 16 \cos 5.2 \text{ grad})°}$

4.   Determine the value of the following trigonometric functions expressed in the °, ', " format. *Hint*: Use the $\boxed{\text{D.MS}}$ key.

a. $\sin 12°13'25''$

b. $\tan 42°1'4''$

c. $\dfrac{1}{\cos 7°7'7''}$

d. $\sec 14°12'5''$

e. $\cot 21°8'4.2''$

15.3 THE INVERSE TRIGONOMETRIC FUNCTIONS

Section 15.2 demonstrated that by knowing the value of an angle, you can determine the ratios of any two sides of a right triangle. Again, it is important to keep in mind that what is known is a ratio, not necessarily the precise lengths of any of the sides. Suppose we reverse the situation: we now know the length of any two sides of a right triangle but not the magnitude of any of the angles. Use Figure 15.9 as an example. The value of h can be easily determined by applying the Pythagorean theorem; it is the value of $\underline{/\theta}$ and/or $\underline{/\phi}$ that is now of interest. Let us arbitrarily determine $\underline{/\theta}$; then $\underline{/\phi}$ must be $90° - \underline{/\theta}$. From the point of view of $\underline{/\theta}$, the side opposite is 3 and the side adjacent is 4. The following equation is therefore established with reference to $\underline{/\theta}$:

$$\frac{\text{Side opposite } \underline{/\theta}}{\text{Side adjacent } \underline{/\theta}} \equiv \tan \theta = \frac{3}{4}$$

or
$$\tan \theta = 0.75 \qquad\qquad (15.1)$$

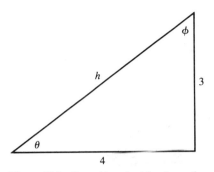

Figure 15.9 Two sides of a triangle are known.

Because 0.75 is a unique ratio, there can be only one exact, unique angle that has 0.75 as its tangent. To determine this value, we must separate the tangent term from θ. The following format then exists:

$$\theta = \text{some mathematical expression}$$

As an analogy for the mathematical symbolism that denotes this separation, recall from the laws of exponents

$$a \cdot a^{-1} = a^{1+(-1)} = a^{\circ} = 1$$

If we multiply both sides of Equation (15.1) by $\tan^{-1}$, the following results:

$$\underbrace{\tan^{-1} \tan\ \theta}_{\text{Equates to 1}} = \tan^{-1} 0.75$$

or

$$\theta = \tan^{-1} 0.75 \tag{15.2}$$

The expression $\tan^{-1}$ is called the *arctan*. Equation (15.2) can be interpreted as follows:

θ is that angle whose inverse tangent function (arctan) has a value of 0.75.

Any trigonometric function that has the -1 associated with it (written as an exponent) is called the arc function. It has the same literal interpretation for Equation (15.2) when any other trigonometric function is substituted for "tan." Note that -1 is specifically not called an exponent because "tan" (or any trigonometric function) is a definition and definitions cannot be raised to powers. However, $\tan \theta$, whether θ is in degrees, radians, or grades, is a finite number that can be raised to an exponent. As of consequence,

$$\tan^{-1} \theta$$

is a unique mathematical notation meaning arctan and should not be interpreted as meaning

$$\frac{1}{\tan \theta}$$

If you really do want the reciprocal of tan θ, it is written as

$$(\tan \theta)^{-1}$$

Example 1

Using your calculator, determine the value of the following angles. Use $\boxed{\text{Fix 3}}$, except for b, which is at $\boxed{\text{Fix 4}}$ due to the format requested.

a. $\sin^{-1} 0.321$ in   ° format

b. $\tan^{-1} 4.83$ in   °, ', " format

c. $\cos^{-1} 0.936$ in   radian format

d. $\cot^{-1} 1.432$ in   ° format

e. $\csc^{-1} 2.155$ in   radian format

f. $\sec^{-1} 3.862$ in   grade format

PRESS	DISPLAY	COMMENT
$\boxed{\text{Deg}}$ .321 $\boxed{\text{INV}}$ $\boxed{\text{sin}}$ [1]	18.723	Answer to a: 18.723°
4.83 $\boxed{\text{INV}}$ $\boxed{\text{tan}}$ $\boxed{\text{INV}}$ $\boxed{\text{D.MS}}$	78.1810	Answer to b: 78°18'10"
$\boxed{\text{Rad}}$ .936 $\boxed{\text{INV}}$ $\boxed{\text{cos}}$	0.360	Answer to c: 0.36 rad
$\boxed{\text{Deg}}$ 1.432 $\boxed{1/x}$ $\boxed{\text{INV}}$ $\boxed{\text{tan}}$	34.928	Answer to d: 34.928°
$\boxed{\text{Rad}}$ 2.155 $\boxed{1/x}$ $\boxed{\text{INV}}$ $\boxed{\text{sin}}$	0.483	Answer to e: 0.483 rad
$\boxed{\text{Grad}}$ 3.862 $\boxed{1/x}$ $\boxed{\text{INV}}$ $\boxed{\text{cos}}$	83.326	Answer to f: 83.326 grad

PROBLEMS

1. Using your calculator, determine the value of the following angles in the units indicated.

    a. $\sin^{-1} 0.427$ in   ° format
    b. $\cos^{-1} 0.563$ in   radian format
    c. $\tan^{-1} 2.15$ in   grad format
    d. $\csc^{-1} 2.17$ in   °, ', " format
    e. $\sec^{-1} 5.12$ in   ° format
    f. $\cot^{-1} 0.532$ in   radian format

---

[1]Note that depressing the $\boxed{\text{INV}}$ key before depressing the trigonometric function key calculates the arc function.

2. Using your calculator, perform the following exercises to develop proficiency in solving problems that intermix both the trigonometric and inverse trigonometric functions. Be careful as to whether your answer is purely numerical or expressed as an angle.

a. $\sin (\cos^{-1} 0.325)°$

b. $\tan (\sin^{-1} 0.721)$ grad

c. $\cot (0.01 \sin^{-1} 0.5)$ rad

d. $\dfrac{\sin (\csc^{-1} 1.55)°}{\cos (\tan^{-1} 2.0)°}$

e. $\dfrac{\cos (\sin^{-1} 0.153)°}{\cot (0.01 \cos^{-1} 0.253) \text{ rad}}$

f. $\dfrac{\tan (\cot^{-1} 1.235) \text{ grad}}{\csc (0.017 \tan^{-1} 2.052) \text{ rad}}$

g. $\tan^{-1} (15 \sin (\tan^{-1} 2.89)° + 23 \sin (\cot^{-1} 1.15)°)$

h. $\sin^{-1} (0.5 \cos (\tan^{-1} 0.96)° + 0.95 \cos (\csc^{-1} 1.2)°)$

15.4 PRACTICAL APPLICATIONS OF THE TRIGONOMETRIC FUNCTIONS

Previous sections discussed the basic trigonometric functions and their inverse (the arc functions). The problems at the end of each section were only exercises for manipulating the calculator to determine the numerical solutions involving these functions. The exercises did not provide the in-depth appreciation for the *practical* applications of the trigonometric functions that this section covers:

1. As purely numerical solutions of right triangles (which will be invaluable when you study vectors)
2. For solving electronic circuits.

Numerical Solutions

1. Knowing the hypotenuse and one angle of a right triangle. See Figure 15.10; determine a, b, and θ.

$$\sin 49.2° = \frac{\text{opposite}}{\text{hypotenuse}} = \frac{b}{11.87}$$
$$\therefore b = 11.87 \sin 49.2° = 8.986$$

$$\cos 49.2° = \frac{\text{adjacent}}{\text{hypotenuse}} = \frac{a}{11.87}$$
$$\therefore a = 11.87 \cos 49.2° = 7.756$$
$$\theta = 90° - 49.2° = 40.8°$$

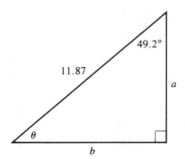

Figure 15.10   Knowing the hypotenuse and one angle.

or using the inverse functions,

$$\tan \theta = \frac{a}{b} = \frac{7.756}{8.986} = 0.863$$

$$\therefore \theta = \tan^{-1} 0.863 = 40.8°$$

2. Knowing one side and one angle of a right triangle. See Figure 15.11; determine h, b, and ϕ.

$$\sin 36.5° = \frac{\text{opposite}}{\text{hypotenuse}} = \frac{17.2}{h}$$

$$\therefore h = \frac{17.2}{\sin 36.5°} = 28.916$$

$$\tan 36.5° = \frac{\text{opposite}}{\text{adjacent}} = \frac{17.2}{b}$$

$$\therefore b = \frac{17.2}{\tan 36.5°} = 23.244$$

$$\phi = 90° - 36.5° = 53.5°$$

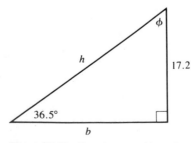

Figure 15.11   Knowing one side and one angle.

or using the inverse function,

$$\tan \phi = \frac{23.244}{17.2} = 1.351$$

$$\phi = \tan^{-1} 1.351 = 53.499°$$

3. Knowing two sides of a right triangle. See Figure 15.12; find h, θ, and ϕ.

$$h = (13.57^2 + 11.89^2)^{1/2} = 18.042$$

$$\theta = \tan^{-1} \frac{11.89}{13.57} = 41.225°$$

$$\phi = \tan^{-1} \frac{13.57}{11.89} = 48.775°$$

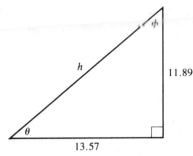

Figure 15.12   Knowing two sides.

4. Knowing one side and the hypotenuse of a right triangle. See Figure 15.13; determine θ, ϕ, and b.

$$b^2 = 27.62^2 - 15.21^2$$

$$\therefore b = \sqrt{27.62^2 - 15.21^2} = 23.055$$

$$\cos \theta = \frac{15.21}{27.62} = 0.551$$

$$\therefore \theta = \cos^{-1} 0.551 = 56.56°$$

$$\phi = 90° - 56.56° = 33.44°$$

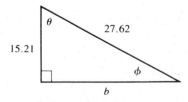

Figure 15.13   Knowing the hypotenuse and one side.

Electronic Circuit Solutions

The electronic theory behind the following three examples is more appropriate for a textbook about electronics. These examples better illustrate the applicability of basic trigonometric theory to electronic circuit solutions.

1. The instantaneous voltage of an alternator is given as

$$v = V_p \sin 2\pi f t$$

An alternator has a maximum output voltage (V_p) of 78 V and a frequency (f) of 500 Hz. What is v when t equals 0.32 ms? *Note: $2\pi f t$ represents radians.*

$$v = 78 \sin 2\pi \times 500 \times 0.32 \times 10^{-3}$$
$$= 78 \sin 1.005 \text{ rad}$$
$$= 65.86 \text{ V}$$

2. In a manner analogous to 1,

$$i = I_p \sin 2\pi f t$$

In the circuit in Figure 15.14, how long will it take for i to equal 4 mA? *Note: V (peak) is 35 V.*

$$I_p = \frac{V_p}{R} = \frac{35 \text{ V}}{5.6 \text{ k}\Omega} = 6.25 \text{ mA}$$
$$4 \times 10^{-3} = 6.25 \times 10^{-3} \sin (2\pi \times 750 \times t)$$
$$\sin 1500 \, \pi t = \frac{4 \times 10^{-3}}{6.25 \times 10^{-3}} = 0.64$$
$$1500 \, \pi t = \sin^{-1} 0.64 \text{ rad} = 0.6945$$
$$t = \frac{0.6945}{1500 \pi} = 147.38 \ \mu s$$

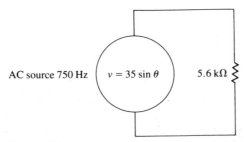

AC source 750 Hz $v = 35 \sin \theta$ 5.6 kΩ

Figure 15.14

3. See Figure 15.15; in a series L/R circuit (Figure 15.15a), the resistance (R) and the reactance (X_L) form the base and altitude, respectively, of a right triangle (Figure 15.15b). The total opposition to current flow is represented by the magnitude of Z. The angle of lag of the current behind the voltage is the magnitude of $\underline{/\theta}$. If $R = 4.7$ kΩ, $f = 4.5$ kHz, and $L = 100$ mH, what is I_p if V_p equals 40 V? What is the angle (θ) between the voltage and the current?

$$X_L = 2\pi fL = 2\pi \times 4.5 \times 10^3 \times 0.1 = 2827.4 \ \Omega$$
$$R = 4700 \ \Omega \quad \text{(given)}$$
$$Z = \sqrt{R^2 + X_L{}^2} = \sqrt{4.7^2 + 2.8274^2} \ \text{k}\Omega$$
$$= 5.485 \ \text{k}\Omega$$
$$\therefore I_p = \frac{V_p}{Z} = \frac{40 \ \text{V}}{5.485 \ \text{k}\Omega} = 7.293 \ \text{mA}$$
$$\tan \theta = \frac{X_L}{R} = \frac{2827.4}{4500} = 0.628$$
$$\therefore \theta = \tan^{-1} 0.628 = 32.14°$$

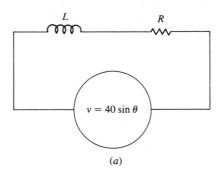

(a)

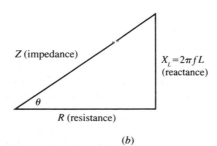

(b)

Figure 15.15

PROBLEMS

1. Use the right triangle in Figure 15.16 to solve the following.

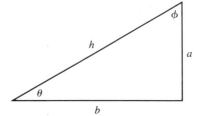

Figure 15.16

   a. $a = 16.2$, $b = 17.9$; find h, θ, ϕ

   b. $h = 3 \times 10^4$, $b = 9.2 \times 10^3$; find a, θ, ϕ

   c. $h = 119.7$, $a = 62.8$; find b, θ, ϕ

   d. $h = 21.89$, $\theta = 72.8°$, find a, b, ϕ

   e. $a = 71.62$, $\phi = 48.9$ grad; find b, h, θ

   f. $b = 36.9$, $\phi = 0.927$ rad; find a, h, θ

   g. $b = 17.82$, $\phi = 19.8°$; find a, h, θ

   h. $a = 125.96$, $\theta = 68.2$ grad; find b, h, ϕ

   i. $h = 3.86 \times 10^5$, $\theta = 1.15$ rad; find a, b, ϕ

2. In a series L/R circuit similar to that in Figure 15.15, $R = 3.3$ kΩ, $L = 100$ mH, and $f = 500$ Hz. What is the magnitude of the impedance Z?

3. In the L/R circuit in Figure 15.15, $R = 4.7$ kΩ and $L = 375$ mH. What frequency f results in an impedance Z of 6.8 kΩ?

4. Given $v = V_p \sin 2\pi f t$. What is v if $V_p = 105$ mV, $f = 150$ Hz, and $t = 750$ μs?

5. Given $i = I_p \sin 2\pi f t$. If $I_p = 350$ mA and $f = 450$ Hz, at what time t will $i = 237.5$ mA?

6. Given $v = V_p \sin 2\pi f t$. If $f = 475$ Hz, what should the peak voltage V_p be so that $v = 750$ mV when $t = 745$ μs?

15.5 RANGE, POLARITY, AND SHAPE OF THE BASIC TRIGONOMETRIC FUNCTIONS

Most of our discussion about the trigonometric functions has been restricted to the first quadrant which was sufficient for purely numerical analysis of right triangles and the L/R circuits. However, subjects to be discussed, such as the generation of AC voltages and currents, the analysis of RC circuits, vector analysis, and the j operator, require an understanding of the sine, cosine, and tangent functions throughout all four quadrants—quadrants I to IV—of a circle if we are to establish a thorough foundation for analyzing future electronic problems.

Figure 15.6 illustrated the nonlinearity of the sine function in increments of 10°
each throughout quadrant I. If we use the calculator to extend the structure of the
sine, cosine, and tangent functions in increments of 10° each, from 0° to 360°, we
get the results shown in Table 15.3. Figures 15.17a to c plot the values of Table 15.3
for these trigonometric functions.

Note that both the sine and cosine magnitudes never exceed an absolute value of
1. This is reasonable because both functions represent a ratio of either side of a right
triangle divided by the hypotenuse of the same triangle. Because the hypotenuse is
always greater than or at least equal to any side, this ratio can never exceed ±1. As

TABLE 15.3   VALUES OF TRIGONOMETRIC FUNCTIONS

Degrees	sin	cos	tan
0	0.0000	1.0000	0.0000
10	0.1736	0.9848	0.1763
20	0.3420	0.9397	0.3640
30	0.5000	0.8660	0.5774
40	0.6428	0.7660	0.8391
50	0.7660	0.6428	1.1918
60	0.8660	0.5000	1.7321
70	0.9397	0.3420	2.7475
80	0.9848	0.1736	5.6713
90	1.0000	0.0000	∞
100	0.9848	−0.1736	−5.6713
110	0.9397	−0.3420	−2.7475
120	0.8660	−0.5000	−1.7321
130	0.7660	−0.6428	−1.1918
140	0.6428	−0.7660	−0.8391
150	0.5000	−0.8660	−0.5774
160	0.3420	−0.9397	−0.3640
170	0.1736	−0.9848	−0.1763
180	0.0000	−1.0000	0.0000
190	−0.1736	−0.9848	0.1763
200	−0.3420	−0.9397	0.3640
210	−0.5000	−0.8660	0.5774
220	−0.6428	−0.7660	0.8391
230	−0.7660	−0.6428	1.1918
240	−0.8660	−0.5000	1.1732
250	−0.9397	−0.3420	2.7475
260	−0.9848	−0.1736	5.6713
270	−1.0000	0.0000	−∞
280	−0.9848	0.1736	−5.6713
290	−0.9397	0.3420	−2.7475
300	−0.8660	0.5000	−1.7321
310	−0.7660	0.6428	−1.1918
320	−0.6428	0.7660	−0.8391
330	−0.5000	0.8660	−0.5774
340	−0.3420	0.9397	−0.3640
350	−0.1736	0.9848	−0.1763
360	0.0000	1.0000	0.0000

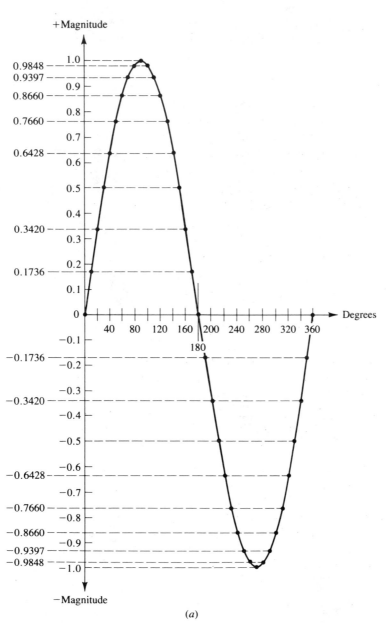

Figure 15.17 (*a*) Sine function.

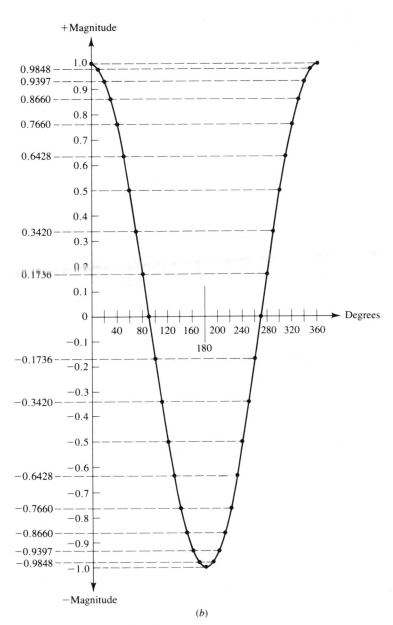

Figure 15.17   (*b*) Cosine function.

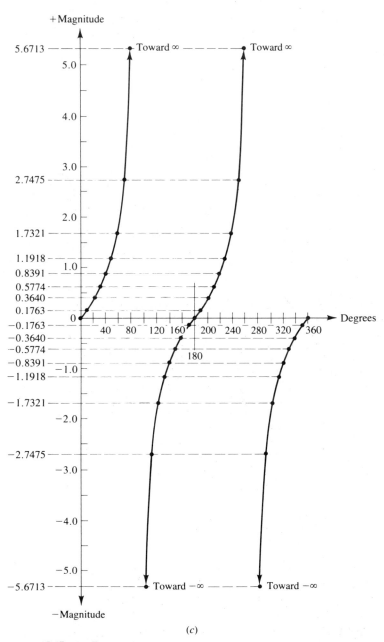

Figure 15.17 (c) Tangent function.

indicated by Figure 15.17*a* and *b*, however, the polarity of the trigonometric functions changes as a line is rotated through all four quadrants of a circle; hence the use of the term "absolute magnitude" of 1. The range of the magnitude of the tangent function, in contrast to the sine and cosine function, reaches the undefined values of $\pm\infty$. To illustrate this concept for quadrant I, you must recognize that as the vertex angle increases from $0°$ to $90°$, the side opposite the vertex angle increases in magnitude as the side adjacent to the vertex angle decreases in magnitude. Therefore, you encounter a ratio whose numerator approaches and finally equals the value of the hypotenuse while the denominator approaches and finally equals zero. At exactly $90°$ (the upper delimiter of quadrant I), the tangent function equals the unusual ratio of

$$\tan 90° = \frac{\text{magnitude of the hypotenuse}}{\text{zero}} = \infty$$

Approaching 90° in a clockwise direction, the opposite tangent value of $-\infty$ is reached by the same reasoning, hence the extreme magnitude fluctuations of the tangent function. The same rationale applies when approaching $270°$ from both directions.

Now that you have acquired a feeling for the magnitude changes of the primary trigonometric functions as a line sweeps through the four quadrants of a circle, it is necessary to investigate the polarity changes. This can be done best by translating the graphs in Figure 15.17 into the quadrant representations in Figures 15.18*a* to *d*. In the *real* world, *h* (the line being rotated) always exists, therefore it is always positive, regardless of its location within the circle. The projections of *h* on the *x* and *y* axes (which we call base *b* and altitude *a*, respectively) account for the polarity changes, according to which quadrant *h* is positioned in. Let us progress through quadrants I to IV:

1. In Figure 15.18*a*, note that $0° \leq \theta \leq 90°$. The base *b* (hypotenuse *h* projected upon the *x* axis) is positive because it is in the $+x$ direction, and the altitude *a* (hypotenuse *h* projected upon the *y* axis) is positive because it is in the $+y$ direction. Triangle *OPQ* is the right triangle formed by these projections, with θ the resulting vertex angle. Therefore, with regard to only the polarities involved,

$$\sin \theta = \frac{+a}{h} = +\text{ value}$$

$$\cos \theta = \frac{+b}{h} = +\text{ value}$$

$$\tan \theta = \frac{+a}{+b} = +\text{ value}$$

2. In Figure 15.18*b*, note that $90° \leq \theta \leq 180°$. Using the same projections of *h* as in step 1, the base *b* is negative because it is in the $-x$ direction and *a* is positive because it is in the $+y$ direction. Triangle *OPQ* is the right triangle

$$\sin\theta = \frac{+a}{h} = +$$

$$\cos\theta = \frac{+b}{h} = +$$

$$\tan\theta = \frac{+a}{+b} = +$$

$$\sin\theta = \sin\phi = \frac{+a}{h} = +$$

$$\cos\theta = \cos\phi = \frac{-b}{h} = -$$

$$\tan\theta = \tan\phi = \frac{+a}{-b} = -$$

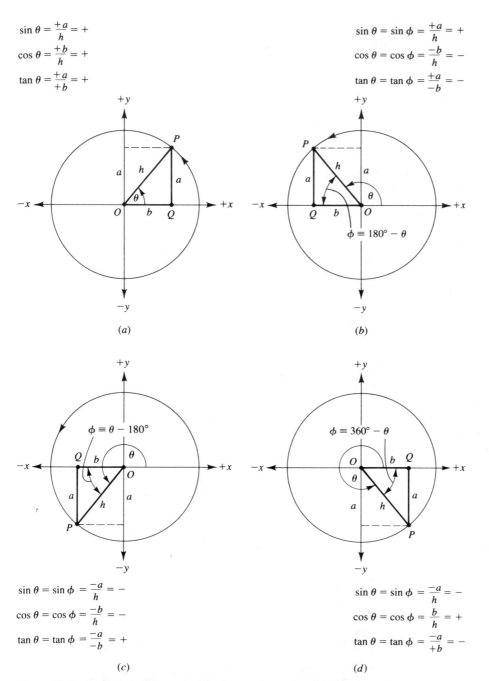

(a)

(b)

$$\sin\theta = \sin\phi = \frac{-a}{h} = -$$

$$\cos\theta = \cos\phi = \frac{-b}{h} = -$$

$$\tan\theta = \tan\phi = \frac{-a}{-b} = +$$

$$\sin\theta = \sin\phi = \frac{-a}{h} = -$$

$$\cos\theta = \cos\phi = \frac{b}{h} = +$$

$$\tan\theta = \tan\phi = \frac{-a}{+b} = -$$

(c)

(d)

Figure 15.18 (*a*) First quadrant projection of *a* onto the *y* axis and *b* onto the *x* axis.
(*b*) Second quadrant projection of *a* onto the *y* axis and *b* onto the *x* axis.
(*c*) Third quadrant projection of *a* onto the *y* axis and *b* onto the *x* axis.
(*d*) Fourth quadrant projection of *a* onto the *y* axis and *b* onto the *x* axis.

formed by these projections, with ϕ $(180° - \theta)$ the resultant vertex angle for determining the values of the basic trigonometric functions. Therefore, with regard to only the polarities involved,

$$\sin \theta = \sin \phi = \frac{+a}{h} = + \text{ value}$$

$$\cos \theta = \cos \phi = \frac{-b}{h} = - \text{ value}$$

$$\tan \theta = \tan \phi = \frac{+a}{-b} = - \text{ value}$$

To substantiate that it is actually the magnitude of the acute angle closest to the x axis which determines the absolute value of the function, perform the following simple exercise. Arbitrarily choose any angle between 90° and 180°, and with your calculator, determine the sine, cosine, and tangent functions for $\angle \theta$ and ϕ, the resultant of $180° - \theta$. Compare the results.

3. In Figure 15.18c, note that $180° \leqq \theta \leqq 270°$. The projections of h on the x and y axes, respectively, result in b being negative because it is in the $-x$ direction and a being negative because it is in the $-y$ direction. Triangle OPQ is the right triangle formed by these projections, with ϕ $(\theta - 180°)$ the resultant vertex angle for determining the values of basic trigonometric functions. Therefore, with regard to only the polarities involved,

$$\sin \theta = \sin \phi = \frac{-a}{h} = - \text{ value}$$

$$\cos \theta = \cos \phi = \frac{-b}{h} = - \text{ value}$$

$$\tan \theta = \tan \phi = \frac{-a}{-b} = + \text{value}$$

4. In Figure 15.18d, note that $270° \leqq \theta \leqq 360°$. The projections of h on the x and y axes, respectively, result in b being positive because it is in the $+x$ direction and a being negative because it is in the $-y$ direction. Triangle OPQ is a right triangle formed by these projections, with $\phi(360° - \theta)$ the resultant vertex angle for determining the values of the basic trigonometric functions. Therefore, with regard to only the polarities involved,

$$\sin \theta = \sin \phi = \frac{-a}{h} = - \text{ value}$$

$$\cos \theta = \cos \phi = \frac{b}{h} = + \text{value}$$

$$\tan \theta = \tan \phi = \frac{-a}{+b} = - \text{ value}$$

Certain angles within any given quadrant tend to occur more frequently than other angles. These special angles and the major axes that encompass them are the x and y axes themselves and the 30°, 45° and 60° angles within each respective quadrant. To establish exacting trigonometric ratios for these values, consider the following:

1. See Figure 15.19, which has both 30° and 60° angles. Hypotenuse h is purposely given a value of 2 so that the side opposite the 30° angle will be 1 ($a = h \sin 30° = 2(0.5) = 1$). By the Pythagorean theorem, $b = \sqrt{2^2 - 1^2} = \sqrt{3}$. After the basic trigonometric functions are applied, the following values result:

$$\sin 30° = \frac{1}{2}$$

$$\cos 30° = \frac{\sqrt{3}}{2}$$

$$\tan 30° = \frac{1}{\sqrt{3}}$$

$$\sin 60° = \frac{\sqrt{3}}{2} = \cos 30°$$

$$\cos 60° = \frac{1}{2} = \sin 30°$$

$$\tan 60° = \sqrt{3} = \frac{1}{\tan 30°} = \cot 30°$$

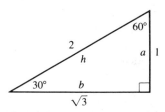

Figure 15.19 Ratios of the 30° and 60° angles.

2. See Figure 15.20. In this right triangle both acute angles are 45°. Each side of the triangle is purposely given a value of 1 so that h will be $\sqrt{2}$ ($h = \sqrt{1^2 + 1^2} = \sqrt{2}$). After the primary trigonometric functions are applied, the following values result:

$$\sin 45° = \frac{1}{\sqrt{2}}$$

$$\cos 45° = \frac{1}{\sqrt{2}}$$

$$\tan 45° = 1$$

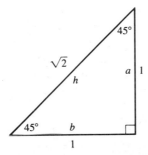

Figure 15.20   Ratios of the 45° angle.

Combining these conclusions for a 30°, 45° and 60° angle with the polarities associated with each of the four quadrants produces Table 15.4, a summary of the magnitudes of the key angles within the entire circle.

TABLE 15.4   MAGNITUDES OF KEY ANGLES

Quadrant	Angle	sin	cos	tan
I	0°	0	1	0
	30°	$\dfrac{1}{2}$	$\dfrac{\sqrt{3}}{2}$	$\dfrac{1}{\sqrt{3}}$
	45°	$\dfrac{1}{\sqrt{2}}$	$\dfrac{1}{\sqrt{2}}$	1
	60°	$\dfrac{\sqrt{3}}{2}$	$\dfrac{1}{2}$	$\sqrt{3}$
	90°	1	0	∞
II	120°	$\dfrac{\sqrt{3}}{2}$	$\dfrac{-1}{2}$	$-\sqrt{3}$
	135°	$\dfrac{1}{\sqrt{2}}$	$\dfrac{-1}{\sqrt{2}}$	-1
	150°	$\dfrac{1}{2}$	$\dfrac{-\sqrt{3}}{2}$	$\dfrac{-1}{\sqrt{3}}$
	180°	0	-1	0
III	210°	$\dfrac{-1}{2}$	$\dfrac{-\sqrt{3}}{2}$	$\dfrac{1}{\sqrt{3}}$
	225°	$\dfrac{-1}{\sqrt{2}}$	$\dfrac{-1}{\sqrt{2}}$	1
	240°	$\dfrac{-\sqrt{3}}{2}$	$\dfrac{-1}{2}$	$\sqrt{3}$
	270°	-1	0	$-\infty$ [1]
IV	300°	$\dfrac{-\sqrt{3}}{2}$	$\dfrac{1}{2}$	$-\sqrt{3}$
	315°	$\dfrac{-1}{\sqrt{2}}$	$\dfrac{1}{\sqrt{2}}$	-1
	330°	$\dfrac{-1}{2}$	$\dfrac{\sqrt{3}}{2}$	$\dfrac{-1}{\sqrt{3}}$
	360°	0	1	0

[1] Approaching 270° clockwise.

PROBLEMS

1. Evaluate the cosecant, secant, and cotangent functions at 0°, 30°, 45°, 60°, and 90°.

2. Draw a graph of the cosecant, secant, and cotangent functions from 0° to 360°.

3. Repeat Problem 1 for 120°, 135°, and 360°.

4. Repeat Problem 1 for 180°, 210°, 225°, and 240°.

5. Repeat Problem 1 for 270°, 300°, 315°, and 330°.

15.6 POWERS OF THE TRIGONOMETRIC FUNCTIONS

When a trigonometric function is raised to a power, there are two ways to indicate this process. For example, if the sine of $\underline{/\theta}$ is to be squared, it can be represented as

$$\sin^2 \theta \quad \text{or} \quad (\sin \theta)^2$$

The first method is preferable because it does not require writing a set of parentheses to represent this operation. Any power can be associated with the function, except the power -1, which exclusively implies the inverse trigonometric function. In summary, $\sin^{-1} \theta$ means the inverse sine function, *not* that $\sin \theta$ is to be raised to the -1 power. If you want to express the reciprocal of $\sin \theta$, you must write it as $(\sin \theta)^{-1}$ and not (again, to stress the point) as $\sin^{-1} \theta$. All other numbers can be written as a power of a trigonometric function. The normal laws of exponents apply equally well to the trigonometric functions.

Example 1

a. $(\tan^2 \theta)(\tan^3 \theta) = \tan^5 \theta$ first law of exponents

b. $\dfrac{\sin^2 \phi}{\sin^3 \phi} = (\sin \phi)^{-1}$ or $\csc \phi$ second law of exponents

c. $(\sec^{1/2} \beta)^4 = \sec^2 \beta$ third law of exponents

d. $\sqrt[3]{\cos^6 \psi} = \cos^2 \psi$ fourth law of exponents

Trigonometric functions, unlike abstract base numbers raised to powers, are a little more versatile when the laws of simplification are applied to them. For example,

$$\frac{\sin \alpha}{\csc \alpha} = \sin \alpha \left(\frac{1}{\csc \alpha} \right) = \sin \alpha \cdot \sin \alpha = \sin^2 \alpha$$

These simplifications are explored more thoroughly in the next section, "Trigonometric Identities."

When using your calculator to evaluate trigonometric functions raised to powers, remember that special functions such as x^2, y^x, and the trigonometric functions themselves all react immediately upon the value in the calculator's display register. Therefore you must evaluate the trigonometric function first and then the power sequence.

Example 2   Fix 3

a.   Evaluate $\sin^2 28° + \tan^3 64°$

PRESS	DISPLAY	COMMENT
Deg 28 sin x^2	0.220	$\sin^2 28°$
+ 64 tan y^x 3 =	8.839	Solution

b.   Evaluate $\dfrac{\sec^{1/2} 1.13 \text{ rad}}{\cos^2 63°}$

PRESS	DISPLAY	COMMENT
Rad 1.13 cos $1/x$ $\sqrt{}$	1.531	$\sec^{1/2} 1.13$ rad
÷ Deg 63 cos x^2	0.206	$\cos^2 63°$
=	7.428	Solution

PROBLEMS

Using your calculator, perform the following exercises to gain proficiency in resolving problems of trigonometric functions raised to powers.

1. $\sin^2 43.3° - \cos^2 11.2°$

2. $17 \tan^3 1.3 \text{ rad} + 8 \sin^3 0.65 \text{ rad}$

3. $2.4 \sec^{1/2} 32.5° + 1.5 \csc^{1/3} 42.1°$

4. $\dfrac{1.95 \cot^{1/2} 1.1 \text{ rad}}{0.87 \tan^2 0.63 \text{ rad}}$

5. $\dfrac{4.82 \sin^{3/2} 55 \text{ grad}}{1.82 \sec^{1/2} 31.6 \text{ rad}}$

6. $(2.13 \csc^{1/2} 0.9 \text{ rad})(4.21 \sec^{1/2} 55°)$

7. $(42.6 \sin^{1/4} 61.8°)^{1.2}$

8. $(5.2 \csc^{1/2} 0.34 \text{ rad})^{\sin 37°}$

9. $\dfrac{(1.11 \cot^{2/7} 56.1°)^{1.52}}{2.187 \sec^{1/2} 72.6 \text{ grad}}$

10. $\dfrac{(8.9 \cos^{1/4} 17.2°)^{\tan 51°}}{(4.96 \csc^{1/3} 42.9°)^{\cos 0.9 \text{ rad}}}$

15.7 TRIGONOMETRIC IDENTITIES

As you study electronic theory, you will encounter many equations that involve the trigonometric functions. Certain combinations of these functions have predictable results that, when substituted into the original equation, greatly reduce the complexity of the equation and hence simplify its solution. These combinations are called *trigonometric identities*. This section develops some commonly encountered identities.

Using the right triangle in Figure 15.16 as reference,

$$a = h \sin \theta$$
$$b = h \cos \theta$$

$$\therefore \tan \theta = \frac{a}{b} = \frac{h \sin \theta}{h \cos \theta} = \frac{\sin \theta}{\cos \theta} \tag{15.3}$$

Because
$$\cot \theta = \frac{1}{\tan \theta}$$

it follows that
$$\cot \theta = \frac{1}{\dfrac{\sin \theta}{\cos \theta}} = \frac{\cos \theta}{\sin \theta} \tag{15.4}$$

These are the simplest trigonometric identities; they are useful when simplifying complex equations.

Again, using the right triangle in Figure 15.16,

$$h^2 = a^2 + b^2 \tag{15.5}$$
$$a = h \sin \theta$$
$$\therefore a^2 = h^2 \sin^2 \theta \tag{15.6}$$
$$b = h \cos \theta$$
$$\therefore b^2 = h^2 \cos^2 \theta \tag{15.7}$$

When substituting Equations (15.6) and (15.7) into Equation (15.5), the results are

$$h^2 = h^2 \sin^2 \theta + h^2 \cos^2 \theta \tag{15.8}$$

Dividing each side of Equation (15.8) by h^2 produces the following:

$$1 = \sin^2 \theta + \cos^2 \theta \tag{15.9}$$

Interesting! Equation (15.9) says that no matter what angle is selected, its sine squared plus its cosine squared is *always* 1.

Try it: Let us arbitrarily use as examples 23.91° and 2.46 rad.

PRESS		DISPLAY	COMMENT
[Deg] 23.91 [sin] [x^2] [+] 23.91 [cos] [x^2] [=]		1	Satisfied?
[Rad] 2.46 [sin] [x^2] [+] 2.46 [cos] [x^2] [=]		1	Satisfied?

Now, if we in turn divide Equation (15.9) by $\sin^2 \theta$, we get

$$\frac{1}{\sin^2} = 1 + \frac{\cos^2 \theta}{\sin^2 \theta}$$

or

$$\csc^2 \theta = 1 + \cot^2 \theta \qquad (15.10)$$

Finally, if we divide Equation (15.9) by $\cos^2 \theta$, we get

$$\frac{1}{\cos^2 \theta} = \frac{\sin^2 \theta}{\cos^2 \theta} + 1$$

or

$$\sec^2 \theta = \tan^2 \theta + 1 \qquad (15.11)$$

PROBLEMS

Using your calculator and the right triangle shown in Figure 15.21, verify Problems 1 to 5.

1. $\sin^2 \alpha + \cos^2 \alpha = 1$

2. $\tan \beta = \dfrac{\sin \beta}{\cos \beta}$

3. $\cot^2 \alpha + 1 = \csc^2 \alpha$

4. $\dfrac{\cos \beta}{\sin \beta} = \cot \beta$

5. $\tan^2 \alpha + 1 = \sec^2 \alpha$

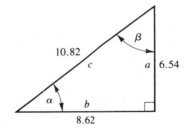

Figure 15.21

Simplify, and then evaluate Problems 6 to 10 without using your calculator.

6. $\sin 45°(\sin^2 15° + \cos^2 15°)$

7. $\sin 30° \cos^2 35° + \sin 30° \sin^2 35°$

8. $\cos 180°(\sec^2 81° - \tan^2 81°)$

9. $\dfrac{\cos 45°}{\cot^2 117° - \csc^2 117°}$

10. $\dfrac{\sin 210° \sin^2 98° + \sin 210° \cos^2 98°}{\cos 0°}$

15.8 LAWS OF SINES AND COSINES

Earlier chapters discussed equations of the second degree, that is, the form $ax^2 + bx + c = 0$. *Nice* equations were studied first for their appropriate solutions. But these nice equations were merely special cases of the generalized format—special because they can be factored easily. The generalized solution was the quadratic equation

$$x_{1,2} = \frac{-b \pm \sqrt{b^2 - 4ac}}{2a}$$

which solves *all* second-degree equations, whether or not they can be easily factored. But why not just teach the quadratic equation from the start and forget all that factoring? One of many valid answers is that it is first necessary to understand the easier types of problems before you can understand the more complex derivations of a generalized solution. But what does the quadratic equation have to do with trigonometric functions? By itself, nothing, but in principle, everything. Right triangles, just like easily factorable equations, are specialized cases of the generalized triangle. They are similar to the nice factorable equations because special, simple rules like the Pythagorean theorem can be applied. This section develops the generalized solution of triangles and illustrates that when applying this solution to right triangles, the Pythagorean theorem results.

The Law of Sines

Figure 15.22 is an irregularly shaped triangle, not a right triangle. Because we no longer have the unique altitude, base, or hypotenuse of the right triangle, we label this triangle's sides as x, y, and z to completely divorce it from our previous nomenclature and we label the angles X, Y, and Z. Now, if we drop a line from point P to point O such that it is perpendicular to z, we can define this line as h, the *height* of the triangle. Keep in mind that in this case h denotes height, not hypotenuse. By virtue of our construction, both triangles PQO and PRO in Figure 15.22 are now right triangles, so the basic trigonometric functions can be applied. From triangle PQO,

$$h = x \sin Y \qquad (15.12)$$

From triangle PRO,

$$h = y \sin X \qquad (15.13)$$

Because things equal to the same things can be equated, Equation (15.12) equals Equation (15.13), or

$$x \sin Y = y \sin X \qquad (15.14)$$

and

$$\frac{x}{\sin X} = \frac{y}{\sin Y} \qquad (15.15)$$

Equation (15.15) suggests a unique verbal description:

Side x is to the sine of the angle opposite x as side y is to the sine of the angle opposite y.

Interesting! Does this same uniqueness extend to z and the sine of z? To answer this question, redraw Figure 15.22 so that the perpendicular originates from either of the other two vertex angles; let us arbitrarily say $\underline{/Y}$. The resultant is Figure 15.23. Now, from triangle PQO

$$h = x \sin Z \tag{15.16}$$

and from triangle RQO

$$h = z \sin X \tag{15.17}$$

Again equating Equations (15.16) and (15.17) (things equal to the same things equal each other), there results

$$x \sin Z = z \sin X$$

or

$$\frac{x}{\sin X} = \frac{z}{\sin Z} \tag{15.18}$$

But notice now that Equation (15.15) can be equated to Equation (15.18), resulting in

$$\frac{x}{\sin X} = \frac{y}{\sin Y} = \frac{z}{\sin Z} \tag{15.19}$$

We may now expand our previous definition:

All the ratios of any side of a triangle to the sine of the angle opposite that side are equal.

The inverse of course is also true:

$$\frac{\sin X}{x} = \frac{\sin Y}{y} = \frac{\sin Z}{z} \tag{15.20}$$

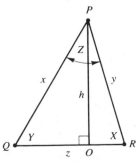

Figure 15.22 Irregularly shaped triangle.

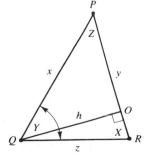

Figure 15.23 Completion of the law of sines solution.

Example 1

Use your calculator and the nomenclature in Figure 15.22. Given one side and two angles: $x = 17.2$, $Y = 68.7°$, and $Z = 41.2°$. Find X, y, and z. Sketch the resulting triangle as Figure 15.24.

a. $X = 180° - Y - Z = 180° - 68.7° - 41.2° = 70.1°$

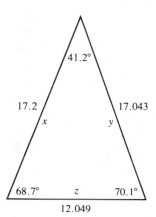

Figure 15.24

Using Equation (15.19),

b. $\dfrac{y}{\sin Y} = \dfrac{x}{\sin X}$

$\therefore y = \dfrac{x \sin Y}{\sin X} = \dfrac{17.2 \sin 68.7°}{\sin 70.1°}$

c. $\dfrac{z}{\sin Z} = \dfrac{x}{\sin X}$

$\therefore z = \dfrac{x \sin Z}{\sin X} = \dfrac{17.2 \sin 41.2°}{\sin 70.1°}$

PRESS	DISPLAY	COMMENT
17.2 $\boxed{\div}$ 70.1 $\boxed{\sin}$ $\boxed{=}$ $\boxed{\text{STO 1}}$ $\boxed{\times}$ 68.7 $\boxed{\sin}$ $\boxed{=}$	17.043	y
$\boxed{\text{RCL 1}}$ $\boxed{\times}$ 41.2 $\boxed{\sin}$ $\boxed{=}$	12.049	z

Example 2

Use your calculator and the nomenclature in Figure 15.22. Given one angle and two sides: $z = 14.83$, $y = 12.4$, $Y = 25.14°$. Find Z, X, and x. Sketch the resulting triangle as Figure 15.25.

Using Equation (15.20),

a. $\dfrac{\sin Z}{z} = \dfrac{\sin Y}{y}$

$\sin Z = \dfrac{z \sin Y}{y}$

$Z = \sin^{-1}\left(\dfrac{z \sin Y}{y}\right) = \sin^{-1}\left(\dfrac{14.83 \sin 25.14°}{12.4}\right)$

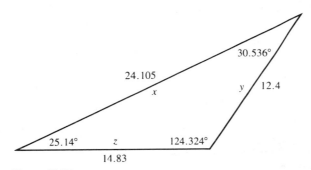

Figure 15.25

PRESS

PRESS	DISPLAY	COMMENT
14.83 $\boxed{\times}$ 25.14 $\boxed{\sin}$ $\boxed{\div}$ 12.4 $\boxed{=}$ $\boxed{\text{INV}}$ $\boxed{\sin}$	30.536	Z in degrees

b. $X = 180° - 25.14° - 30.536° = 124.324°$

Using Equation (15.19),

c. $\dfrac{x}{\sin X} = \dfrac{y}{\sin Y}$

$x = \dfrac{y \sin X}{\sin Y} = \dfrac{12.4 \sin 124.324°}{\sin 25.14°}$

PRESS	DISPLAY	COMMENT
12.4 $\boxed{\times}$ 124.324 $\boxed{\sin}$ $\boxed{\div}$ 25.14 $\boxed{\sin}$ $\boxed{=}$	24.105	x

The Law of Cosines

In Example 2, if $\underline{/X}$ (the angle between the two known sides) had been the angle given rather than either $\underline{/Y}$ or $\underline{/Z}$, we could not have used the law of sines. Under these conditions, the first equation would have been

$$\frac{\sin Y}{y} = \frac{\sin X}{x}$$

$$\therefore \sin Y = \frac{y \sin X}{x}$$

This would have resulted in both Y and x as unknowns: This is an impossible situation when you have only one equation. To get around this restriction and not limit ourselves to which three of the six quantities of a triangle are required for solution, we use the law of cosines. This law also illustrates the development of the Pythagorean theorem as a special application of the law of cosines to right triangles.

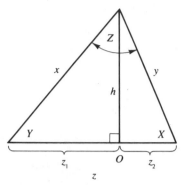

Figure 15.26   Triangle used to develop
the law of cosines.

Redraw Figure 15.22 (used to develop the law of sines) as Figure 15.26. The major difference is that we now divide side z, the side to which the perpendicular h is drawn, into two parts: z_1 and z_2. From the Pythagorean theorem,

$$x^2 = h^2 + z_1^2$$
$$h^2 = x^2 - z_1^2 \tag{15.21}$$

Similarly, $$h^2 = y^2 - z_2^2 \tag{15.22}$$

Equating Equations (15.21) and (15.22) results in

$$x^2 - z_1^2 = y^2 - z_2^2$$

Progressing algebraically,

$$x^2 = y^2 - z_2^2 + z_1^2$$
$$= y^2 - z_2^2 + (z - z_2)^2$$
$$= y^2 - z_2^2 + z^2 - 2zz_2 + z_2^2$$
$$= y^2 + z^2 - 2zz_2$$

But $$z_2 = y \cos X$$
$$\therefore x^2 = y^2 + z^2 - 2yz \cos X \tag{15.23}$$

This expression forms a pattern for determining side x when given the other two sides y and z and the angle between them. This is precisely the nature of the restriction stated at the beginning of this section. If the perpendicular of Figure 15.26 is dropped from Y or X to edges y or x, respectively, similar reasoning produces the following:

$$y^2 = x^2 + z^2 - 2xz \cos Y \tag{15.24}$$
$$z^2 = x^2 + y^2 - 2xy \cos Z \tag{15.25}$$

Generalizing Equations (15.23) to (15.25) gives us the following verbal description:

The square of any side of *any* triangle equals the sum of the squares of the other two sides minus twice their product times the cosine of the angle between them.

For a simple right triangle,

$$h^2 = a^2 + b^2 - 2ab \cos 90°$$
$$= a^2 + b^2 - 2ab(0)$$
$$= a^2 + b^2$$

Now you can see how the famous Pythagorean theorem evolved. It is merely a simplified, special application of the general solution of all triangles, just as factoring second-degree equations is merely a special application of the general solution of all second-degree equations provided by the quadratic equation.

Example 3

Using your calculator at $\boxed{\text{Fix 3}}$, solve for the unknowns in Figure 15.27. Note that the shape suggests all angles are acute.

$$y^2 = 15.26^2 + 14.15^2 - 2(15.26)(14.15) \cos 19.6°$$

PRESS	DISPLAY	COMMENT
15.26 $\boxed{x^2}$ $\boxed{+}$ 14.15 $\boxed{x^2}$ $\boxed{-}$ 2 $\boxed{\times}$ 15.26		
$\boxed{\times}$ 14.15 $\boxed{\times}$ 19.6 $\boxed{\cos}$ $\boxed{=}$ $\boxed{\sqrt{}}$	5.124	Solution of y

Now that we have solved for y, we may apply the law of sines or cosines to determine X or Z. We use the law of sines because it is simpler.

$$X = \sin^{-1}\left(\frac{15.26 \sin 19.6°}{5.124}\right)$$

PRESS	DISPLAY	COMMENT
15.26 $\boxed{\times}$ 19.6 $\boxed{\sin}$ $\boxed{\div}$ 5.124 $\boxed{=}$ $\boxed{\text{INV}}$ $\boxed{\sin}$	87.466	$X = 87.466°$

$$\therefore Z = 180° - 19.6° - 87.466° = 72.934°$$

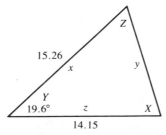

Figure 15.27

Example 4

Solve for the unknowns in Figure 15.28.

$$z = (29.32^2 + 12.7^2 - 2(29.32)(12.7)(\cos 21.5°))^{1/2} = 18.112$$

At this point we can first solve for X (the acute angle) and then substract both this value and 21.5° from 180° to obtain Y. Or we can first solve for Y (the obtuse angle) and subtract both this value and 21.5° from 180° to obtain X. Let us illustrate both approaches to show how to detect an obtuse angle (and triangle) if the dimensions given do not make this apparent.

a.   Solve for X (the acute angle) first. Use the law of sines:

$$\frac{\sin X}{12.7} = \frac{\sin 21.5°}{18.112}$$

$$X = \sin^{-1}\left(\frac{12.7 \sin 21.5°}{18.112}\right) = 14.891°$$

$$\therefore Y = 180° - 21.5° - 14.891° = 143.609°$$

Knowing that 29.32 is the dimension of the longest side of Figure 15.28 tells us that Y must be the largest angle; hence the fact that it is obtuse seems reasonable.

b.   Solve for Y (the obtuse angle) first. Use the law of sines:

$$\frac{\sin Y}{29.32} = \frac{\sin 21.5°}{18.112}$$

$$Y = \sin^{-1}\left(\frac{29.32 \sin 21.5°}{18.112}\right) = 36.391°$$

$$\therefore X = 180° - 21.5° - 36.391° = 122.109°$$

Wait a minute—something is wrong. Angle Y is supposed to be the largest angle because it is opposite the longest side. Could it be that because the sine function is positive in both quadrants I and II, Y is actually in quadrant II and 36.391° is really the angle between side x and the horizontal axis? Try it.

$$Y = 180° - 36.391° = 143.609°$$

X of course is now $180° - 21.5° - 143.609° = 14.891°$

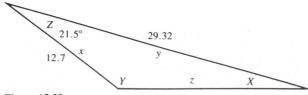

Figure 15.28

Because the sine function has the same polarity in both quadrants I and II, but the cosine function changes polarity, can we use the law of cosines to solve for both X and Y, thereby eliminating the "intuition" needed to correct for obtuse triangles? Let us try.

Example 5

Repeat Example 4, using the law of cosines to solve for X and Y. Given from Example 4 that $z = 18.112$. Now, using the law of cosines, the following results:

$$y^2 = x^2 + z^2 - 2xz \cos Y$$

$$Y = \cos^{-1}\left(\frac{x^2 + z^2 - y^2}{2xz}\right)$$

$$= \cos^{-1}\left(\frac{12.7^2 + 18.112^2 - 29.32^2}{2 \times 12.7 \times 18.112}\right)$$

PRESS	DISPLAY	COMMENT
$($ 12.7 $\boxed{x^2}$ $\boxed{+}$ 18.112 $\boxed{x^2}$ $\boxed{-}$ 29.32 $\boxed{x^2}$ $)$	-370.328	Negative number means quadrant II angle
$\boxed{\div}$ $($ 2 $\boxed{\times}$ 12.7 $\boxed{\times}$ 18.112 $)$ $\boxed{=}$ $\boxed{\text{INV}}$ $\boxed{\cos}$	143.609	$= 143.609°$

The calculator automatically adjusts for an obtuse angle when the cosine is detected as negative.

$$x^2 = y^2 + z^2 - 2yz \cos X$$

$$\therefore X = \cos^{-1}\left(\frac{y^2 + z^2 - x^2}{2yz}\right) = \cos^{-1}\left(\frac{29.32^2 + 18.112^2 - 12.7^2}{2 \times 29.32 \times 18.112}\right) = 14.891°$$

We also could have determined X by subtracting Y and Z from $180°$.

So it appears that a trade-off occurs. If it is intuitively obvious that an angle is obtuse, you can use the simpler law of sines to determine the angle's value. If it is not intuitively obvious that this is true, it is more prudent to use the slightly more complex law of cosines to determine the unknown angles.

PROBLEMS

All the problems in this section refer to a generalized triangle ABC whose angles are A, B, and C. The side opposite $\angle A$ is a, the side opposite $\angle B$ is b, and the side opposite $\angle C$ is c. In Problems 1 to 4 use the law of sines.

1. In triangle ABC $a = 193.8$, $b = 236.3$, and $B = 63.25°$. Find A.

2. In triangle ABC $c = 28.7$, $a = 36.3$, and $A = 48.6°$. Find C.

3. In triangle ABC $A = 66.5°$, $B = 41.8°$, and $c = 12.6$. Find a and b.

4. In triangle ABC $A = 69°12'$, $b = 261.5$, and $C = 51°34'$. Find c.

In Problems 5 to 8 use the law of cosines.

5. Find the third side in triangle ABC, where $b = 55.6$, $c = 43.3$, and $A = 37°45'$.

6. Find the third side in triangle ABC, where $a = 28.5$, $c = 44.3$, and $B = 108.54°$.

7. In triangle ABC $c = 38.62$, $a = 32.51$, and $B = 49.6°$. Find b.

8. In triangle ABC $c = 18.6$, $a = 31.2$, and $b = 48.6$. find C.

In Problems 9 to 15 solve triangle ABC, finding the remaining sides and angles using the law of sines, the law of cosines, or both laws.

9. $\underline{/A} = 41.5°$, $a = 86.6$, $b = 59.5$

10. $a = 16.3$, $b = 11.5$, $c = 20.45$

11. $\underline{/B} = 61°35'$, $b = 48.65$, $c = 42.8$

12. $\underline{/B} = 38°15'$, $\underline{/C} = 56°25'$, $a = 64.5$

13. $\underline{/A} = 74.6°$, $b = 2.54$, $c = 4.54$

14. $\underline{/A} = 36°18'$, $a = 45.28$, $b = 30.65$

15. $\underline{/A} = 53°25'$, $b = 154.6$, $c = 287.4$

15.9 THE SUM AND DIFFERENCE IDENTITIES

Often in electronic circuits the sine or cosine of a double angle is required. For example, in simple series circuits the instantaneous current takes the form of

$$i(t) = I_p \sin(\theta - \phi) \tag{15.26}$$

for an L/R circuit and

$$i(t) = I_p \sin(\theta + \phi) \tag{15.27}$$

for an RC circuit.

Because the trigonometric functions are not linear, it is incorrect to state, for example, that $\sin 30° + \sin 45° = \sin 75°$. It is therefore advantageous to develop these formats into simpler trigonometric relationships.

The Sum Identities

To develop the sine and cosine functions of the sum identities, it is convenient to redraw our generalized triangle with a different set of labels; see Figure 15.29. When the perpendicular h is drawn from point T to side t, it divides $\underline{/STR}$ into $\underline{/\beta}$ and $\underline{/\alpha}$ and side t into segments u and v. Two right triangles are formed: OST and ORT. Because the area of any triangle is one-half the product of its altitude times its base, it easily follows that

$$A_{\triangle OST} = \frac{uh}{2} = \frac{1}{2}hr \sin \alpha \qquad (15.28)$$

and

$$A_{\triangle ORT} = \frac{vh}{2} = \frac{1}{2}hs \sin \beta \qquad (15.29)$$

Additionally, because the whole is the sum of its parts, it follows that

$$A_{\triangle OST} + A_{\triangle ORT} = A_{\triangle SRT} \qquad (15.30)$$

The key now is to determine the area of the generalized triangle SRT and then appropriately substitute into Equation (15.30). We apply our previously learned knowledge of the law of sines to develop $A_{\triangle SRT}$.

From Figure 15.29,

$$\frac{\sin (\alpha + \beta)}{t} = \frac{\sin \omega}{s}$$

or

$$s \sin (\alpha + \beta) = t \sin \omega$$

Multiply both sides of this equation by $r/2$:

$$\frac{1}{2}rs \sin (\alpha + \beta) = \frac{1}{2}t(r \sin \omega)$$

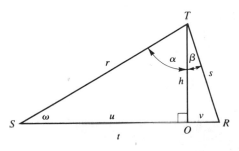

Figure 15.29   Figure for determining the sum angles.

But $r \sin \omega = h$; therefore

$$\frac{1}{2} rs \sin (\alpha + \beta) = \frac{ht}{2} \qquad (15.31)$$

Because h and t are the altitude and base of triangle SRT, respectively, then $ht/2$ equals $A_{\triangle SRT}$. Substituting Equations (15.28), (15.29), and (15.31) into Equation (15.30) results in

$$\frac{1}{2} rs \sin (\alpha + \beta) = \frac{1}{2} hr \sin \alpha + \frac{1}{2} hs \sin \beta \qquad (15.32)$$

Multiplying both sides of Equation (15.32) by $2/rs$ results in

$$\sin (\alpha + \beta) = \frac{h}{s} \sin \alpha + \frac{h}{r} \sin \beta$$

but $h/s = \cos \beta$ and $h/r = \cos \alpha$

$$\therefore \sin (\alpha + \beta) = \sin \alpha \cos \beta + \cos \alpha \sin \beta \qquad (15.33)$$

Just as we used the law of sines to develop the sine *sum* identity, we can apply the law of cosines to Figure 15.29 to develop the cosine *sum* identity. From Figure 15.29,

$$t^2 = r^2 + s^2 - 2rs \cos (\alpha + \beta)$$
$$(u + v)^2 = r^2 + s^2 - 2rs \cos (\alpha + \beta)$$
$$u^2 + 2uv + v^2 = r^2 + s^2 - 2rs \cos (\alpha + \beta)$$

Solving for $\cos (\alpha + \beta)$

$$\cos (\alpha + \beta) = \frac{r^2 + s^2 - u^2 - 2uv - v^2}{2rs}$$

$$= \frac{r^2 - u^2}{2rs} + \frac{s^2 - v^2}{2rs} - \frac{2uv}{2rs}$$

But from the two right triangles in Figure 15.29,

$$h^2 = r^2 - u^2 = s^2 - v^2$$

$$\therefore \cos (\alpha + \beta) = \frac{h^2}{2rs} + \frac{h^2}{2rs} - \frac{2uv}{2rs}$$

$$= \frac{h^2}{rs} - \frac{uv}{rs}$$

$$= \frac{h}{r} \cdot \frac{h}{s} - \frac{u}{r} \cdot \frac{v}{s} \qquad (15.34)$$

Again using Figure 15.29 as reference, the four fractions in Equation (15.24) convert to

$$\cos (\alpha + \beta) = \cos \alpha \cos \beta - \sin \alpha \sin \beta \qquad (15.35)$$

Example 1

The angle between voltage and current in an RC circuit is 30°. Develop an expression for the instantaneous current $i(t)$ in milliamps at any $\underline{/\theta}$ if the peak current (I_p) equals 10 mA.

From Equation (15.27),

$$
\begin{aligned}
i(t) &= I_p \sin(\theta + \phi) \\
&= 10 \sin(\theta + 30°) \\
&= 10(\sin\theta \cos 30° + \cos\theta \sin 30°) \\
&= 10\left(\frac{\sqrt{3}}{2}\sin\theta + \frac{1}{2}\cos\theta\right) \\
&= 5(\sqrt{3}\sin\theta + \cos\theta)
\end{aligned}
$$

The Difference Identities

To determine the sine and cosine functions of angle differences, it is necessary to draw triangles such that at least one angle is the difference of two other angles. Essentially, Figure 15.30 begins as the obtuse triangle TOS. A perpendicular is dropped from point T to an extension of base v, thereby defining the height (h) of triangle TOS. If α is the angle from r to h and β is the angle from s to h, then $\underline{/STO}$ is now the difference angle $(\alpha - \beta)$.

To develop the sine *difference* identify, we use the law of sines, much as in the same fashion as when we developed the sine *sum* identity. From Figure 15.30,

$$A_{\triangle TRO} + A_{\triangle TOS} = A_{\triangle TRS}$$

$$\frac{1}{2}hs\sin\beta + \frac{1}{2}vh = \frac{1}{2}hr\sin\alpha \qquad (15.36)$$

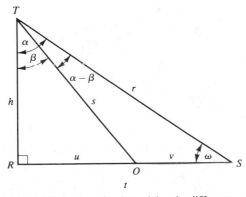

Figure 15.30   Figure for determining the difference angles.

The trick is to develop an expression for $1/2\ vh$.
From Figure 15.30,

$$\frac{\sin(\alpha - \beta)}{v} = \frac{\sin \omega}{s}$$
$$s\ \sin(\alpha - \beta) = v\ \sin \omega$$

Multiplying by $r/2$ results in

$$\frac{1}{2}rs\ \sin(\alpha - \beta) = \frac{1}{2}v(r\ \sin \omega)$$

But $r\ \sin \omega = h$; therefore

$$\frac{1}{2}rs\ \sin(\alpha - \beta) = \frac{1}{2}vh \qquad (15.37)$$

Substituting Equation (15.37) into Equation (15.36) results in

$$\frac{1}{2}hs\ \sin \beta + \frac{1}{2}rs\ \sin(\alpha - \beta) = \frac{1}{2}hr\ \sin \alpha$$

Solve algebraically for $\sin(\alpha - \beta)$:

$$\sin(\alpha - \beta) = \frac{hr\ \sin \alpha - hs\ \sin \beta}{rs}$$

$$= \frac{h}{s}\ \sin \alpha - \frac{h}{r}\ \sin \beta$$

But from Figure 15.30,

$$\frac{h}{s} = \cos \beta \qquad \text{and} \qquad \frac{h}{r} = \cos \alpha$$

$$\therefore\ \sin(\alpha - \beta) = \sin \alpha \cos \beta - \cos \alpha \sin \beta \qquad (15.38)$$

To develop the cosine *difference* identity, use Figure 15.30 and the law of cosines. From Figure 15.30,

$$v^2 = s^2 + r^2 - 2sr\ \cos(\alpha - \beta)$$
$$(t - u)^2 = s^2 + r^2 - 2sr\ \cos(\alpha - \beta)$$
$$t^2 - 2tu + u^2 = s^2 + r^2 - 2sr\ \cos(\alpha - \beta)$$

Solve for $\cos(\alpha - \beta)$

$$\cos(\alpha - \beta) = \frac{s^2 + r^2 - t^2 - u^2 + 2tu}{2sr}$$

$$= \frac{s^2 - u^2}{2sr} + \frac{r^2 - t^2}{2sr} + \frac{2tu}{2sr}$$

From Figure 15.30,

$$h^2 = s^2 - u^2 = r^2 - t^2$$

$$\therefore \cos(\alpha - \beta) = \frac{h^2}{2sr} + \frac{h^2}{2sr} + \frac{2tu}{2sr}$$

$$= \frac{h^2}{sr} + \frac{tu}{sr}$$

$$= \frac{h}{r} \cdot \frac{h}{s} + \frac{t}{r} \cdot \frac{u}{s}$$

Again using Figure 15.30 as reference, convert the ratios of the previous equation to the following:

$$\cos(\alpha - \beta) = \cos \alpha \cos \beta + \sin \alpha \sin \beta \qquad (15.39)$$

Example 2

The angle between voltage and current in an L/R circuit is 45°. Develop an expression for the instantaneous current $i(t)$ in milliamps at any $\underline{/\theta}$ if the peak current (I_p) equals 17.5 mA. From Equation (15.26):

$$i(t) = I_p \sin(\theta - \phi)$$

$$= 17.5 \sin(\theta - 45°)$$

$$= 17.5(\sin \theta \cos 45° - \cos \theta \sin 45°)$$

$$= 17.5\left(\frac{1}{\sqrt{2}} \sin \theta - \frac{1}{\sqrt{2}} \cos \theta\right)$$

$$= \frac{17.5}{\sqrt{2}}(\sin \theta - \cos \theta)$$

PROBLEMS

1. Expand the identity.

    a. $\sin(A - B) =$
    b. $\cos(\phi + \theta) =$
    c. $\sin(r + s) =$
    d. $\cos(x - y) =$

2. Rewrite as a single trigonometric function.

    a. $\sin 45° \cos 20° + \cos 45° \sin 20°$
    b. $\cos 137° \cos 47° + \sin 137° \sin 47°$
    c. $\cos 50° \cos 10° - \sin 50° \sin 10°$
    d. $\sin 26.5° \cos 94° + \cos 26.5° \sin 94°$
    e. $\cos 157° \cos 173° - \sin 157° \sin 173°$

3.   Use the identity cos $(\alpha + \beta)$ to prove cos $(\theta + 90°) = -\sin \theta$

4.   Use the identity cos $(\alpha - \beta)$ to prove cos $(270° - \theta) = -\sin \theta$

5.   Use the identity sin $(\alpha - \beta)$ to prove sin $(2\pi \text{ rad} - \theta) = -\sin \theta$

6.   Use the identity sin $(\alpha + \beta)$ to prove sin $(90° + \theta) = \cos \theta$

7.   The angle between voltage and current in an L/R circuit is 90°. Using $i(t) = I_p$ sin $(\theta - \phi)$, find $i(t)$ in milliamps at any peak angle θ if I_p equals 15.5 mA.

8.   The angle between voltage and current in an RC circuit is 45°. Develop an expression for $i(t)$ in milliamps using $i(t) = I_p$ sin $(\theta - \phi)$ if I_p equals 12.5 mA.

9.   Use the identity sin $(\alpha - \beta)$ to find the value of the following in radical form.
     a.   sin 315°   :   let $\alpha = 360°$ and $\beta = 45°$
     b.   sin 15°   :   let $\alpha = 60°$ and $\beta = 45°$

10.   Use cos $(\alpha - \beta) = \cos \alpha \cos \beta + \sin \alpha \sin \beta$ to find the value of the following in radical form:
$$\cos 15°   :   \text{let } \alpha = 45° \text{ and } \beta = 30°$$

SUMMARY

The trigonometric functions are a means for the solution of right triangles when any side and any acute angle are known. The primary functions are the sine (ratio of the opposite side to the hypotenuse), cosine (ratio of the adjacent side to the hypotenuse), and tangent (ratio of the opposite side to the adjacent side). These ratios define unique angles. The reciprocal of each ratio is designated as the cosecant, secant, and cotangent, respectively.

The inverse trigonometric functions are designated by associating a (-1) with the function in a position similar to that of an exponent. In effect, a (-1) identifies the angle having that value as a trigonometric ratio. The range of values of the primary functions are ± 1 for the sine and cosine functions and $\pm\infty$ for the tangent function.

Although the primary trigonometric functions (or their reciprocals) are restricted to the right triangle, a more generalized approach to the solution of any triangle is provided by the law of sines or the law of cosines:

1.   The law of sines states that all the ratios of any side of a triangle to the sine of the angle opposite that side are equal.
2.   The law of cosines states that the square of any side of any triangle equals the sum of the squares of the other two sides minus twice their product times the cosine of the angle between them.

Whether to apply the law of sines or the law of cosines when determining the value of an angle depends upon one's intuitive size of the angle. If it is apparent that the angle is obtuse, it is more expeditious to apply the law of sines; otherwise it is more prudent to apply the slightly more complex law of cosines.

Electronic equations frequently require the solution of the sum and/or difference of the sine and/or cosine of multiple angles. These equations are useful for determining the instantaneous values of current or voltage in resistive/reactive circuits.

16 COMPLEX ALGEBRA

16.1 INTRODUCTION

Electronics, like many other technical disciplines, often requires mathematics other than simple algebra. Electronics has evolved to become sophisticated enough to use long-established mathematical principles, such as the algebra of logic (called Boolean algebra). Boolean algebra, which is applied to computers, existed several hundred years before computers were developed. Boolean algebra is discussed in Part V; here we discuss other long-established electronic principles, such as the phase shift between voltage and current in simple L/R or RC circuits, which require the application of a form of mathematics known as *complex algebra*.

When you study a new mathematical concept, often at first the material seems to contradict previously learned laws of polarity manipulation. Hence the following short discussion illustrates the applicability of and need for complex algebra by

1. Showing the need for applying complex algebraic principles to even the most simple AC circuits
2. Encouraging and/or whetting your appetite to continue your studies of mathematics so as to better comprehend electronic phenomena

See Figure 16.1a. Suppose the instantaneous circuit (i) through a perfect inductor is defined as

$$i = I_m \sin \omega t \tag{16.1}$$

where I_m = maximum current
ω = radian frequency
t = time

Electronic theory states that the voltage (V_L) across the coil (L) is $V_L = L\, di/dt$. The expression di/dt is called the derivative of i (current) with respect to t (time). This involves the use of calculus and as such is beyond the level of this text; the important point is that when executed, Equation (16.1) reduces to

$$V_L = I_m \, \omega L \cos \omega t \tag{16.2}$$

Note that the current is a sine function, whereas the voltage is a cosine function. Referring to the sine curve in Figure 15.17a, note that the maximum positive value of 1 is reached at 90°. On the cosine curve in Figure 15.17b, the maximum positive

362

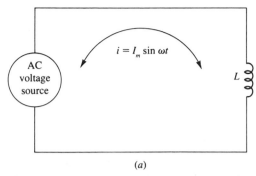

Figure 16.1 (*a*) Simple inductive circuit to determine *I*-*V* relationship.

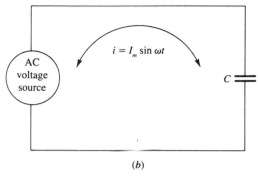

Figure 16.1 (*b*) Simple capacitive circuit to determine *I*-*V* relationship

value of 1 occurs at 0°. The cosine function therefore leads the sine function by 90°, or stated another way, the voltage across the coil leads the current through the coil by 90°.

This relationship is most easily illustrated by returning to the axis of a graph. Refer to Figure 16.2. Part *a* (the axes for now are unnamed) shows the *I*-*V* relationship for a purely resistive circuit. Both *V* and *I* are drawn in a direction of 0°. However, for the inductive circuit, *V* should be at 90° when *I* is at 0°. To accomplish this, we now introduce the *j* operator to effect the necessary rotation. If we multiply *V* in Figure 16.2*a* by *j*, then *V* is rotated 90° counterclockwise, resulting in Figure 16.2*b*, the necessary *I*-*V* relationship for an inductive circuit.

In Figure 16.1*b* there is a purely capacitive circuit. If the series current is again given as

$$i = I_m \sin \omega t$$

electronic theory states that $V_c = (1/C) \int i \, dt$. The expression $\int i \, dt$ is called the integral of i (current) with respect to t (time). Again, this is a calculus expression beyond the level of this textbook. When resolved, however, the expression reduces to

$$V_C = -\frac{I_m}{\omega C} \cos \omega t \qquad (16.3)$$

Note that similar to the inductive circuit, the current is a sine function and the voltage is the negative cosine function, which means that *V* should be 90° behind *I*, or at an angle of 270° in relation to the reference. To effect this, *V* in Figure 16.2*a* must be multiplied by $-j$ to swing it 90° clockwise. The result is Figure 16.2*c*. Thus electronic phenomena need the *j* operator to effect 90° rotations.

At this point we can now label the axes in Figure 16.2*a*: The axis in the direction of 90° (formerly $+Y$) is *j* and the axis in the direction of 270° (formerly $-Y$) is $-j$. Furthermore, because this implies $1j$ and $-1j$, respectively, the axis in the direction of 0° (formerly $+X$) is labeled $+1$ and the axis in the direction of 180° (formerly

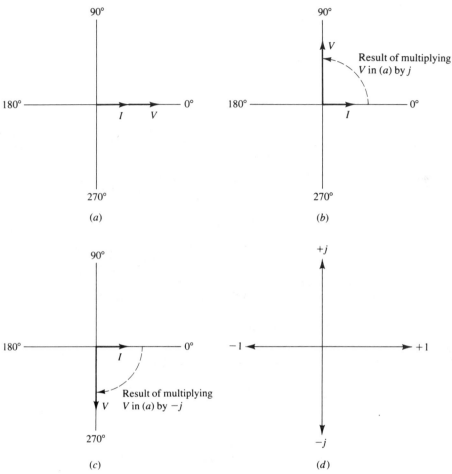

Figure 16.2 (*a*) *I*-*V* relationship for a purely resistive element.
(*b*) *I*-*V* relationship for a purely inductive element.
(*c*) I-*V* relationship for a purely capacitive element.
(*d*) Labeling the complex plane.

$-X$) is labeled -1. The result of this labeling is Figure 16.2*d*, referred to as the *complex plane*.

By convention, the ± 1 axes are the *real* axes and the $\pm j$ axes are the *imaginary* axes. This is an unfortunate choice of words because certainly the *I*-*V* inductive and capacitive relationships are not imaginary. However, if you encounter the terminology *real* axes or *imaginary* axes, remember that it refers to the complex plane in Figure 16.2*d*.

This new plane produces some interesting results and provides a solution for a group of problems that up to this point we have not had the means to handle. If we multiply $+1$, the axis in the $0°$ direction, by j, it then rotates $90°$ counterclockwise and becomes the j axis. If in turn we again multiply the j axis by j, the effect is another 90° counterclockwise rotation, in the direction of the -1 axis. Summarizing this:

$$\text{First 90° counterclockwise rotation}\quad 1 \times j = j^1 = j \qquad (16.4)$$
$$\text{Second 90° counterclockwise rotation}\quad j \times j = j^2 = -1 \qquad (16.4a)$$

If we extract the square root of both sides of Equation (16.4a), the following results:

$$\sqrt{j^2} = \sqrt{-1}$$

or
$$j \equiv \sqrt{-1} \qquad (16.5)$$

This means that the square root of -1 is defined as j. We have never been able before to take the square root of a negative number, but it now appears that Equation (16.5) eliminates this obstacle.

To summarize, the j operator

1. Mathematically facilitates extracting the square root of negative numbers
2. Electronically facilitates the representation of phasors to depict the I-V relationships of L/R or RC circuits

Example 1

Extract the square root of the following.

a. $\sqrt{49}$
   7

b. $\sqrt{-49}$
   $\sqrt{-1}\sqrt{49} = j7$

c. $\sqrt{75}$
   $\sqrt{25}\sqrt{3} = 5\sqrt{3}$

d. $\sqrt{-75}$
   $\sqrt{-1}\sqrt{25}\sqrt{3} = j5\sqrt{3}$

Custom requires that j *precede* the number. This is another of those specially reserved mathematical notations, similar to $\sin^{-1} \theta$. For example, the notation $j3$, implies a 90° rotation of the number 3 and should not be confused with $3j$, which implies three times the variable j. Furthermore, mathematics uses i as the definition of $\sqrt{-1}$. Because electronics reserves i for current and j has not yet been designated as an electronic notation, j then is the electronically accepted symbol for both a 90° counterclockwise rotation and/or the definition of $\sqrt{-1}$.

Continuing the analysis of 90° rotations,

$$j^3 = j^2 \times j = -1 \times j = -j \qquad (16.6)$$
$$j^4 = j^2 \times j^2 = -1 \times -1 = 1 \qquad (16.6a)$$

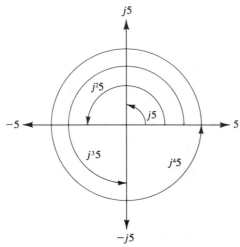

Figure 16.3   Successive application of j on 5.

Equation (16.6) represents three 90° rotations, to a final value of 270°. This is the $-j$ axis. Equation (16.6a) represents four 90° rotations, which is a return to the original +1 axis direction. Figure 16.3 summarizes all these rotations by illustrating the repeated application of j on the number 5.

PROBLEMS

Determine the square root in Problems 1 to 15.

1. $\sqrt{-121}$

2. $\sqrt{-169}$

3. $\sqrt{-\alpha^2}$

4. $-\sqrt{-\alpha^2}$

5. $-\sqrt{-49\beta^4}$

6. $\sqrt{-196i^2R^2}$

7. $-7\sqrt{49}$

8. $-7\sqrt{-49}$

9. $-5\sqrt{150}$

10. $-5\sqrt{-150}$

11. $j\sqrt{-100}$

12. $-j^2\sqrt{-100}$

13. $j^2\sqrt{\dfrac{-121}{25}}$

14. $-j^5\sqrt{\dfrac{-36}{49}}$

15. $-(-j)^2\sqrt{-169}$

16.2 THE MATHEMATICS OF j

Section 16.1 demonstrated the need for additional rules to the basic algebraic laws to solve special electronic needs and to determine the square root of negative numbers. These new rules seem to defy our previous knowledge of the laws of polarities; the balance of this chapter demonstrates that they actually add a new dimension to our ability to solve electronic problems. To apply these principles to electronic phenomena, we have to be able to implement the basic arithmetic operations that involve j.

Addition or Subtraction

When a binomial (or any polynominal) contains j as a term, treat j just as though it is an independent variable. For example, if two binomials each contain the variables x and y, a typical summation is

$$\begin{aligned} 2x + 3y \\ + \underline{5x - 6y} \\ 7x - 3y \end{aligned}$$

Each variable is added or subtracted independently, as indicated by the operators. When one term contains j, this term is as uniquely different from all other terms of the expression, including pure numbers, as the x terms were from the y terms in the preceding example.

Example 1

Perform the following operations.

a. $(2 + j3) + (4 - j5)$
$2 + 4 + j3 - j5 = 6 - j2$

b. $(-2 - j4) + (6 - j9)$
$-2 + 6 - j4 - j9 = 4 - j13$

c. $-(j6) - (3 - j4)$
$-3 - j6 + j4 = -3 - j2$

d. $-(2 - j3) - (j4)$
$-2 + j3 - j4 = -2 - j$

Multiplication

The laws of exponents apply to a series of j terms multiplied together just as they apply to any other variables. Unlike other variables, however, j terms are usually replaced with simpler expressions. If the j rotations in Figure 16.3 are extended for two full rotations, the following results:

$$\begin{aligned} j^1 &= +j & j^5 &= +j \\ j^2 &= -1 & j^6 &= -1 \\ j^3 &= -j & j^7 &= -j \\ j^4 &= +1 & j^8 &= +1 \end{aligned}$$

Whatever the power of j within the first full rotation, that evaluation of j repeats itself for any j having that power plus an integer multiple of 4.

For example,

$$j^1 = j^5 = j^9 = j^{1+4n} = j$$

This simple rule now facilitates the multiplication of any monomials or polynomials containing j. Initially develop the product in the conventional form, and then go back and simplify the j terms using the preceding rule. Example 2 illustrates the rule.

Example 2

a. $(j5)(j) = j^2 5 = (-1)(5) = -5$

b. $(-j3)(j7) = -j^2 21 = -(-1)21 = 21$

c. $(j5)(1 + j2) = j5 + j^2 10 = j5 + (-1)10 = -10 + j5$

Note that when a binominal consists of a real number and a j term, the real number is written first.

d. $(3 - j4)(5 - j) = 15 - j3 - j20 + j^2 4$
$$= 15 - j23 + (-1)4$$
$$= 15 - j23 - 4$$
$$= 11 - j23$$

e. $(-1 + j5)(-4 - j6) = 4 + j6 - j20 - j^2 30$
$$= 4 - j14 - (-1)30$$
$$= 4 - j14 + 30$$
$$= 34 - j14$$

f. $(3 - j4)^3 = 3^3 - 3(3)^2(j4) + 3(3)(j4)^2 - (j4)^3$
$$= 27 - j108 + j^2 144 - j^3 64$$
$$= 27 - j108 + (-1)144 - (-j)64$$
$$= 27 - j108 - 144 + j64$$
$$= -117 - j44$$

Division

Before we illustrate the division process, it is beneficial to first discuss what happens when *only* a j term is in the divisor. Given an expression such as

$$\frac{1}{j}$$

Both the numerator and denominator can be multiplied by j, and, just as is the case with ordinary fractions, still retain the original value of the fraction:

$$\frac{1}{j} \cdot \frac{j}{j} = \frac{j}{j^2} = \frac{j}{-1} = -j$$

Conversely, $$\frac{1}{-j} = \frac{1}{-j} \cdot \frac{j}{j} = \frac{j}{-j^2} = \frac{j}{-(-1)} = \frac{j}{1} = j$$

This is a somewhat interesting result. It appears that a single j term can be moved *up* or *down* within a fraction if only the polarity preceding j changes. This is rather analogous to moving a variable raised to a power *up* or *down* within a fraction if only the polarity of the power changes.

Example 3

Solve the following division problems when the divisor is only a singular term.

a. $\dfrac{j10}{2}$

$j\left(\dfrac{10}{2}\right) = j5$

b. $\dfrac{10}{j2}$

$-j\left(\dfrac{10}{2}\right) = -j5$

c. $\dfrac{j12}{j3}$

$\dfrac{12}{3} = 4$

d. $\dfrac{25 + j15}{5}$

$\dfrac{25}{5} + j\dfrac{15}{5} = 5 + j3$

e. $\dfrac{25 + j15}{j5}$

$-j\left(\dfrac{25}{5}\right) + \left(\dfrac{15}{5}\right) = 3 - j5$

f. $\dfrac{36 - j12}{-j4}$

$j\left(\dfrac{36}{4}\right) + \left(\dfrac{12}{4}\right) = 3 + j9 = 3(1 + j3)$

If the divisor is not a singular term but a binomial containing a pure number and a j term, you can reduce the divisor to a singular real term by a process known as multiplication by the *conjugate*. The conjugate of a complex binominal is another binominal that contains the same magnitudes of the original two terms but is separated by the opposite polarity sign. For example:

$$\text{Conjugate of } a + jb = a - jb$$

and $$\text{Conjugate of } a - jb = a + jb$$

Observe the interesting result when a complex number is multiplied by its conjugate: What is left is only a real number.

$$
\begin{array}{r}
a + jb \\
\times\ a - jb \\
\hline
a^2 + jab \\
-\ jab - j^2b^2 \\
\hline
a^2 \qquad -\ (-1)b^2 = a^2 + b^2
\end{array}
$$

Summarizing, the result of multiplying a complex number by its conjugate is always a pure number equal to the sum of the squares of the absolute values of the coefficients of each term in the original number. The expression "absolute value" is valid because any number, negative or positive, is always positive when squared.

Example 4

Find the resultant after multiplying each of the following by its conjugate.

a. $j2$

$$(j2)(-j2) = -j^2 4 = -(-1)4 = 4$$

b. $-j5$

$$(-j5)(j5) = -j^2 25 = -(-1)25 = 25$$

c. $1 + j$

$$(1 + j)(1 - j) = 1 - j^2 = 1 - (-1) = 2$$

d. $3 - j4$

$$(3 - j4)(3 + j4) = 9 - j^2 16 = 9 - (-1)16 = 25$$

e. $-5 - j6$

$$(-5 - j6)(-5 + j6) = 25 - j^2 36 = 25 - (-1)36 = 61$$

Example 5

Perform each of the following divisions.

a. $\dfrac{1}{1 + j}$

$$\frac{1}{1 + j} \cdot \frac{1 - j}{1 - j} = \frac{1 - j}{1 + 1} = \frac{1 - j}{2} = 0.5 - j0.5$$

b. $\dfrac{5}{-2 - j2}$

$$\frac{5}{-2 - j2} \cdot \frac{-2 + j2}{-2 + j2} = \frac{-10 + j10}{4 + 4} = \frac{-10 + j10}{8} = -1.25 + j1.25$$

c. $\dfrac{10 - j12}{3 + j2}$

$$\frac{10 - j12}{3 + j2} \cdot \frac{3 - j2}{3 - j2} = \frac{30 - j36 - j20 - 24}{13} = \frac{6 - j56}{13}$$

$$= 0.462 - j4.308$$

d. $\dfrac{-25 - j36}{4 + j2}$

$\dfrac{-25 - j36}{4 + j2} \cdot \dfrac{4 - j2}{4 - j2} = \dfrac{-100 - j144 + j50 - 72}{16 + 4} = \dfrac{-172 - j94}{20}$

$\qquad\qquad\qquad\qquad\qquad\qquad\qquad = -8.6 - j4.7$

e. $\dfrac{16.5 + j12.8}{3 - j4}$

$\dfrac{16.5 + j12.8}{3 - j4} \cdot \dfrac{3 + j4}{3 + j4} = \dfrac{49.5 + j38.4 + j66 - 51.2}{9 + 16} = \dfrac{-1.7 + j104.4}{25}$

$\qquad\qquad\qquad\qquad\qquad\qquad\qquad = -0.068 + j4.176$

PROBLEMS

Evaluate the complex operations in Problems 1 to 27.

1. $(2.5 - j3) + (1.7 + j7.2)$

2. $(1.38 + j6.9) + (2.4 - j5.6)$

3. $(-5.12 - j1.6) - (2.1 + j5.1)$

4. $-(2.12 - j5.3) - (1.89 - j3.25)$

5. $(3.6)(j5.6)$

6. $(j1.5)(j3.2)$

7. $(3 - j6.1)^2$

8. $(4.2 + j8.6)^2$

9. $(5.3 - j1.8)^3$

10. $(1.1 + j2.2 + j^2 3.3)^2$

11. $-(j1.8)(-j3.6)$

12. $5.7(1.6 - j2.9)$

13. $-j1.6(2.55 + j3.86)$

14. $(1.21 + j8.6)(2.89 - j3.42)$

15. $(8.16 - j1.19)(-3.21 - j5.17)$

16. $-(-2.17 + j8.6)(9.3 - j2.16)$

17. $\dfrac{1}{j5}$

18. $\dfrac{-5}{j1.2}$

19. $\dfrac{j3.5}{1.25}$

20. $\dfrac{-j2.7}{1 + j}$

21. $\dfrac{1.5 + j3}{j2}$

22. $\dfrac{3.62 + j2.5}{3 - j4}$

23. $\dfrac{21.6 - j5.7}{2.1 + j4.5}$

26. $\dfrac{(2.1 + j5.6)(1.7 - j3.6)}{2 + j5}$

24. $\dfrac{3.16 + j8.9}{2.1 - j3.5}$

27. $\dfrac{(1.6 - j2.7)(3.2 + j1.8)}{(0.5 - j1.2)(1.9 - j8.6)}$

25. $\dfrac{8.21 + j3.62}{3.14 + j5.62}$

16.3 SCALARS, VECTORS, AND PHASORS

Section 16.1 illustrated that in a series L/R or RC circuit, both voltage and current have not only a finite value, but each has a definite direction relative to the other. Current was chosen as a reference because it is the same in all elements of a series circuit. Given this magnitude in a reference direction of $0°$, voltage was drawn in a direction of $90°$ for the L/R circuit and at $270°$ ($-90°$) for the RC circuit. Thus there is a need not only to define magnitudes but to specify a direction for these magnitudes in relationship to a given reference direction. This section establishes the definitions of scalars, vectors, and phasors.

Scalars

Scalars are merely magnitudes. For example, $21\,°C$ or $\$19.95$ are numbers expressing relative magnitudes. Trying to associate a direction with these numbers is meaningless.

Vectors

Vectors indicate both magnitude and direction. When a direction is associated with a scalar, the scalar has a more significant meaning. For example, one airplane is flying at 600 mph due north and another is flying at 550 mph due east. If both planes leave the ground from the same airport and at the same time, to determine the distance between the planes at any given time, you have to know both their respective speeds and their general direction to obtain an accurate answer.

In physics, a vector is the magnitude of a force acting in a given direction from a given origin or point of reference. As will be shown in subsequent sections of this chapter, it is the resolution of two or more vectors into one net solution that is important. As another example of vectors, suppose you are stationary on a skateboard and two friends are going to pull you along. Each friend takes hold of one of your arms. Now, with reference to where you are, one friend pulls one of your arms with a force equal to 50 lb at an angle of $0°$, and the other friend pulls your other arm with a force of 50 lb at an angle of 180°. What is your net movement? You would remain stationary if the two friends pulled you with equal forces (magnitudes) but in opposite directions. If they both pull you in the same direction, or at least only 90° apart, you might move off at an equal angle between the two forces in the latter case, somewhere between your stationary speed and a speed equal to their efforts if they are working together. Do you now get the idea? Vectors do not just involve magnitudes like scalars, they also involve the directions in which they are heading.

Phasors

When we analyzed resistive and reactive circuits, we demonstrated that directional displacement occurred between voltage and current. Technically, either voltage or current, in a purely isolated sense, is a scalar. However, electronics requires that a definite angular displacement exist between the scalars. Phasors represent electronic scalars requiring angular directions to satisfy electronic theory and thus are subject to the same mathematical analysis as vectors. Henceforth, whenever we discuss the mathematics of vectors, the same rationale exists for phasors.

Vectors versus Phasors

At this point you may think that there is little difference between a vector and a phasor. Indeed, the difference is subtle. In recent years the trend has been to restrict the definition of vectors to a representation of amplitude versus direction and the definition of a phasor to the representation of amplitude versus time (such as a sine wave whose parameters are amplitude and time). Because the mathematics involving the resolution of either a vector or phasor are identical, this text follows this distinction: When a problem illustrated is purely mathematical, the term *vector* is used; when a problem illustrated is electronic, the term *phasor* is used.

16.4 VECTOR FORMATS

Polar

Section 16.3 defined a vector as having a magnitude and a direction. The simplest way to describe a vector is to express its magnitude and then use the Cartesian coordinate system to describe its direction. Figure 16.4 illustrates a vector of magnitude 15 and at an angle of 30° $((\pi/6)$ rad). Mathematically, this is expressed as:

$$15\underline{/30°} \quad \text{or} \quad 15\underline{/(\pi/6)} \text{ rad}$$

Note that the *tail* of the vector is the origin and the *head* of the vector is denoted by an arrow. Figure 16.5 illustrates vectors of $27\underline{/79°}$, $16\underline{/127°}$, and $21\underline{/-155°}$. When designated in this fashion, vectors are in the *polar* format.

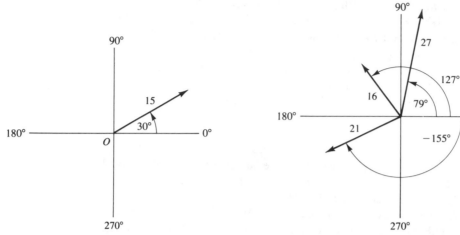

Figure 16.4   Vector of 15 at an angle of 30°.

Figure 16.5   Vectors of 27/79°, 16/127°, and 21/−155°.

Rectangular

Suppose you walk off distances on a map that correspond to the Cartesian coordinate system. That is, thinking of yourself as at the origin, east is at 0°, north is at 90°, west is at 180°, and south is at 270°. Walking due east 5 mi and then turning north for another 5 mi may be thought of as two separate operations. With reference to Figure 16.6, this can be represented first as a vector along the horizontal axis and then as a 90° turn upward and a vector parallel to the vertical axis. When vectors are placed head to tail, the ending point is equivalent to a singluar vector (the resultant) originating at the origin, with its *head* terminating at the *head* of the last vector. Logic indicates that the resultant of this example must lie at a 45° angle to the horizontal, and from the solution of right triangles, it must have a magnitude equal to $5\sqrt{2}$ mi. The importance of this discussion is not the resolution of right triangles—

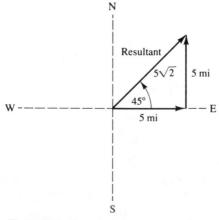

Figure 16.6   Rectangular components of a vector.

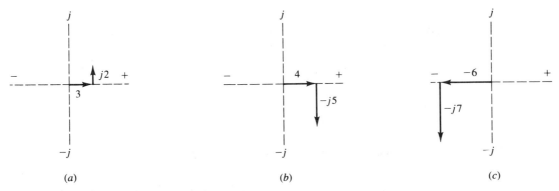

Figure 16.7 Examples of rectangular coordinate vectors.

that already has been investigated thoroughly. It is that a vector, such as $5\sqrt{2}\ \underline{/45°}$, can also be represented as the sum of two components: the projection of the vector on the horizontal and vertical axes, respectively.

By convention the horizontal component is specified first. From Figure 16.6, this is 5. Second, the vertical component, which is rotated by 90°, is preceded by $+j$. Third, the magnitude of the vertical component is now described; this is also 5. When all parts are combined, the vector $5\sqrt{2}\ \underline{/45°}$ in Figure 16.6 can be expressed in its rectangular format as

$$5 + j5$$

Figures 16.7a to c, respectively, illustrate vectors of $3 + j2$, $4 - j5$, and $-6 - j7$. Vectors in a format consisting of a *real* part and an *imaginary* part are called *rectangular coordinates*.

16.5 VECTOR FORMAT CONVERSIONS

Converting from the polar (P) format into the rectangular (R) format, or vice versa, involves applying the simple trigonometric functions, the Pythagorean theorem, and the polarities of the Cartesian coordinate system. We illustrate conversions between both formats by using two major categories of calculators: those having $P \rightarrow R$ and $R \rightarrow P$ capabilities and those that do not.

Conversions Using Basic Trigonometric Principles

Polar to Rectangular Format

Figure 16.8 illustrates vector $25\ \underline{/42°}$; we want to convert this polar format into rectangular components. Using the cosine and sine functions, the real component can be expressed as 25 cos 42° and the imaginary component as 25 sin 42°. Although it is obvious that both the real and imaginary components are in the $+1$ and $+j$ directions, respectively, calculators automatically display the polarity whenever the angular magnitude is expressed with reference to 0°.

Example 1

Using the generalized conversion of

$$Z\underline{/\theta} = Z \cos \theta + jZ \sin \theta \qquad (16.7)$$

convert the following four polar vectors (one in each of the four quadrants, respectively) into rectangular format. Use [Fix 3].

a. $25\underline{/42°}$

b. $17.3\underline{/105.7°}$

c. $16.2\underline{/1.15\pi \text{ rad}}$

d. $22.8\underline{/-37.9 \text{ grad}}$

PRESS	DISPLAY	COMMENT
[Deg] 25 [×] 42 [cos] [=]	18.579	Real component
25 [×] 42 [sin] [=]	16.728	Solution of a is $18.579 + j16.728$
17.3 [×] 105.7 [cos] [=]	-4.681	Real component
17.3 [×] 105.7 [sin] [=]	16.655	Solution of b is $-4.681 + j16.655$
[Rad] 16.2 [×] [(] 1.15 [×]		
[π] [)] [STO 1] [cos] [=]	-14.434	Real component
16.2 [×] [RCL 1] [sin] [=]	-7.355	Solution of c is $-14.434 - j7.355$
[Grad] 22.8 [×] 37.9 [+/−] [cos] [=]	18.878	Real component
22.8 [×] 37.9 [+/−] [sin] [=]	-12.786	Solution of d is $18.878 - j12.786$

Rectangular to Polar Format

Figure 16.9 illustrates the vector $21.3 + j27.5$; we want to convert this rectangular format into polar format. The resultant Z, using Pythagorean's theorem, is always expressed as

$$Z = \sqrt{\left| \text{real component} \right|^2 + \left| \text{imaginary component} \right|^2}$$

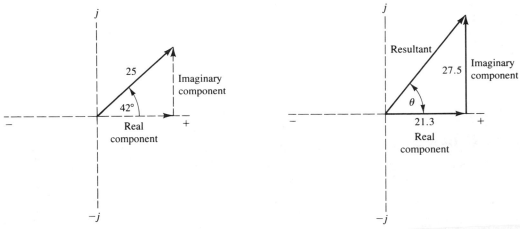

Figure 16.8   Converting $25\underline{/42°}$ into an $a + jb$ format.   **Figure 16.9**   Converting $21.3 + j27.5$ into an $R\underline{/\theta}$ format.

Associating the absolute value symbol with each rectangular component saves keying in the $\boxed{+ / -}$ calculator function because any number, positive or negative, is always positive when squared.

The major thing to watch for is the quadrant in which $\underline{/\theta}$ lies. When using your calculator, the magnitude of the angle is always expressed as the smallest angle between the resultant (R) and the real axis. This magnitude is calculated as

$$\theta = \tan^{-1}\left(\frac{\text{imaginary component}}{\text{real component}}\right)$$

To illustrate the process, let the real component be of absolute magnitude 5 and the imaginary component of absolute magnitude 10. Table 16.1 lists this vector as it appears in quadrants I to IV, the inverse tangent function, the correction necessary to express θ with reference to $0°$, and the corrected angle magnitude with reference to $0°$. As you can see, you have to carefully observe the polarities of the rectangular components to "sense" the correct quadrant value of θ. Corrections are necessary only for quadrants II and III and are accomplished by adding and/or subtracting $180°$ to the calculated display.

TABLE 16.1   COMPONENTS OF VECTOR $|5| + j|10|$

Quadrant	Rectangular Format	$\tan^{-1}\left(\dfrac{\text{imaginary comp.}}{\text{real comp.}}\right)$	Correction	θ
I	$5 + j10$	$63.435°$	None	$63.435°$
II	$-5 + j10$	$-63.435°$	$+180$	$116.565°$
III	$-5 - j10$	$63.435°$	-180	$-116.565°$
IV	$5 - j10$	$-63.435°$	None	$-63.435°$

Example 2

Convert the following four rectangular vectors (one in each of the four quadrants, respectively) into polar format. Use [Fix 3] .

a. $2.5 + j3.6$

b. $-1.7 + j0.9$

c. $-5.2 - j3.6$

d. $3.65 - j1.42$

PRESS	DISPLAY	COMMENT
2.5 $\boxed{x^2}$ $\boxed{+}$ 3.6 $\boxed{x^2}$ $\boxed{=}$ $\boxed{\sqrt{}}$	4.383	Resultant vector of a
3.6 $\boxed{\div}$ 2.5 $\boxed{=}$ $\boxed{\tan^{-1}}$	55.222	Solution of a is $4.383\underline{/55.222°}$
1.7 $\boxed{x^2}$ $\boxed{+}$ .9 $\boxed{x^2}$ $\boxed{=}$ $\boxed{\sqrt{}}$	1.924	Resultant vector of b
.9 $\boxed{\div}$ 1.7 $\boxed{+/-}$ $\boxed{=}$ $\boxed{\tan^{-1}}$ $\boxed{+}$ 180 $\boxed{=}$	152.103	Solution of b is $1.924\underline{/152.103°}$
5.2 $\boxed{x^2}$ $\boxed{+}$ 3.6 $\boxed{x^2}$ $\boxed{=}$ $\boxed{\sqrt{}}$	6.325	Resultant vector of c
3.6 $\boxed{+/-}$ $\boxed{\div}$ 5.2 $\boxed{+/-}$		
$\boxed{=}$ $\boxed{\tan^{-1}}$ $\boxed{-}$ 180 $\boxed{=}$	-145.305	Solution of c is $6.325\underline{/-145.305°}$
3.65 $\boxed{x^2}$ $\boxed{+}$ 1.42 $\boxed{x^2}$ $\boxed{=}$ $\boxed{\sqrt{}}$	3.916	Resultant vector of d
1.42 $\boxed{+/-}$ $\boxed{\div}$ 3.65 $\boxed{=}$ $\boxed{\tan^{-1}}$	-21.258	Solution of d is $3.916\underline{/-21.258°}$

Conversions Using P → R and R → P

Many scientific calculators have built-in algorithms that convert from the polar into the rectangular format and from the rectangular into the polar format. Unfortunately, these algorithms are built in by the manufacturer. As a consequence, although the procedural steps used by each manufacturer are very similar to one another, some algorithms display the x coordinate first and the y coordinate second when going from the polar into the rectangular format, whereas other algorithms display the y coordinate first and the x coordinate second. The same type of differences exists for displaying Z (the resultant) and θ (the angle) when converting from rectangular into polar formats. For this reason, here we present both the TI-55 and fx-80 algorithms side by side to illustrate these differences. The EL-5813 does not have a built-in algorithm for direct P → R or R → P conversions, so you should thoroughly acquaint yourself with your particular calculator. Note too that when the basic trigonometric functions are used along with the Pythagorean theorem, all calculators react identically. Only specialized, built-in algorithms designed to reduce the number of steps involving format conversions can result in slightly different sequences of displayed answers.

Converting from a rectangular into a polar format requires both a real component and an imaginary component as inputs into the calculator before you can evaluate an output consisting of a resultant (Z) and an angle (θ). The reverse is also true. Therefore, for calculators with built-in algorithms, the first value should be input into the display or x register and then transferred into a temporary storage area (often referred to as the y register). The second value can then be input into the display register. The calculator now has both values of the known coordinate system. After you key in the appropriate request ($P \rightarrow R$ or $R \rightarrow P$), the built-in algorithm evaluates both unknowns of the desired coordinate system, displays one value in the display or x register and stores the other value in the y register for subsequent transfer into the display register. Follow the sequences illustrated for the **TI-55** and *fx*-80 calculators, and note the similarity of the input sequence and the order of display of the evaluated results. Know *your* calculator when using these shortcuts.

Example 3

Convert the following rectangular vector into a polar vector, using the built-in algorithms of the indicated calculator.

$$3.5 + j5.6$$

	TI-55			*fx*-80	
PRESS	**DISPLAY**	**COMMENT**	**PRESS**	**DISPLAY**	**COMMENT**
3.5	3.5	Real component	3.5	3.5	Real component
$x \rightleftarrows y$	0	Transfer real component into y register	$R \rightarrow P$ [2]	3.5	Transfer real component into y register
5.6	5.6	Imaginary component	5.6	5.6	Imaginary component
$R \rightarrow P$ [1]	57.995	Does an $R \rightarrow P$ conversion; display θ in degrees	$=$	6.604	Display Z
$x \rightleftarrows y$	6.604	Display Z	$x \leftrightarrow y$	57.995	Display θ in degrees

[1] The actual key sequence (specifically for the TI-55) is 2nd INV P→R
[2] The actual key sequence (specifically for the *fx*-80) is INV R→P

Example 4

Convert the following polar vector into a rectangular vector using the built-in algorithms of the indicated calculator.

$$31.62\underline{/-151.23°}$$

TI-55			fx-80		
PRESS	**DISPLAY**	**COMMENT**	**PRESS**	**DISPLAY**	**COMMENT**
31.62	31.62	Polar magnitude	31.62	31.62	Polar magnitude
$x \rightleftarrows y$	0	Transfer R into y register	$\boxed{P \rightarrow R}$ [2]	31.62	Sets up algorithm
151.23			151.23		
$\boxed{+/-}$	-151.23	Set angle into x register	$\boxed{+/-}$	-151.23	Set angle into x register
$\boxed{P \rightarrow R}$ [1]	-15.219	Display imaginary component	$\boxed{=}$	-27.717	Display real component
$x \rightleftarrows y$	-27.717	Display real component	$\boxed{x \leftrightarrow y}$	-15.219	Display imaginary component

Again note the similarities of the inputs and outputs. Subsequent problems involving format conversions present only the results; you should either apply the algorithms of your particular calculator or use the basic trigonometric functions for format conversions.

PROBLEMS

Convert the polar vectors in Problems 1 to 6 into rectangular vectors.

1. $6\underline{/30°}$

2. $12.5\underline{/99.2°}$

3. $16.2\underline{/-161°}$

4. $21.63\underline{/0.91 \text{ rad}}$

5. $8.96\underline{/-0.81 \text{ rad}}$

6. $1.03 \times 10^3\underline{/255 \text{ grad}}$

---

[1] The actual key sequence (specifically for the TI-55) is $\boxed{\text{2nd}}$ $\boxed{P \rightarrow R}$.

[2] The actual key sequence (specifically for the fx-80) is $\boxed{\text{INV}}$ $\boxed{P \rightarrow R}$.

Convert the rectangular vectors in Problems 7 to 12 into polar vectors.

7. $2.35 + j6.81$

10. $-19.63 - j20.12$

8. $1.03 \times 10^3 + j2.065 \times 10^3$

11. $82.21 - j75.03$

9. $-8.72 + j10.12$

12. $-0.81 \times 10^{-5} - j9.42 \times 10^{-6}$

16.6 THE MATHEMATICS OF VECTORS

The basic arithmetic operations for vectors follow the same format as the algebraic hierarchy of the calculator. That is, to add and subtract vectors (priority one), you have to break down vectors into their rectangular components if they are not already in that format. It is easy to multiply and divide vectors (priority two) if they are in the polar format, although you can also perform the operations (with a little more labor) when vectors are in the rectangular format. We now illustrate each of the four basic arithmetic operations. Learning these academic principles, in conjunction with the principles of AC electronics discussed in the following chapters, will give you the background for solving the AC electronic circuits developed in Chapter 18.

Addition and/or Subtraction

When vectors are *flying off* in different directions, you cannot readily add or subtract them. However, if you break down each vector into its *real* and *imaginary* components, you can algebraically combine them because respective elements are then acting along the same *straight-line* axis. This means that when all vectors are in rectangular format, you can directly add or subtract them. If any one vector is in polar format, you must first reduce it into rectangular format before you can perform the addition or subtraction processes. Examples 1 and 2 illustrate all the possible combinations.

Example 1

Add and/or subtract the following vectors, all of which are in rectangular format. Express all your answers in rectangular format.

a. $(3 + j4) + (5 - j7)$
    $(3 + 5) + j(4 - 7) = 8 + j(-3) = 8 - j3$

b. $(-2.1 - j6.5) - (3.6 - j1.8)$
    $(-2.1 - 3.6) + j(-6.5 + 1.8)$
    $= -5.7 + j(-4.7)$
    $= -5.7 - j4.7$

c. $(-1.15 + j6.3) + (2.1 - j5.06) - (-7.6 - j1.9)$
    $(-1.15 + 2.1 + 7.6) + j(6.3 - 5.06 + 1.9)$
    $= 8.55 + j3.14$

Reminder: Before you can add and/or subtract one or more vectors in polar format, you must first convert all the vectors into rectangular format.

Example 2

Add and/or subtract the following vectors, some or all of which are in polar format. Express all your answers in rectangular format.

a. $(2.87 + j6.91) + 12.3\underline{/20.6°}$
    $(2.87 + j6.91) + (11.51 + j4.33) = (2.87 + 11.51) + j(6.91 + 4.33)$
    $= 14.38 + j11.24$

b. $-(11.82 - j10.19) - 6.82\underline{/155.7°}$
    $(-11.82 + j10.19) - (-6.216 + j2.807)$
    $= (-11.82 + 6.216) + j(10.19 - 2.807)$
    $= -5.604 + j7.383$

c. $13.21\underline{/15.62°} + 17.85\underline{/-49.7°} - 11.8\underline{/129°}$
    $(12.722 + j3.557) + (11.545 - j13.614) - (-7.426 + j9.170)$
    $= 31.693 - 19.227$

Multiplication and/or Division

You can multiply or divide vectors in either rectangular or polar format. We already illustrated rectangular operations; polar multiplication and/or division is much simpler because the resultant of a multiplication operation is

$$A\underline{/\theta} \cdot B\underline{/\phi} = A \cdot B\underline{/\theta + \phi} \qquad (16.8)$$

and the resultant of a division operation is

$$\frac{A\underline{/\theta}}{B\underline{/\phi}} = \frac{A}{B}\underline{/\theta - \phi} \qquad (16.9)$$

Equation (16.8) states that the resultant of the multiplication process is the product of the vectors' magnitudes directed at an angle equal to the sum of the angles of the operands. Equation (16.9) states that the resultant of the division process is the quotient of the vectors' magnitudes directed at an angle equal to the angle of the dividend minus the angle of the divisor.

Example 3

Perform the following operations. Note that you must first convert those vectors in rectangular format into polar format before the operation can take place. Express all your answers in polar format.

a. $16.1\underline{/13.2°} \cdot 12.1\underline{/47.5°}$
    $(16.1)(12.1)\underline{/13.2° + 47.5°} = 194.81\underline{/60.7°}$

b. $3.8\underline{/-42.1°} \cdot 7.9\underline{/63.4°}$
    $(3.8)(7.9)\underline{/-42.1° + 63.4°} = 30.02\underline{/21.3°}$

c. $14.87\underline{/161.8°} \cdot (13.89 - j12.47)$
    $(14.87)(18.666)\underline{/161.8° + (-41.916°)} = 277.563\underline{/119.884°}$

d. $(-21.6 + j13.8) \cdot (17.5 - j11.91)$
    $(25.632)(21.168)\underline{/147.426° + (-34.238°)} = 542.578\underline{/113.188°}$

e. $\dfrac{42.3\ \underline{/27.4°}}{13.4\ \underline{/11.7°}}$

    $\dfrac{42.3}{13.4}\underline{/27.4° - 11.7°} = 3.157\underline{/15.7°}$

f. $\dfrac{46.3\underline{/-147.2°}}{39.6\underline{/-57.6°}}$

    $\dfrac{46.3}{39.6}\underline{/-147.2° - (-57.6°)} = 1.169\underline{/-89.6°}$

g. $\dfrac{15.72\underline{/31.42°}}{6.17 - j5.82}$

    $\dfrac{15.72\underline{/31.42°}}{8.482\underline{/-43.328°}} = \dfrac{15.72}{8.482}\underline{/31.42° - (-43.328°)}$
    $= 1.853\underline{/74.748°}$

h. $\dfrac{8.21 + j11.16}{-9.37 - j16.2}$

    $\dfrac{13.855\underline{/53.659°}}{18.715\underline{/239.955°}} = \dfrac{13.855}{18.715}\underline{/53.659° - 239.955°}$
    $= 0.74\underline{/-186.296°}$

Vectors Raised to Powers

Vectors raised to powers are most easily solved when they are in polar format. Observe the following:

$$(5\underline{/30°})^2 = (5\underline{/30°})(5\underline{/30°}) = (5)(5)\underline{/30° + 30°} = 25\underline{/60°}$$

Raising any vector to a power is merely repeated products of the vector, so the rules of multiplication can be applied. Generally, raise the vector magnitude to the indicated power and multiply the vector angle by the indicated power.

Example 4

Raise the following vectors to the indicated power.

a. $(4\underline{/50°})^3$
    $4^3\underline{/3 \times 50°} = 64\underline{/150°}$

b. $(2\underline{/-27.5°})^4$
    $2^4\underline{/-27.5° \times 4} = 16\underline{/-110°}$

c. $(j5)^2$
    $(j5)^2 = (5\underline{/90°})^2 = 5^2\underline{/90° \times 2} = 25\underline{/180°}$

If the power to which a vector is raised is a fraction, that is, if the nth root is required, you must implement the reverse procedure: Extract the indicated root of the magnitude and divide the angle of the vector by the degree of the indicated root.

Example 5

Extract the indicated root of the following vectors.

a. $(49\underline{/25°})^{1/2}$
    $49^{1/2}\underline{/25°/2} = 7\underline{/12.5°}$

b. $(125\underline{/-156°})^{1/3}$
    $125^{1/3}\underline{/-156°/3} = 5\underline{/-52°}$

c. $(j16)^{1/2}$
    $(j16)^{1/2} = (16\underline{/90°})^{1/2} = 16^{1/2}\underline{/90°/2} = 4\underline{/45°}$

If vectors are in rectangular format, first convert them into polar format before applying the preceding rules.

Example 6

Perform the indicated operations on the following vectors.

a. $(3 + j4)^2$
$(5\underline{/53.13°})^2 = 25\underline{/106.26°}$

b. $(-4 - j6.2)^{1/3}$
$(7.378\underline{/237.171°})^{1/3} = 1.947\underline{/79.057°}$

c. $(-2.1 + j6.3)^3$
$(6.641\underline{/108.435°})^3 = 292.887\underline{/325.305°}$

d. $(2.7 - j1.9)^{1/4}$
$(3.302\underline{/-35.134})^{1/4} = 1.348\underline{/-8.784°}$

PROBLEMS

Perform the following operations in Problems 1 to 20. Express all answers in both rectangular and polar formats.

1. $(2 + j3.5) + (6 - j8.2)$

2. $(-4.9 - j5.6) + (7.7 - j1.6)$

3. $(12.39 + j18.2) - (21.6 - j10.1)$

4. $(2.5 - j3.6) + (1.8 + j2.9) - (-5.1 - j13.6)$

5. $8.2 - j11.1 + 6\underline{/37.2°}$

6. $7.9\underline{/-42.3°} - (8.2 - j5.61)$

7. $2.75\underline{/105.6°} + 3.2\underline{/0.9\pi \text{ rad}}$

8. $8.41\underline{/-21.4 \text{ grad}} + 9.2\underline{/251.3°} - 10.7\underline{/0.52\pi \text{ rad}}$

9. $(13.1\underline{/56°})(7.4\underline{/-17.1°})$

10. $(18.6\underline{/-43.1 \text{ grad}})(6.2\underline{/(\pi/3) \text{ rad}})$

11. $(2.62 - j3.19)(4\underline{/102.3°})$

12. $\dfrac{22.5\underline{/51.2°}}{31.8\underline{/18.7°}}$

13. $\dfrac{42.4\underline{/-71.8°}}{6.4\underline{/-51.3°}}$

14. $\dfrac{2.17 - j6.3}{5.9\underline{/0.72\pi \text{ rad}}}$

15. $\dfrac{11.42\underline{/-271.8°}}{6.41 + j8.92}$

16. $\dfrac{32.7\underline{/126.8°}}{2.9 - j6.3} - (18.2 + j10.11)$

17. $(27.9\underline{/-15.40})^3$

18. $(21.4 + j30.7)^2$

19. $(17.3\underline{/\pi \text{ rad}})^{1/2}$

20. $(18.2 - j11.9)^{1/4}$

16.7 ALL THE ROOTS OF A NUMBER

Recall from the discussion in Chapters 3 and 6, which dealt with the roots of numbers, that there are as many different answers to the problem as the magnitude of the root. For example,

$$\sqrt{16}$$

in which the implied magnitude of the root to be extracted is 2, has answers of ± 4. Thus there are two answers because the root is 2. By the same rationale,

$$\sqrt[3]{8}$$

must have three solutions because this is the cube root. One of the three solutions is obviously 2. But what are the other two solutions? Our knowledge of the complex plane and the trigonometric functions can now provide the missing solutions.

To approach the problem, let us first tackle the easy root of a problem, such as the root of 16. First we extract the absolute magnitude of $\sqrt{16}$, which is 4. Second, visualize that 4 is initially directed at an angle of 0°; see Figure 16.10. This results in $4\underline{/0°}$, which has already been calculated as one of the answers, namely $+4$. Third, divide 360° (one complete circle) by 2, the magnitude of the root being extracted. This results in an angle of 180°. Therefore, the second and final solution of the root of 16 is another vector of magnitude 4, displaced 180° from $4\underline{/0°}$, or $4\underline{/180°}$. However, $4\underline{/180°}$ is also referred to as -4 in the complex plane. The two roots therefore are $+4$ and -4.

To continue the process, $\sqrt[3]{8}$ equates to $2\underline{/0°}$, $2\underline{/120°}$, and $2\underline{/240°}$. Note that the magnitude 2 results from extracting the cube root of 8, and the angular displacements result from starting at 0° and rotating the vector two increments of $(360/3)°$,

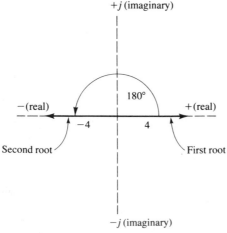

Figure 16.10 Determining the two solutions of $\sqrt{16}$.

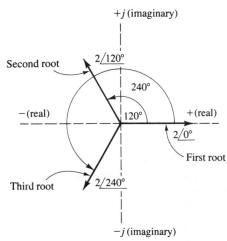

Figure 16.11   Determining the three solutions to $\sqrt[3]{8}$.

or 120°, until returning to the reference point. Figure 16.11 illustrates the three roots of $\sqrt[3]{8}$. From the complex plane and the trigonometric functions, the results are

$$2\underline{/0°} = 2$$
$$2\underline{/120°} = -1 + j\sqrt{3}$$
$$2\underline{/240°} = -1 - j\sqrt{3}$$

Therefore the three solutions of $\sqrt[3]{8}$ are 2, $-1 + j\sqrt{3}$, and $-1 - j\sqrt{3}$.

   To substantiate these solutions, cube each of the three answers and see if the resultant is exactly 8:

1. $2 \times 2 = 4$ and $4 \times 2 = 8$
2. $(-1 + j\sqrt{3})(-1 + j\sqrt{3}) = -2 - j2\sqrt{3}$ and
    $(-2 - j2\sqrt{3})(-1 + j\sqrt{3}) = 8$
3. $(-1 - j\sqrt{3})(-1 - j\sqrt{3}) = -2 + j2\sqrt{3}$ and
    $(-2 + j2\sqrt{3})(-1 - j\sqrt{3}) = 8$

PROBLEMS

In Problems 1 to 6 determine all the roots.

1. $\sqrt{25}$ 3. $\sqrt[3]{125}$ 5. $\sqrt[5]{32}$

2. $\sqrt[3]{64}$ 4. $\sqrt[4]{16}$ 6. $\sqrt[6]{\pi}$

SUMMARY

The phenomenon of phase shift between voltage and current in an L/R or RC circuit requires the use of the complex plane. This shift is mathematically accomplished by use of the j operator, which

1. Electronically represents I-V phasors
2. Mathematically facilitates extracting the square root of negative numbers

The four basic arithmetic functions (add, subtract, multiply, and divide) may be applied to j. If j is part of a polynomial (especially a binomial), it is easier to add or subtract in rectangular format and to multiple or divide in polar format.

In general, conversions between formats are

$$Z\underline{/\theta} = Z \cos \theta + j \sin \theta \qquad \text{for P} \to \text{R}$$

and
$$a + jb = \sqrt{a^2 + b^2}\,\underline{/\tan^{-1} b/a} \qquad \text{for R} \to \text{P}$$

Scalars are purely numeric magnitudes. Vectors are magnitudes with direction. Phasors are special forms of vectors in that they represent magnitudes versus time as opposed to magnitudes versus direction.

Many calculators have built-in algorithms for P $\to$ R and R $\to$ P conversions. Because these algorithms are specifically designed by each manufacturer for their calculator, precise sequencing of the input and output data can vary from manufacturer to manufacturer.

Raising a vector to a power or extracting a root is most easily accomplished when the vector is in the polar format. Raising the vector to a power takes the format

$$(Z\underline{/\theta°})^n = Z^n\underline{/n \cdot \theta°}$$

Extracting the root of a vector takes the format

$$(Z\underline{/\theta°})^{1/n} = \sqrt[n]{Z}\,\underline{/\theta°/n}$$

17 GENERATION OF THE PERIODIC SINE FUNCTION

17.1 INTRODUCTION

Chapter 15 investigated the shape of the sine, cosine, and tangent functions. These functions are cyclic in nature; that is, they are exactly repetitive after a given unit period of time, and so they are called *periodic* functions. Although all three basic trigonometric functions and their reciprocals are periodic, only the sine and cosine functions are continuous—they do not exhibit severe discontinuities in their graphs. Therefore, only the sine and cosine functions have practical electronic application. This chapter investigates the generation of both the single-phase and the polyphase sine curve, the fundamental terminology associated with the sine curve, and the applications of the single- and three-phase system. Chapter 18 expands upon the application of the principles of complex algebra to the solution of AC circuits.

17.2 GENERATION OF THE SINE CURVE

Figure 17.1 depicts a simple, three-dimensional coil rotating in a magnetic field. Electronic textbooks adequately develop the theories that interrelate flux density, length of the conductors, speed of rotation, and magnitude of the induced voltage. From a purely mathematical viewpoint, it is sufficient for us to consider only a two-dimensional view of the electronic principles involved, based upon accepted theoretical formulas.

When a coil of wire is rotated through a magnetic field, a voltage is induced in the coil. In Figure 17.1, the plane of the coil is aligned at a reference of 0°. The voltage across R changes from a value of 0 V to a maximum positive voltage as the coil rotates counterclockwise to 90°. The nature of this change and subsequent changes that occur during the remaining 270° of rotation concern us in this chapter.

The value of the induced voltage (v_i) is given as

$$V_i = N\frac{d\phi}{dt} \tag{17.1}$$

where N represents the number of turns of the coil (which for Figure 17.1 is 1) and $d\phi/dt$ represents the instantaneous rate of change of flux lines cut by the coil's rods per unit length of time. As stated in earlier mathematical expressions, the term $d\phi/dt$ is related to calculus and concerns us only as to its conceptual interpretations.

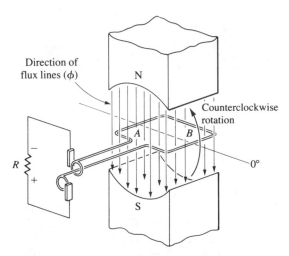

Figure 17.1   Fundamental circuit to generate AC voltage.

Figure 17.2a is a two-dimensional rendering of the three-dimensional drawing in Figure 17.1. The plane formed by rods A and B is shown at 0°. The tangential projection of the direction of rotation of rods A and B illustrates that when at 0°, *no* lines of flux (ϕ) are cut; therefore Equation (17.1) predicts that v_i is 0 V.

Figure 17.2b is a two-dimensional end view of the coil in Figure 17.1, after rotating 90°. Note that the tangential direction of each rod of the coil is perpendicular to the flux field, which means that the maximum number of lines of flux are being cut per unit of time. Equation (17.1) predicts that v_i is now a maximum. As indicated in previous sections, the absolute value of this maximum voltage (V_M) is part of electronic theory. Our concern is only with the trigonometric relationships that occur as the coil rotates between the values of 0° and 90°. Is this cutting of flux lines linear? For example, when halfway between 0° and 90°, is the value of v_i halfway between 0 V and the maximum voltage generated? Let us investigate this question.

Figure 17.2c illustrates the coil after a rotation of θ°. Points a and b are the end views of rods A and B, respectively. Line ns represents the direction of the central flux line between the magnetic pole faces; the flux line emanates from the north pole and terminates at the south pole. The remaining flux lines between the pole faces are drawn as dashed lines. Line ns by its nature is perpendicular to the 0° axis. Both vectors bn and as represent the tangential directions of rods B and A, respectively. Vectors bn and as are cutting some flux lines, but the actual numbers cut can be counted only by observing the perpendicular movement of rods B and A through the flux field; hence line bb' is drawn perpendicular to the flux field. At this point it is only necessary to analyze the magnitude of bb' to determine the number of flux lines being cut because the identical effect is also happening to the perpendicular direction of rod A through the flux field, which is identified as vector aa'. Because Figure 17.1 shows that rods B and A are series connected and therefore series additive, the same conclusion can be reached by analyzing any one rod.

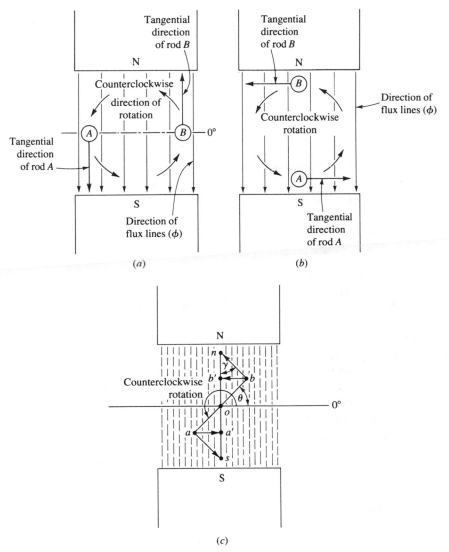

Figure 17.2   (*a*) Zero flux lines cut by coil at 0°.
(*b*) Maximum number of flux lines being cut by coil at 90°.
(*c*) Determination of the sine voltage curve.

Line *ob* (one-half the plane formed by the closed coil *AB*) is at an angle θ to the 0° axis. Line *ob* is also perpendicular to *bn*, the tangential direction of rotation of rod *B* (this is the definition of tangential). Line *nb'* in turn completes our picture by being perpendicular to the beginning reference line, the 0° axis. From the rules of plane geometry, this makes $\angle \gamma$ equal to $\angle \theta$. So now the stage is set. We are interested in *bb'*, which is that portion of *bn* actually perpendicular to and cutting the lines of flux between the pole faces. For simplicity, consider bn a unit vector, that is, of magnitude 1. Then, from triangle nbb',

$$\frac{d\phi}{dt} \equiv bb' = (bn) \sin \gamma = (1) \sin \gamma = \sin \gamma$$

But $\underline{/\gamma} = \underline{/\theta}$; therefore

$$\frac{d\phi}{dt} = \sin \theta \qquad (17.2)$$

It appears then that v_i does not follow a linear function as coil AB rotates through the magnetic field; it follows the sine function relationship. Consequently, we can state that the coil is generating a *sine voltage curve*. Substituting Equation (17.2) into Equation 17.1 and using the theories of electronics to develop a value for the maximum voltage generated, we can rewrite Equation (17.1) as

$$v_i = V_m \sin \theta \qquad (17.3)$$

The instantaneous voltage developed at any given point in the rotation of a coil through a magnetic field equals the product of the maximum magnitude of the developed voltage (at 90° or 180°) and the sine of the angle between the coil plane and the 0° reference axis.

We can use a rotating vector to represent the sine function. Figure 17.3a shows a unit vector starting at 0° and progressing in 30° increments. The height of the vector above or below the horizontal axis (the vertical component) represents the sine function. It is projected along the horizontal axis in equal increments of angular displacement. The horizontal axis, for convenience, is incremented in both degree and radians. Notice that the maximum amplitude can never exceed the length of the unit vector, and the maximum positive value occurs at 90° ($\sin 90° = 1$) and the maximum negative value at 270° ($\sin 270° = -1$). The cyclic nature of the sine function is obvious, thereby classifying it as a continuous, periodic function.

17.3 NOMENCLATURE OF THE SINE FUNCTION

When a vector rotates at a constant speed about an origin O, it generates a sine wave such as that in Figure 17.3. Assume the vector initially starts at 0°; after a lapse of time (t) it has transversed through an angular displacement of θ degrees and/or radians. This is represented in Figure 17.4; this figure is the basis for the following electronic terminology.

Frequency

We showed that a periodic function, such as a sine wave, is repetitive; it repeats itself after a given unit of time. The number of times a function repeats itself per unit length of time is the *frequency* (f) of the function. The specific unit of time used

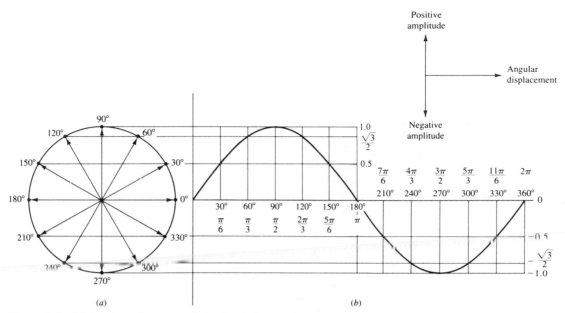

Figure 17.3   (*a*) Rotating unit vector; (*b*) sine (vertical) projection.

generally depends upon the science to which it has reference. For example, in astronomy the rotation of the earth around the sun is defined as

1 revolution (r)/year

In automotive mechanics, the tachometer in a car might read

2500 rpm   or   2500 r/min

In electronics, a periodic function seldom repeats itself less than once a second. Therefore, if the vector in Figure 17.4 completes 25r/s, its frequency is defined as

25 r/s

A revolution per second used to be called a cycle, or 25 cps or 25 cycles/s. Current electronic terminology now refers to a cycle per second as *hertz*, abbreviated Hz.

Example 1

Define the frequency (*f*) for the following number of revolutions of the vector in Figure 17.4.

a.   325 r/s
     $f = 325$ Hz

b.   1750 r/s
     $f = 1750$ Hz or 1.75 kHz

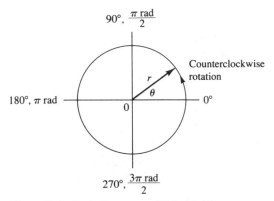

Figure 17.4   Basic figure to establish definitions.

c.   2,540,000 r/s
   $f = 2,540,000$ Hz or 2.54 MHz

d.   3600 r/m
   $f = 3600$ r/min $\times$ 1 min/60 s = 60 r/s = 60 Hz

Period

The period is the time (T) it takes to complete 1 Hz of a periodic function. It is mathematically expressed as

$$T = \frac{1}{f} \qquad\qquad (17.4)$$

The period of a function is equal to the reciprocal of its frequency.

Example 2

Determine the period of the following frequencies.

a.   60 Hz

   $$T = \frac{1}{f} = \frac{1}{60} = 0.0167 \text{ s} = 16.7 \text{ ms}$$

b.   3250 Hz

   $$T = \frac{1}{3250} = 0.000308 \text{ s} = 308 \ \mu s$$

c.   21.6 MHz

   $$T = \frac{1}{21.6 \times 10^6} = 0.0463 \ \mu s = 46.3 \text{ ns}$$

Velocity

Velocity, or the speed of motion, is either (1) linear or (2) angular. Most of us are familiar with *linear* velocity, such as the speed of a car traveling down a straight highway. This velocity measures distance traveled per unit length of time. Typical examples are 55 mph or 88 km/h.

A form of *angular* velocity is the hertz. The expression 35 Hz means 35 r (distance traveled)/s (unit of time). Distance traveled per unit of time is the definition of either angular or linear velocity. Another very important measurement of angular velocity in electronics is expressed in units of radians per second, symbolized as ω (omega). Because there are 2π rad in a circle, radian velocity is simply 2π times the frequency of revolutions in hertz or

$$\omega = 2\pi f \text{ rad/s} \tag{17.5}$$

Example 3

Determine the radian velocity of the following frequencies.

a.   0.1 Hz

$$\omega = 2\pi f = 2\pi(0.1) = 0.2\pi \text{ rad/s}$$

b.   245 Hz

$$\omega = 2\pi(245) = 490\pi \text{ rad/s}$$

c. $\dfrac{3260}{\pi}$ Hz

$$\omega = 2\pi\left(\frac{3260}{\pi}\right) = 6520 \text{ rad/s}$$

Displacement

Displacement, which refers to distance traveled, directly corresponds to velocity and can be *linear* or *angular*. Refer to the linear velocity (v) of the car in the "velocity" section. If the velocity is 88 km/h and the car travels for a period of time (t) of 2 h, then the distance (d) is given as

$$d = vt \tag{17.6}$$

or
$$d = 88 \text{ km/h} \times 2 \text{ h} = 176 \text{ km}$$

Distance or displacement is the product of the velocity and the period of time the velocity was in effect.

By the same analogy, the angular displacement (θ) in radians is

$$\theta = \omega t \tag{17.7}$$

Example 4

Determine the angular displacements in both radians and degrees for the following.

a. $\omega = 3$ rad/s for 0.1 s

$\theta = \omega t = 3(0.1) = 0.3$ rad $= 0.3\left(\dfrac{180}{\pi}\right) = 17.19°$

b. $\omega = 50\pi$ rad/s for 0.005 s

$\theta = 50\pi(0.005) = 0.25\pi$ rad $= 45°$

c. $\omega = 1500\pi$ rad/s for 10 μs

$\theta = 1500\pi(10^{-5}) = 0.015\pi$ rad $= 2.7°$

Instantaneous Values

Let vector r in Figure 17.4 represent the maximum voltage (V_M) that can be generated by an AC source. This maximum value can occur only at 90° for the positive
excursion and at 270° for the negative excursion. How can instantaneous voltage be
expressed at any point between these extremes? We showed that the magnitude of
the generated voltage was directly proportional to the sine of the angle between the
rotating vector and the 0° axis. Because the vector has a magnitude equal to V_M, we
may conclude that

$$v = V_M \sin \theta$$

However, the "velocity" section indicated that the angular displacement equaled ωt.
Therefore,

$$v = V_M \sin \omega t \qquad (17.8)$$

If you consider only a resistive circuit, where the magnitude of current is directly
proportional to the voltage, then

$$i = I_M \sin \omega t \qquad (17.9)$$

Example 5

Determine the instantaneous value of v or i under the following conditions.

a. v if $V_M = 15$ V, $\omega = 60\pi$ rad, $t = 6.3$ ms

$v = V_M \sin \omega t$

$= 15 \sin ((60\pi)(6.3 \times 10^{-3}))$ rad

$= 15 \sin 1.1875$ rad

$= 13.912$ V

b. v if $V_M = 23.5$ V, $\omega = 1750\pi$ rad, $t = 350$ μs

$v = 23.5 \sin ((1750\pi)(350 \times 10^{-6}))$ rad

$= 23.5 \sin 1.924$ rad

$= 22.049$ V

c. i if $I_M = 18.2$ A, $\omega = 2530\pi$ rad, $t = 485$ μs
$i = 18.2 \sin ((2530\pi)(485 \times 10^{-6}))$ rad
$= 18.2 \sin 3.855$ rad
$= -11.909$ A

d. i if $I_M = 39.75$ μA, $\omega = 16{,}250\pi$ rad, $t = 101$ μs
$i = 39.75 \sin ((16250\pi)(101 \times 10^{-6}))$ rad μA
$= 39.75 (\sin 5.156$ rad$)$ μA
$= -35.9$ μA

Variations of these examples are possible. Suppose we want to determine the length of time required for the instantaneous value to reach some fractional part of its maximum. Now we use the $\sin^{-1}$ function.

Example 6

Assume Figure 17.5 is a series circuit where $R = 4.7$ kΩ, $v = 20 \sin \theta$, and $f = 1$ kHz. How long will it take for i to reach a value of 1.2 mA?
  From Equation (17.9),

$$i = I_m \sin \omega t = I_m \sin 2\pi ft$$

$$I_m = \frac{20}{4.7 \text{ k}\Omega} = 4.255 \text{ mA} = 4.255 \times 10^{-3} \text{A}$$

$$1.2 \times 10^{-3} = 4.255 \times 10^{-3} \sin ((360°)(10^3)t)$$

$$\sin^{-1} \left(\frac{1.2 \times 10^{-3}}{4.255 \times 10^{-3}} \right) = (360°)(10^3)t$$

$$\sin^{-1} 0.282 = (360°)(10^3)t$$

$$16.381° = (360°)(10^3)t$$

$$t = \left(\frac{16.381}{360} \right)(10^{-3})\text{s}$$

$$t = 46 \text{ }\mu\text{s}$$

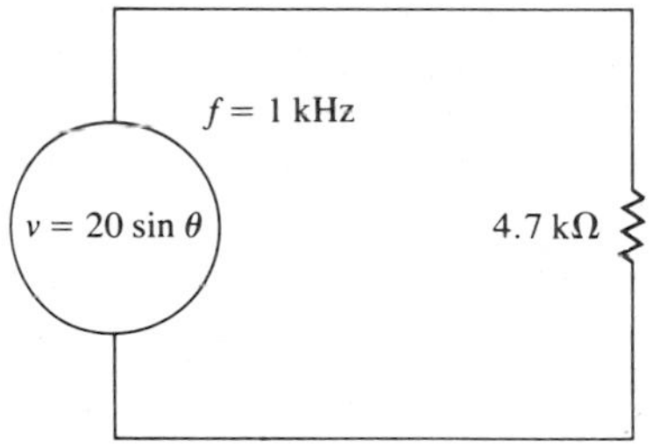

Figure 17.5

$f = 1$ kHz

$v = 20 \sin \theta$

4.7 kΩ

Phase Relationships

Chapter 16 demonstrated that a phase shift of 90° occurs between the voltages and currents in a purely inductive or capacitive circuit. Because some resistance is always present, this shift is actually less than 90°. To illustrate, consider Figure 17.6a, a simple RC circuit. The current is the same everywhere in a series circuit, so the current phasor will be the reference (Figure 17.6b). There are two voltage drops, v_R and v_C, the sum of which must at any given time add to v. The voltage phasor across R must be *in phase* with the current because there is no phase shift through a resistor. The voltage phasor across the capacitor is rotated clockwise 90° $(-j)$ from the $I_{\max}$ phasor.

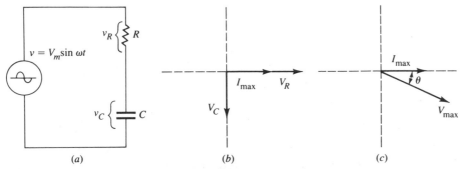

Figure 17.6   (*a*) Circuit to illustrate phase relationships.
(*b*) Phasor diagram of (*a*).
(*c*) Summing the voltage phasors of (*b*).

Applying the rules of vector addition to the two voltage phasors results in Figure 17.6*c*. Because this is an AC circuit, the phasors I_{max} and V_{max} rotate counterclockwise at a radian frequency specified by ω in Figure 17.6*a*. At any given moment, I_{max} *leads* V_{max} by θ, or, worded another way, V_{max} *lags* I_{max} by θ.

Assume V_R is 10 V and V_C is $(10/\sqrt{3})$ V. These numbers were conveniently chosen so that $\underline{/\theta}$ between I_{max} and V_{max} results in

$$\theta = \tan^{-1}\frac{(10/\sqrt{3})}{10} = \tan^{-1}\frac{1}{\sqrt{3}} = 30°$$

Constructing a sine wave graph (similar to Figure 17.3) of the I_{max} and V_{max} phasors for $\theta = 30°$ results in Figure 17.7. Convenient reference points for measuring the angular displacement are where the phasors peak or cross the 0° axis.

Although Figure 17.7 illustrates *I* leading *V* for an *RC* circuit, an analogous graph results for an *L/R* circuit if the labels are reversed, that is, if *V* leads *I*.

Example 7

For Figure 17.7, assume $V_M = 150$ V, $I_M = 230$ mA, and $f = 1.1$ kHz.
a.   Write the equation for *i* and *v*.
$i = I_M \sin\theta$
$= 0.23 \sin\omega t$
$= 0.23 \sin 2\pi ft$
$= 0.23 \sin (2200\pi t)$
Note that 2200π is the radian frequency.
$v = V_M \sin(\theta - 30°) = 150 \sin(2200\pi t - 30°)$

b.   What is the instantaneous value of *i* and *v* when $t = 1$ ms?
$i = 0.23 \sin(2200\pi \times 10^{-3})$ rad $= 0.23 \sin(2.2\pi)$ rad
$= 0.23 (0.588) = 135$ mA
$v = 150 \sin(2.2\pi$ rad $- 30°)$

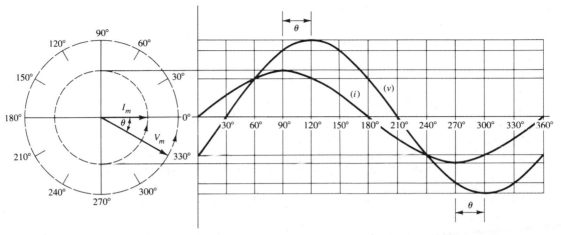

Figure 17.7 *I-V* phase relationship for $\theta = 30°$.

Note the mix of both radians and degrees.

$$v = 150 \sin (36° - 30°)$$
$$= 150 \sin 6°$$
$$= 15.679 \text{ V}$$

c. Starting with $t = 0$ when I_M is at $0°$, how long will it take for i to equal 110 mA?

$$i = I_m \sin \omega t$$
$$110 = 230 \sin 2\pi ft$$
$$\frac{110}{230} = \sin ((360°)(1.1 \times 10^3)t)$$
$$\sin^{-1}\left(\frac{110}{230}\right) = 28.572° = (360°)(1.1 \times 10^3)t$$
$$t = \left(\frac{28.572°}{360}\right)\left(\frac{1}{1.1 \times 10^3}\right)$$
$$t = 72.15 \ \mu s$$

d. In regards to part c, how long will it take for v to reach -127 V?

$$-127 = 150 \sin (2\pi ft - 30°)$$
$$= 150 \sin ((360°)(1.1 \times 10^3)t - 30°)$$
$$\sin^{-1}\left(\frac{-127}{150}\right) = 302.149° = (360°)(1.1 \times 10^3)t - 30°$$
$$(1.1 \times 10^3)t = \frac{302.149° + 30°}{360°} = 0.923$$
$$t = \frac{0.923}{1.1 \times 10^3} = 839 \ \mu s$$

PROBLEMS

1. A motor rotates at 7200 rpm. What is its frequency of rotation per

   a. hour   b. second   c. millisecond

2. What is the period per second of the following frequencies?

   a. 60 Hz   c. 3,700,000 Hz   e. 142 MHz
   b. 2100 Hz   d. 18.6 kHz

3. A jet plane is flying in a straight line at 975 km/h. What distance does it travel in

   a. 1 s   b. 8 min, 15 s   c. 2 h, 14 min, 3.6 s

4. How long will it take the plane in Problem 3 to travel

   a. 1.92 km   c. 182 mi
   b. 1,000,000 m   d. 420 mi, 3050 ft

5. A coil rotating in a magnetic field generates a sine wave of frequency 675 Hz. How many revolutions and radians does it travel in

   a. 1 ms   c. 182.6 μs
   b. 0.54 s   d. 13.2 min

6. The maximum voltage generated by a rotating coil is 42.5 V. If the coil rotates at a frequency of 60 Hz,

   a. Write an equation for the instantaneous voltage.
   b. What is the value of the instantaneous voltage at the following displacements into the cycle:
      14°
      105°
      1.27 rad
      1.68 rad

7. For the generator in Problem 6, assume $v = 0$ when $t = 0$. How long will it take for the instantaneous voltage to equal

   a. 14.3 V (quadrant I)   c. -26.4 V (quadrant III)
   b. 14.3 V (quadrant II)   d. -26.4 V (quadrant IV)

8. In a series RC circuit, $I_M = 55$ mA, $V_{RM} = 4.8$ V, and $V_{CM} = 6.1$ V. If $f = 2.5$ kHz, write an equation for the instantaneous value of i and v. Consider current as reference.

9. In Problem 8, how long will it take for i to equal 21 mA?

10. In Problem 8, how long will it take for v to equal $V_M/2$?

11. In a series L/R circuit, $I_M = 127$ mA, $V_{RM} = 16.2$ V, and $V_{LM} = 20.5$ V. If $f = 4.5$ kHz, write an equation for the instantaneous value of i and v. Consider current as reference.

12. In Problem 11, how long will it take for i to equal 50 mA?

13. In Problem 11, how long will it take for v to equal $0.85\ V_M$?

17.4 THREE-PHASE GENERATION

Former sections discussed the basic mathematics and theories of a single-phased system, symbolized as (1ϕ). Three-phase systems (3ϕ), however, display a significant advantage over single-phase systems, especially when a balanced load is connected, because then the power delivered is a constant. This offers many advantages; as one example, a three-phase system will develop a constant torque. Thus most commercial power distribution systems are three-phase. Our interest in these systems is the unusual extent to which we can use phasors and complex algebra to solve electronic problems and develop electronic theory. To establish a common ground for this analysis, the following notations, methodologies, and definitions are developed.

Figure 17.8a shows three sets of coils equally spaced and capable of rotating between the poles of a magnet. In practice, the points labeled A', B', and C' are common (Figure 17.8b); hence we can show the vector representation of these coils as emanating from a common point (Figure 17.8c). Each individual coil generates its own unique phase voltage. If we plot each of these three phases individually in 30° increments for one-half cycle, starting from the positions in Figure 17.8b, Fig-

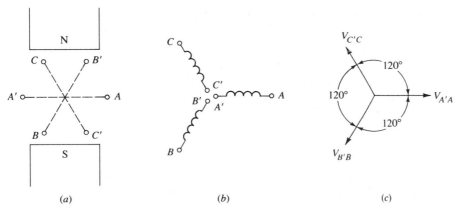

(a)   *(b)*   *(c)*

Figure 17.8   Basic 3ϕ representations.

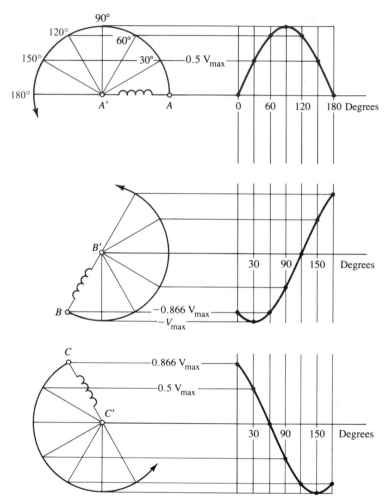

Figure 17.9 Voltage-generated sequence for Figure 17.8.

ure 17.9 results. For each coil, the amplitude of voltage at the unprimed end of the coil is measured with respect to its prime (common) end. When the three-phase voltages in Figure 17.9 are superimposed one upon the other for a full cycle, Figure 17.10 results.

We can make an interesting observation when analyzing Figure 17.10. At any given time, the sum of the three-phase voltages is always zero. For example, if $V_{AA'}$, $V_{BB'}$, and $V_{CC'}$ are respectively summed at 0°, 30°, 60°, and 90°, the following results:

$$0° \quad V_T = (0 + (-0.866) + (0.866))V_{max} = 0 \text{ V}$$
$$30° \quad V_T = (0.5 + (-1) + 0.5)V_{max} = 0 \text{ V}$$
$$60° \quad V_T = (0.866 + (-0.866) + 0)V_{max} = 0 \text{ V}$$
$$90° \quad V_T = (1 + (-0.5) + (-0.5))V_{max} = 0 \text{ V}$$

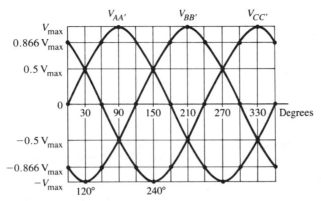

Figure 17.10 Superimposed voltage sequence in Figure 17.9.

17.5 CHARACTERISTICS OF 3ϕ SYSTEMS

The Y Configuration

Up to this point we have discussed only a Y generator. A generator is of little value unless connected, via transmission lines, to a load. When this connection is accomplished, the power developed by the generator can be delivered to a factory to illuminate lights or drive the motors of machinery, and so on. Visualize a great hydroelectric station, where the force of falling water turns huge turbines. These 3ϕ generators transmit this power to the customer. Figure 17.11 illustrates this concept. First, notice that a neutral wire is added symbolically as a dashed line; whether it is used depends upon the load.

Second, note that the load, whatever its configurations, is connected to the 3ϕ generator via three transmission lines. Elements of the load are therefore affected by the voltage difference between any two lines, such as V_{AC}, V_{CB}, or V_{BA}. Additionally, the magnitude and characteristics of the load elements determine the current through each line, such as I_{LA}, I_{LB}, and I_{LC}.

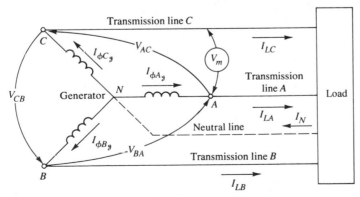

Figure 17.11 Y generator with load connected.

For convenience of analysis, currents are assumed as emanating from the generator's neutral point, through each coil and its respective line to the load, and back to the generator via the neutral line (obviously, when present). Because each coil is in series with its own line to the load, it is clear that

$$I_{\phi A_g} = I_{LA} \tag{17.10a}$$
$$I_{\phi B_g} = I_{LB} \tag{17.10b}$$
$$I_{\phi C_g} = I_{LC} \tag{17.10c}$$

Each phase current of a Y generator equals its respective line current. Note that when referring to the line characteristics, a capital L subscript is used; a script l subscript is used to identify load characteristics. A script g subscript is used to identify generator characteristics.

Voltages between lines do not have quite so simple a relationship to their respective phase voltages. For example, the voltmeter in Figure 17.11 is measuring V_{AC}, the voltage of line C with respect to line A. Remember that some element(s) of the load are connected between lines A and C. Figure 17.12a illustrates this measurement. By vector addition,

$$V_{AC} = V_{AN} + V_{NC} = -V_{NA} + V_{NC}$$

Figure 17.12b combines the vectors in question in solid lines, leaving those phase vectors not needed in the calculation as dashed lines. Because the magnitudes of the phase voltages are equal, triange ANC is an isosceles triangle and V_{AC} bisects $\underline{/ANC}$. Furthermore, by symmetry, the distance from N to P is $1/2\,(V_{AC})$ and NPC is a right triangle. Therefore,

$$\left(\frac{1}{2}\right) V_{AC} = NP = V_{NC} \cos 30° = \left(\frac{\sqrt{3}}{2}\right) V_{NC}$$
$$V_{AC} = 2\left(\frac{\sqrt{3}}{2}\right) V_{NC} = \sqrt{3}\, V_{NC}$$

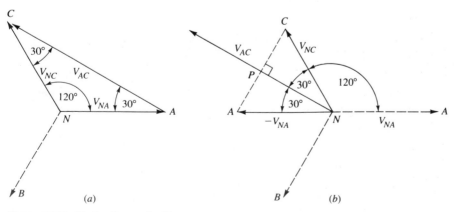

Figure 17.12 Vector diagram for V_{AC}.

V_{AC} is called a *line voltage*, to distinguish it from a *phase voltage*. Therefore, to extend our notation format,

$$V_{LAC} = V_{LCB} = V_{LBA} = \sqrt{3}\, V_{\phi g} \qquad (17.11)$$

The voltage between any two transmission lines (line voltage) of a Y generator equals $\sqrt{3}$ times any alternator phase voltage.

Note that by convention generators are rated by their line voltage. If the Y generator developed 120 V per phase, it would be rated as $120\sqrt{3}$, or 208 V (line).

The Δ Configuration

If the unprimed letters of each coil in Figure 17.8b are connected to the next sequential primed letters, that is if A and B', B and C', and C and A' are connected, Figure 17.13a results. The Y-connected generator is now structured like a delta (Δ). The sum of the voltages forms a closed loop ($V_{CA} + V_{AB} + V_{BC} = 0$). Because they represent 0 V, there is zero short circuit current. As with the Y configuration, however, the isolated generator is not as important as the effects produced by the generator when connected to a load.

Figure 17.13b illustrates the Δ generator connected to a load. It should be apparent that no fourth neutral wire is possible in the Δ configuration. For simplicity, each common connection is identified by its unprimed letter. If a voltmeter is again connected between transmission lines A and C, as in Figure 17.11 for the Y generator, you can see that

$$V_{LAC} = V_{\phi AC g}$$

Similarly, if a voltmeter is connected between any two transmission lines of a Δ generator, the line voltage equals the phase voltage:

$$V_L = V_{\phi g} \qquad (17.12)$$

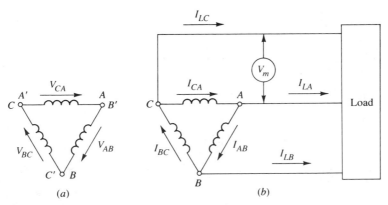

Figure 17.13   (a) Δ coil and (b) load configurations.

The voltage between any two transmission lines of a Δ generator equals the phase voltage of the coil between these same two lines.

Any given line current, such as I_{LA}, however, does not have such a simple relationship to its corresponding phase current. From Figure 17.13b,

$$I_{CA} = I_{LA} + I_{AB}$$
$$I_{LA} = I_{CA} - I_{AB} = I_{CA} + I_{BA}$$

If the relevant vectors in Figure 17.13b are restructured as solid-line vectors, leaving the unneeded vectors in this calculation as dashed lines, Figure 17.14 results. This is analogous to Figure 17.12b, which relates the Y-generator line voltage to its phase voltage. By similar logic, for a Δ generator,

$$I_L = \sqrt{3}\,I_{\phi_g}$$

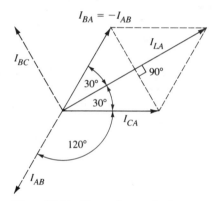

Figure 17.14   Vector diagram for I_{LA}.

PROBLEMS

1. Each phase of a Y generator generates 4.5 A and 120 V.

    a. What is the line current?

    b. What is the line voltage?

2. If the line voltage of a three-phase Y generator is 225 V, what is each individual phase voltage?

3. Each phase of a Δ generator generates 6 A and 220 V.

    a. What is the line current?

    b. What is the line voltage?

4. If the current of a three-phase Δ generator is 12.5 A, what is each individual phase current?

17.6 THREE-PHASE APPLICATIONS

The mathematics of complex algebra are ideally suited for solving the problems presented by the 120° phase displacements of a 3ϕ power distribution system. The level of complexity is determined by whether the load is *balanced* or *unbalanced*. Furthermore, under each condition there can be four configurations. To present a mixture of examples, we illustrate the following four types of problems:

1. $Y \to Y$ (unbalanced load)

2. $Y \to \Delta$ (balanced load)

3. $\Delta \to Y$ (balanced load)

4. $\Delta \to \Delta$ (unbalanced load)

Example 1

Assume the loads on a four-wire, three-phase alternator are unbalanced. The circuit is shown in Figure 17.15. If the loads are

$$Z_A = 10 + j0 \qquad Z_B = 4 + j3 \qquad Z_C = 6 + j8$$

calculate the line currents when the reference voltage is $120 \underline{/0°}$. Calculate I_N (the neutral line current).

The alternator phase voltages equal their respective load phase voltages and the alternator phase currents equal their respective line currents.

$$Z_{A\ell} = 10 + j0 = 10\underline{/0°}$$

$$Z_{B\ell} = 4 + j3 = 5\underline{/36.87°}$$

$$Z_{C\ell} = 6 + j8 = 10\underline{/53.13°}$$

$$I_{LA} = \frac{V_{\phi A g}}{Z_{A\ell}} = \frac{120\underline{/0°}}{10\underline{/0°}} = 12\underline{/0°} = 12 + j0$$

$$I_{LB} = \frac{V_{\phi B g}}{Z_{B\ell}} = \frac{120\underline{/-120°}}{5\underline{/36.87°}} = 24\underline{/-156.87°} = -22.07 - j9.43$$

$$I_{LC} = \frac{V_{\phi C g}}{Z_{C\ell}} = \frac{120\underline{/120°}}{10\underline{/53.13°}} = 12\underline{/66.87°} = 4.71 + j11.04$$

$$I_N = I_{LA} + I_{LB} + I_{LC}$$

$$= (12 + j0) + (-22.07 - j9.43) + (4.71 + j11.04)$$

$$= -5.36 + j1.61$$

$$= 5.60\underline{/163.28°}$$

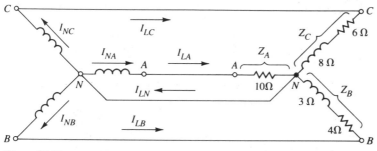

Figure 17.15

Example 2

Assume the Y generator in Figure 17.16 generates 120 V per phase.

a. Find the phase currents of the load.

b. Show that when the load is balanced, the circulating load current is zero.

c. Find the line currents.

$$V_{LAC} = \sqrt{3}\, V_{\phi g} \underline{/150°} \qquad \text{see Figure 17.12}b$$
$$= 208 \underline{/150°}$$

Similarly,

$$V_{LCB} = 208 \underline{/270°}$$
$$V_{LBA} = 208 \underline{/30°}$$
$$Z_{\theta\ell} = 6 + j8 = 10 \underline{/53.13°}$$

a. $I_{\phi AC\ell} = \dfrac{V_{LAC}}{Z_{AC\ell}} = \dfrac{208 \underline{/150°}}{10 \underline{/53.13°}} = 20.8 \underline{/96.87°}$

$\qquad = -2.49 + j20.65$

$\quad I_{\phi CB\ell} = \dfrac{V_{LCB}}{Z_{CB\ell}} = \dfrac{208 \underline{/270°}}{10 \underline{/53.13°}} = 20.8 \underline{/216.87°}$

$\qquad = -16.64 - j12.48$

$\quad I_{\phi BA\ell} = \dfrac{V_{LBA}}{Z_{BA\ell}} = \dfrac{208 \underline{/30°}}{10 \underline{/53.13°}} = 20.8 \underline{/-23.13°}$

$\qquad = 19.13 - j8.17$

b. $\Sigma I_{\phi\ell} = (-2.49 + j20.65) + (-16.64 - j12.48) + (19.13 - j8.17)$

$\qquad = 0 + j0$

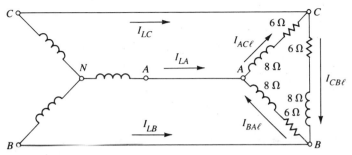

Figure 17.16

c. $I_{LA} = I_{\phi AC\ell} - I_{\phi BA\ell} = -2.49 + j20.65 - (19.13 - j8.17)$
$= -21.62 + j28.82 = 36.03\underline{/126.87°}$

$I_{LB} = I_{\phi BA\ell} - I_{\phi CB\ell} = (19.13 - j8.17) - (-16.64 - j12.48)$
$= 35.77 + j4.31 = 36.03\underline{/6.87°}$

$I_{LC} = I_{\phi CB\ell} - I_{\phi AC\ell} = (-16.64 - j12.48) - (-2.49 + j20.65)$
$= -14.15 - j33.13 = 36.03\underline{/246.87°}$

Note that a balanced load produces equal line currents (36.03 A) and equal phase displacements of 120°.

Example 3

For the circuit in Figure 17.17 assume $I_{LA} = 10\underline{/0°}$, $I_{LB} = 10\underline{/-120°}$, and $I_{LC} = 10\underline{/120°}$. Find $V_{\phi AC\ell}$, $V_{\phi CB\ell}$, and $V_{\phi BA\ell}$.

$$Z_{\phi\ell} = 5 - j12 = 13\underline{/-67.38°}$$

$$V_{\phi NA\ell} = I_{LA}Z_{\phi\ell} = (10\underline{/0°})(13\underline{/-67.38°}) = 130\underline{/-67.38°}$$
$$= 50 - j120$$

$$V_{\phi NC\ell} = I_{LC}Z_{\phi\ell} = (10\underline{/120°})(13\underline{/-67.38°}) = 130\underline{/52.62°}$$
$$= 78.92 + j103.3$$

$$V_{\phi NB\ell} = I_{LB}Z_{\phi\ell} = (10\underline{/-120°})(13\underline{/-67.38°}) = 130\underline{/172.62°}$$
$$= -128.92 + j16.7$$

$$V_{\phi AC\ell} = V_{\phi AN\ell} + V_{\phi NC\ell} = -V_{\phi NA\ell} + V_{\phi NC\ell}$$
$$= -50 + j120 + 78.92 + j103.3$$
$$= 28.92 + j223.3$$
$$= 225.16\underline{/82.62°}$$

$$V_{\phi CB\ell} = V_{\phi CN\ell} + V_{\phi NB\ell} = -V_{\phi NC\ell} + V_{\phi NB\ell}$$
$$= -78.92 - j103.3 - 128.92 + j16.7$$
$$= -207.84 - j86.6$$
$$= 225.16\underline{/-157.38°}$$

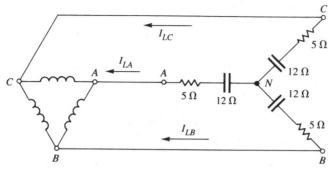

Figure 17.17

$$V_{\phi BA\ell} = V_{\phi BN\ell} + V_{\phi NA\ell} = -V_{\phi NB\ell} + V_{\phi NA\ell}$$
$$= 128.92 - j16.7 + 50 - j120$$
$$= 178.92 - j136.7$$
$$= 225.16\underline{/-37.38°}$$

Note the 120° phase displacement between each set of line voltages.

Example 4

In the Δ configuration generator and load in Figure 17.18, which line current will conduct the greatest current if the load impedances are $Z_1 = 200$ Ω, $Z_2 = 300 \underline{/40°}$ Ω, $Z_3 = 400 \underline{/-60°}$ Ω? Construct a phasor diagram showing the line currents in relation to the phase voltages.

$$I_{CA} = 120\underline{/0°} / 200 = 0.6 \text{ A } \underline{/0°}$$

$$I_{BC} = 120 \underline{/-120°} / 400 \underline{/-60°} = 0.3 \text{ A } \underline{/-60°}$$

$$I_{AB} = 120\underline{/120°} / 300 \underline{/40°} = 0.4 \text{ A } \underline{/80°}$$

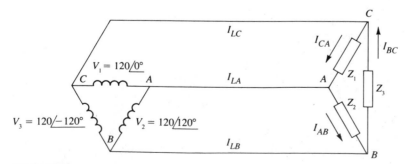

Figure 17.18

Converting the polar formats into rectangular formats results in

$$I_{CA} = 0.6 + j0$$

$$I_{BC} = 0.15 - j0.26$$

$$I_{AB} = 0.07 + j0.39$$

Combining phase currents to calculate line currents results in

$$I_{LC} = I_{BC} - I_{CA} = (0.15 - j0.26) - (0.6 + j0) = -0.45 - j0.26$$
$$= 0.52\underline{/-149.98°}\ \text{A}$$

$$I_{LA} = I_{CA} - I_{AB} = (0.6 + j0) - (0.07 + j0.39) = 0.53 - j.39$$
$$= 0.658\underline{/-36.35°}\ \text{A}$$

$$I_{LB} = I_{AB} - I_{BC} = (0.07 + j0.39) - (0.15 - j0.26) = -0.08 + j0.65$$
$$= 0.655\underline{/97.02°}\ \text{A}$$

Therefore, the greatest line current flows from terminal A. Figure 17.19 is a phasor diagram relating the line currents to the phase voltages.

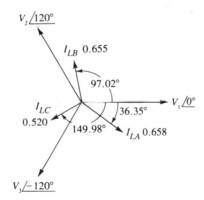

Figure 17.19

PROBLEMS

1. Assume a circuit similar to that in Figure 17.15 having a balanced load. If $V_{\phi_9} = 150$ V and each phase load is 10 Ω in parallel with a 15-Ω capacitive reactance, find each line current and show that I_N is 0 A.

2. Assume the circuit in Figure 17.15 has an unbalanced load. $V_{\phi_9} = 175$ V, $Z_A = 10 + j15$, $Z_B = 8 - j6$, and $Z_C = 5$-Ω resistance in parallel with 8-Ω capacitive reactance. Find each line current and I_N.

3. Assume a circuit similar to that in Figure 17.16 with a balanced load. If $V_{\phi_g} = 200$ V and each phase load is $16.5 + j8.9$, find each load phase current and line current and show that the sum of the circulating load currents is zero.

4. Assume a circuit similar to that in Figure 17.16 having an unbalanced load. $V_{\phi_g} = 155$ V, $Z_{AC} = 7.5 - j3.6$, $Z_{CB} = 10.2 + j5.9$, and $Z_{BA} = 12 + j0$. Find each load phase current and line current.

5. Assume a circuit similar to that in Figure 17.17. $I_{LA} = 12.5\underline{/0°}$, $I_{LB} = 12.5\underline{/-120°}$, and $I_{LC} = 12.5\underline{/120°}$. If each load phase impedance is $3.5 + j6.2$, find $V_{\phi AC\ell}$, $V_{\phi CB\ell}$, $B_{\phi BA\ell}$.

6. Assume a circuit similar to that in Figure 17.20. $V_{\phi AC_g} = 100\underline{/0°}$, $V_{\phi CB_g} = 100\underline{/120°}$, and $V_{\phi BA_g} = 100\underline{/-120°}$. Find each load phase current and line current.

7. Assume a circuit similar to that in Figure 17.18. $Z_1 = 200 + j150$, $Z_2 = 150 - j125$, and $Z_3 = 95$ in parallel with $j105$. Find each load phase current and line current.

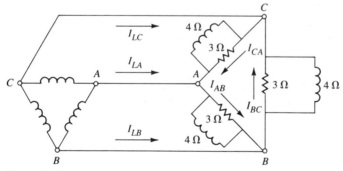

Figure 17.20

SUMMARY

The laws of electromagnetics dictate that when a closed-loop coil is rotated within a uniform magnetic field, the voltage induced in the coil is directly proportional to the sine of the angle the coil makes with a plane that is perpendicular to the direction of the magnetic field.

Nomenclature has been developed to define certain characteristics of a sine wave, including

*   *Frequency*   The number of times a cyclic effect repeats itself; usually expressed in hertz per second.

- *Period* The time it takes to complete 1 Hz of a periodic function.

- *Radian frequency* Designated ω (omega), this defines the number of radians the periodic function transverses per unit length of time.

- *Displacement* With reference to a rotating vector, defines the number of degrees or radians the rotating vector has transversed from the reference.

When a phase shift occurs, resolution of electric phenomena is vectorial as opposed to an algebraic solution.

Most commercial power systems are three-phase (3ϕ) in nature. The characteristics are:

Y Configuration

1. Each phase current equals its corresponding line current.
2. Line voltages equal $\sqrt{3}$ times any phase voltage.

Δ Configuration

1. Each line voltage equals its corresponding phase voltage.
2. Line currents equal $\sqrt{3}$ times any phase current.

18 APPLICATIONS 3: AC ELECTRONIC CIRCUITS

This is the third chapter dedicated exclusively to applying the more recently learned mathematics to electronic circuits. Chapter 13, "Applications 2: Network Theorems," illustrated the application of the rules of simultaneous equations to developing electronic theorems. Chapter 9, "Applications 1: DC Electronic Circuits," applied simple linear and fractional equations to the solution of simple DC circuits and discussed such electronic laws as Ohm's law, Kirchhoff's laws, the proportionate voltage theorem, and the inverse current ratio theorem. All these laws and theorems are equally applicable to AC circuits. However, because of the phasor relationships in alternating current, complex mathematics must replace the simpler approach used for DC circuit analysis. Examples 1 to 3 illustrate more applications of complex mathematics to electronic circuits. Keep in mind that inductive reactance is expressed as jX_L, capacitive reactance as $-jX_C$.

18.1 SERIES CIRCUITS

Proportionate Voltage Ratios

Example 1

See the circuit in Figure 18.1. The general expression for the voltage (v_{Za}) across a given impedance (Z_a) when in series with a second impedance (Z_b) is

$$v_{Za} = \frac{Z_a}{Z_a + Z_b} v_S \qquad v_S \text{ is total voltage across both impedances} \qquad (18.1)$$

Apply these principles to determine

a. The maximum voltage across R

$$v_R = \frac{R}{R - jX_c} v_S = \frac{3.3}{3.3 - j2.5}(25)$$

$$= \frac{82.5\underline{/0°}}{4.14\underline{/-37.147°}} = 19.927\underline{/37.147°}$$

$$= 15.884 + j12.033$$

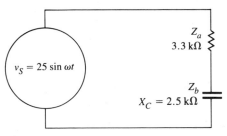

Figure 18.1

b. The maximum voltage across C

$$v_C = \frac{-jX_c}{R - jX_C} v_s = \frac{-j2.5}{3.3 - j2.5}(25)$$

$$= \frac{62.5\underline{/-90°}}{4.14\underline{/-37.147°}} = 15.097\underline{/-52.853°}$$

$$= 9.117 - j12.034$$

Now apply the principles to

c. Verify Kirchhoff's voltage law by showing that $v_R + v_C = v_S$

$$v_S = v_R + v_C$$
$$= (15.884 + j12.033) + (9.117 - j12.034)$$
$$= 25.001 - j0.001$$
$$\approx 25$$

Example 2

What should X_L in Figure 18.2 be so that v_L maximum equals 19.25 V?
  From the principles set by Equation (18.1),

$$v_{Zb} = \frac{jX_L}{4.7 + jX_L}(v_s)$$

$$19.25 = \frac{jX_L}{4.7 + jX_L}(47.5)$$

$$90.475 + j19.25X_L = j47.5X_L$$

$$90.475 = j28.25X_L$$

$$jX_L = \frac{90.475}{28.25}$$

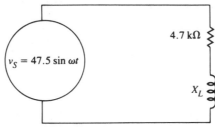

Figure 18.2

You can drop the j term because this problem only indicates that the reactance is inductive. In addition, X_L is automatically in kΩ because all values in Figure 18.2 are also in kΩ.

$$\therefore X_L = 3.203 \text{ k}\Omega$$

Ohm's Law

Example 3

See the circuit in Figure 18.3

a. Find i_T.

Because Z_T is in kΩ, i_T is automatically in mA.

$$Z_T = 2.2 + j4.45 - j6.15$$
$$= 2.2 - j1.7$$
$$= 2.78\underline{/-37.694°} \text{ k}\Omega$$
$$\therefore i_T = \frac{v_T}{Z_T} = \frac{32\underline{/0°}}{2.78\underline{/-37.694°}} = 11.51\underline{/37.694°}$$
$$= (9.107 + j7.037)\text{mA}$$

Note that if voltage is taken as reference, current leads voltage by slightly more than 37°, making this circuit capacitive.

b. Find v_R, v_L, and v_C
$$v_R = i_T R = (11.51\underline{/37.694°})(2.2\underline{/0°})$$
$$= 25.322\underline{/37.694°}$$
$$= (20.037 + j15.483) \text{ V}$$

$$v_L = i_T X_L = (11.51\underline{/37.694°})(4.45\underline{/90°})$$
$$= 51.22\underline{/127.694°}$$
$$= (-31.318 + j40.53) \text{ V}$$

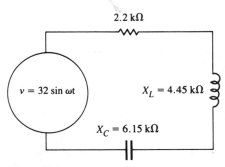

Figure 18.3

$$v_C = i_T X_C = (11.51 \underline{/37.694°})(6.15 \underline{/-90°})$$
$$= 70.787 \underline{/-52.306°}$$
$$= (43.282 - j56.013) \text{ V}$$

c. Verify Kirchhoff's laws of voltages in a closed loop.

$$v_S = v_R + v_L + v_C$$
$$= (20.037 + j15.483) + (-31.318 + j40.53) + (43.282 - j56.013)$$
$$= 32.001 + j0$$
$$\approx 32 \text{ V}$$

d. Draw a phasor diagram of the values requested in parts a and b. The resultant is Figure 18.4.

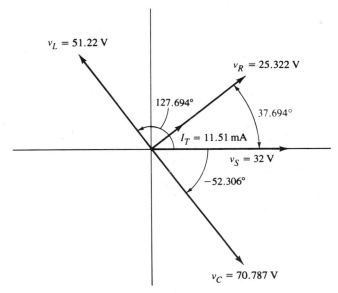

Figure 18.4

PROBLEMS

1. In a series circuit $R = 6.8 \text{ k}\Omega$ and $X_L = 5.6 \text{ k}\Omega$. The circuit is connected across a source of $22.5 \sin \omega t$ V. What is the maximum voltage across R? What is the maximum voltage across L? Show that $v_R + v_L = 22.5$ V.

2. A series RC circuit is connected across a voltage source of $56.15 \sin \omega t$ V. If $R = 2.2 \text{ k}\Omega$ and $X_C = 3.3 \text{ k}\Omega$, what is the maximum value of v_R and v_C?

3. What should X_C in Problem 2 be so that v_C maximum is 37.5 V?

4. In a series RC circuit $R = 1.5 \text{ k}\Omega$ and $X_C = 3.3 \text{ k}\Omega$. What is the total circuit impedance? If $v_S = 43.75$ V, what is i_T?

5. In a series RCL circuit $R = 10 \text{ k}\Omega$ and $X_L = 4.7 \text{ k}\Omega$. What should X_C be so that $Z_T = 16.1 \text{ k}\Omega$?

6. In a series RCL circuit $R = 1.5 \text{ k}\Omega$, $X_L = 4.7 \text{ k}\Omega$, and $X_C = 2.2 \text{ k}\Omega$. The circuit is connected across a source voltage of $29.35 \sin \omega t$ V. What is the maximum current? What is the maximum value of v_R, v_L, and v_C? Show that $v_R + v_L + v_C = 29.35$ V.

18.2 PARALLEL CIRCUITS

Equivalent Resistance

Example 1

For the circuit shown in Figure 18.5, determine Z_T. (Z_T is in kΩ.)

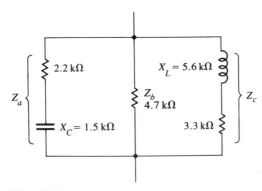

Figure 18.5

Using the product-over-sum method for three impedances in parallel, we get

$$Z_T = \frac{Z_a Z_b Z_c}{Z_a Z_b + Z_a Z_c + Z_b Z_c} = \frac{N}{D}$$

$$Z_a = 2.2 - j1.5 = 2.663\underline{/-34.287°}$$
$$Z_b = 4.7 + j0 = 4.7\underline{/0°}$$
$$Z_c = 3.3 + j5.6 = 6.5\underline{/59.49°}$$
$$Z_a Z_b = (2.663\underline{/-34.287°})(4.7\underline{/0°}) = 12.516\underline{/-34.287°}$$
$$= 10.341 - j7.051$$
$$Z_a Z_c = (2.663\underline{/-34.287°})(6.5\underline{/59.49°}) = 17.310\underline{/25.203°}$$
$$= 15.662 + j7.371$$
$$Z_b Z_c = (4.7\underline{/0°})(6.5\underline{/59.49°}) = 30.55\underline{/59.49°}$$
$$= 15.51 + j26.32$$
$$D = (10.341 - j7.051) + (15.662 + j7.371) + (15.51 + j26.32)$$
$$= 41.513 + j26.64$$
$$= 49.326\underline{/32.689°}$$
$$N = (2.663\underline{/-34.287°})(4.7\underline{/0°})(6.5\underline{/59.49°})$$
$$= 81.355\underline{/25.203°}$$
$$Z_T = \frac{N}{D} = \frac{81.355\underline{/25.203°}}{49.326\underline{/32.689°}} = 1.649\underline{/-7.486}$$
$$= 1.635 - j0.215$$

Thus all the elements in Figure 18.5 can be replaced with a series combination of $R = 1.635$ kΩ and $X_C = 215$ Ω.

Inverse Current Ratio

Example 2

See the circuit in Figure 18.6. The general expression for the current (i_{Za}) through a given impedance (Z_a) in parallel with a second impedance (Z_b) is

$$i_{Za} = \frac{Z_b}{Z_a + Z_b} i_T \qquad i_T = \text{total current entering junction} \qquad (18.2)$$

a.   Determine the maximum value of i_{Za}. Applying the principles of the inverse current ratio law results in the following:

$$i_{Za} = \frac{0.56 + j1}{(0.47 - j0.68) + (0.56 + j1)} (185)$$

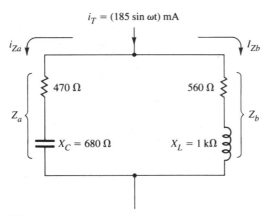

$i_T = (185 \sin \omega t)$ mA

Figure 18.6

$$= \frac{(1.146\underline{/60.751°})(185\underline{/0°})}{1.079\underline{/17.259°}}$$
$$= 196.487\underline{/43.492°}$$
$$= (142.546 + j135.233) \text{ mA}$$

b. Determine the maximum value of i_{Zb}. Using data from part a

$$i_{Zb} = \frac{0.47 - j0.68}{1.079\underline{/17.259°}}(185)$$
$$= \frac{(0.827\underline{/-55.349°})(185\underline{/0°})}{1.079\underline{/17.259°}}$$
$$= 141.793\underline{/-72.608°}$$
$$= (42.383 - j135.311) \text{ mA}$$

c. Show that the sum of parts a and b equals i_T.
$$i_{Za} + i_{Zb} = (142.546 + j135.233) + (42.383 - j135.311)$$
$$= 184.929 - j0.078$$
$$\approx 185 \text{ mA}$$

PROBLEMS

1. Determine the equivalent resistance of $Z_a = (2.54 + j1.68)$ kΩ in parallel with $Z_b = (4.76 - j3.52)$ kΩ.

2. See Problem 1. If a maximum of 327.5 mA enter the junction formed by Z_a and Z_b, determine the current through each branch using the method of inverse currents. Show that the sum of the currents is 327.5 mA.

3. The following three impedances are in parallel: $Z_a = 4.7 - j3.6$, $Z_b = 5.2 + j3.9$, and $Z_c = -j3.2$. Determine their equivalent resistance.

4. A maximum current of 210 mA enter the junction of the three parallel impedances in Problem 3. Determine how the current splits among the three impedances. Show that the sum of these currents equals 210 mA.

18.3 SERIES-PARALLEL CIRCUITS

The previous examples mostly sought solutions when the input voltage was at its maximum, that is, when $\omega t = 90°$ for a sine input. As a variation of the previous examples, the following inputs are in polar format when the angle is other than 90°.

Example 1

In Figure 18.7, determine i_T, i_1, and i_2. Express all three answers in both rectangular and polar formats. In this example, all impedances are in kΩ, so all currents automatically result in mA. Let $Z_1 = 4.7 + j0 = 4.7\underline{/0°}$, $Z_2 = 3.3 + j2.5 = 4.14\underline{/37.147°}$, and $Z_3 = -j1.68 = 1.68\underline{/-90°}$.

$$Z_T = Z_1 + Z_2\|Z_3 = 4.7 + \frac{(4.14\underline{/37.147°})(1.68\underline{/-90°})}{(3.3 + j2.5) + (-j1.68)}$$

$$= 4.7 + \frac{6.955\underline{/-52.853°}}{3.4\underline{/13.955°}}$$

$$= 4.7 + 2.046\underline{/-66.808°}$$

$$= 4.7 + 0.806 - j1.881$$

$$= 5.506 - j1.881 = 5.818\underline{/-18.862°}$$

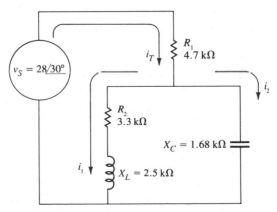

Figure 18.7

a. $i_T = \dfrac{V_S}{Z_T} = \dfrac{28\underline{/30°}}{5.818\underline{/-18.862°}}$

$= 4.813\underline{/48.862°} = (3.166 + j3.625)$ mA

By the inverse current ratios,

b. $i_1 = \dfrac{Z_3}{Z_2 + Z_3} i_T$

$= \dfrac{1.68\underline{/-90°}}{(3.3 + j2.5) + (-j1.68)}(4.813\underline{/48.862°})$

$= \dfrac{8.086\underline{/-41.138°}}{3.3 + j0.82} = \dfrac{8.086\underline{/-41.138°}}{3.4\underline{/13.995°}}$

$= 2.378\underline{/-55.093°}$ mA

$= (1.361 - j1.950)$ mA

c. $i_2 = i_T - i_1$

$= (3.166 + j3.625) - (1.361 - j1.950)$

$= (1.805 + j5.575)$ mA

$= 5.860\underline{/72.060°}$ mA

Example 2

In Figure 18.8 determine the values of i_T, i_1, i_2, v_{ab}, v_{ac}, and v_{bc}. All currents are in mA because all impedances are in kΩ. Let $Z_0 = 1.5 + j2.2 = 2.663\underline{/55.713°}$, $Z_1 = 5.6 - j4.15 = 6.97\underline{/-36.541°}$, and $Z_2 = 3.3 + j(5.1 - 1.75) = 3.3 + j3.35 = 4.702\underline{/45.431°}$.

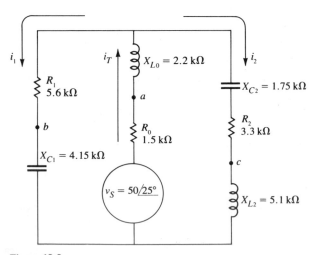

Figure 18.8

Define $Z_3 = Z_1 \| Z_2$

$$= \frac{(6.97\underline{/-36.541°})(4.702\underline{/45.431°})}{(5.6 - j4.15) + (3.3 + j3.35)}$$

$$= \frac{32.773\underline{/8.89°}}{8.936\underline{/-5.136°}}$$

$$= 3.668\underline{/14.026°} = 3.559 + j0.889$$

$$\therefore Z_T = Z_0 + Z_3 = (1.5 + j2.2) + (3.559 + j0.889)$$

$$= 5.059 + j3.089 = 5.928\underline{/31.408°}$$

a. $i_T = \dfrac{v_S}{Z_T} = \dfrac{50\underline{/25°}}{5.928\underline{/31.408°}} = (8.435\underline{/-6.408°})$ mA

$$= (8.382 - j0.941) \text{ mA}$$

Because v_S is at 25° and i_T at −6.408°, voltage leads current, so the circuit is capacitive.

b. $i_1 = \dfrac{Z_2}{Z_1 + Z_2}(i_T)$

$$= \frac{4.702\underline{/45.431°}}{(5.6 - j4.15) + (3.3 + j3.35)}(8.435\underline{/-6.408°})$$

$$= \frac{39.661\underline{/39.023°}}{8.936\underline{/-5.136°}}$$

$$= 4.438\underline{/44.159°} = 3.184 + j3.092$$

c. $i_2 = i_T - i_1$

$$= (8.382 - j0.941) - (3.184 + j3.092)$$

$$= 5.198 - j4.033 = 6.579\underline{/-37.807°}$$

d. $v_{ab} = i_T X_{L0} + i_1 R_1$

$$= (8.435\underline{/-6.408°})(2.2\underline{/90°}) + (4.438\underline{/44.159°})(5.6)$$

$$= 18.557\underline{/83.592°} + 24.853\underline{/44.159°}$$

$$= (2.071 + j18.441) + (17.830 + j17.314)$$

$$= 19.901 + j35.755 = 40.92\underline{/60.9°}$$

e. $v_{ac} = i_T X_{L0} + i_2(R_2 - jX_{C2})$

$$= 18.557\underline{/83.592°} + (6.579\underline{/-37.807°})(3.735\underline{/-27.937°})$$

$$= 2.071 + j18.441 + 24.573\underline{/-65.744°}$$

$$= 2.071 + j18.441 + 10.095 - j22.404$$

$$= 12.166 - j3.963 = 12.795\underline{/-18.043°}$$

f. $v_{bc} = i_1 X_{c1} - i_2 X_{L2}$

$$= (4.438\underline{/44.159°})(4.15\underline{/-90°}) - (6.579\underline{/-37.807°})(5.1\underline{/90°})$$

$$= 18.418\underline{/-45.841°} - 33.553\underline{/52.193°}$$

$$= (12.831 - j13.213) - (20.568 + j26.51)$$

$$= -7.737 - j39.723 = 40.469\underline{/-101.022°}$$

PROBLEMS

1. Determine the values of i_T, i_1, and i_2 in Figure 18.9, expressing all your answers in both rectangular and polar formats.

2. Determine the values of i_T, i_1, and i_2 in Figure 18.10, expressing all your answers in both rectangular and polar formats.

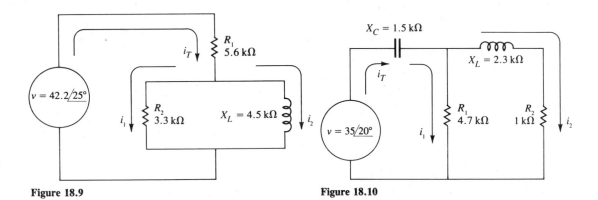

Figure 18.9   **Figure 18.10**

3. Determine all the currents in Figure 18.11.

4. In Figure 18.12 determine the magnitudes of all the currents, expressing all your answers in the rectangular and polar formats.

5. Using the results of Problem 4, determine v_{ab}, v_{ac}, and v_{bc} in Figure 18.12.

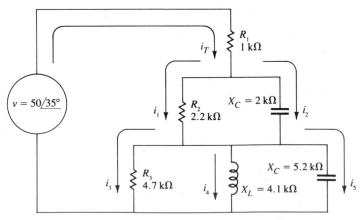

Figure 18.11

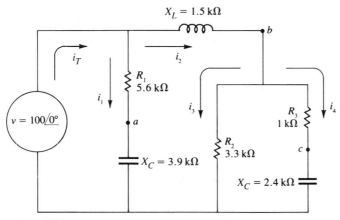

Figure 18.12

SUMMARY

All the basic laws utilized for DC circuit analysis apply equally to AC circuit analysis. The major difference is the introduction of the j operator into the calculations.

When all voltages (including accounting for phase angles) in a series circuit are summed, they total the applied voltage. When all currents (including accounting for phase angles) in a parallel circuit are summed, the result is the total source current.

REVIEW PROBLEMS, PART III

The following problems are a review of Part III of the text and directly relate to the numbered sections.

Section 14.2

1. What is the complement of
   a. $17.6°$
   b. $52°12'54"$

2. What is the supplement of
   a. $141.31°$
   b. $161°21'35"$

3. Express as an angle greater than 0° but less than 360°.
   a. $516°$
   b. $1154°16'$

4. Represent as a negative angle.
   a. $432°$
   b. $125°$

5. Convert to decimal format.
   a. $56°42'$
   b. $12°5'8"$

6. Convert to the degree, minute, second format.
   a. $15.1685°$
   b. $217.0617°$

7. Each angle should be represented in degrees, radians, and grades. Supply the missing two formats.
   a. 1.652 rad
   b. $65°5'$
   c. 0.621 grad

8. Determine B in the format represented
   a. $62°3' + 162.41° = B° ' "$
   c. $21.3° + 2.78$ rad $= B$ rad
   b. 1.65 rad $+ 231.59$ grad $= B° ' "$
   d. 92.6 grad $+ 1.48$ rad $= B$ grad

Section 14.3

9. Refer to Figure 14.10 and determine the values requested.
   a. $OP = 15.2$ in, $PQ = 6.7$ in, and $OP' = 25.6$ in; find $P'Q'$.
   b. $PQ = 6.1$ m, $OQ = 8.2$ m, and $P'Q' = 10.5$ m; find OQ'.

10. Refer to Figure 14.12 and determine the values requested.

   a. $\theta = 43.1°$; find ϕ.

   b. $a = 3.21$ m and $b = 4.17$ m; find h.

   c. $b = 5.12$ ft and $h = 13.1$ ft; find a.

Section 15.2

Refer to Figure 15.8 and calculate the missing variables in Review Problems 11 to 14 for the following linear functions.

11. $x_1 = 5.321$ $y_1 = 21.621$ **13.** $x_1 = -2.041$ $y_1 = 10.876$
   $x_2 = 2.654$ $y_2 = ?$ $x_2 = 1.256$ $y_2 = 5.721$
   $x_3 = 0.865$ $y_3 = 5.627$ $x_3 = ?$ $y_3 = 3.613$

12. $x_1 = -4.213$ $y_1 = 6.782$ **14.** $x_1 = 10.876$ $y_1 = 4.321$
   $x_2 = -1.461$ $y_2 = ?$ $x_2 = 9.672$ $y_2 = -2.685$
   $x_3 = 0.987$ $y_3 = 19.642$ $x_3 = 3.061$ $y_3 = ?$

Using your calculator at ⌐Fix 4⌐ , determine the value of Review Problems 15 to 20.

15. $(\sin 21.8°)(\tan 1.7 \text{ rad})$

18. $(\cos 55°35' + \sec 80°44') \tan 68°50'$

16. $\sec 2.31 \text{ rad} + \cot 35.6 \text{ grad}$

19. $\dfrac{1}{\tan 16°42'35''}$

17. $\dfrac{\tan \frac{\pi}{3} \text{rad} - \cot 67.8 \text{ grad}}{\csc 86°}$

20. $\tan 15°15' \cdot \cot 15°15'$

Section 15.3

21. Using your calculator, determine the value of the following angles in the units indicated.

   a. $\cos^{-1} 0.872$ in ° format

   b. $\cot^{-1} 3.6$ in grade format

   c. $\sin^{-1} 0.213$ in ° ' '' format

   d. $\sec^{-1} 6.31$ in radian format

22. Using your calculator, evaluate the following.

   a. $\tan (\sin^{-1} 0.863)°$

   b. $\sin (2 \cos^{-1} 0.621) \text{ rad}$

   c. $\cot \left(\sin^{-1} \dfrac{\sqrt{3}}{2} \right)°$

   d. $\dfrac{\cos (\cot^{-1} 0.623) \text{ grad}}{\tan 35.6 \text{ grad}}$

Section 15.4

Refer to Figure 15.2 for Review Problems 23 to 26 and find the value of the unknown quantities for the values given.

23. $\theta = 35°$ and $h = 20.7$; find b.

24. $b = 12.8$ and $a = 5.8$; find $\theta°$.

25. $\phi = 0.489$ rad and $h = 30.5$; find θ, a, and b.

26. $b = 15.2$ and $a = 25.6$; find h, θ, and ϕ.

27. Given $i = I_p \sin 2\pi ft$; find i if $I_p = 310$ mA, $f = 520$ Hz, and $t = 10$ mS.

28. Given $v = V_p \sin 2\pi ft$; what is V_p if $v = 120$ mV, $f = 160$ Hz, and $t = 650$ μS?

Section 15.5

Without using your calculator, evaluate the resultant of Review Problems 29 to 34. Note the special nature of these angles.

29. $\sec 60° + \tan 45°$

30. $\sin 120° + \sec 210°$

31. $\csc 300° - \tan 180°$

32. $\dfrac{10 \sin 30°}{\sin 210°}$

33. $(\tan 225° + \sec 120°) \cot 240°$

34. $\dfrac{\cos 315°}{\cos 120°} + \sin 90° - \cos 180°$

Section 15.6

In Review Problems 35 to 39, use your calculator to evaluate the following expressions.

35. $\sin^2 25.5° + \sin^2 14.6°$

36. $2.3 \csc^2 1.6$ rad $+ 5.1 \sec^2 0.5$ rad

37. $(0.5 \sin^2 75$ grad$)^2(0.3 \cos^{1/2} 10$ grad$)^3$

38. $\dfrac{(2.57 \cot^{1/3} 47.2°)^{1.3}}{5.61 \tan^2 98.6°}$

39. $\dfrac{\sec^2 2.61 \text{ rad}}{\sin^2 1.32 \text{ rad} - \cos^2 0.67 \text{ rad}}$

Section 15.7

In Review Problems 40 to 47, express as a single trigonometric function or a real number without using your calculator.

40. $\sin^2 53°15' + \cos^2 53°15'$

41. $\dfrac{\cos 115 \text{ grad}}{\sin 115 \text{ grad}}$

42. $\cot^2 1.21 \text{ rad} + 1$

43. $\dfrac{\sin 143.2°}{\cos 143.2°}$

44. $1 + \tan^2 0.31 \text{ rad}$

45. $(\sin^2 29° + \cos^2 29°) \sin 30°$

46. $\cos 225°(\sec^2 231° - \tan^2 231°)$

47. $\dfrac{\cot^2 310° - \csc^2 310°}{\cos 60°}$

Section 15.8

For Review Problems 48 to 50, assume a generalized triangle ABC whose angles are labeled A, B, and C. The side opposite $\underline{/A}$ is a, the side opposite $\underline{/B}$ is b, and the side opposite $\underline{/C}$ is c.

48. Solve using the law of sines.

   a. In triangle ABC $a = 30.24$, $A = 30.2°$, $C = 103.5°$; find b.

   b. In triangle ABC $a = 274.2$, $c = 55.6$, $A = 115.3°$; find C.

49. Solve using the law of cosines.

   a. In triangle ABC $a = 78.41$, $b = 46.98$, $c = 62.31$; find B.

   b. In triangle ABC $a = 85.4$, $b = 41.2$, $c = 62.3$; find A.

50. Solve the triangle ABC, finding the remaining sides and the remaining angles using the law of sines, the law of cosines, or both laws.

   a. In triangle ABC $A = 28°45'$, $B = 81°30'$, $a = 42.5$.

   b. In triangle ABC $a = 82.1$, $b = 63.8$, $c = 35.4$.

Section 15.9

51. Using either the sum or difference identities, rewrite the following expressions as single trigonometric functions.

   a. $\sin 40° \cos 85° - \cos 40° \sin 85°$

   b. $\cos 65.5° \cos 11.5° + \sin 65.5° \sin 11.5°$

    c. $\sin 134.5° \cos 80° - \cos 134.5° \sin 80°$

    d. $\sin 115° \cos (-12°) + \cos 115° \sin (-12°)$

52. Use the identity for $\sin (\alpha + \beta)$ to prove $\sin (180° + \theta) = -\sin \theta$.

53. The angle between the voltage and current in an L/R circuit is $30°$. Using $i(t) = I_p \sin (\theta - \phi)$, find $i(t)$ in milliamps at any peak angle θ if I_p equals 10 mA.

54. The angle between the voltage and current in a RC circuit is $60°$. Develop an expression for $i(t)$ in milliamps using $i(t) = I_p \sin (\theta + \phi)$ if I_p equals 18.5 mA.

Section 16.1

Determine the square root of Review Problems 55 to 59.

55. $\sqrt{-144}$ **58.** $j\sqrt{-28}$

56. $-\sqrt{-64}$ **59.** $-j^3\sqrt{-225}$

57. $-9\sqrt{-81}$

Section 16.2

In Review Problems 60 to 75, evaluate the complex operations.

60. $(3.2 - j5) + (1.3 + j4.9)$ **68.** $-(8.1 + j5)(3 + j6.2)$

61. $(8.1 + j5.2) - (3.8 - j6.5)$

69. $\dfrac{1}{j3}$

62. $(3.5)(j2.6)$

63. $(3.6 + j5.7)^2$ **70.** $\dfrac{-3}{j2.3}$

64. $(2.5 - j4.1)^2$

71. $\dfrac{-j8.6}{1 - j}$

65. $-(j3.7)(-j4.9)$

66. $-j2.2(3.61 + j4.39)$ **72.** $\dfrac{2.3 + j4}{j2}$

67. $(3.4 - j4.1)(6.6 + j3.3)$

73. $\dfrac{2.6 + j5.4}{3 - j2}$

74. $\dfrac{8.2 + j6.7}{3.1 + j4.6}$

75. $\dfrac{(2.3 + j3.3)(-4.7 - 6.2j)}{2 + j6}$

Section 16.5

For Review Problems 76 to 80, convert the polar formats into rectangular formats.

76. $4\underline{/60°}$

79. $9.6\underline{/325 \text{ grad}}$

77. $13.4\underline{/-134°}$

80. $3.2 \times 10^4\underline{/126.2°}$

78. $26.3\underline{/0.86 \text{ rad}}$

For Review Problems 81 to 85, convert the rectangular formats into polar formats.

81. $3.12 + j5.67$

84. $34.6 - j63.1$

82. $-6.32 + j5.31$

85. $-0.6 \times 10^{-3} + j3.6 \times 10^{-4}$

83. $-18.2 - j11.1$

Section 16.6

In Review Problems 86 to 98, perform the indicated operations. Express all your answers in both rectangular and polar format.

86. $(3.2 - j5.7) + (-8.4 + j2.1)$

93. $\dfrac{64.2\underline{/-27.2°}}{9.8\underline{/-45.8°}}$

87. $(-10.8 + j8.7) - (19.6 - j12.7)$

94. $\dfrac{12.6\underline{/-215.6°}}{6.3 + j9.61}$

88. $6.4 - j5.1 + 5\underline{/41.3°}$

89. $4.62\underline{/115°} + 8.2\underline{/264°}$

95. $(31.2\underline{/-20.6°})^3$

90. $(9.3\underline{/42°})(5.3\underline{/-15°})$

96. $(8.8 + j16.5)^2$

91. $(3.12 - j4.62)(6.72\underline{/194°})$

97. $(16.2\underline{/1.5\pi \text{ rad}})^{1/2}$

92. $\dfrac{35.6\underline{/44.6°}}{39.5\underline{/23.4°}}$

98. $(3.2 - j4.8)^{1/3}$

Section 16.6

In Review Problems 99 to 103, determine all the roots indicated.

99. $\sqrt{49}$

100. $\sqrt[3]{27}$

101. $\sqrt[4]{625}$

102. $\sqrt[5]{243}$

103. $\sqrt[4]{256}$

Section 17.3

104. A motor rotates at 60 r/s. What is its frequency of revolution per

    a.   minute   **b.**   hour   **c.**   millisecond

105. What is the period per second of the following frequencies?

    a.   337 Hz   **b.**   2.5 MHz   **c.**   125 kHz

106. Given a 1250-Hz sine wave. How many revolutions in radians does it travel in

    a.   0.75 ms   **b.**   562 μs   **c.**   1.3 min

107. The maximum voltage generated by a rotating coil is 56.5 V. If the coil rotates at 250 Hz, determine the instantaneous voltage at the following displacements into the cycle:

    a.   13.5°   **b.**   96.2°   **c.**   1.34 rad

Section 17.5

108. If the line voltage of a three-phase Y generator is 117.5 V, what is each individual phase voltage?

109. Each phase of a Δ generator generates 12.35 A and 275 V.

    a.   What is the line current?

    b.   What is the line voltage?

Section 17.6

110. Assume a circuit similar to the one in Figure 17.15 and having a balanced load. If $V_{\phi_g} = 220$ V and each phase load is 7.5 Ω in series with a 10-Ω inductive reactance, find each line current and show that I_N is 0 A.

111. Assume a circuit similar to the one in Figure 17.16 and having an unbalanced load. $V_{\phi_9} = 205$ V, $Z_{AC} = 3.2 + j1.8$, $Z_{CB} = 5.1 - j3.2$, and $Z_{BA} = 6.1 + j5.2$. Find each load phase current and each line current.

Section 18.1

112. A series RL circuit consists of $R = 2.2$ kΩ and $X_L = 4.7$ kΩ. What is the total circuit impedance? If $V_S = 56.2$ V, what is i_T?

113. A series RCL circuit consists of $R = 2.2$ kΩ, $X_L = 1.5$ kΩ, and $X_C = 4.7$ kΩ. If this circuit is connected across a source voltage of $30.15 \sin \omega t$ V, what is the maximum value of v_R, v_L, and v_C?

Section 18.2

114. What is the equivalent series impedance of $3.2 + j5.6$ in parallel with $1.7 - j3.2$?

115. A 525-mA current enters a parallel junction of $Z_A = 3.5 - j1.7$, $Z_B = 1.6 + j2.9$, and $Z_C = j3.2$. Determine how the current splits among these three impedances. Show that the sum of these currents equals 525 mA.

Section 18.3

116. If v in Figure 18.9 equals $36.62\underline{/41.3°}$, determine i_T, i_1, and i_2. Express all currents in both rectangular and polar format.

117. If v in Figure 18.11 equals $73.2\underline{/42.5°}$, determine all currents. Express each current in both rectangular and polar format.

LOGARITHMS, GRAPHS, AND EXPONENTS

19 LOGARITHMS

19.1 DEFINITION OF THE LOGARITHM

Many electrical problems, especially transients (discussed in application sections 19.6 and 19.7) use a mathematical term called the *logarithm*. Before you can work with logarithms, you first have to know what they are. Basically, a logarithm is a *power*. More explicitly, a logarithm is the power to which a given number (the base) must be raised to equal another number. For example, we all recognize that

$$10^2 = 100$$

Therefore, the logarithm of 100 in the base 10 is 2 because 2 is the power that 10 must be raised to to equal 100. If this definition is extrapolated to a more generalized format, called the exponential format, then

$$a^x = N \qquad (19.1)$$

Mathematically, the abbreviation for the term logarithm is *log*. As a further refinement, if no subscript is associated with the term log, the base is implied as 10. The alternate style for logarithmic equations is the logarithmic format. Expressed verbally, Equation (19.1) is: x is the power the base a must be raised to to equal the number N. The following chart illustrates how to mathematically express logarithms in both the exponential and logarithmic formats for different bases.

Exponential Formats	Logarithmic Formats
$10^3 = 1000$	$3 = \log 1000$
$5^2 = 25$	$2 = \log_5 25$
$7^3 = 343$	$3 = \log_7 343$
$64^{1/2} = 8$	$\frac{1}{2} = \log_{64} 8$
$x^y = Z$	$y = \log_x Z$
$\epsilon^p = M$	$p = \log_\epsilon M$

19.2 SPECIAL PROPERTIES OF LOGARITHMS

Logarithms have special properties that make them most useful for solving certain types of equations. These properties further reinforce the generalized laws of exponents. The development of the calculator has virtually eliminated dependency upon extensive logarithmic charts for performing difficult arithmetic operations. For example, prior to the calculator, a numerical problem like

$$\sqrt[3]{3261.34} \times 0.5663^{2.2}$$

was almost impossible to solve without the use of logarithmic tables. Now, using the calculator, a sequence like

PRESS   **DISPLAY**

3261.34 $\boxed{y^x}$ $\boxed{(}$ 3 $\boxed{1/x}$ $\boxed{)}$ $\boxed{\times}$ .5663 $\boxed{y^x}$ 2.2 $\boxed{=}$ 4.2445774

renders the answer to many significant digits and requires but a few seconds. Also rendered obsolete by calculators is the inherent necessity to interpolate logarithmic numbers not expressly specified by the tables. By the way, because logarithmic functions are not linear, the interpolation process has the same built-in errors as interpolation of trigonometric angles not expressly specified by tables. Here we discuss the four special logarithm properties: logarithms of products, quotients, powers, and roots.

Logarithms of Products

Suppose any two numbers are being multiplied. To generalize the procedure, let us refer to this process as

$$M \cdot N \tag{19.2}$$

Now let us choose a base number; for convenience and practicality we select 10, the base of the decimal numbering system. By the way, although any base number could have been chosen, there are only two *practical* bases: base 10—the Naperian system, designated on the calculator as $\boxed{\log x}$—and base ϵ—the natural system, designated on the calculator as $\boxed{\ln x}$. Now, whatever the values assigned to M and N, there is some power to which 10 can be raised such that

$$10^x = M \quad \text{and} \quad 10^y = N \tag{19.3}$$

If expressed in logarithmic format, this is the same as $x = \log M$ and $y = \log N$ (remember that the base 10 is implied). From our knowledge of the laws of exponents,

$$M \cdot N = 10^x \cdot 10^y = 10^{x+y} \tag{19.4}$$

Before we can fully implement Equation (19.4), we have to illustrate one more concept concerning logarithms. We begin with the fundamental idea that equal exponentials can be expressed as equal to each other:

$$x^y = x^y$$

Now, if we convert the left side of this identity into logarithmic format and we define the right side as the value N, there results

$$y = \log_x N = \log_x x^y \tag{19.5}$$

Interesting! The log to the base x of a number N, defined as magnitude x^y, is itself y, the power of the number x. This means that if we take the log of both sides of Equation (19.4), the results are

$$\begin{aligned} \log_{10}(M \cdot N) &= \log 10^{x+y} \\ &= x + y \\ &= \log_{10} M + \log_{10} N \end{aligned} \tag{19.6}$$

Because we placed no restriction on what the value of the base number can be, or for that matter on how many numbers form a product, we can generalize Equation (19.6) as

$$\log_a(M \cdot N \cdots Z) = \log_a M + \log_a N \cdots \log_a Z \tag{19.7}$$

Expressed verbally,

The logarithm of a product of a series of numbers equals the sum of the logarithm of each number of the series.

Proof: Using your calculator, show that

$$\log(23.1 \times 76.42 \times 186.7) = \log 23.1 + \log 76.42 + \log 186.7$$

(In this and the rest of the examples in this chapter, we make exceptions to the number of significant digits to which examples are carried out so as to more accurately illustrate the solutions.) For the left side of the equation,

PRESS	DISPLAY
23.1 $\boxed{\times}$ 76.42 $\boxed{\times}$ 186.7 $\boxed{=}$ $\boxed{\log}$	5.517963

For the right side of the equation,

PRESS	DISPLAY	COMMENT
23.1 $\boxed{\log}$ $\boxed{+}$	1.363612	
76.42 $\boxed{\log}$ $\boxed{+}$	3.246819	
186.7 $\boxed{\log}$ $\boxed{=}$	5.517963	Log of products rule is verified

Logarithms of Quotients

Let two numbers represent a quotient, such as

$$\frac{M}{N} \tag{19.8}$$

Using the principles of logarithms of products, we can express the dividend M as a base raised to a power (a^x) and the divisor as the same base raised to a power (a^y). Using the laws of exponents, there results

$$\frac{M}{N} = \frac{a^x}{a^y} = a^{x-y} \tag{19.9}$$

Keeping in mind that $a^x = M$ is only the exponential format of $x = \log_a M$ and $a^y = N$ is only the exponential format of $y = \log_a N$, we can apply the principles of Equation (19.5) to develop

$$\log_a \frac{M}{N} = \log_a a^{x-y} = x - y = \log_a M - \log_a N$$

or

$$\log_a \frac{M}{N} = \log_a M - \log_a N \tag{19.10}$$

Expressed verbally,

The logarithm of the quotient of two numbers equals the logarithm of the dividend minus the logarithm of the divisor.

Proof: Using your calculator, show that:

$$\log \frac{249.13}{72.6} = \log 249.13 - \log 72.6$$

For the left side of the equation,

PRESS	DISPLAY
249.13 $\div$ 72.6 $=$ $\boxed{\log}$	0.535489

For the right side of the equation,

PRESS	DISPLAY	COMMENT
249.13 $\boxed{\log}$ $\boxed{-}$	2.396426	
72.6 $\boxed{\log}$ $\boxed{=}$	0.535489	Log of quotients rule is verified

Logarithms of Powers

Start with the generalized exponential format

$$M = a^x \qquad\qquad (19.11)$$

which translates into

$$\log_a M = x \qquad\qquad (19.12)$$

If we raise both sides of Equation (19.11) to the nth power, the results are

$$M^n = (a^x)^n = a^{nx} \qquad\qquad (19.13)$$

Taking the $\log_a$ of both sides of Equation (19.13) results in

$$\log_a M^n = \log_a a^{nx} \qquad\qquad (19.14)$$

However, from Equation (19.5),

$$\log_a a^{nx} = nx \qquad\qquad (19.15)$$

Substituting the value of x from Equation (19.12) and the results of Equation (19.15) into Equation (19.14) results in

$$\log_a M^n = nx = n \log_a M$$

or
$$\log_a M^n = n \log_a M \qquad\qquad (19.16)$$

Expressed verbally,

> The logarithm of a number raised to a power equals the product of the power and the logarithm of the number.

Proof: Using your calculator, show that:

$$\log(2.519^{1.783}) = 1.783 \log 2.519$$

For the left side of the equation,

PRESS	DISPLAY
2.519 $\boxed{y^x}$ 1.783 $\boxed{=}$ $\boxed{\log}$	0.7153898

For the right side of the equation,

PRESS	DISPLAY	COMMENT
1.783 $\boxed{\times}$ 2.519 $\boxed{\log}$ $\boxed{=}$	0.7153898	Log of powers rule is verified

Logarithms of Roots

Repeat Equations (19.11) and (19.12):

$$M = a^x \tag{19.11}$$

$$\log_a M = x \tag{19.12}$$

Extract the nth root of both sides of Equation (19.11):

$$M^{1/n} = (a^x)^{1/n} = a^{x/n} \tag{19.17}$$

Taking the $\log_a$ of both sides of Equation (19.17) and applying the principles of Equation (19.5) results in

$$\log_a M^{1/n} = \log_a a^{x/n} = \frac{x}{n} = \frac{\log_a M}{n}$$

or

$$\log_a M^{1/n} = \frac{\log_a M}{n} \tag{19.18}$$

Expressed verbally,

The logarithm of a number extracted to a root is the quotient of the logarithm of the number divided by the root.

Proof: Using your calculator, show that

$$\log \sqrt[5]{4379.6} = \frac{\log 4379.6}{5}$$

For the left side of the equation,

PRESS   DISPLAY

4379.6 $\boxed{y^x}$ $\boxed{(}$ 5 $\boxed{1/x}$ $\boxed{)}$ $\boxed{=}$ $\boxed{\log}$ 0.728287

For the right side of the equation,

PRESS   DISPLAY   COMMENT

4379.6 $\boxed{\log}$ $\boxed{\div}$ 5 $\boxed{=}$ 0.728287   Log of roots
                                        rule is verified

PROBLEMS

Using your calculator, in Problems 1 to 5 find the logs by direct application and by applying the rules of logarithms.

1. $\log \dfrac{23.7 \times 64.82}{17.61}$ 2. $\log \dfrac{2^{5.7} \times 3^{1.9}}{148.65}$ 3. $\log \dfrac{23.19^{1.7}}{1.73 \times 7.16^{1.5}}$

4. $\log \dfrac{\sqrt[4]{317.16} \times 1.49^{4.87} \times 15.2^{3.1}}{694.2}$

5. $\log \dfrac{\sqrt[5]{754.32} \times 3.43^{4.17}}{1.62^{3.15} \times \sqrt[4]{1063.72}}$

19.3 LOGARITHMIC EQUATIONS

Section 19.2 developed special logarithmic principles to illustrate that some types of problems must use these principles to achieve solutions. Consider if a problem is

$$x^2 = 13.129$$

The solution is simple: Take the square root of both sides of the equation:

$$\sqrt{x^2} = \sqrt{13.129}$$

$$x = 3.6234$$

Now suppose that the problem is

$$2^x = 13.129$$

This is a new type of problem. The power is x, and that is the part we do not know. So how do we know what root to extract? The principles learned in Section 19.2 help us find the solution. First take the log (implied base 10) of both sides of the equation and then apply the rules for the logarithm of powers:

$$2^x = 13.129$$

$$\log 2^x = \log 13.129$$

$$x \log 2 = \log 13.129$$

$$x = \frac{\log 13.129}{\log 2} = \frac{1.1182316}{0.30103}$$

$$= 3.7146851$$

To check, use the y^x function on your calculator and see if indeed

$$2^{3.7146851} = 13.129$$

Notice how a slight change from the term x^2 to the term 2^x radically changed the approach necessary to solve the problem. This section describes how to apply the special rules of logarithms to solve special types of equations like those just illustrated. However, before we can proceed, we must first review the *arc* (inverse) function for logarithms.

Recall that when finding inverse trigonometric functions, such as finding θ when we know that $\sin \theta = 0.5$, the inverse sine function cancels the sine function when they are multiplied, resulting in

$$\sin^{-1} \sin \theta = \sin^{-1} 0.5$$

$$\theta = \sin^{-1} 0.5$$

Translated, the inverse sine function asks the question, "What angle has as the sine value the numeric ratio 0.5?"

Similarly, given that

$$\log x = 1.453$$

if we apply the inverse log (abbreviated $\log^{-1}$) to each side of this equation, the results are

$$\log^{-1} \log x = \log^{-1} 1.453$$

$$x = \log^{-1} 1.453$$

Literally, what number has as its log (base 10 implied) the value 1.453? The answer is obviously between 10 and 100 because $10^1 = 10$ and $10^2 = 100$.

On most scientific calculators the value of x is determined by the following sequence:[1]

$$1.453 \; \boxed{10^x} \; \boxed{=}$$

When you execute this sequence, the answer is 28.37919. To check, use the $\boxed{y^x}$ key on your calculator and see if indeed

$$10^{1.453} = 28.37919$$

Now that we understand the application of the inverse log and that the $\boxed{10^x}$ key on the calculator makes this evaluation, let us solve several logarithmic equations.

Example 1

Solve for x if
$$2 \log x + \log 5x = 3.91$$

Applying the logarithmic rule for products,

$$2 \log x + \log 5 + \log x = 3.91$$
$$3 \log x = 3.91 - \log 5 = 3.21103$$
$$\log x = \frac{3.21103}{3} = 1.07034$$
$$\log^{-1} \log x = \log^{-1} 1.07034$$
$$x = \log^{-1} 1.07034 = 11.758177 \qquad \boxed{\text{STO 1}}$$

Check: Does $2 \log(11.758177) + \log(5(11.758177)) = 3.91$?

---

[1] Some calculators use the key sequence $\boxed{\text{INV}}$ $\boxed{\text{log}}$ instead of the $\boxed{10^x}$ key.

PRESS	DISPLAY	COMMENT
2 ☐× RCL 1 log ☐+	2.14068	First term
((5 ☐× RCL 1)) log ☐=	3.90999	Solution verified

Example 2

Solve for x.

$$3 \log \frac{x}{2} + \log \frac{x}{5} = 1.7235$$

$$3 \log x - 3 \log 2 + \log x - \log 5 = 1.7235$$

$$4 \log x = 1.7235 + 3 \log 2 + \log 5 = 3.32556$$

$$\log x = \frac{3.32556}{4} = 0.83139$$

$$\log^{-1} \log x = \log^{-1} 0.83139$$

$$x = \log^{-1} 0.83139 = 6.7825 \qquad \boxed{\text{STO 1}}$$

Check: Does $3 \log 6.7825 / 2 + \log 6.7825 / 5 = 1.7235$?

PRESS	DISPLAY	COMMENT
3 ☐× ((RCL 1 ☐÷ 2)) log ☐+	1.59108	First term
((RCL 1 ☐÷ 5)) log ☐=	1.7235	Solution verified

Example 3

Solve for x.

$$4.3^{2.6x} = 507.6$$

$$2.6x \log 4.3 = \log 507.6$$

$$x = \frac{\log 507.6}{2.6 \log 4.3}$$

$$= 1.6426789 \qquad \boxed{\text{STO 1}}$$

Check: Does $4.3^{(2.6)(1.6426789)} = 507.6$?

PRESS	DISPLAY	COMMENT
4.3 $\boxed{y^x}$ ((2.6 ☐× RCL 1)) ☐=	507.6	Solution verified

Example 4

Solve for x.

$$\sqrt[x]{12.7^5} = 23.974781$$

$$\sqrt[x]{12.7^5} = 12.7^{5/x} = 23.974781$$

$$\frac{5}{x} \log 12.7 = \log 23.974781$$

$$x = \frac{5 \log 12.7}{\log 23.974781} = 4$$

Check: Does $\sqrt[4]{12.7^5} = 23.974781$?

PRESS	DISPLAY	COMMENT
12.7 $\boxed{y^x}$ $\boxed{(}$ 5 $\boxed{\div}$ 4 $\boxed{)}$ $\boxed{=}$	23.974781	Solution verified

PROBLEMS

Solve the logarithmic equations in Problems 1 to 10.

1. $x = \log 346.19$

2. $\log x + 4 \log x = 13.6$

3. $3 \log x + \log 3x = 9.7$

4. $3 \log \frac{x}{3} + \log \frac{3}{x} = 8.66$

5. $\log x^3 - \log 17 = 6.1$

6. $x^{3.62} = 172.7$

7. $6.1^x = 325$

8. $\log 2.7^x + \log 3.1^x = 4.2$

9. $4.2^{x+1.5} = 109.36$

10. $5.6^{x/3} = 96.22$

19.4 NEGATIVE LOGARITHMS

To this point we have discussed only positive logarithms because only the log of numbers greater than 1 have been calculated. This concept becomes clearer as you observe the following progression:

Exponential Format	Logarithm Format
$10^2 = 1000$	$3 = \log 1000$
$10^2 = 100$	$2 = \log 100$
$10^1 = 10$	$1 = \log 10$
$10^0 = 1$	$0 = \log 1$

Now what happens when the power to which 10 is raised becomes less than 0? Continuing the chart:

Exponential Format	Logarithm Format
$10^{-1} = 0.1$	$-1 = \log 0.1$
$10^{-2} = 0.01$	$-2 = \log 0.01$
$10^{-3} = 0.001$	$-3 = \log 0.001$
.	.
.	.
.	.
$10^{-\infty} = 0$	$-\infty = \log 0$

When $0 \leq n \leq 1$, then $-\infty \leq \log N \leq 0$. Because a power cannot be less negative than $-\infty$ and raising a number to a power of $-\infty$ imposes a limiting value of 0, the conclusion must be that it is impossible to find the log of a negative number. This is true, $\log(-N)$ is not possible. Try finding the log of a negative number on your calculator; your calculator will indicate an error. However, do not confuse our conclusion with the expression $(-\log N)$, the negative of the log of N—this expression *is* quite valid.

PROBLEMS

In Problems 1 to 6, which are valid and which are invalid expressions?

1. $-\log(0.15)$ 3. $-\log(-0.3)$ 5. $\log(\log 0.75)$

2. $\log(-0.3)$ 4. $\log(\log 5.3)$ 6. $\log(-(\log 0.75))$

19.5 THE BASE OF A LOGARITHM

The base of a logarithm is that number which when raised to a power equal to the logarithm equates to another number. Up until now we have used only base 10, the base of the decimal numbering system. When the base is 10, the logarithm is simply designated as $\log N$. It is also correct to designate this as $\log_{10} N$, although the subscript is implied, just as the number 1 in $1x$ is implied when only the term x is written.

There is only one other practical base, *epsilon*, identified as ϵ. Because most calculators depict epsilon as e, such as the $\boxed{e^x}$ key, we will use the same symbol rather than the Greek letter. Theoretically, logarithms can use any base greater than 1, but as stated, only the bases 10 and e are practical. Base e is practical because it represents that unique number which when used as a base exactly predicts natural electronic phenomena. This includes all voltage and current transients in all RC or L/R circuits.

The derivation of the value of e is basically a calculus-oriented problem beyond the scope of this text. However, its numerical value is calculated as

$$e = \lim_{n \to \infty} \left(1 + \frac{1}{n}\right)^n \tag{19.19}$$

Equation (19.19) states that as n increases beyond bounds, the exact value of e is approached. To see that e does indeed approach a limiting value, use the $\boxed{y^x}$ key of your calculator to verify the following:

n	$\left(1 + \dfrac{1}{n}\right)$	$e = \left(1 + \dfrac{1}{n}\right)^n$
1	2	2
2	1.5	2.25
5	1.2	2.48832
10	1.1	2.5937425
50	1.02	2.691588
100	1.01	2.7048138
1000	1.001	2.716924
10000	1.0001	2.7181463

To obtain a still more exact value of e, as evaluated by the calculator, key in 1 and $\boxed{e^x}$ and note the results: This effectively evaluates e^1, or simply e as 2.7182818. However, the exact value of e is similar to the value of π in that it can never be precisely defined. The $\log_e N$ is called the *natural* logarithmic base because it follows natural electronic phenomena. This is analogous to calling the base 10 the Naperian system. Just as base 10 is indicated on the calculator as $\boxed{\log x}$, the base e is indicated on the calculator as $\boxed{\ln x}$.

Now, because any logarithmic base greater than 1 can theoretically be utilized, there must be a method to evaluate

$$\log_b N$$

when b is neither the value 10 nor the value e, the only two bases provided for on the calculator. To convert from any logarithmic base (whether 10, e, or any other number), we have to first verify the following identity:

$$\log_b a = \frac{1}{\log_a b} \tag{19.20}$$

To verify Equation (19.20), first evaluate the right side of the equation by substituting a number purposely selected for its ease of evaluation, such as

$$\frac{1}{\log_a b} = \frac{1}{\log_{10} 4} = \frac{1}{0.60206} = 1.660964$$

Note that we selected a as 10 and b as 4. (Again, keep in mind that this choice was for convenience; the same results occur regardless of the number selected.) Equation (19.20) now becomes

$$\log_4 10 = 1.660964 \qquad \text{or} \qquad 4^{1.660964} = 10$$

To verify this,

PRESS   **DISPLAY**

4 $\boxed{y^x}$ 1.660964 $\boxed{=}$ 10

Now that we have substantiated Equation (19.20), we proceed to develop a generalized relationship between logarithmic bases. In exponential form, if

$$N = a^x \tag{19.21}$$

then in logarithmic form

$$x = \log_a N \tag{19.22}$$

Taking the $\log_b$ of both sides of Equation (19.21) results in

$$\log_b N = \log_b a^x = x \log_b a \tag{19.23}$$

Substituting Equation (19.22) into Equation (19.23) results in

$$\log_b N = \log_a N \cdot \log_b a$$

From Equation (19.20),

$$\log_b N = \frac{\log_a N}{\log_a b} \tag{19.24}$$

If we select base a as either 10 or e, the calculator can evaluate the logarithm of any number to any other base.

Example 1

Use Equation (19.24) to reverify that $\log_4 10 = 1.660964$.

PRESS   **DISPLAY**

10 $\boxed{\log x}$ $\boxed{\div}$ 4 $\boxed{\log x}$ $\boxed{=}$ 1.660964

or

10 $\boxed{\ln x}$ $\boxed{\div}$ 4 $\boxed{\ln x}$ $\boxed{=}$ 1.660964

Example 2

Determine $\log_7 54.1$. Verify your answer by using the y^x function.

PRESS	DISPLAY	COMMENT
54.1 $\boxed{\ln x}$ $\boxed{\div}$ 7 $\boxed{\ln x}$ $\boxed{=}$ $\boxed{\text{STO 1}}$	2.0508833	Evaluates $\log_7 54.1$
7 $\boxed{y^x}$ $\boxed{\text{RCL 1}}$ $\boxed{=}$	54.1	Answer verified

Example 3

Using your calculator, evaluate

$$2.73 \log(\ln 23.6^{2.41})$$

Because of the number of special functions used, it is frequently easier to start at the innermost parentheses and work outward.

PRESS	DISPLAY
23.6 $\boxed{y^x}$ 2.41 $\boxed{=}$ $\boxed{\ln x}$ $\boxed{\log x}$ $\boxed{\times}$ 2.73 $\boxed{=}$	2.4075

PROBLEMS

1. Use Equation (19.24) to verify that $\ln N = 2.3026 \log N$.

2. Use Equation (19.24) to verify that $\log N = 0.4343 \ln N$.

Using your calculator, evaluate Problems 3 to 17 to develop proficiency in using the $\boxed{e^x}$, $\boxed{10^x}$, $\boxed{\ln x}$, and $\boxed{\log x}$ keys. Remember that the $\boxed{e^x}$ key is effectively the same as $\ln^{-1}$ and the $\boxed{10^x}$ key is effectively the same as $\log^{-1}$.

3. $3 \log 34.2$

4. $(4.7 \log 39.6)(2.3 \ln 17.9)$

5. $\dfrac{8.4 \ln 17.6}{e^{3.5}}$

6. $(10^{1.52})(\log 94.6)(e^{1.4})$

7. $\dfrac{2.7 \ln(\log 594.6)}{10^{1.91}}$

8. $\dfrac{1.8 \log(\log 15691.8)}{\ln 1.3^{2.6}}$

9. $(\log^{-1} 2.6834)(\ln^{-1} 3.6427)$

10. $(\log^{-1} 1.72^{1.69})(4.7 \ln^{-1} 1.63^{0.94})$

11. $\dfrac{4.25 \log 17.39}{8.62 \ln^{-1} (\log 10.35)}$

12. $\dfrac{\ln^{-1}\,(\log(\ln\,83521))}{\ln(\log^{-1}(\ln\,9.63))}$

13. $(\log_5\,39.6)(\log_7\,63.52)$

14. $\ln(\log_3\,105.7)$

15. $\log_6(\log_5\,5342.87)$

16. $\ln^{-1}(\log_8\,83.61)$

17. $\dfrac{\log^{-1}(\ln\,3.62^{3.17})}{\log_\pi\,39.42}$

19.6 ELECTRONIC APPLICATIONS OF LOGARITHMS—BASE 10

Many types of electronic problems, from comparison of different antenna gains to the distributed capacitance between parallel transmission lines, require the application of the log functions for their solution. Perhaps the most common application is measuring the power gains of different amplifying devices, especially as represented by the high fidelity industry.

The basis for the relationships of power gains to the logarithmic functions is really a human characteristic. For example, studies show that when listening to sound at a given level of volume (loudness), the human ear cannot discern only a minute amount of change in the volume. To be discernible, the change must be about 25 percent before the average hearing ability of an individual can even perceive that a change has taken place. This means that if you are listening to a loudspeaker delivering only 1 W of power, on the average the amount of power output has to increase to 1.25 W before you notice any change in volume. By the same rationale, if you are listening to a loudspeaker delivering 10 W of power, the power has to increase to 12.5 W before you notice the change. These changes represent equal displacements on a logarithmic scale, so human responses and power gains more readily adapt to measurements on a logarithmic scale.

The basic unit of power gain is the *bel*. By definition,

$$1\ \text{bel} = \log\frac{P_o}{P_i} \qquad (19.25)$$

where P_o is the power output from a device and P_i is the power input to the device.

Example 1

How many bels are necessary for the average person's hearing to just perceive a change in volume?

For any level of input P_i, the output must be 25 percent greater (or less) to perceive a change. Thus P_o must be $1.25\,P_i$, or

$$\text{Bel} = \log\frac{1.25\,P_i}{P_i} = \log\,1.25 = 0.09691$$

This represents a rather small decimal number. Therefore a different unit is desirable so that practical measurements of power gains or losses are more nearly whole numbers rather than fractional numbers. If we multiply the previous number of bels by 10, we have a unit of measurement of power gain nearly equal to unity (1). Such a unit is a decibel (abbreviated dB). As you can see, a just-perceivable human recognition of change then becomes 0.9691 dB, or very nearly equal to a nice even 1 (dB). From this we then conclude that

$$1 \text{ dB} = 10 \text{ bel} = 10 \log \frac{P_o}{P_i} \qquad (19.26)$$

Example 2

If a power amplifier's output is 30 W when its input is 4.5 mW, what is its power gain (P_g) in decibels?

$$P_g = 10 \log \frac{30}{4.5 \times 10^{-3}} = 10 \log 6666.67 = 38.24 \text{ dB}$$

Example 3

If $P_i = 3.5$ mW, what is P_o if the gain is 40 dB?

$$40 = 10 \log \frac{P_o}{3.5 \times 10^{-3}}$$

$$\log^{-1} \frac{40}{10} = \log^{-1} \log \frac{P_o}{3.5 \times 10^{-3}} = \frac{P_o}{3.5 \times 10^{-3}}$$

$$P_o = 3.5 \times 10^{-3} \times \log^{-1} 4 = 3.5 \times 10^{-3} \times 10^4$$

$$= 35 \text{ W}$$

Notice that to determine the power gain, both an input power and an output power were required. If a series of different amplifiers from several manufacturers are compared for gain, each amplifier has to be judged upon its P_o, for a given reference P_i, which means that there must be a reference input for comparison. Throughout the audio-amplifier industry, this reference is usually accepted as 6 mW.

Example 4

Compare the power gain in decibels for three different amplifier outputs: $P_o = 40$ W, $P_o = 45$ W, and $P_o = 50$ W.

$$P_{40} = 10 \log \frac{40}{6 \times 10^{-3}} = 38.24 \text{ dB}$$

$$P_{45} = 10 \log \frac{45}{6 \times 10^{-3}} = 38.75 \text{ dB}$$

$$P_{50} = 10 \log \frac{50}{6 \times 10^{-3}} = 39.21 \text{ dB}$$

Notice that for a spread of 10-W output power between the lowest and highest amplifiers there is only about 1-dB difference in their ratings. This is to be expected because, as described earlier, to just perceive a change in volume above 40 W takes a change of about 25 percent of 40 W, or a new value of 50 W. As calculated earlier, a just-perceivable change represents about 1 dB. Sometimes P_o is much less than P_i, such as when a signal is attenuated while passing through a coupling device. In such a case the decibels is a negative number representing a loss in power.

Example 5

What is the output power represented by -60 dB if the input power is the 6-mW reference level?

$$-60 = 10 \log \frac{P_o}{6 \times 10^{-3}}$$

$$\log^{-1} \frac{-60}{10} = \log^{-1} \log \frac{P_o}{6 \times 10^{-3}} = \frac{P_o}{6 \times 10^{-3}}$$

$$P_o = 6 \times 10^{-3} \times \log^{-1}(-6) = 6 \times 10^{-3} \times 10^{-6}$$

$$= 6 \text{ nW}$$

When a series of electronic devices are cascaded—that is, when the output of one device is used as the input to the next device—the overall gain is the product of the gains of each device. For example, in Figure 19.1, the overall gain is $20 \times 30 \times 25 = 15,000$. If 6 mW is the input, then $P_o = 6 \times 10^{-3} \times 15 \times 10^3 = 90$ W. Now, if the gain of each device is expressed in decibels, then the logarithmic law of products states that the overall gain equals the sum of the logarithmic gains of each device. In short, when gain is expressed in decibels, the overall gain of all the devices is merely the sum of the algebraic gain of each device.

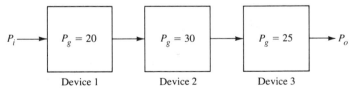

$P_i \longrightarrow$ | $P_g = 20$ | $\longrightarrow$ | $P_g = 30$ | $\longrightarrow$ | $P_g = 25$ | $\longrightarrow P_o$

Device 1   Device 2   Device 3

Figure 19.1

Example 6

Figure 19.2 depicts a microphone (high-loss device) coupled to a preamplifier coupled to a matching pad (another loss device) coupled to a power amplifier.

a. Determine the overall gain.

$$P_g = -80 + 62 - 12 + 85 = 55 \text{ dB}$$

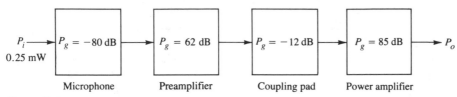

Figure 19.2

b. Determine P_o.

$$55 = 10 \log \frac{P_o}{0.25 \times 10^{-3}}$$

$$\log^{-1} \frac{55}{10} = \log^{-1} \log \frac{P_o}{0.25 \times 10^{-3}} = \frac{P_o}{0.25 \times 10^{-3}}$$

$$P_o = 0.25 \times 10^{-3} \times \log^{-1} 5.5 = 79.06 \text{ W}$$

Figure 19.3 depicts an electronic device whose input and output characteristics are known in terms of their impedances, voltages, and currents. Because P_i can be expressed as V_i^2/Z_i or $I_i^2 Z_i$ and P_o as V_o^2/Z_o or $I_o^2 Z_o$, we can develop a relationship between decibels and impedances, voltages, and/or currents:

$$P_i = i_i^2 Z_i \qquad \text{and} \qquad P_o = i_o^2 Z_o$$

Using the logarithmic rules of products and powers, the results are

$$\text{dB} = 10 \log \frac{P_o}{P_i} = 10 \log \frac{I_o^2 Z_o}{I_i^2 Z_i}$$

$$= 10 \log \left(\frac{I_o}{I_i}\right)^2 \left(\frac{\sqrt{Z_o}}{\sqrt{Z_i}}\right)^2$$

$$= 20 \log \frac{I_o \sqrt{Z_o}}{I_i \sqrt{Z_i}} \qquad (19.26)$$

Similarly,

$$\text{dB} = 10 \log \frac{P_o}{P_i} = 10 \log \frac{E_o^2/Z_o}{E_i^2/Z_i}$$

$$= 10 \log \frac{E_o^2 Z_i}{E_i^2 Z_o} = 10 \log \left(\frac{E_o}{E_i}\right)^2 \left(\frac{\sqrt{Z_i}}{\sqrt{Z_o}}\right)^2$$

$$= 20 \log \frac{E_o \sqrt{Z_i}}{E_i \sqrt{Z_o}} \qquad (19.27)$$

Example 7

Figure 19.4 is an electronic amplifying device, the common-emitter transistor (less the external resistances and voltages necessary for its operation). If $I_i = 40 \ \mu\text{A}$, $Z_i = 1 \text{ k}\Omega$, $I_o = 3.2 \text{ mA}$, and $Z_o = 20 \text{ k}\Omega$, determine the power gain in decibels.

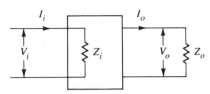

Figure 19.3   **Figure 19.4**

Using Equation (19.26):

$$dB = 20 \log \frac{I_o \sqrt{Z_o}}{I_i \sqrt{Z_i}}$$

$$= 20 \log \frac{3.2 \times 10^{-3} \times \sqrt{2 \times 10^4}}{40 \times 10^{-6} \times \sqrt{10^3}}$$

$$= 20 \log 357.771$$

$$= 51.07$$

PROBLEMS

1. Determine the decibel gain if

   a. $P_o = 3.4$ W, $P_i = 150$ mW

   b. $P_o = 32$ mW, $P_i = 5$ μW

   c. $P_o = 160$ μW, $P_i = 320$ mW

   d. $I_o = 3.4$ mA, $Z_o = 13.5$ kΩ, $I_i = 50$ μA, $Z_i = 500$ Ω

   e. $E_o = 20$ V, $Z_o = 6.4$ kΩ, $E_i = 300$ mV, $Z_i = 350$ Ω

2. What is the output power represented by -75 dB if $P_i = 10$ mW?

3. A series of electronic devices have power gains of 30 dB, -18 dB, and 50 dB. If $P_o = 62.5$ W, determine P_i to the overall system.

Two of the more common forms of transmission lines are the two-wire open air line and the coaxial cable, represented as Figure 19.5(a) and (b), respectively. In (a), d is the distance between centers of the transmission lines and r is the radius of a line. In (b), d_i is the outside diameter of the inside wire and d_o is the inside diameter of the outer wire.

In Problems 4 to 7, refer to Figure 19.5a.

4. The characteristic impedance of the two-wire open air line is given as

$$Z = 276 \log \frac{d}{r}$$

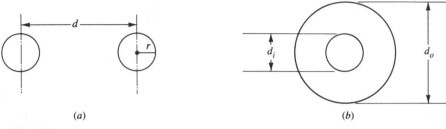

(a)   (b)

Figure 19.5

Determine this impedance if $d = 7$ cm and $r = 0.8$ mm. Note that d and r must be in the same units.

5. The capacitance of the transmission line (in microfarads) is given as

$$C = \frac{0.0121\ \ell}{\log d/r}$$

where ℓ is in kilometers and d and r must be in similar units. Determine C if $\ell = 25$ km, $d = 50$ cm, and $r = 0.25$ cm.

6. See Problem 5. What is the spacing between wires if a 1-mi length of the transmission line has a capacitance of 0.025 μF and the radius remains 0.25 cm?

7. The inductance of the wire (in millihenries) is given as

$$L = \ell\left(0.01 + 0.92 \log \frac{d}{r}\right)$$

where ℓ is in kilometers and d and r must be in similar units. Determine the inductance of the transmission line in Problem 5.

For Problems 8 and 9, refer to Figure 19.5b.

8. The characteristic impedance of an air-insulated coaxial transmission line is given as

$$Z = 138 \log \frac{d_o}{d_i}$$

Calculate Z if $d_o = 2$ cm and $d_i = 5.5$ mm. Note that as in previous problems, d_o and d_i must be converted into the same units.

9. The transmission of a signal from a house antenna to a television set frequently occurs on a 75-Ω characteristic impedance coaxial cable. If the radius (to the outer surface) of the center conductor is 0.815 mm, what is the value of d_o?

19.7 ELECTRONIC APPLICATIONS OF LOGARITHMS—BASE ϵ

The most common application of the $\ln x$ or e^x function occurs with electrical transients in RC or L/R circuits. For example, in Figure 19.6, if switch S is moved to position 1 (the capacitor is initially assumed uncharged), the instantaneous voltage across C is expressed as

$$v_C = V(1 - e^{-t/\tau}) \qquad \tau = RC$$

and the instantaneous voltage across the resistor R equals

$$v_R = Ve^{-t/\tau} \qquad \tau = RC$$

After the capacitor has been fully charged to V volts, if switch S is moved to position 2, then conditions completely reverse:

$$v_C = Ve^{-t/\tau}$$

and

$$v_R = V(1 - e^{t/t})$$

The same types of changes take place in L/R circuits.

It appears then that for an RC circuit you must compile a chart of equations for both charging and discharging conditions, for the voltages across C and R, and for currents under all these conditions. In addition, these equations change if the capacitor is precharged in either an aiding or opposing condition. As usual, all these complications also apply to L/R circuits. But rather than having to compile charts or memorize equations for all the possible conditions just described, you need only *one* equation, specified under only *two* different conditions. This solves all possible transient problems. Because there are two conditions, this is essentially the mathematical equivalent of solutions of simultaneous equations.

The following electronic phenomena (the theory of which is left to textbooks about electronic circuits) are germane to the subsequent mathematical analysis:

1. In resistive capacitive circuits, the time constant $\tau = RC$.

2. In resistive inductance circuits, the time constant $\tau = L/R$.

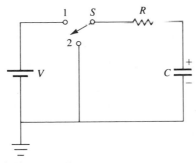

Figure 19.6

3. The voltage across C cannot change instantaneously, but if changing, it changes from the initial to the final point in infinite time (for practical purposes, after 5τ).

4. The current through L cannot change instantaneously, but if changing, it changes from the initial to the final point in infinite time (for practical purposes, after 5τ).

The basic equation utilizing the natural logarithmic function for all electrical transient conditions is

$$A + Be^{-t/\tau} \qquad\qquad (19.28)$$

Note that there are two unknowns: A and B. Therefore, two time (t) conditions are required to evaluate A and B. We can select any times, but $t = 0$ and $t = \infty$ are the two times when circuit conditions are best known. Before proceeding, keep in mind that if $t = 0$,

$$e^{-t/\tau} = e^{-0/\tau} = e^0 = 1$$

and if $t = \infty$

$$e^{-\infty/\tau} = e^{-\infty} = \frac{1}{e^\infty} = \frac{1}{\infty} = 0$$

Examples 1 to 4 illustrate the universal application of Equation (19.28) to electrical transients.

Example 1

Figure 19.7. What is the voltage across the capacitor 75 ms after switch S is closed? The capacitor is initially assumed uncharged.

$$\tau = RC = 2.2 \times 10^3 \times 22 \times 10^{-6}$$
$$= 48.4 \times 10^{-3}$$
$$= 48 \text{ ms}$$

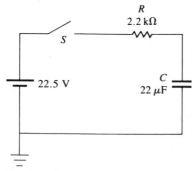

Figure 19.7

Using Equation (19.28),

$$v_C = A + Be^{-t/\tau} \tag{19.29}$$

When $t = 0$, $v_c = 0$; therefore

$$v_C(t = 0) = 0 = A + Be^{-0/\tau} = A + B(1) = A + B \tag{19.30}$$

When $t = \infty$, $v_C = 22.5$; therefore

$$v_C(t = \infty) = 22.5 = A + B^{-\infty/\tau} = A + B(0) = A \tag{19.31}$$

Equations (19.30) and (19.31) establish the two simultaneous equations necessary to solve for the values of A and B in Equation (19.29):

$$A + B = 0$$
$$A = 22.5$$
$$\therefore B = -A = -22.5$$

Substituting into Equation (19.29),

$$v_C = 22.5 - 22.5\,e^{-75\text{ ms}/48\text{ ms}}$$
$$= 22.5(1 - e^{-1.5496})$$

PRESS	DISPLAY	COMMENT
1.5496 $\boxed{+/-}$ $\boxed{e^x}$ $\boxed{+/-}$		
$\boxed{+}$ 1 $\boxed{=}$ $\boxed{\times}$ 22.5 $\boxed{=}$	17.723	Voltage across C

Note the order of solving for v_C: 10 steps were required. To solve the same equation in a straightforward, left-to-right fashion, proceed as follows:

PRESS	DISPLAY
22.5 $\boxed{\times}$ $\boxed{(}$ 1 $\boxed{-}$ 1 $\boxed{e^x}$ $\boxed{y^x}$ 1.5496 $\boxed{+/-}$ $\boxed{)}$ $\boxed{=}$	17.723

This latter process takes 12 steps; the extra two steps involved developing e and then raising it to the x power via the $\boxed{y^x}$ key.

Example 2

See Figure 19.8. How long will it take for the voltage across C (that is, the voltage of the top plate of C with respect to ground) to reach -9.5 V? Note that the top plate of C is initially precharged to $+5$ V.

$$\tau = RC = 0.47 \times 10^6 \times 0.02 \times 10^{-6} = 9.4 \text{ ms}$$
$$v_C = A + Be^{-t/\tau} \tag{19.32}$$
$$v_C(t = 0) = 5 = A + B(1) = A + B$$
$$v_C(t = \infty) = -17.75 = A + B(0) = A$$
$$\therefore B = 5 - A = 5 - (-17.75) = 22.75$$

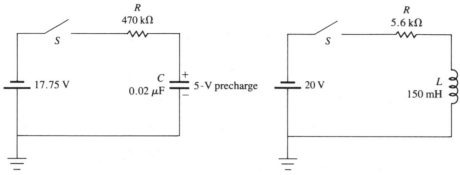

Figure 19.8   **Figure 19.9**

From Equation (19.32),

$$-9.5 = -17.75 + 22.75\ e^{-t/9.4\ \text{ms}}$$
$$22.75\ e^{-t/9.4\ \text{ms}} = -9.5 + 17.75 = 8.25$$
$$e^{-t/9.4\ \text{ms}} = \frac{8.25}{22.75} = 0.3626$$
$$\ln e^{-t/9.4\ \text{ms}} = \ln 0.3626$$
$$-\frac{t}{9.4\ \text{ms}} = -1.0144$$
$$t = 1.0144 \times 9.4\ \text{ms}$$
$$= 9.535\ \text{ms}$$

Example 3

See Figure 19.9. What is the voltage across R 40 μs after switch S is closed?

$$\tau = \frac{L}{R} = \frac{150 \times 10^{-3}}{5.6 \times 10^{3}} = 26.786\ \mu\text{s}$$
$$v_R = A + Be^{-t/\tau} \tag{19.33}$$
$$v_R(t = 0) = 0 = A + B$$
$$v_R(t = \infty) = 20 = A$$
$$\therefore B = -A = -20$$

From Equation (19.33),

$$v_R = 20 - 20e^{-40\ \mu\text{s}/26.786\ \mu\text{s}}$$
$$= 20(1 - e^{-1.4933})$$
$$= 15.51\ \text{V}$$

Example 4

In Figure 19.9, how long will it take for the value of the current to reach 2 mA after switch S is closed?

From Example 3,

$$\tau = 26.786 \ \mu s$$
$$i = A + Be^{-t/\tau} \tag{19.34}$$
$$i(t = 0) = 0 = A + B$$
$$i(t = \infty) = \frac{20}{5.6 \ k\Omega} = 3.57 \ mA = A$$
$$\therefore B = -A = -3.57 \ mA$$

From Equation (19.34),

$$2 \ mA = 3.57(1 - e^{-t/26.786 \ \mu s}) \ mA$$
$$\frac{2}{3.57} = 0.56 = 1 - e^{-t/26.786 \ \mu s}$$
$$e^{-t/26.786 \ \mu s} = 1 - 0.56 = 0.44$$
$$\ln e^{-t/26.786 \ \mu s} = \ln 0.44$$
$$-\frac{t}{26.786 \ \mu s} = -0.821$$
$$t = (26.786)(0.821)\mu s$$
$$= 21.991 \ \mu s$$

PROBLEMS

1. See Figure 19.10. What is the voltage across C 200 μs after switch S is closed?

2. If the capacitor in Figure 19.10 is precharged to 6.5 V (grounded side positive and the other plate negative), how long will it take in microseconds for C to reach 18.63 V after switch S is closed?

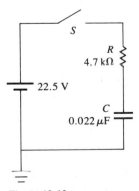

Figure 19.10

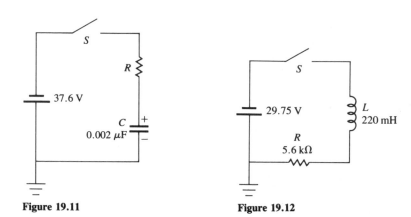

Figure 19.11   **Figure 19.12**

3. If C in Figure 19.11 is precharged to 5 V in the polarity direction shown, what value of R is required to result in a voltage across C of 23.897 V 2.6 μs after switch S is closed?

4. See Figure 19.12. What is the voltage across R 45 μs after switch S is closed?

5. Semiconductor theory expresses the current (I_d) through a diode as

$$I_d = I_s(e^{(v_d/(kT/q))} - 1) \qquad (19.35)$$

where I_s = leakage current under reverse bias conditions
   V_d = voltage across diode
   k = Boltzmann's constant (1.38×10^{-23} J/K)
   T = temperature in K
   q = electron charge (1.6×10^{-19} C)

At 25°C, kT/q equals approximately 0.026 (V). Determine the diode current (at 25°C) if $I_s = 1$ μA and there is 0.2 V across the diode.

6. See Equation (19.35). What voltage is required across the diode for an I_d of 4 mA? Assume $I_s = 850$ nA.

7. Figure 19.13 represents a logarithmic amplifier, where v_o is expressed as

$$v_o = \frac{kT}{q} \left(\ln \frac{v_i}{R} - \ln I_s \right) \qquad (19.36)$$

If $R_1 = 220$ kΩ and $I_s = 20$ nA, calculate v_o if $v_i = 4.75$ V. Assume T is such that $kT/q = 0.026$ V.

8. See Equation (19.36). Determine the value of v_i if $v_o = 0.215$ V, $R = 200$ kΩ, $I_s = 22.5$ nA, and $kT/q = 0.026$ V.

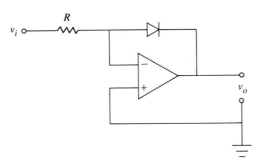

Figure 19.13 Logarithmic amplifier

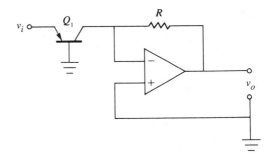

Figure 19.14 Antilog amplifier

9. Figure 19.14 represents an antilog amplifier. The output voltage (v_o) is expressed as

$$v_o = RI_s \ln^{-1}\left(\frac{v_i}{kT/q}\right) \tag{19.37}$$

Find v_o if $R = 100$ kΩ, $I_s = 5$ nA, and $v_i = 0.15$ V. Assume that $kT/q = 0.026$ V.

10. See Equation (19.37). Find v_i if $v_o = 25$ mV, $R = 150$ kΩ, $I_s = 8.5$ nA, and $kT/q = 0.026$ V.

SUMMARY

A logarithm is a power to which a given number (called the base) must be raised to equal another number. Logarithms possess four properties that facilitate their application to the solution of certain types of problems:

1. The logarithm of a product of a series of numbers equals the sum of the logarithm of each number of the series.

2. The logarithm of the quotient of two numbers equals the logarithm of the dividend minus the logarithm of the divisor.

3. The logarithm of a number raised to a power equals the product of the power times the logarithm of the number.

4. The logarithm of a number extracted to a root is the quotient of the logarithm of the number divided by the root.

The inverse log (designated by $\log^{-1}$) is analogous in its function, as are the inverse trigonometric functions. The designation $x = \log^{-1} N$ asks, "What number x has as its log the value N?"

The two most practical base numbers for logarithms are base 10 and ϵ. Both bases have many direct applications in the field of electronics. Base 10 is especially useful for calculating power ratios; base ϵ is useful for calculating electronic transients.

20 GRAPHS

20.1 BASIC DEFINITIONS

Graphs are visual pictures of the relationships between two (or more) variables. Such a relationship is usually expressed as an equation. Several two-dimensional, linear graphs were shown in preceding chapters, such as the magnitude of the sine, cosine, or tangent functions versus the magnitude of an angle. The term *linear* means that equal displacements along the horizontal and vertical axes are represented as equal changes in the variables.

As stated in earlier chapters, one variable is called the dependent variable and the other the independent variable. A typical format is

$$y = \underbrace{}_{\substack{\text{Dependent} \\ \text{variable}}} \quad \underbrace{3x + 5}_{\substack{\text{Independent} \\ \text{variable}}}$$

Values are assigned to x; depending upon this assignment, y can then be determined. Hence y is dependent upon the chosen value of x. Customarily, although not mandatory, the dependent variable is plotted along the vertical axis (called the ordinate) and the independent variable is plotted along the horizontal axis (abscissa).

Plotting a graph is essentially a mechanical process. Within the range of interest of the independent variable, select enough increments to accurately display the relationship between the dependent variable and the independent variable. The word "sufficient" as used here is not a fixed number; it is arrived at by common sense. For example, if you are plotting a straight line, as few as two points can be sufficient to get an accurate plot of the function. If there are extreme fluctuations in the dependent variables over the range of interest of the independent variable, many dozens of points may be necessary to plot their interrelationship.

Speaking of points, let us digress for a moment and identify a point. Figure 20.1 is a simple linear graph. The location of point P_1 is to be identified. The intersection of the major axis (origin) should be used as a reference point when referring to P_1.

Because P_1 is 4 units in the direction of the $+x$ axis and 5 units in the direction of the $+y$ axis, it can uniquely be identified as (4, 5). (Custom requires that the x-axis direction displacement be specified first and the y-axis displacement be specified second). Using this same relationship, P_2 in Figure 20.1 is identified as $(-9, -3)$.

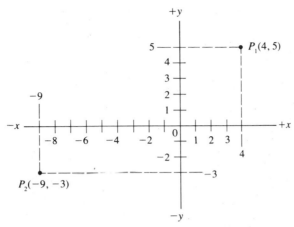

Figure 20.1

The last major definition to consider in a linear graph is the *slope*. The easiest graph for determining the slope is a straight line. Refer to Figure 20.2, a graph of the equation

$$y = 3x$$

Assume we are interested only in the values of the dependent variable y over the range of the independent variable x from $x = -1$ to $x = 4$, in increments of 1. Developing a chart that plots this graph requires merely assigning the values of -1, 0, 1, 2, 3, and 4 to x and calculating the resulting value of y:

If x equals	Then y equals	Plot Point
-1	-3	$P_{-1}(-1, -3)$
0	0	$P_0(0, 0)$
1	3	$P_1(1, 3)$
2	6	$P_2(2, 6)$
3	9	$P_3(3, 9)$
4	12	$P_4(4, 12)$

Thus the slope of a graph is the number of units the graph rises on the ordinate (or falls for a negative slope) to the number of units along the abscissa over which this measurement is taken. Mathematically, the slope is the ratio of the distance between *any* two points measured vertically to the corresponding distance between the *same* two points measured horizontally. From Figure 20.2, after arbitrarily choosing points P_4 and P_2,

$$\text{Slope} = \frac{\triangle y}{\triangle x} = \frac{y_4 - y_2}{x_4 - x_2} = \frac{12 - 6}{4 - 2} = \frac{6}{2} = 3$$

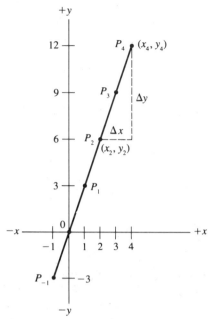

Figure 20.2

This equation states that any incremental change in x results in an incremental change in y three times the magnitude of the change in x. Because this slope is positive, y increases if x increases and y decreases if x decreases. Figure 20.3 is the graph of a straight line with a negative slope. (Incidentally, the equation of this straight line is $y = -x - 4$; we discuss this equation in the next section.) Specifically, from Figure 20.3,

$$\text{Slope} = \frac{y_2 - y_1}{x_2 - x_1} = \frac{3 - (-5)}{-7 - 1} = -1$$

Thus we can conclude that positive slopes angle *upward* when progressing from left to right and negative slopes angle *downward* when progressing from left to right. The following two observations are immediately apparent:

1. The equation of the straight-line graph in Figure 20.2 is

$$y = 3x$$

and its slope is 3. The equation of the straight-line graph of Figure 20.3 was

$$y = -x - 4$$

and its slope was -1. At least for straight lines, the slope is equal to the number preceding the independent variable.

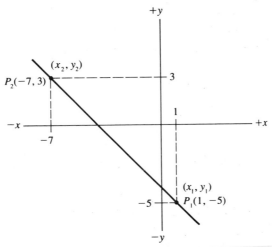

$$+y$$

(x_2, y_2)

$P_2(-7, 3)$

3

1

$-x$

$+x$

-7

(x_1, y_1)

-5 $P_1(1, -5)$

$-y$

Figure 20.3

Example 1

What is the slope of the following straight-line equations? Where necessary, rearrange the equation so that the dependent variable is expressed as a unit variable.

a. $y = 6x + 3$
   Slope $= 6$

b. $s = 3.5t - 7$
   Slope $= 3.5$

c. $u = \dfrac{2}{3}v + 5.6$

   Slope $= \dfrac{2}{3}$

d. $4a = 5b + 4.8$
   Rearranging by dividing through by 4 results in

   $a = \dfrac{5}{4}b + 1.2$

   $\therefore$ Slope $= \dfrac{5}{4}$

2. If the straight line passes through the origin, only one point (say, P_2) is needed to determine the slope because the second point (say, P_1) can always be chosen as the origin, and

$$\text{Slope} = \frac{y_2 - y_1}{x_2 - x_1} = \frac{y_2 - 0}{x_2 - 0}$$

which is the same thing as

$$\text{Slope} = \frac{y_2}{x_2}$$

Example 2

See Figure 20.2. Show that if any single point (except 0, 0) is selected, the slope is still 3.

Arbitrarily choose P_{-1} and P_2:

$$\text{For } P_{-1} \quad \text{Slope} = \frac{-3}{-1} = 3$$

$$\text{For } P_2 \quad \text{Slope} = \frac{6}{2} = 3$$

Until now, only straight lines have been used for determining the slope. The slope of a *curve* can also be determined, but with these two restrictions:

1. If we use the Δy (incremental change along the ordinate) to the Δx (incremental change along the abscissa) ratio to determine the slope between any two points on a curve, this ratio represents only an average slope between the two points.

2. The distance between two points can be reduced to lessen the error between the exact slope and the average slope. When the distance between the two points is so reduced that the points coincide, we can determine the exact slope of a curve at that specific point. Determining the slope at a singular point on a curve is called taking the derivative of the curve at that point. However, this process is calculus oriented and therefore beyond the scope of this text.

Figure 15.6, reproduced as Figure 20.4, is nonlinear and represents the value of the sine function for angles between 0° and 90°, with the added features of P_2 at 90° and P_1 at 30°.

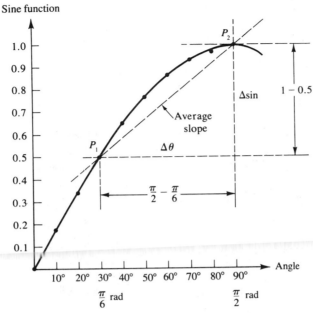

Figure 20.4

Example 3

Using Figure 20.4, determine the *average* slope between $\pi/2$ rad (90°) and $\pi/6$ rad (30°).

$$\text{Average slope} = \frac{\sin{(P_2)} - \sin{(P_1)}}{\pi/2 - \pi/6}$$

$$= \frac{6(1 - 0.5)}{2\pi}$$

$$= \frac{1.5}{\pi}$$

PROBLEMS

Determine each slope in Problems 1 to 8.

1. $a = 2b$

2. $x = -3z + 7$

3. $s = 2t + 1.5$

4. $3c = 6d + 7$

5. $P_2 = (5, 0),\ P_1 = (3, 4)$

6. $P_2 = (-1, 4),\ P_1 = (3, 5)$

7. $P_2 = (4.5,\ 6.9),\ P_1 = (1.2,\ 7.8)$

8. $P_2 = (-3.4,\ -1.9),\ P_1 = (-5.2,\ 6.5)$

20.2 GRAPHING A STRAIGHT LINE

Section 20.1 illustrated that the equation of a straight line has the format

$$y = mx + b \qquad (20.1)$$

The value of m represents the slope of the line. If the line passed through the origin, the value of b would have been zero; otherwise b represents that value at which the line intersects the ordinate axis.

Because any two points on a straight line represent the projection of all points on the line, we are now in a situation analogous to that of solving the general exponential equation of $A + B\epsilon^{-t/\tau}$ in that to develop the graph of a straight line it is necessary to just determine the location of any two points. The two most convenient points are when $x = 0$ and $y = 0$. To illustrate, consider the equation

$$y = 3x + 4 \qquad (20.2)$$

When x in Equation (20.2) equals zero, then

$$y = 3(0) + 4 = 4$$

therefore one point is (0, 4). Similarly, when $y = 0$, then

$$0 = 3x + 4$$

or

$$3x = -4$$

and

$$x = \frac{-4}{3}$$

Therefore another point is $(-4/3, 0)$.

Plotting the two points and projecting an extension beyond each results in Figure 20.5.

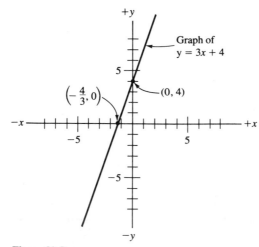

Figure 20.5

Example 1

Graph the following straight lines.

a. $s = 2t - 7$
     When $s = 0$, then $2t = 7$ and $t = 3.5$.
     $\therefore P_2 = (3.5, 0)$
     When $t = 0$, then $s = -7$.
     $\therefore P_1 = (0, -7)$
     See Figure 20.6a.

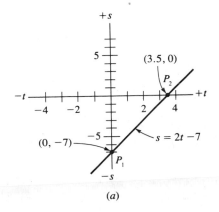

(a)

b. $u = -3t + 6$
     When $u = 0$, then $3t = 6$ and $t = 2$.
     $\therefore P_2 = (2, 0)$
     When $t = 0$, then $u = 6$.
     $\therefore P_1 = (0, 6)$
     See Figure 20.6b

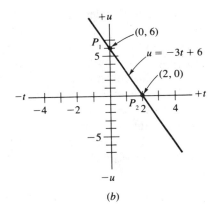

(b)

c. $3a = 6b + 12$
     Rearranging: $a = 2b + 4$
     When $a = 0$, then $2b = -4$ and $b = -2$.
     $\therefore P_2 = (-2, 0)$
     When $b = 0$, then $a = 4$.
     $\therefore P_1 = (0, 4)$
     See Figure 20.6c

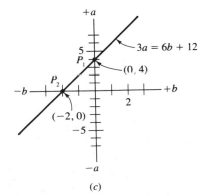

(c)

Figure 20.6

PROBLEMS

In Problems 1 to 10 graph the straight lines.

1. $y = x$

2. $c = 2d$

3. $\alpha = 3\beta + 4$

4. $u = -5v + 6$

5. $2e = 3f - 5$

6. $3y = -6z + 12$

7. $-3x = 2y + 7$

8. $5s - 3t = 4$

9. $2.1d = 7.6g - 9.5$

10. $8.62a + 7.12b - 3.4 = 0$

20.3 LINEAR GRAPHS WITH APPLICATIONS

Linear graphs are ideally suited for the solution of simultaneous equations, the nature of which can sometimes be difficult to solve academically. Here we apply the principles of graphs to solving simple simultaneous equations. Once you learn these principles, you will be able to solve more complex simultaneous equations.

Figure 20.7 illustrates two resistors in series across an 18-V source. We want to graphically determine the voltage across each resistor. For such a simple problem it would be more practical to apply the proportionate voltage ratio law. But remember that at this point we are developing a methodology for attacking such problems. As you will soon see, when the mathematics become more complex, graphic solution is the only practical approach.

To solve the problem, first consider each resistor independently, as though the other resistor did not exist. If only R_1 is present, the I-V equation is

$$I_1 = \frac{1}{R_1}V \qquad (20.3)$$

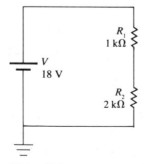

Figure 20.7

If only R_2 is present, the I-V equation is

$$I_2 = \frac{1}{R_2}V \tag{20.4}$$

Notice that both Equations (20.3) and (20.4) are of the straight-line format

$$y = mx + b$$

where $b = 0$

m = slope of the line

This means that the slope of an I-V graph, where only resistance is present, is $1/R$.

Back to the problem. Graphically determined, what is the value of V_{R1} and V_{R2} in Figure 20.7? To answer this question, construct a mirror image graph, such as that in Figure 20.8. Note that the abscissa of the independent variable goes from left to right for V_{R1} and from right to left for V_{R2}. The two ordinate axes represent I_1 and I_2, respectively. All axes are in equal increments. The I_1 and V_{R1} axes are used to plot Equation (20.3), left to right; the I_2 and V_{R2} axes are used to plot Equation (20.4), right to left. Now all that remains is to determine any two points for both Equations 20.3 and 20.4 and then connect these points with a straight line.

For Equation (20.3):

When $V = 0$, $I_1 = 0$ therefore $P_1 = (0, 0)$

When $V = 10$, $I_1 = 10$ therefore $P_2 = (10, 10)$

For Equation (20.4):

When $V = 0$, $I_2 = 0$ therefore $P_3 = (0, 0)$

When $V = 18$, $I_2 = 9$ therefore $P_4 = (18, 9)$

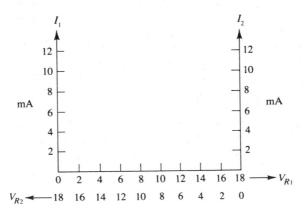

Figure 20.8

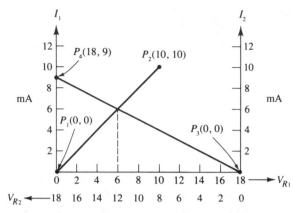

Figure 20.9

Plotting these two straight lines results in Figure 20.9. Projecting straight down from their intersection gives the voltage across each resistor:

$$V_{R1} = 6 \text{ V} \qquad \text{reading left to right}$$

$$V_{R2} = 12 \text{ V} \qquad \text{reading right to left}$$

The accuracy of these answers is quickly verified by the proportionate voltage ratio law.

Using these principles, let us illustrate a more practical application of graphical solution. See Figure 20.10, which is similar to Figure 20.7 except that one of the resistors is replaced with a diode whose characteristics are nonlinear. We want to determine the voltage across the diode. This is the difference between the source voltage (0.5 V) and the voltage drop across the resistor, which in turn is determined by the total current. The problem is that the current through the diode is determined by the voltage across the diode. At the same time, the voltage across the resistor is not independent of the current through the diode, which is also the current through the resistor because both the diode and resistor are in series. It seems we are chasing our tail! Here are the equations. Current through the diode, which is also the total

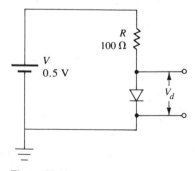

Figure 20.10

series current, is (at room temperature) approximated as

$$I = I_s \epsilon^{38.46\,V_d} \qquad (20.5)$$

where I is synonymous with I_d (current through diode and hence through circuit)
I_s is saturation current (characteristic of diode)
V_d = voltage across diode

The voltage across R is

$$I = \frac{V_R}{R} \qquad (20.6)$$

Get the idea? These simultaneous equations are not so nice anymore—here is where graphs prove advantageous.

To solve the problem, first select a range of values for I, V_d, and V_R. Start with Equation (20.6) because it is the easier of the two equations. Certainly, if the diode was not present,

$$I_{max} = \frac{0.5 \text{ V}}{100 \text{ } \Omega} = 5 \text{ mA}$$

Therefore, the ordinate scale need not exceed 5 mA in height. The abscissa cannot exceed 0.5 V, the applied voltage. The resultant of these conclusions is Figure 20.11. Note that on the abscissa axis V_R reads from right to left, V_d from left to right, as in the previous example. Additionally, because both ordinate axes are identical (as in Figure 20.9), only one such axis is necessary, hence the more simplified graph in Figure 20.11.

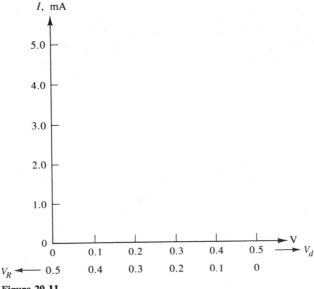

Figure 20.11

Next, determine the location of points. For R (assuming the diode not present):

If $I = 0$ mA, then $V_R = 0$ V   therefore $P_1 = (0, 0)$.

If $I = 5$ mA, then $V_R = 0.5$ V   therefore $P_2 = (0.5, 5)$

For the diode, we must plot several points because the relationship is not so simple (certainly it is not a straight line). Assume $I_s = 0.1 \ \mu A = 10^{-7}$ A:

If $V_d =$	then I_d(mA) $=$	Point
0.05	0	$P_4(0.05, 0)$
0.10	0.0047	$P_5(0.1, 0.0047)$
0.15	0.032	$P_6(0.15, 0.032)$
0.20	0.219	$P_7(0.2, \ 0.219)$
0.25	1.499	$P_8(0.25, 1.499)$
0.275	3.92	$P_9(0.275, 3.92)$

When both plots are superimposed on Figure 20.11, Figure 20.12 results. It appears that $I = 2.4$ mA, $V_d = 0.26$ V, and $V_R = 0.24$ V. To test these results:

1. $V_R = (2.4 \text{ mA})(100 \ \Omega) = 0.24$ V

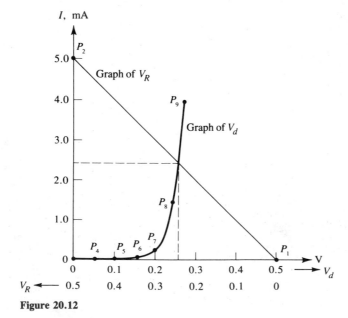

Figure 20.12

2. From Equation (20.5):

$$2.4 \times 10^{-3} = 10^{-7} \epsilon^{38.46 V_d}$$

$$\epsilon^{38.46 V_d} = 2.4 \times 10^4$$

$$\therefore V_d = \frac{\ln 2.4 \times 10^4}{38.46} = 0.262 \text{ V}$$

Note that $V_R + V_d$ (graphically) is 0.498 V, almost exactly 0.5 V. This indicates that graphs can be quite but not *extremely* accurate. However, graphs frequently do provide the only practical solution to complex sets of simultaneous equations. Note too that the straight-line graph of V_R is called (in electronic jargon) the *load line*. The load line represents all possible combinations of the current through R and the voltage across R. The interaction of this line with the curve of an electronic device (generally a very complex mathematical expression) represents the solutions of the net series current through and the voltages across each device.

PROBLEMS

In Problems 1 to 5 use a linear graph to solve the sets of simultaneous equations.

1. $y = 2x - 8$
$y = 3x + 7$

3. $2a + 3b = -10$
$-a + 2b = -9$

5. $y = x^2$
$y = 2x + 3$

2. $s = -6t + 33$
$s = 2t - 15$

4. $2u - 3v = -1$
$-u + 4v = 8$

20.4 LOGARITHMIC GRAPHS AND THE BODE PLOT

In electronics you often have to graphically depict data when one or both variables extends over a range too difficult to portray on linear graph paper. For example, suppose the manufacturer of high fidelity equipment wants to depict the linearity of the response of their device over the normal human hearing range (generally from 20 Hz to 20 kHz). Because this linearity represents a 1000:1 ratio of the higher to the lower frequency, it is impractical to linearly graph this data and simultaneously have it accurate at both ends of the frequency range. What is practical, however, is to span this range logarithmically. Because the log of 1000 is 3, only three adjacent *decade* scales are needed for the graph paper. Figure 20.13 is an example of such a graph. As implied, a decade is a range of frequencies in which the ratio of the highest to the lowest frequency is 10, such as a range from 1000 to 100.

Study Figure 20.13: Note the spacing of the lines (say, from 100 to 1000 Hz). The space from 100 to 200 is longer than the space from 200 to 300, which is longer than the spacing from 300 to 400, and so on because the log of each successive ratio gets

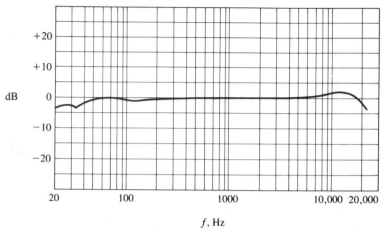

Figure 20.13   Frequency response as measured via Dolby with CRO_2 tape.

smaller and smaller, thereby requiring a smaller representative distance displacement. For example, 200 to 100 Hz is a ratio of 2. The log of 2 is 0.30103. At 800 Hz, the ratio to 100 Hz is 8. The log of 8 is 0.90309. Now the ratio magnitude has jumped by a factor of 4 (from 2 to 8). The log, however, has jumped by only a factor of 3 (from 0.30103 to 0.90309). This effect becomes more pronounced as the ratios get even higher. The net result is a compression of the spacing of the graph lines, thereby facilitating the graphing of great expanses of the range of values of either or both variables.

From a practical viewpoint, one of the most frequently encountered independent variables in electronics is frequency. Frequency is plotted along the abscissa and often ranges through several decades (or cycles, in nonelectronic terms). The ratio of an output voltage (v_o) to an input voltage (v_i) is plotted along the ordinate axis. The ratio of v_o/v_i represents a transfer function, which is the basis of the Bode plot. The Bode plot depicts both the transfer function and phase shift of a device as it is affected by frequency. Data can be experimentally determined over a series of these ratios and logarithmic graphs then plotted to show these relationships. The ordinate axis is in units of positive decibels for amplifiers or negative decibels for high- or low-pass filter networks.

Simple mathematics that involve the j operator can facilitate a rapid reconstruction of these Bode graphs using *straight-line approximations*. These graphs are remarkably accurate for more detailed experimental analysis. Both the detailed and the straight-line approximations occur so frequently in electronics that the balance of this section discusses their analysis. To portray the extent of their utilization, Figure 20.14a to c illustrates their application to operational amplifiers. These figures respectively represent:

1.   A three-stage operational amplifier
2.   The frequency response of each individual stage
3.   The overall response of the composite amplifier

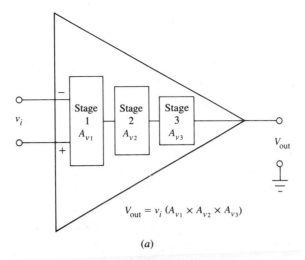

(a)

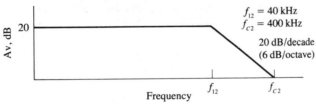

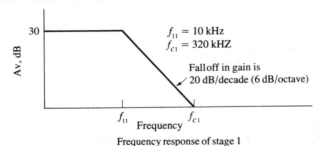

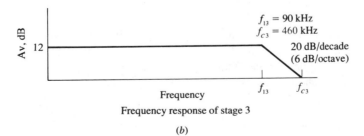

(b)

Figure 20.14 (a) Three-stage operational amplifier.
(b) Bode plot (frequency response) of each stage of the amplifier in (a).

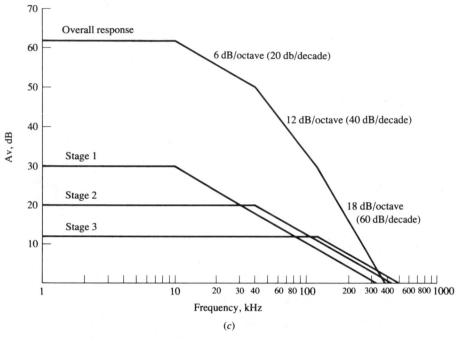

(c)

Figure 20.14 (*continued*)   (*c*) Overall frequency response of the cascaded amplifiers in (*a*) and (*b*).

Note that the frequencies in this figure refer to specific characteristics not specified in the drawings. Also note that the straight-line segmentations (when properly directed) can so accurately represent the actual detailed response curves that they dramatically reduce the work necessary to depict the transfer ratio-frequency relationships. We use a simple *RC* filter network to develop this methodology and graph a detailed Bode plot. Then we illustrate the simpler straight-line approximation method for comparison.

Consider the low-pass filter in Figure 20.15. By simple voltage divider ratios,

$$v_o = \frac{-jX_C}{R - jX_C} v_i \tag{20.7a}$$

or

$$\frac{v_o}{v_i} = \frac{-jX_C}{R - jX_C} \tag{20.7b}$$

At very low input frequencies, $v_o/v_i \to 1$ because $X_C >> R$. At very high input frequencies, $v_o/v_i \to 0$ because $X_C \to 0$. Hence this circuit passes low frequencies and shorts out high frequencies. At one critical frequency (f_c), $v_o/v_i = 0.5$; that is, the output voltage diminishes to one-half its input value when $X_C = R$, or

$$\frac{1}{2\pi f_c C} = R \tag{20.8a}$$

$$\therefore f_c = \frac{1}{2\pi RC} \tag{20.8b}$$

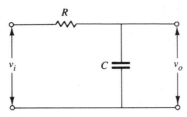

Figure 20.15

Inasmuch as the abscissa is in units of frequency, it behooves us to reduce Equation (20.7b) into units of f and f_c. Substituting R in Equation (20.8a) into Equation (20.7b) and further substituting $X_C = 1/2\pi fC$ results in

$$\frac{v_o}{v_i} = G = \frac{1}{\sqrt{1 + (f/f_c)^2}} \; \underline{/-\tan^{-1}(f/f_c)}$$

This means that the gain (G) is

$$G = \frac{1}{\sqrt{1 + (f/f_c)^2}} \tag{20.9a}$$

or, when expressed in decibels, the gain is

$$G_{dB} = 20 \log \frac{1}{\sqrt{1 + (f/f_c)^2}} \tag{20.9b}$$

You can see from these equations that for a fixed f_c, the gain diminishes as f increases and the phase angle becomes increasingly negative (approaching $-90°$ as f becomes much larger than f_c) and is equal to

$$\theta = -\tan^{-1}\frac{f}{f_c} \tag{20.10}$$

First, let us do the detailed analysis.

Example 1

Suppose $C = 0.01\ \mu F$. Determine the value of R necessary for f_c (the cutoff frequency) to equal 1 kHz. Graph G_{dB} and the phase angle for a range of frequencies from 10 Hz to 10 kHz.

$$R = \frac{1}{2\pi f_c C}$$

$$= \frac{1}{2\pi (10^3)(10^{-8})}$$

$$= 15.92\ k\Omega$$

f	G_{dB}	$\theta°$
10	-4.34×10^{-4}	-0.57
20	-1.74×10^{-3}	-1.15
30	-3.91×10^{-3}	-1.72
50	-0.011	-2.86
70	-0.021	-4.0
100	-0.043	-5.7
200	-0.17	-11.3
300	-0.374	-16.7
500	-0.97	-26.6
700	-1.732	-35
1000 (f_c)	-3	-45
2000	-6.99	-63.4
3000	-10	-71.6
5000	-14.15	-78.7
7000	-16.99	-81.9
10000	-20	-84.3

It is apparent from the data in Example 1 that this is a low-pass filter: Well below f_c there is little attenuation or phase shift. As the frequency increases, attenuation increases significantly, as does the phase shift. Plotting the data results in Figure 20.16. We can make the following two observations from both the data and Figure 20.16 to establish the bases for the straight-line approximation method:

1. When starting from both the high and low frequencies and approaching f_c, data is asymptotic to two straight lines that intersect at 0 dB and f_c. The dashed lines in Figure 20.16 depict this relationship.

   a. The line below f_c is horizontal to the abscissa and at a level of 0 dB. Its slope is zero.

   b. The slope of the second line is -20 dB/decade. This slope may also be expressed as -6 dB/octave (an octave is defined as a frequency ratio of 2:1).

   c. From the data table, at f_c the actual data point is -3 dB below the straight-line intersection point.

   d. Also from the data table, at 1 octave below f_c (in this specific example, 500 Hz) the actual value is approximately -1 dB and at 1 octave above f_c (in this specific example, 2000 Hz) the actual value is approximately -7 dB.

2. The phase-angle component is approximately 0° when $f \ll f_c$; it equals 45° at f_c and approaches 90° as $f \gg f_c$. To help facilitate these approximations, consider the following:

When	$f \ll f_c$	$f = 0.5 f_c$	$f = f_c$	$f = 2 f_c$	$f \gg f_c$
$\theta°$ approximates	0	-25	-45	-65	-90

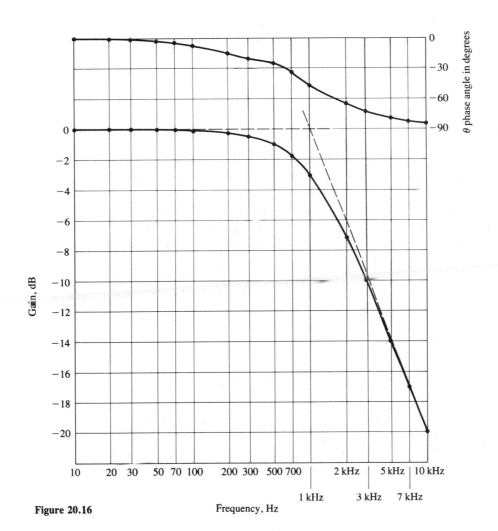

Figure 20.16

Frequency, Hz

Example 2 illustrates how rapidly a transfer function can be constructed using these approximations.

Example 2

Graph the transfer function of Figure 20.17. Reference is made to Figure 20.18 during this solution.

$$R = \frac{1}{2\pi f_c C}$$

$$\therefore f_c = \frac{1}{2\pi RC} = \frac{1}{2\pi(18 \times 10^3)(0.0022 \times 10^{-6})}$$

$$= 4 \text{ kHz}$$

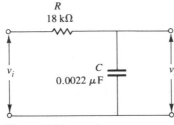

Figure 20.17

1. From the corner frequency (f_c = 4 kHz), construct two straight lines, one parallel to the abscissa and drawn at a 0-dB level. For the second straight line, connect the two points (4 kHz, 0 dB) and (40 kHz, −20 dB). The first of these two points is f_c; the second point is 1 decade higher in frequency and represents the required −20 dB attenuation slope.

2. Place three points at

   a. 2 kHz, −1 dB (1 octave below f_c)

   b. 4 kHz, −3 dB

   c. 8 kHz, −7 dB (1 octave above f_c)

3. Draw a curve through the three points placed in step 2 and extend it asymptotically toward the two straight lines.

4. For the phase shift, select

   a. the lowest frequency graphed at 0°

   b. 2 kHz, −25°

   c. 4 kHz, −45°

   d. 8 kHz, −65°

   e. the highest frequency graphed at −90°

   Connect these points with a smooth curve.

   When more than one transfer function is involved, such as in Figure 20.14, the total transfer function of the device naturally becomes more complex. Generally, however, the total decibel loss (or gain) is additive. Hence each stage contributes another 6-dB/octave loss (or gain) slope, as shown in Figure 20.14c.

PROBLEMS

Construct a Bode transfer function for Figure 20.19a to c. Note that b and c are high-pass filters and therefore represent a positive-slope transfer function.

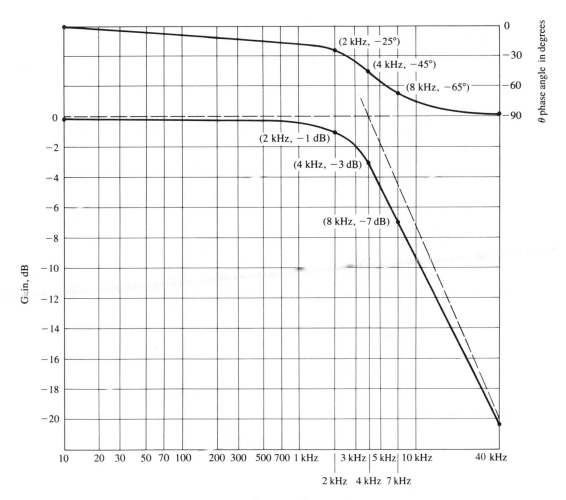

Figure 20.18

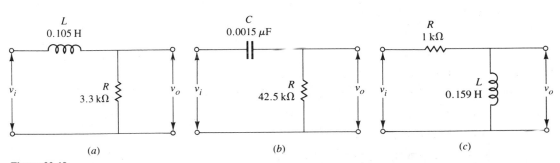

Figure 20.19

20.5 POLAR GRAPHS

Polar graphs are used in electronics to depict radiation and/or reception strengths of antennae. They are unusual in that the magnitude and angle are the variables. Usually, the magnitude (expressed as a radius r from the origin) is a function of $\underline{/\theta}$. Typically polar graphs are aligned in such a manner that what was formerly the abscissa of the linear graph is the 0° to 180° axis and what was formerly the ordinate of the linear graph is the 90° to 270° axis. Most polar graphs have 10 major, equally spaced concentric circles radiating outward from the center. The easiest way to demonstrate how to plot a polar graph is by the following example.

Example 1

See Figure 20.20. Plot $r = |\cos 2\theta|$. Note the absolute magnitude signs. Because the cosine function can have both positive and negative values and we have arbitrarily selected that any radii from the origin is positive, we can avoid negative distances by using the absolute magnitude sign. The data is as follows:

$\theta°$	$2\theta°$	$r = |\cos 2\theta|$	Point #
0	0	1	1
30	60	$\frac{1}{2}$	2
45	90	0	3
60	120	$\frac{1}{2}$	4
90	180	1	5
120	240	$\frac{1}{2}$	6
135	270	0	7
150	300	$\frac{1}{2}$	8
180	360	1	9
210	420	$\frac{1}{2}$	10
225	450	0	11
240	480	$\frac{1}{2}$	12
270	540	1	13
300	600	$\frac{1}{2}$	14
315	630	0	15
330	660	$\frac{1}{2}$	16

Each point of the data table is plotted in Figure 20.20.

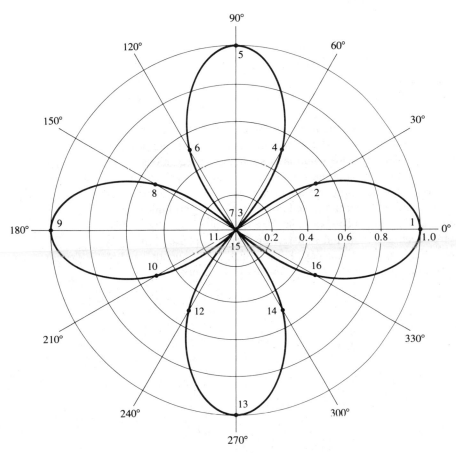

Figure 20.20

PROBLEMS

In Problems 1 to 9 plot the polar graphs.

1. $r = |\cos \theta|$

2. $r = |\sin 2\theta|$

3. $r = |\sin 3\theta|$

4. $r = 1 + \cos \theta$

5. $r = 1 - \sin \theta$

6. $r = 2 - \sin \theta$

7. $r = 2 + \cos \theta$

8. $r = 2 + 2 \cos \theta$

9. $r = 2 - 2 \sin \theta$

SUMMARY

Graphs are visual pictures of the relationships between two (or more) variables. The dependent variable is usually plotted along the ordinate, and the independent variable is plotted along the abscissa. When identifying the location of a point on a graph, the values of the independent variable are specified first and the dependent variable second.

The equation of a straight line takes the format $y = mx + b$, where m specifies the slope of the line and b specifies that point where the line intersects the ordinate axis. Graphs can be used to determine the solution of simultaneous equations. They are especially useful when one or both of the simultaneous equations are not linear.

Besides linear graphs, two other forms of graphs are useful for portraying electronic phenomena. These are the logarithmic and polar graphs. Logarithmic graphs usually have frequency as the independent variable. The Bode plot is ideally applied to logarithmic graphs because it plots the transfer function and phase shift of an electronic device (or circuit) against frequency. Polar graphs are frequently used in electronics to depict radiation and/or reception strengths of antennae.

REVIEW PROBLEMS, PART IV

The following problems are a review of Part IV of the text and directly relate to the numbered sections.

Section 19.2

In Review Problems 1 to 5, use your calculator to find the logs of the indicated quantities by both direct application and applying the rules of logarithms.

1. $\log \dfrac{25.6 \times 42.9}{16.34}$

2. $\log \dfrac{3^{6.1} \times 2^{5.3}}{16.5}$

3. $\log \dfrac{19.8^{2.1}}{1.64 \times 7.12^{0.6}}$

4. $\log \dfrac{\sqrt[3]{21.6} \times 1.51^{6.2}}{3.42}$

5. $\log \dfrac{\sqrt[5]{321.2} \times 3.6^{3.6}}{2.4^{4.21} \times \sqrt[4]{986.2}}$

Section 19.3

In Review Problems 6 to 12, solve each logarithmic equation for x.

6. $x = \log 253.61$

7. $5 \log x + 2 \log x = 21.86$

8. $2 \log x + \log 4x = 16.2$

9. $2 \log (x/4) + \log x = 42.62$

10. $x^{5.82} = 9876.21$

11. $3 \log x - \log 5x - \log \dfrac{x}{2} = 10.2$

12. $8.4^{x/2} = 642.61$

Section 19.4

For Review Problems 13 to 16, indicate which are valid and which are invalid expressions.

13. $\log (-5.2)$

14. $-\log (9.8)$

15. $\log (-\log 8.3)$

16. $\log (\log 6.2)$

Section 19.5

In Review Problems 17 to 23, use your calculator to evaluate the indicated operations.

17. $(3.1 \log 42.1)(4.7 \ln 6.3)$

18. $(\log 32.6)e^{-2.1}$

19. $\dfrac{2.3 \log (\log 6421.6)}{\ln 4.6^{2.1}}$

20. $\dfrac{(\log^{-1} 3.6214)(\ln^{-1} 2.9876)}{\log^{-1} 2.5431}$

21. $\log_4 62$

22. $\log_4 (\log_2 42.317)$

23. $\log^{-1} (\log_7 61.872)$

Section 19.6

In Review Problems 24 to 28, determine the decibel gain using the given information.

24. $P_o = 28$ mW, $P_i = 10$ μW

25. $P_o = 6.8$ W, $P_i = 250$ mW

26. $P_o = 9.8$ μW, $P_i = 400$ mW

27. $I_o = 5.2$ mA, $Z_o = 14.5$ kΩ, $I_i = 40$ μA, $Z_i = 450$ Ω

28. $E_o = 25$ V, $Z_o = 6.8$ kΩ, $E_i = 350$ mV, $Z_i = 400$ Ω

29.   Three electronic devices in series have 35, -16, and 55 dB power gains. If $P_i = 0.9$ μW, determine P_o for the overall system.

Section 19.7

For Review Problems 30 and 31, refer to Figure IV.1.

30.   What is the current in the circuit 1 μs after switch S is closed?

31.   How long will it take for the voltage across the coil to equal 15 V?

For Review Problems 32 to 34, refer to Figure IV.2.

32.   What is the voltage at point A (with respect to ground) 4.8 ms after switch S is closed?

33.   How long will it take after switch S is closed for the voltage across the capacitor to equal 10 V?

34.   If point A is initially precharged to $+6$ V, how long will it take after switch S is closed before the voltage at point A reaches -2.5 V?

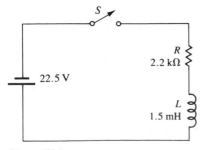

Figure IV.1

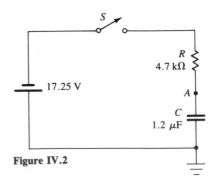

Figure IV.2

Section 20.1

For Review Problems 35 to 39, find the slope of the line specified by the given equation or through the given points.

35. $y = -3x$

36. $s = 2t - 3$

37. $3y = 6.3x - 9$

38. $P_2 = (4, 0), P_1 = (2, 5)$

39. $P_2 = (4.3, -6.2), P_1 = (-5.6, -3.9)$

Section 20.2

For Review Problems 40 to 44, graph the straight lines.

40. $a = b$

41. $y = -4.5x$

42. $3x - 6y + 3$

43. $3.2s - 2.5t = 9.8$

44. $3.1a + 7.12b - 4.6 = 0$

Section 20.3

In Review Problems 45 to 49, use a linear graph to solve the indicated sets of simultaneous equations.

45. $y = 2x$
$y = 3x - 3$

46. $2x - y = -1$
$x = y + 1$

47. $3s + t = 13$
$s + 6t = -7$

48. $a + 2b + 6 = 0$
$a = 2b$

49. $y = x^2$
$y = 2x - 1$

Section 20.4

50. In Figure 20.15, suppose $C = 0.0033$ μF. Determine what R should be for f_c to equal 3 kHz. Graph G_{dB} and the phase angle for a range of frequencies from 10 Hz to 30 kHz.

Section 20.5

In Review Problems 51 to 55, use polar graph paper to plot the indicated equations.

51. $r = \sin \theta$

52. $r = 2 \cos \theta$

53. $r = 3 + 2 \cos \theta$

54. $r = 1 + \sin \theta$

55. $r = \cos 5\theta$

NUMBERING SYSTEMS AND LOGIC

21 NONDECIMAL NUMBERING SYSTEMS

21.1 FUNDAMENTAL IDEAS

Prior to the development of the broad field of digital electronics in general, and the computer/calculator in particular, there was little need for any numbering system other than decimal. However, the enormous impact of digital electronics now mandates a thorough understanding of nondecimal numbering systems. This chapter thoroughly explores these new systems and demonstrates how a calculator can be readily used to convert from one system to another.

Students beginning a study of the nondecimal numbering system frequently ask "Why can't these computing devices use electronic circuitry capable of operating directly in ten separate states; that way the decimal system could be directly applicable? For example, why can't we take an IC [integrated circuit] device and break up its operational voltage levels into ten separate states?" Because the typical IC might operate between 0 and 3.5 V, why can't

$$0.0 \text{ V} = \text{decimal } 0$$
$$0.388 \text{ V} = \text{decimal } 1$$
$$0.777 \text{ V} = \text{decimal } 2$$
$$1.167 \text{ V} = \text{decimal } 3$$
$$\cdot$$
$$\cdot$$
$$\cdot$$
$$3.5 \text{ V} = \text{decimal } 9$$

The answer to these otherwise "ideal" questions is that we live in a practical world. For example, commercial resistors have ±10 percent or ±5 percent tolerance. The gain of an amplifying device, even within the same family, may easily have a 2:1 ratio factor. If the output of an IC device is used as the input to only one other device, its output voltage might ideally be 3.5 V. But if its output is used as the input to, say, five other devices, this "fan-out" effect might drop the voltage level to 2 V. In short, practical circuitry cannot reliably define enough different voltage levels to accurately depict any one of 10 different decimal digits. We could say that the actual value of the digit being depicted is so ambiguous that practical circuitry is totally impractical. What is reliable is that the manufacturer might typically specify that either the *high* output voltage of the IC device is between 2 to 3.5 V or the *low* output voltage is between 0 to 1 V. Within this constraint there seems to be only two regions of predictable voltage: one high or one low. Because these are the only reliable "regions" of voltage output, logically we can surmise that there are only 2 digits (not 10 as in the decimal system) which can be practically represented by commercial circuitry. Hence you must understand a new two-digit system of numbers for digital electronics: the *binary* system.

Before you begin this study, you must first recognize certain basic definitions and mathematical laws. We illustrate these by using the decimal system.

1. The *base* of a numbering system equals the number of digits in that system. In the decimal system there are 10 digits (0 through 9); therefore the base is 10. This concept is important because it is the base that is raised to various powers when the decimal value of a nondecimal number is being determined. Extrapolating this ideal: The base in binary is 2 (digits 0 through 1); the base in octal is 8 (digits 0 through 7); and so on. Also note that the highest digit in any system is one less than the base of that system.

2. A number such as 724 is not a digit; it is a number consisting of three separate digits. Each digit has two meanings: (*a*) magnitude, such as the digit 4, which is one more than 3 and one less than 5, and (*b*) a digit's position within the number. Certainly, the 4 in 247 (same three digits as before) has the same magnitude, but it is *weighted* differently because of its location within the number.

3. Starting with the extreme right-hand digit position, which is always weighted as unity, each successive digit position progressing left is as many times greater in weighted value as is the base of that numbering system. The full meaning of the 4 in 724, considering both its magnitude and position, is 4×1, or 4. In 247, it is 4×10, or 40.

We can express the full meaning of a decimal number (let us arbitrarily choose 3567) by the following breakdown of its components:

$$3567 = 7 \times 1 + 6 \times 10 + 5 \times 100 + 3 \times 1000$$
$$= 7 \times 10^0 + 6 \times 10^1 + 5 \times 10^2 + 3 \times 10^3$$

If the weight of a position is displayed directly below its respective digit, such as

3	5	6	7	digits
10^3	10^2	10^1	10^0	weighted

it is quickly apparent that

1. It is indeed the base of a numbering system that is raised to a power as part of the process of evaluating the full meaning of a number.

2. The powers always start with 0 (at the unit position) and increment by 1 progressing left for each digit position.

With these basic concepts as background, let us proceed to investigate binary, the most fundamental nondecimal numbering system.

21.2 THE BINARY NUMBERING SYSTEM

As stated in Section 21.1, the base of the binary numbering system is 2. Thus the system uses only the two digits 0 and 1. Although this is the most limiting system as far as the availability of digits, it does represent the only system where *every* digit can correspond to predictable regions (levels) of voltage outputs from digital circuitry. In a common form of logic, digit 0 is represented by a low voltage output (typically 0 to 1 V) and the digit 1 is represented by a high voltage output (typically 2 to 3.5 V).

Everyday mathematics use the digits 0 through 9, but binary uses only a 0 or 1. Therefore there must be a method for translating between the two systems. This section develops these methods, but before we start we must be able to recognize the system we are talking about or we will be confused. For example, is 101 a binary number, or is it one-hundred and one in decimal? To avoid this confusion, any numbering system other than decimal has a subscript associated with it. Therefore, binary 101 is written as 101_2. If the number is in decimal, it may or may not have a subscript. If it does, then decimal is *explicitly* indicated; if not, it is implied to be decimal, in which case it is said to be *implicitly* indicated.

Binary to Decimal—Integer Numbers

Consider a binary number such as 10101_2. When applying the weighting factor for each digit position, there results

$$
\begin{array}{cccccl}
1 & 0 & 1 & 0 & 1_2 & \text{digits} \\
2^4 & 2^3 & 2^2 & 2^1 & 2^0 & \text{weights}
\end{array}
$$

This evaluates to

$$
1 \times 2^4 + 0 \times 2^3 + 1 \times 2^2 + 0 \times 2^1 + 1 \times 2^0
$$
$$
= 16 + 0 + 4 + 0 + 1
$$
$$
= 21
$$

Now in practice you would not even take the time to evaluate those positions where the digit is zero because 0×2^n always results in zero and as such contributes nothing to the final summation.

Example 1

Evaluate (that is, find the decimal equivalent of) 10010101_2.

$$
\begin{array}{cccccccc}
1 & 0 & 0 & 1 & 0 & 1 & 0 & 1_2 \\
2^7 & 2^6 & 2^5 & 2^4 & 2^3 & 2^2 & 2^1 & 2^0
\end{array}
$$
$$
\begin{aligned}
N_{10} &= 1 \times 2^7 + 1 \times 2^4 + 1 \times 2^2 + 1 \times 2^0 \\
&= 128 \quad + \quad 16 \quad + \quad 4 \quad + \quad 1 \\
&= 149
\end{aligned}
$$

Study this example: There are further shortcuts for evaluating *binary* numbers. Because only digit positions that contain a 1 are evaluated and 1×2^n always results in 2^n, the process is merely the sum of a series of powers:

$$N_{10} = 2^7 + 2^4 + 2^2 + 2^0$$

If you record (or mentally visualize) only the power to which the base is raised below each digit position, you should quickly see how this example is derived:

1	0	0	1	0	1	0	1_2	binary number
7	6	5	4	3	2	1	0	power of base
2^7			$+$ 2^4	$+$	2^2	$+$	$2^0 = 149$	resultant

The y^x function on a calculator can rapidly convert binary to decimal (especially large numbers) using this pattern.

Example 2

Using your calculator, evaluate 1011110110110_2.

1	0	1	1	1	1	0	1	1	0	1	1	0_2	
12	11	10	9	8	7	6	5	4	3	2	1	0	base powers

Recognizing that $2^1 = 2$ and $2^2 = 4$, we can shorten the last two y^x steps by merely adding 4 and 2 to the pending summation.

PRESS	DISPLAY	COMMENT
2 $\boxed{y^x}$ 12 $\boxed{+}$ 2 $\boxed{y^x}$ 10 $\boxed{+}$	5120	
2 $\boxed{y^x}$ 9 $\boxed{+}$ 2 $\boxed{y^x}$ 8 $\boxed{+}$ 2 $\boxed{y^x}$ 7 $\boxed{+}$	6016	
2 $\boxed{y^x}$ 5 $\boxed{+}$ 2 $\boxed{y^x}$ 4 $\boxed{+}$ 4 $\boxed{+}$ 2 $\boxed{=}$	6070	Final solution

Binary to Decimal—Fractional Numbers

Suppose a binary number is represented as 101.1011_2; that is, a decimal point separates parts of the number. This decimal point has the same meaning in binary as it does in decimal: It divides the number into an integer portion and a fractional portion. Because the powers to which the base is raised decrements by 1 for each position (when progressing from left to right) we can extrapolate this same progression into negative powers to the right of the decimal point. The base powers for 101.1011_2 then look like

1	0	1.	1	0	1	1_2	
2	1	0	-1	-2	-3	-4	base power

Example 3

Using your calculator, evaluate 101.1011_2.

The full procedure is now illustrated, although you can easily mentally evaluate 101_2 as 5.

PRESS	DISPLAY	COMMENT
2 $\boxed{y^x}$ 2 $\boxed{+}$ 2 $\boxed{y^x}$ 0 $\boxed{+}$	5	Positions to left of decimal point evaluated
2 $\boxed{y^x}$ 1 $\boxed{+/-}$ $\boxed{+}$	5.5	2^{-1} summed to above
2 $\boxed{y^x}$ 3 $\boxed{+/-}$ $\boxed{+}$	5.625	2^{-3} summed to above
2 $\boxed{y^x}$ 4 $\boxed{+/-}$ $\boxed{=}$	5.6875	2^{-4} summed to above
		Problem completed

PROBLEMS

Convert the binary numbers in Problems 1 to 7 into their decimal equivalent.

1. 1101

2. 1110011

3. 10101010101

4. 0.11

5. 0.1001

6. 11011.101

7. 110000001.1111

Decimal to Binary—Integer Numbers

The previous two sections used the approach "Given a binary number, what does it equal in decimal?" We now turn the problem around: What is the binary representation of a given decimal number? The trick to determining this value (and incidentally, converting a decimal number into any other numbering system) is to divide successively by the base (in this case 2), carry down the remainder, and continue the process until the quotient is less than the base. One further note: Although we frequently perform division left to right, here we perform our conversion division right to left because this latter method results in the binary integers *falling out* into their proper position. Examples 4 and 5 illustrate this concept.

Example 4

Convert 37 into binary.

Study the following sequence, remembering that the division and remainder directions are as follows:

$\leftarrow$ Divide to the left

$\downarrow$

Carry remainders straight down,
placing them under their quotient

$$0 \leftarrow 1 \leftarrow 2 \leftarrow 4 \leftarrow 9 \leftarrow 18 \leftarrow 37(2$$
$$1 \quad 0 \quad 0 \quad 1 \quad 0 \quad 1_2 \quad \text{Remainders equal the binary answer}$$

Note the small triangle represented by the dashed lines. This sequence *always* occurs when converting decimal into binary. Translated, it says that the base 2 divides into 1 zero whole number of times, thereby leaving a remainder of 1. Mathematically,

$$\frac{1}{2} = 0 \text{ quotient, } 1 \text{ remainder}$$

Any binary number calculated in this fashion always starts with a 1 because leading zeros add nothing to its meaning.

Example 5

Convert 125 into binary.

Dropping off the arrows:

0	1	3	7	15	31	62	125(2
1	1	1	1	1	0	1_2	

We now learn how to use the calculator to perform the preceding two operations. Actually, it is probably just as easy to *manually* convert into binary. However, because the principles learned are equally applicable to any other base conversions—and these bases are much more prone to division error (especially base 16)—it behooves us to learn the methodology for a simple system for more realistic applications to the more complex systems.

Because the calculator displays a quotient (when it is not an even division) as an integer part, decimal point, and fractional part, we need to save the integer part for later division and convert the fractional part into an integer remainder. For example, in Example 5, 125 divided by 2 equals 62.5 on the calculator. The 62 must be preserved for later division, and the .5 must be converted to a remainder of 1. Observe the following sequence in a simple conversion.

Example 6

Convert 13 into binary.

PRESS	DISPLAY	COMMENT
13 $\div$ 2 $=$	6.5	Initial division
STO 1 $-$ 6 $=$	0.5	Store first quotient (6.5) in memory; separate out the fractional part
$\times$ 2 $=$	1*	Remainder of 1 $\rightarrow$ 2^0 position
RCL 1 $-$ .5 $=$	6**	Reestablish integer part of initial quotient for next division sequence

Now we repeat this short sequence until the final solution. Whenever the fractional part of a quotient is zero, that is, whenever there is an even division, no store or recall sequence is necessary, merely place a zero into the 2^n position affected. Also, whenever the dividend is less than the base (in binary, whenever it is a 1), merely place it into the highest binary digit position affected. Remember, the 6 from the previous sequence is still in the display register.

PRESS	DISPLAY	COMMENT
$\div$ 2 $=$	3.0	0 into the 2^1 position
$\div$ 2 $=$	1.5	
STO 1 $-$ 1 $=$ $\times$ 2 $=$	1	1 into the 2^2 position
RCL 1 $-$ .5 $=$	1	Final dividend less than base, therefore 1 into the 2^3 position

Answer is 1101_2.

As stated earlier, we illustrated this calculator procedure for a simple system (which could just as easily have been done mentally) to learn a methodology for converting more complex numbering systems.

---

*Multiplying the fractional portion of a quotient by the base (the same number you just divided by) *should* result in an integer. Most of the time this happens. But on same calculators, dividing by an unusual base results in 1.1111111 rather than a "nice" fractional part (such as $10 \div 9$). When subtracting out the integer part and multiplying by 9 again, the resultant *might* appear as 0.9999999. In situations like this, always round up to the next highest whole integer. For example, 6.9999998 $\rightarrow$ 7.

**On some calculators with the following functions, the INT (integer part) and its associated INV INT (fractional part) greatly simplify the entire process by separating out the components of a mixed number.

Decimal to Binary—Fractional Numbers

Suppose we want to convert a decimal number such as 23.625 into binary. We calculate the integer portion (23) as in the previous section. We must also convert the fractional portion (.625) into binary; however, this latter method requires successive multiplication rather than successive division. After each multiplication, observe the integer to the left of the decimal point: If the product equals 1, that 1 is placed into the 2^{-n} position affected, otherwise a zero is placed there. To illustrate:

$0.625 \times 2 = 1.25$ place the 1 of product 1.25 into the 2^{-1} position
$0.25 \times 2 = 0.5$ place the 0 of product 0.5 into 2^{-2} position
$0.5 \times 2 = 1.0$ place the 1 of product 1.0 into 2^{-3} position

At this point the process is completed because any further multiplication would result in a zero, and zeros following the last nonzero digit to the right of the decimal point are meaningless. One further comment. Some fractional numbers, for example, 0.691, continue indefinitely when multiplied by 2 and never have a final terminating product of 1.0. (This is analogous to how accurately you want to express π.) What do you do? Simply carry out the accuracy to the point desired and stop the process.

Example 7

Convert 95.8125 into binary.
   First, convert 95 by successive division:

0	1	2	5	11	23	47	95(2
1	0	1	1	1	1	1_2	

Second, convert 0.8125 by successive multiplication.

$0.8125 \times 2 = 1.625$ 2^{-1} position = 1
$0.625 \times 2 = 1.25$ 2^{-2} position = 1
$0.25 \times 2 = 0.5$ 2^{-3} position = 0
$0.5 \times 2 = 1.0$ 2^{-4} position = 1

Therefore $95.8125 = 1011111.1101_2$.

PROBLEMS

Convert the decimal numbers in Problems 1 to 10 into binary.

1. 86
2. 117
3. 2516
4. 63.5
5. 151.3125
6. 1000.125
7. 847.1875
8. 505.5625
9. 351.9375
10. 191.15625

21.3 OCTAL, HEXADECIMAL, AND OTHER SYSTEMS

The principles you learned for converting binary to decimal and decimal to binary apply equally well to numbering systems in other than base 2. The two most popular (and practical) bases used by the computer industry are octal (base 8) and hexadecimal (base 16). Theoretically we can manipulate numbers in any base, so although we concentrate upon bases 2, 8, and 16, other bases are presented as exercise problems.

Decimal to Base N

Recall that converting from decimal to any base requires successive division by that base, carrying down the remainder, and continuing the process until the final quotient is less than the base. This is for integer numbers. Successive multiplication by the base, using the methodology presented in Section 21.2, is used for fractional parts of a numbering system. Examples 1 to 6 illustrate the principles.

Example 1

Convert 1256 to base 8.

$$0 \leftarrow 2 \leftarrow 19 \leftarrow 157 \leftarrow 1256(8$$
$$\downarrow \quad \downarrow \quad \downarrow \quad \downarrow$$
$$2 \quad 3 \quad 5 \quad 0_8$$

Therefore $1256_{10} = 2350_8$.

Notice that the collective magnitude of the digits of a number in a base less than decimal always appears larger than the decimal equivalent. The opposite is true for bases larger than decimal. Note too that no digit in octal can exceed 7 because the eight digits of octal are 0, 1, 2, 3, 4, 5, 6, and 7.

Example 2

Using your calculator, convert 2694 to octal.

PRESS	DISPLAY	COMMENT
2694 $\div$ 8 $=$	336.75	Initial quotient
STO 1 $-$ 336 $=$ $\times$ 8 $=$	6	$6 \rightarrow 8^0$ position
RCL 1 $-$ .75 $=$	336	
$\div$ 8 $=$	42	$0 \rightarrow 8^1$ position
$\div$ 8 $=$	5.25	

PRESS						DISPLAY	COMMENT
$\boxed{\text{STO 1}}$	$\boxed{-}$	5	$\boxed{=}$	$\boxed{\times}$	8 $\boxed{=}$	2	$2 \to 8^2$ position
$\boxed{\text{RCL 1}}$	$\boxed{-}$	.25	$\boxed{=}$			5	$5 \to 8^3$ position

Therefore $2694_{10} = 5206_8$

Example 3

Convert 916 into base 7.

$$0 \leftarrow 2 \leftarrow 18 \leftarrow 130 \leftarrow 916(7$$
$$\downarrow \quad \downarrow \quad \downarrow \quad \downarrow$$
$$2 \quad 4 \quad 4 \quad 6_7$$

Therefore $916_{10} = 2446_7$.

Again note that no digit in a system of base 7 can exceed 6. Why?

Before proceeding to bases higher than decimal, specifically base 16, you must become accustomed to some unusual digits. For example, base 16, (the hexadecimal system) must have 16 digits. The first 10 digits are 0 through 9 inclusive. But this accounts for only 10 of the 16 required digits. What are the remaining 6 digits? At this point we must create six new digits; the easiest to remember are the first six letters of the alphabet. Thus the 16 digits of hexadecimal and their equivalent value in decimal are

Decimal	Hexadecimal	Decimal	Hexadecimal
0	0	8	8
1	1	9	9
2	2	10	A
3	3	11	B
4	4	12	C
5	5	13	D
6	6	14	E
7	7	15	F

Therefore it is possible for a hexadecimal number to look like $1B3_{16}$, or even ACE_{16}.

Example 4

Convert 3517 to base 16.

$$0 \leftarrow 13 \leftarrow 219 \leftarrow 3517(16$$
$$\downarrow \quad \downarrow \quad \downarrow$$
$$D \quad B \quad D_{16}$$

Therefore $3517_{10} = DBD_{16}$.

Example 5

Using your calculator, convert 2919 into base 16.

PRESS	DISPLAY	COMMENT
2919 $\div$ 16 $=$	182.4375	Initial quotient
STO 1 $-$ 182 $=$ $\times$ 16 $=$	7	$7 \to 16^0$ position
RCL 1 $-$ .4375 $=$ $\div$ 16 $=$	11.375	
STO 1 $-$ 11 $=$ $\times$ 16 $=$	6	$6 \to 16^1$ position
RCL 1 $-$ .375 $=$	11	$B \to 16^2$ position

Therefore $2919_{10} = B67_{16}$.

Example 6

Convert 0.455_{10} into octal (stop at the 8^{-4} position)

$$0.455 \times 8 = 3.64 \qquad 8^{-1} \text{ position} = 3$$
$$0.64 \ \times 8 = 5.12 \qquad 8^{-2} \text{ position} = 5$$
$$0.12 \ \times 8 = 0.96 \qquad 8^{-3} \text{ position} = 0$$
$$0.96 \ \times 8 = 7.68 \qquad 8^{-4} \text{ position} = 7$$

Therefore $0.455_{10} = 0.3507_8$.

Because the answer can go on indefinitely, use your calculator to check the accuracy by stopping at the 8^{-4} position.

PRESS	DISPLAY
3 $\times$ 8 $1/x$ $+$ 5 $\times$ 8 y^x 2 $+/-$	
$+$ 7 $\times$ 8 y^x 4 $+/-$ $=$	0.4548

PROBLEMS

In Problems 1 to 13 convert the decimal numbers into the bases indicated. Do not carry fractional numbers past the fourth decimal place. Use your calculator.

1. $3120 = ?_8$ **5.** $3976 = ?_{16}$ **9.** $343 = ?_5$ **13.** $621.485 = ?_6$

2. $9763 = ?_8$ **6.** $10,715 = ?_{16}$ **10.** $562 = ?_9$

3. $131.575 = ?_8$ **7.** $357.15 = ?_{16}$ **11.** $1743 = ?_7$

4. $417.6355 = ?_8$ **8.** $593.375 = ?_{16}$ **12.** $421.35 = ?_5$

Base *N* to Decimal

When converting binary to decimal, it is necessary to sum only powers of 2 because any nonzero digit in binary can be only 1 and multiplying any power of 2 by 1 contributes nothing to its value. In any other numbering system, most likely a nonzero digit will be greater than 1, so you have to sum the products of digits and powers.

Example 7

Convert 3456_8 into decimal.

$$N_{10} = 3 \times 8^3 + 4 \times 8^2 + 5 \times 8^1 + 6 \times 8^0$$
$$= 1838$$

Example 8

Convert $3A7_{16}$ into decimal.

$$N_{10} = 3 \times 16^2 + 10 \times 16 + 7 = 935$$

Example 9

Using your calculator, convert 7354.54_8 into decimal.

PRESS	DISPLAY	COMMENT
7 $\boxed{\times}$ 8 $\boxed{y^x}$ 3 $\boxed{+}$ 3 $\boxed{\times}$		
8 $\boxed{x^2}$ $\boxed{+}$ 5 $\boxed{\times}$ 8 $\boxed{+}$ 4 $\boxed{+}$	3820	Integer portion
5 $\boxed{\times}$ 8 $\boxed{1/x}$ $\boxed{+}$ 4 $\boxed{\times}$ 8 $\boxed{y^x}$ 2 $\boxed{+/-}$ $\boxed{=}$	3820.6875	Integer portion plus fractional portion

PROBLEMS

In Problems 1 to 10 use your calculator to convert the indicated bases into decimal.

1. 654_8 3. $2BC_{16}$ 5. 123.456_8 7. 456_7 9. 728.16_9

2. 7001_8 4. ACE_{16} 6. $1C6.AB_{16}$ 8. 3442_5 10. 555.25_6

Base *N* to Base *M*

Bases that are multiples of 2, namely, binary (2^1), octal (2^3), and hexadecimal (2^4), can be easily converted from one into the other by pairing off selected numbers of binary bits. The power of 2 associated with a given numbering system indicates the number of binary digits necessary to represent a digit in that system. See Table 21.1. For example, if we consider each block of digits in octal and hexadecimal a single

TABLE 21.1

Decimal	Binary	Octal	Hexadecimal
0	0	000	0000
1	1	001	0001
2		010	0010
3		011	0011
4		100	0100
5		101	0101
6		110	0110
7		111	0111
8			1000
9			1001
10			1010 (A)
11			1011 (B)
12			1100 (C)
13			1101 (D)
14			1110 (E)
15			1111 (F)

entity (that is, a block of binary digits) and they are weighted in the standard fashion, then we obtain the decimal equivalent value.

Example 10

Convert 10110010111_2 into octal.

Pair off the binary number in blocks of three digits, beginning at the right and progressing left:

10	110	010	111	binary
2	6	2	7	octal

Note that even though there are only two digits in the leftmost bracket, you can always fill in leading zeros to bring the total to three digits, which adds nothing to the value. If there is a decimal point, the pairing progresses in both directions from the decimal point. If the pairing on the extreme right does not contain a full compliment of digits, then a zero *does* have to be filled in.

Example 11

Convert 11001010.1_2 into base 16.

Pair off in blocks of four binary digits:

1100	1010	.	1000	binary
C	A	.	8	hexadecimal

See why zeros must be filled in for the rightmost bracket? The hexadecimal number is 8 (filling in the zeros) and not the 1 that would be obtained if you did not fill in the zeros.

Example 12

Convert 345.6_8 into base 16.
First convert octal into binary and then into hexadecimal:

$$\underbrace{\quad 3 \quad}_{} \quad \underbrace{\quad 4 \quad}_{} \quad \underbrace{\quad 5 \quad}_{} \quad . \quad \underbrace{\quad 6 \quad}_{} \quad \text{octal}$$

$$\underbrace{\emptyset\emptyset\emptyset 0 \; 11}_{0} \quad \underbrace{10 \; 0}_{E} \quad \underbrace{101}_{5} \quad . \quad \underbrace{110\emptyset}_{C} \quad \text{binary}$$

$$\quad\; 0 \qquad\quad E \qquad\quad 5 \qquad . \qquad C \qquad \text{hexadecimal}$$

When the bases are not both multiples of 2, we must go through the decimal system as the common link between the two bases concerned. The cancelled zeros ($\emptyset$) have been inserted above to complete the brackets.

Example 13

Convert from 543_6 into octal.

PRESS	DISPLAY	COMMENT
5 $\boxed{\times}$ 6 $\boxed{x^2}$ $\boxed{+}$ 4 $\boxed{\times}$		
6 $\boxed{+}$ 3 $\boxed{=}$	207	207_{10}; convert this into octal
$\boxed{\div}$ 8 $\boxed{=}$	25.875	
$\boxed{\text{STO 1}}$ $\boxed{-}$ 25 $\boxed{=}$		
$\boxed{\times}$ 8 $\boxed{=}$	7	$7 \rightarrow$ position 8^0
$\boxed{\text{RCL 1}}$ $\boxed{-}$ .875 $\boxed{=}$		
$\boxed{\div}$ 8 $\boxed{=}$	3.125	
$\boxed{\text{STO 1}}$ $\boxed{-}$ 3 $\boxed{=}$		
$\boxed{\times}$ 8 $\boxed{=}$	1	$1 \rightarrow$ position 8^1
$\boxed{\text{RCL 1}}$ $\boxed{-}$ .125 $\boxed{=}$	3	$3 \rightarrow$ position 8^2

Therefore $543_6 = 317_8$.

PROBLEMS

Convert the numbers in Problems 1 to 10 into the bases indicated.

1. $101100110_2 = ?_8 = ?_{16}$

2. $1011000010110.101 = ?_8 = ?_{16}$

3. $4563_8 = ?_2 = ?_{16}$

4. $1234.56_8 = ?_2 = ?_{16}$

5. $12A3_{16} = ?_2 = ?_8$

6. $3D4.F_{16} = ?_2 = ?_8$

7. $1432_5 = ?_8$

8. $3148_9 = ?_6$

9. $3256.1_7 = ?_5$

10. $2BC1.D3_{16} = ?_9$

21.4 BASIC ARITHMETIC OPERATIONS ON THE NONDECIMAL NUMBERING SYSTEMS

The four basic arithmetic operators $(+, -, \times, \div)$ can be applied to nondecimal numbering systems just as they can be to decimal systems. The same fundamental principles apply, such as *carrying* in addition or *borrowing* in subtraction. When the bases of the two operands are alike, you can operate upon them directly. When the bases are different, you must change one base to the other before you can operate upon them. Otherwise, you can change both bases to decimal and then perform the required operation.

Addition and multiplication tables for binary and octal are in Tables 21.2 to 21.6. Hexadecimal tables are omitted because they are quite extensive and hexadecimal numbers can be easily converted into octal. The resultant octal can then be easily converted back into hexadecimal. Two-digit resultants, such as $1 + 1 = 10$ in binary or $5 + 6 = 13$ in octal, have the same meaning (within their own respective base system) as $9 + 7 = 16$ does in decimal. That is, the sum of $9 + 7$ in decimal is 6, with a carry of 1 into the next higher-order position.

Example 1

Add $10110_2 + 10111_2$.

	1		1	1			carry
		1	0	1	1	0	augend
+		1	0	1	1	1	addend
	1	0	1	1	0	1	sum
	2^5	2^4	2^3	2^2	2^1	2^0	position

Note that in the 2^2 position we have a carry from the 2^1 position plus a 1 in both operands. This results in three 1s, or 11_2, meaning a sum of 1 for the 2^2 position and a carry of 1 into the 2^3 position.

TABLE 21.2 BINARY ADDITION OF TWO OPERANDS

0	0	1	1
+ 0	+ 1	+ 0	+ 1
0	1	1	10

Example 2

Subtract 1011_2 from 10101_2.

$$
\begin{array}{ccccc}
 & 1 & 0 & 1 & 0 & 1 \\
- & & 1 & 0 & 1 & 1 \\
\hline
 & - & - & - & - & 0 \\
 & 2^4 & 2^3 & 2^2 & 2^1 & 2^0
\end{array}
$$

In the 2^0 position, 1 from 1 was easy, resulting in a difference of 0. In the 2^1 position, however, 1 from 0 cannot occur without a borrow from the 2^2 position. Because this is binary, we can borrow only from the next highest position containing a 1. This results in that position of the minuend being reduced to 0 and "passing down" a 2 (similar to passing down a 10 in decimal). Continuing the problem:

$$
\begin{array}{ccccc}
0 & 2 & 0 & 2 & \\
\not{1} & 0 & \not{1} & 0 & 1 \\
- & 1 & 0 & 1 & 1 \\
\hline
 & 1 & 0 & 1 & 0
\end{array}
$$

Example 3

Multiply 1011_2 by 101_2.

$$
\begin{array}{ccccccc}
 & & & 1 & 0 & 1 & 1 \\
\times & & & & 1 & 0 & 1 \\
\hline
 & & & 1 & 0 & 1 & 1 \\
 & & 0 & 0 & 0 & 0 & \\
 & 1 & 0 & 1 & 1 & & \\
\hline
 & 1 & 1 & 0 & 1 & 1 & 1
\end{array}
$$

Example 4

Divide 100011_2 by 101_2.

Proceed as you would for long division in decimal:

$$
\begin{array}{r}
111 \\
101 \overline{)100011} \\
-\underline{101} \\
111 \\
-\underline{101} \\
101 \\
-\underline{101} \\
0
\end{array}
$$

TABLE 21.3 BINARY ADDITION OF TWO OPERANDS PLUS A CARRY

0	0	0	0	1	1	1	1	carry
0	0	1	1	0	0	1	1	two
0	1	0	1	0	1	0	1	operands
0	1	1	10	1	10	10	11	

TABLE 21.4 BINARY MULTIPLICATION OF TWO OPERANDS

0	0	1	1
$\times 0$	$\times 1$	$\times 0$	$\times 1$
0	0	0	1

Example 5

Add 4325_8 to 1647_8.

Using Table 21.5 directly:

$$
\begin{array}{cccc}
1 & & 1 & \text{carry} \\
4 & 3 & 2 & 5_8 \\
+\,1 & 6 & 4 & 7_8 \\
\hline
6 & 1 & 7 & 4_8
\end{array}
$$

Example 6

Subtract 3164_8 from 7516_8.

$$
\begin{array}{cccc}
7 & 5 & 1 & 6 \\
-\,3 & 1 & 6 & 4 \\
\hline
- & - & - & 2 \\
8^3 & 8^2 & 8^1 & 8^0
\end{array}
$$

Note that at the 8^1 position we are trying to take 6 from 1. Obviously we must go to the 8^2 position to borrow, reducing that 5 of the minuend to a 4 and passing down 8. This 8, when added to the 1 of the 8^1 position of the minuend, gives it a new value of 9. We can now readily subtract the 6 from that 9, resulting in 3. Continuing the problem:

$$
\begin{array}{cccc}
 & 4 & 9 & \\
7 & \cancel{5} & \cancel{1} & 6 \\
-\,3 & 1 & 6 & 4 \\
\hline
4 & 3 & 3 & 2
\end{array}
$$

Example 7

Multiply 435_8 by 26_8.

Use Table 21.6. First, multiply the multiplicand by the 6 of the multiplier:

```
    2  3        carry
    4  3  5     multiplicand
×      2  6     multiplier
---------------
 3  2  5  6
```

Second, multiply the multiplicand by the 2 of the multiplier:

```
       1        carry
    4  3  5     multiplicand
×      2  6     multiplier
---------------
 3  2  5  6
1  0  7  2
---------------
1  4  1  7  6₈
```

Example 8

Divide $3C1_{16}$ by $1F_{16}$.

Convert the hexadecimal numbers into octal, then use Tables 21.5 and 21.6:

$$3C1_{16} = 1701_8$$
$$1F_{16} = 37_8$$

TABLE 21.5   OCTAL ADDITION

+	0	1	2	3	4	5	6	7
0	0	1	2	3	4	5	6	7
1	1	2	3	4	5	6	7	10
2	2	3	4	5	6	7	10	11
3	3	4	5	6	7	10	11	12
4	4	5	6	7	10	11	12	13
5	5	6	7	10	11	12	13	14
6	6	7	10	11	12	13	14	15
7	7	10	11	12	13	14	15	16

TABLE 21.6   OCTAL MULTIPLICATION

×	0	1	2	3	4	5	6	7
0	0	0	0	0	0	0	0	0
1	0	1	2	3	4	5	6	7
2	0	2	4	6	10	12	14	16
3	0	3	6	11	14	17	22	25
4	0	4	10	14	20	24	30	34
5	0	5	12	17	24	31	36	43
6	0	6	14	22	30	36	44	52
7	0	7	16	25	34	43	52	61

$$\begin{array}{r} 3\ 7 \\ 37 \overline{\smash{\big)}\ 1\ 7\ 0\ 1} \\ -1\ 3\ 5 \\ \hline 3\ 3\ 1 \\ -3\ 3\ 1 \\ \hline 0 \end{array}$$

$$37_8 = 1F_{16}$$
$$\therefore 3C1_{16} \div 1F_{16} = 1F_{16}$$

When you perform operations on mixed bases, it is simpler to use your calculator to convert numbers into decimal, perform the required operations, and then convert back into the desired base. If more than two operands are involved, the memory plus (M+), memory minus (M−), and, if your calculator has them, memory times (M×) and memory divide (M÷) functions can be very useful. Some calculators use the $\boxed{\text{SUM}}$ and $\boxed{\text{INV}}$ $\boxed{\text{SUM}}$ keys to call the (M+) and (M−) functions. We use the M+ and M− symbol to imply adding or subtracting the concepts of the display register to the contents of a memory in which data had been previously stored.

Example 9

Perform the indicated operation using your calculator.

$$3154_6 + 1576_8 - 885_9 = ?_{16}$$

PRESS	DISPLAY	COMMENT
3 $\boxed{\times}$ 6 $\boxed{y^x}$ 3 $\boxed{+}$ 6 $\boxed{x^2}$ $\boxed{+}$ 5		
$\boxed{\times}$ 6 $\boxed{+}$ 4 $\boxed{=}$ $\boxed{\text{STO 1}}$	718	$3154_6 = 718_{10}$, stores → 1
$\boxed{\text{CLR}}$ 8 $\boxed{y^x}$ 3 $\boxed{+}$ 5 $\boxed{\times}$ 8 $\boxed{x^2}$ $\boxed{+}$		
7 $\boxed{\times}$ 8 $\boxed{+}$ 6 $\boxed{=}$ $\boxed{\text{M+}}$	894	$1576_8 = 894_{10}$, add to above
$\boxed{\text{CLR}}$ 8 $\boxed{\times}$ 9 $\boxed{x^2}$ $\boxed{+}$		
8 $\boxed{\times}$ 9 $\boxed{+}$ 5 $\boxed{=}$ $\boxed{\text{M−}}$	725	$889_9 = 725_{10}$, subtract from above
$\boxed{\text{RCL 1}}$ $\boxed{\div}$ 16 $\boxed{=}$ $\boxed{\text{STO 1}}$	55.4375	
$\boxed{-}$ 55 $\boxed{=}$ $\boxed{\times}$ 16 $\boxed{=}$	6.9999999	$7 \rightarrow 16^0$ position
$\boxed{\text{RCL 1}}$ $\boxed{-}$ .4375 $\boxed{=}$ $\boxed{\div}$ 16 $\boxed{=}$ $\boxed{\text{STO 1}}$	3.4375	
$\boxed{-}$ 3 $\boxed{=}$ $\boxed{\times}$ 16 $\boxed{=}$	7	$7 \rightarrow 16^1$ position
$\boxed{\text{RCL 1}}$ $\boxed{-}$ .4375 $\boxed{=}$	3	$3 \rightarrow 16^2$ position

Answer is 377_{16}.

PROBLEMS

Perform Problems 1 to 15 by direct applications of Tables 21.2 to 21.6. Do not use your calculator. Note that it may be necessary to convert among binary, octal, and hexadecimal.

1. $101101_2 + 100111_2 = ?_2$

2. $101111_2 + 456_8 = ?_2$

3. $11110011_2 + AD_{16} = ?_{16}$

4. $532_8 - 1001110_2 = ?_2$

5. $1BC_{16} - 561_8 = ?_8$

6. $11001101_2 - A3_{16} = ?_8$

7. $1101_2 \times 101_2 = ?_2$

8. $111010_2 \times 42_8 = ?_8$

9. $326_8 \times 154_8 = ?_8$

10. $10010101_2 \times 3D_{16} = ?_8$

11. $1110110_2 \times 357_8 = ?_{16}$

12. $1120_8 \div 20_8 = ?_8$

13. $2466_8 \div 35_8 = ?_{16}$

14. $3C9_{16} \div 13_{16} = ?_{16}$

15. $1EE_{16} \div 23_8 = ?_2$

Perform Problems 16 to 25 by using your calculator on the mixed bases.

16. $354_6 + 127_8 = ?_{10}$

17. $2567_8 + 1183_9 = ?_{10}$

18. $2332_4 + 157_8 = ?_9$

19. $647_8 - 353_7 = ?_6$

20. $447_8 + 1443_5 - 166_7 = ?_{10}$

21. $3535_6 \times 144_5 = ?_8$

22. $1588_9 \times 562_7 = ?_{16}$

23. $1553_8 \times 61_9 - 3541_6 = ?_7$

24. $10052_6 \div 20_8 = ?_9$

25. $719_{16} \div 43_5 = ?_7$

SUMMARY

Advancements in the field of digital electronics in recent decades have necessitated an understanding of nondecimal numbering systems. The most basic system upon which all other practical systems are ultimately built is the binary numbering system, which uses only the digits 0 and 1 to represent all data.

Mathematical processes exist to go back or forth between the decimal numbering system and any other numbering system. These processes can be broken down into the following four rules:

1. *Decimal into base N* (integer part).   Divide the decimal number consecutively by the base of the numbering system into which the conversion is taking place. Carry down the remainder. Continue this process until the quotient is less than the divisor.

2. *Base N into decimal* (integer part).   Starting at the right and progressing left, form the sum of the products of each digit of the number multiplied by the base system raised to a power that represents its weighted value. Starting with the rightmost position, the weighted value is the base raised to the power zero, with the power incrementing by one for each digit progressing to the left.

3. *Decimal into base N* (fractional part).   Multiply the fractional part of the decimal number by the base of the numbering system into which the conversion is taking place. The whole number integer created by this multiplication then becomes the integer for base system N for that position. Continue multiplying the remaining fractional part while applying this process until you attain the desired degree of accuracy.

4. *Base N into decimal* (fractional part).   Starting at the decimal point and progressing right, form the sum of the products of each digit of the number multiplied by its weighted value. This weighted value is the base raised to a power of minus one when beginning with the digit immediately to the right of the decimal point. Thereafter, each power decrements by one when progressing to the right.

The calculator, with its y^x function, is ideally suited for conversions between different numbering systems.

22 LOGIC GATES AND BOOLEAN ALGEBRA

Chapter 21 demonstrated that electronic digital circuitry operates around only two voltage regions: *high* and *low*. Typically, this could represent a 1 or 0, respectively. Any specifically designed combination of components, such as transistors, diodes, and resistors (usually as part of an IC chip), that responds to these voltages in a predetermined manner is referred to as a *gate*. Electronic gates either prohibit or permit the passage of a voltage level much the same way as a fence gate blocks or permits entry onto property. As complex as logic circuitry may appear, there are only three basic gates that comprise all digital circuitry. This chapter examines these gates (from a logic point of view), discusses the principles of Boolean algebra, and applies simplification procedures to reduce the complexity of equations. Finally, we demonstrate how to implement simple logic design to perform required tasks, such as adding binary numbers.

22.1 THE BASIC GATES

AND Gate

To begin our analysis of gates, it is convenient to compare a gate to a wall switch. See Figure 22.1. As shown, the gate (switch) is open. In this position the flow of current from the voltage source to the light is blocked. If we take the liberty of referring to current as a signal, then we can say that the gate blocks the signal. If closed, the gate passes the signal and the light is illuminated.

To be practical, most gates consist of two or more switches; some gates, such as inverters or level shifters, are exceptions. If a second switch is placed in series with the gate in Figure 22.1, the result is Figure 22.2. We now adopt the following con-

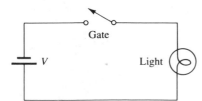

Figure 22.1

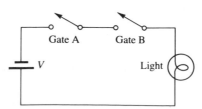

Figure 22.2

TABLE 22.1   TRUTH TABLE OF THE AND GATE

Switch status	Switch A	Switch B	Light
Both switches open	0	0	0
A open, B closed	0	1	0
A closed, B open	1	0	0
Both switches closed	1	1	1

vention: If a switch is open, refer to it as being in a state of LOGIC 0; if a switch is closed, refer to it as being in a state of LOGIC 1. Furthermore, if we refer to the light as in the LOGIC 1 state when lit and the LOGIC 0 state when not lit, we can construct a table that relates the states of the light to the four possible conditions gates A and B can collectively represent.

Table 22.1 depicts all possible conditional states under which Figure 22.2 can exist. Such a table is called a *truth table*. Table 22.1 is the truth table for an AND gate. The term AND is derived from the fact that both switch A *and* switch B must be in a state of LOGIC 1 (closed) for the light to be in a state of LOGIC 1 (lit).

It becomes quite impractical to construct complex logic diagrams as a series of mechanical switches, so there is a convention for representing an AND gate. This symbol is depicted in Figure 22.3; A and B represent the inputs, and C represents the output. An AND gate can have any number of inputs, but it can have only one output. The definition of an AND gate can be stated as

All inputs must be in a state of LOGIC 1 (sometimes called HIGH or TRUE) for the output to be a LOGIC 1.

Keep in mind when you are tracing through a maze of electronic logic symbols that typically LOGIC 1 might represent 3.5 V and LOGIC 0 might represent 0 V. Then, when viewing the AND gate symbol in Figure 22.3 and correlating with the truth table, we might have a case where input A equals 3.5 V and input B equals 0 V; therefore output C equals 0 V. Henceforth, for simplicity and to make the analysis more general, we refer to LOGIC 1 or LOGIC 0 as states instead of voltages.

OR Gate

If the two series gates (switches) in Figure 22.2 are rearranged and placed in parallel, Figure 22.4 results. Because there are now two paths through which current can flow to illuminate the light, there is a new set of logic conditions. Table 22.2 represents

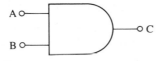

Figure 22.3   AND gate symbol.

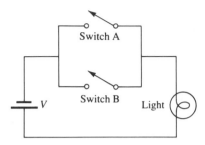

Figure 22.4

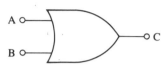

Figure 22.5   OR gate symbol.

this logic; it is the truth table for an OR gate. The term OR is derived from the fact that if either switch A *or* switch B is closed (LOGIC 1), the light is said to be in a state of LOGIC 1 (lit). The symbol of an OR gate is shown in Figure 22.5. An OR gate can have any number of inputs, but it can have only one output. The definition of an OR gate can be stated as

If any input is in a state of LOGIC ONE, the output is a LOGIC 1.

INVERTER Gate

An INVERTER is the last fundamental gate and also the simplest. As its name implies, it merely *inverts* the input. If the input is LOGIC 1, the output is LOGIC 0, or vice versa. The mechanical switch analogy that depicts an INVERTER is shown in Figure 22.6*a*. Note that unlike the AND or OR gate switch, the INVERTER switch is positioned to act as a *short circuit* across the light. When the switch is open (LOGIC 0 input), the light is lit (LOGIC 1 output). Conversely, when the switch is closed (LOGIC 1 input), the light is shorted out and not lit (LOGIC 0 output). The INVERTER symbol is shown in Figure 22.6*b*.

Note several points. First, there is only one input and one output to an INVERTER. Second, an INVERTER's only function is to alter (invert) the LOGIC state of its input. Third, the output has a bar drawn over the A. This convention has been adopted by the computer industry to denote a LOGIC state inverted or opposite to that of A. Fourth, the small circle at the output of the INVERTER logic triangle is the actual symbol denoting the inversion.

TABLE 22.2   TRUTH TABLE OF THE OR GATE

Switch status	Switch A	Switch B	Light
Both switches open	0	0	0
A open, B closed	0	1	1
A closed, B open	1	0	1
Both switches closed	1	1	1

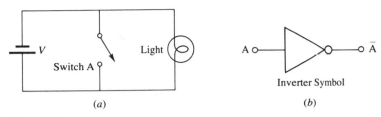

Figure 22.6

We stress this fourth point because when an INVERTER is drawn as an isolated gate, the entire symbol in Figure 22.6*b* is used. When an INVERTER is drawn as part of another gate, such as an AND or OR gate, only the circle portion of Figure 22.6*b* is used. Table 22.3 is the truth table of an inverter.

TABLE 22.3 INVERTER TRUTH TABLE

	Input A	Output A
Input A is LOGIC 0, output $\overline{A}$ is LOGIC 1	0	1
Input A is LOGIC 1, output $\overline{A}$ is LOGIC 0	1	0

The following two gates are combinations of the three basic gates. The simplest way to analyze their logic is to think of them as two separate gates.

NAND Gate

Figure 22.7 depicts the logic symbol of a NAND gate. The term NAND is a combination of the two words *n*egated *and*. In effect, it is nothing more than the output of an AND gate, which in turn is then run through an INVERTER. Notice that symbolically it is the AND gate connected to the INVERTER circle. Only when both (all) inputs are LOGIC 1 is the output LOGIC 0. Under any other input conditions, the output is LOGIC 1. Table 22.4 is the truth table of a NAND gate.

NOR Gate

Figure 22.8 depicts the logic symbol of a NOR gate. The term NOR is a combination of the two words *n*egated *or*. In effect, it is nothing more than the output of an OR gate, which is then run through an INVERTER. Notice that symbolically it is an OR

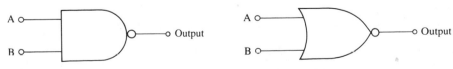

Figure 22.7 NAND gate symbol.   **Figure 22.8** NOR gate symbol.

TABLE 22.4   TRUTH TABLE OF THE NAND GATE

	Input A	Input B	Output
A is LOGIC 0, B is LOGIC 0	0	0	1
A is LOGIC 0, B is LOGIC 1	0	1	1
A is LOGIC 1, B is LOGIC 0	1	0	1
A is LOGIC 1, B is LOGIC 1	1	1	0

gate connected to the INVERTER circle. Only when all inputs are LOGIC 0 is the output LOGIC 1. Under all other conditions, the output is LOGIC 0. Table 22.5 is the truth table of a NOR gate.

As in the case of the basic AND or OR gates, the inputs of the NAND and NOR gates can have more than two inputs. However, their truth table logic remains the same.

PROBLEMS

In Problems 1 to 5, assume an AND, OR, NAND, and NOR gate each have three inputs, respectively referred to as A, B, and C. The output is D. For the inputs shown, what is D for each of the four gates?

	A	B	C
1.	0	1	0
2.	1	0	1
3.	1	1	0
4.	0	0	0
5.	0	1	1

TABLE 22.5   TRUTH TABLE FOR THE NOR GATE

	Input A	Input B	Output
A is LOGIC 0, B is LOGIC 0	0	0	1
A is LOGIC 0, B is LOGIC 1	0	1	0
A is LOGIC 1, B is LOGIC 0	1	0	0
A is LOGIC 1, B is LOGIC 1	1	1	0

22.2 CONVERTING LOGIC DIAGRAMS TO BOOLEAN EQUATIONS

Recall that certain algebraic processes, such as factoring, make it possible to reduce complex expressions into simpler expressions. For example,

$$\frac{x^2 + 5x + 6}{x + 3} = \frac{(x + 2)(x + 3)}{(x + 3)} = x + 2$$

In a similar manner, it is possible to reduce complex logic diagrams into simpler logic diagrams; both types of diagrams are capable of performing the same functions. To facilitate this transformation, it is necessary to reduce a logic diagram to an algebraic expression, perform the appropriate simplification processes, and then reconstruct simpler logic diagram from the reduced algebraic expression. This and the following sections illustrate these collective processes.

We stated that logic diagrams can be reduced to algebraic expressions. This algebra is referred to as *Boolean* algebra, the algebra of logic. It somewhat resembles standard algebra, but it has its own unique properties. Let us discuss the Boolean signs of the AND, OR, INVERTER, NAND, and NOR gates.

AND Sign

The AND sign is the same as the multiplication sign of standard algebra. Figure 22.3 can be *written* as

$$C = AB \qquad \text{or} \qquad C = A \cdot B \tag{22.1}$$

Either AND sign is valid, although the absence of the *dot* is usually the preferred version only because it is just one less item to write. If there are more than two inputs, such as A, B, and C, and D is the output, then

$$D = ABC \tag{22.1a}$$

OR Sign

The OR sign is the same as the plus sign of standard algebra. Figure 22.5 can be *written* as

$$C = A + B \tag{22.2}$$

If there is a three-input OR gate, such as inputs A, B, and C, and D is the output, then

$$D = A + B + C \tag{22.2a}$$

INVERTER Sign

We already discussed the INVERTER sign; it is shown in Figure 22.6*b*. If A is the input, then the output is $\overline{A}$. Note that the sign or symbol of the INVERTER is the *bar* over the variable.

NAND Sign

The NAND sign is a combination of the AND Equation (22.1) and the *bar*. Figure 22.7 is represented as

$$C = \overline{AB} \qquad\qquad (22.3)$$

NOR Sign

The NOR sign is a combination of the OR Equation (22.2) and the *bar*. Figure 22.8 is represented as

$$C = \overline{A + B} \qquad\qquad (22.4)$$

Each of the preceding five discussions shows the sign of an individual gate. When a collection of gates are combined, as they are in most practical cases, the resultant output of each gate is carried along (in a left-to-right fashion) as the input into the next sequential gate, until the final resultant is obtained. Now we illustrate this process, although we make no attempt at this time to simplify the final result.

Consider Figure 22.9. Note that gate 1 is an AND gate, with A and B as inputs. Gate 2 is also an AND gate, but its inputs are the *invert* of C and the output of gate 1. The total picture is further complicated by both gate 1's and gate 2's output serving as inputs to gate 3 (an OR gate). Just looking at such a picture in total makes it difficult to express output D. But if we analyze individually and carry along the results step by step, we can quickly express output D, as we now demonstrate.

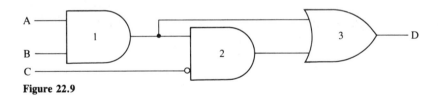

Figure 22.9

First, label the output of gate 1, carrying its output up to every other gate that it influences. Figure 22.10 results. Next, label all inputs and outputs to gate 2, as in Figure 22.11*a*. Note that it is convenient to label C's input within the gate diagram itself because space does not permit otherwise. During the process of redrawing a

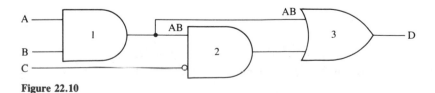

Figure 22.10

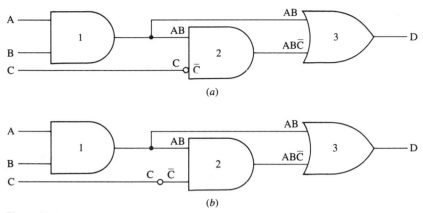

Figure 22.11

diagram for analysis it is sometimes convenient to separate the circle negating C from gate 2 and then label both sides of the circle, as in Figure 22.11*b*. In either case you get the same results. Draw the output of gate 2 as the input to gate 3.

As a final step, the two inputs to gate 3 may be ORed, resulting in the Boolean expression for D. The resultant output is shown in Figure 22.12.

This approach undertakes just one gate at a time, analyzing each output before progressing to the next stage.

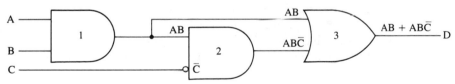

Figure 22.12

PROBLEMS

Develop the Boolean equation for the output of the figures in Problems 1 to 4. Use the method outlined in previous discussion.

1.

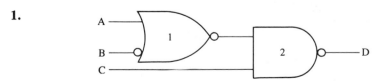

2.

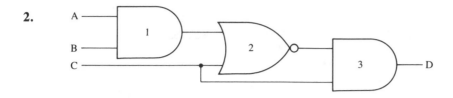

3.

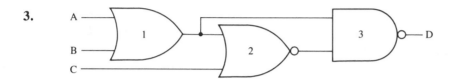

4.

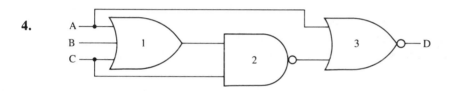

22.3 CONVERTING BOOLEAN EQUATIONS TO LOGIC DIAGRAMS

Section 22.2 related the Boolean algebraic expression to each of the five major gates. Most complex logic is made up of many such gates, all intricately connected to achieve the desired result. Outputs of a given gate usually serve as the inputs of still other gates, which in turn are inputs to still more gates, and so on until a subsystem or even a complete system is formed. For example, the expression

$$AB + C = D$$

says that there are at least two gates involved in the formation of D. AB represents two inputs to an AND gate. The output of this AND, combined with input C through an OR gate, produces the resultant D. Figure 22.13 is the logic diagram of $AB + C = D$.

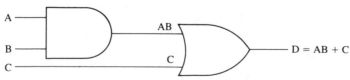

Figure 22.13   Logic diagram of $AB + C = D$.

You simply observe the gates indicated by the equation, draw the gates one at a time, and then interconnect each gate. Having done that, what do you now have? At this point we have a situation akin to putting the horse before the cart. We have simply written an arbitrary equation and drawn its diagram; now we analyze the conditions under which its output is HIGH (or LOW). Although this is approaching practical problems in reverse, the methodology of this analysis is the requisite foundation for subsequent sections in which we first specify the problem and then structure the logic that answers the problem. Let us proceed with seeing what we have; we begin by structuring a truth table.

The truth table associated with this diagram must involve three input variables: A, B, and C. There are therefore eight possible input combinations. An easy way to determine this number is to raise the number 2 to a power equal to the number of variables. As in this example, there are three input variables—A, B, and C—so 2^3 equals 8 possible input combinations.

First, construct all eight combinations. One simple method is to consider A as the most significant digit (MSD) and C as the least significant digit (LSD). Then write the decimal numbers 0 through 7 in binary, as follows:

Decimal	Binary (MSD) A	B	(LSD) C
0	0	0	0
1	0	0	1
2	0	1	0
3	0	1	1
4	1	0	0
5	1	0	1
6	1	1	0
7	1	1	1

Second, structure the function AB by ANDing the eight combinations of A and B:

Decimal	A	B	C	AB
0	0	0	0	0
1	0	0	1	0
2	0	1	0	0
3	0	1	1	0
4	1	0	0	0
5	1	0	1	0
6	1	1	0	1
7	1	1	1	1

Third, OR the function AB with C:

Decimal	A	B	C	AB	AB + C = D
0	0	0	0	0	0
1	0	0	1	0	1
2	0	1	0	0	0
3	0	1	1	0	1
4	1	0	0	0	0
5	1	0	1	0	1
6	1	1	0	1	1
7	1	1	1	1	1

We have now determined the resultant D; its status is known for all possible input conditions of A, B, and C. If we are interested in when D is high, we can state that this occurs when the collective inputs represent a decimal 1, 3, 5, 6, and 7.

As another example, consider the Boolean equation $(\overline{AB})$ C = D. Note that AB forms an AND gate, but note further that this is inverted—this output is really a NAND gate. The output of the NAND is now ANDed with C. Figure 22.14 is the logic diagram and the resultant truth table for D. If we again consider A as the MSD and C as the LSD, then D is HIGH when the inputs represent a decimal 1, 3, and 5.

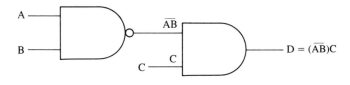

Decimal	A	B	C	$\overline{AB}$	$(\overline{AB})$C = D
0	0	0	0	1	0
1	0	0	1	1	1
2	0	1	0	1	0
3	0	1	1	1	1
4	1	0	0	1	0
5	1	0	1	1	1
6	1	1	0	0	0
7	1	1	1	0	0

Figure 22.14   Format for structure of $(\overline{AB})$C = D.

PROBLEMS

Convert the Boolean equations in Problems 1 to 8 into logic diagrams. If A is the MSD and C the LSD, define the decimal number when output D is HIGH.

1. $D = (A + \overline{B})C$ **3.** $D = \overline{\overline{AB}\,C}$ **5.** $D = \overline{A(B + \overline{C})}$ **7.** $D = \overline{A + \overline{B}} + \overline{AC}$

2. $D = \overline{\overline{A}B} + C$ **4.** $D = \overline{A + \overline{B} + C}$ **6.** $D = A\overline{B} + \overline{A}B$ **8.** $D = A + \overline{BC} + \overline{AB}$

22.4 BOOLEAN AND DE MORGAN'S THEOREMS

In geometry or algebra, rules exist that govern the manipulation of equations. These rules make solving equations easier, but you must thoroughly understand them to obtain the correct results. So it is with Boolean algebra; there is a series of rules, called postulates and theorems, that govern the simplification and manipulation of Boolean equations.

One characteristic you will soon notice is that sometimes there is great similarity between the rules of mathematical algebra and Boolean algebra, but other times, depending upon the nature of the equation, this similarity does not exist. It is therefore important to distinguish when this similarity exists and when it is nonexistent. Consider the three postulate laws presented in the following three tables.

COMMUTATIVE LAW

Mathematical Algebra[a]	Boolean Algebra
A plus B = B plus A A + B = B + A	A or B = B or A A + B = B + A } OR gate
A times B = B times A A · B = B · A	A and B = B and A A · B = B · A } AND gate

[a]Both forms, mathematical algebra and Boolean algebra, are identical.

ASSOCIATIVE LAW

Mathematical Algebra[a]	Boolean Algebra
(A plus B) plus C = A plus (B plus C) (A + B) + C = A + (B + C)	(A or B) or C = A or (B or C) (A + B) + C = A + (B + C) } OR gate
(A times B) times C = A times (B times C) (A · B) · C = A · (B · C)	(A and B) and C = A and (B and C) (A · B) · C = A · (B · C) } AND gate

[a]Both forms, mathematical algebra and Boolean algebra, are identical.

DISTRIBUTIVE LAW

Mathematical Algebra[a]	Boolean Algebra
Case 1[a]	
$A(B + C) = AB + AC$	$A(B + C) = AB + AC$
Case 2[b]	
$(A + B)(A + C) = A^2 + AB + AC + BC$	$(A + B)(A + C) = A + BC$

[a]Both forms, mathematical algebra and Boolean algebra, are identical.
[b]The two forms, mathematical algebra and Boolean algebra, are not identical; the reason for this breakdown in identity will become apparent when the theorems are analyzed.

To summarize, let us emphasize the following four points:

1. The plus sign (+) in algebra is analogous to the OR sign (+) in Boolean algebra.
2. The times sign (·) in algebra is analogous to the AND sign (·) in Boolean algebra.
3. Some rules in mathematical algebra are identical to those rules used in manipulating Boolean algebra. In these cases they are equally applicable.
4. Those rules in mathematical algebra that are not valid in Boolean algebra are explained by application of the truth theorems.

The truth theorems, or Boolean theorems as they are frequently called, provide an analytical tool for logic-equation reduction. The Boolean theorems are as follows.

OR gate theorems

1. $A + A = A$

2. $A + \overline{A} = 1$

3. $A + 1 = 1$

4. $A + 0 = A$

AND gate theorems

5. $AA = A$

6. $A\overline{A} = 0$

7. $A1 = A$

8. $A0 = 0$

Special theorems

9. $A + \overline{A}B = A + B$

10. $\overline{A} + AB = \overline{A} + B$

11. $\overline{\overline{A}} = A$

Example 1 illustrates the analytical reduction of logic equations.

Example 1

Simplify the following equations by using Boolean reduction theorems.

a. A + AB

Factor out A:

A(1 + B)

From theorem 3, (1 + B) = 1, therefore

A1

From theorem 7, A1 = A, therefore

A

The answer is A + AB = A.

b. $ABC + A\overline{B}C$

Factor out AC:

$AC(B + \overline{B})$

From theorem 2, $(B + \overline{B}) = 1$, therefore

AC1

From theorem 7, AC1 = AC, therefore

AC

The answer is $ABC + A\overline{B}C = AC$.

c. $(A + \overline{A}C + \overline{C})B$

From theorem 9, $A + \overline{A}C = A + C$, therefore

$(A + C + \overline{C})B$

From theorem 2, $C + \overline{C} = 1$, therefore

$(A + 1)B$

From theorem 3, A + 1 = 1, therefore

1B

From theorem 7, 1B = B, therefore

B

The answer is $(A + \overline{A}C + \overline{C})B = B$.

Besides the generalized Boolean theorems, De Morgan's reduction theorem facilitates simplification of a series of logic terms with a common negation. Before investigating this theorem, however, let us review multiple negations over a single term because these principles apply to De Morgan's theorem.

In its simplest form, consider one variable, A, with two bars drawn above it:

$$\overline{\overline{A}}$$

This says that A is twice negated. Such a process returns A to its original state. For example, if A is originally LOGIC 1, the first inversion makes it a LOGIC 0 and the second inversion returns it to a LOGIC 1. We can apply a simple rule: An even number of negate bars over a quantity cancels their effects, so they can be eliminated:

$$\overline{\overline{A}} = A$$

If there are three bars, two cancel, with the following result:

$$\overline{\overline{\overline{A}}} = \overline{A}$$

Now consider $\overline{A + B}$. First, there are two variables, A and B. Second, they are commonly negated. As a matter of fact, the expression is nothing more than a NOR gate. De Morgan's theorem says that the common negate bar can be eliminated if:

1.   Each variable term, as shown, is negated.
2.   Each operational sign is changed; that is, each OR sign is changed to an AND sign and each AND sign is changed to an OR sign.

When these two rules are applied, the following results:

$$\overline{A + B} = \overline{A} \cdot \overline{B} \qquad \text{and} \qquad \overline{AB} = \overline{A} + \overline{B}$$

This application provides for the interchangeability of gates. In the first example, three gates (an INVERTER for A, an INVERTER for B, and an AND gate) can be replaced with a single NOR gate. In the second example, three gates (an INVERTER for A, an INVERTER for B, and an OR gate) can be replaced with a single NAND gate. This is all part of a simplification process that is analyzed more thoroughly in subsequent sections.

An another example, consider the expression

$$\overline{(A + \overline{B}) \cdot C}$$

To apply the rules of De Morgan, first consider the OR gate within the parentheses:

$$\overline{A + \overline{B}} = \overline{A} \cdot \overline{\overline{B}}$$

Next, variable C is merely negated to become $\overline{C}$. Now, because the OR function of $(A + \overline{B})$ was ANDed with C, the AND (dot) function must be changed to an OR (plus) function. Combining all the previous steps,

$$\overline{(A + \overline{B}) \cdot C} = (\overline{A} \cdot B) + \overline{C}$$

An alternate method for De Morganizing multiple-variable Boolean expressions is to consider how many negate bars are above each element of the expression. For the example of

$$\overline{(A + B) \cdot C}$$

A has one bar over it, the OR symbol has one bar over it, B has two bars over it, the AND symbol has one bar over it, and C has one bar over it. When broken down into its components, the following representation of the expression can be written as

$$\overline{(\overline{A} + \overline{\overline{B}}) \cdot \overline{C}}$$

Because an even number of bars over an element cancels each other and an odd number of bars over an element negates that variable and/or operational function, the previous expression reduces to

$$(\overline{A} \cdot B) + \overline{C}$$

Note that the parentheses are retained and transpose into the new expression.

Example 2

De Morganize the following.

a. $\overline{\overline{\overline{ABC}}}$

   If we consider $\overline{AB} = Y$, then our expression becomes $\overline{\overline{YC}}$. This easily reduces to $Y + \overline{C}$ when applying De Morgan's rules. Substituting the value of Y results in

$$\overline{\overline{\overline{ABC}}} = Y + \overline{C} = \overline{AB} + \overline{C}$$

b. $\overline{\overline{\overline{A\overline{B}C}}}$

   If we consider $A\overline{B}$ equal to Y, then our expression becomes $\overline{\overline{YC}}$. This easily reduces to $Y + \overline{C}$ when applying De Morgan's rules. Substituting the value of Y results in

$$\overline{\overline{\overline{A\overline{B}C}}} = Y + \overline{C} = A\overline{B} + \overline{C}$$

PROBLEMS

In Problems 1 to 11, simplify the Boolean expressions.

1. $\overline{\overline{AB}} + C$

2. $AB + \overline{\overline{AB}} + C$

3. $\overline{(A + \overline{BC})B}$

4. $(A + \overline{B})(\overline{A} + B)$

5. $(A + B + C)(A + B + \overline{C})$

6. $\overline{\overline{\overline{ABCD}}}$

7. $\overline{\overline{A + \overline{BC}}}$

8. $\overline{(AB + C)C}$

9. $\overline{\overline{\overline{ABCDE}}}$

10. $\overline{((A + B) + \overline{\overline{C}}))(A + B)}$

11. $\overline{\overline{(A + B + C)\overline{C} + A}}$

22.5 LOGIC DESIGN PROCEDURES

In the previous sections you learned the fundamentals necessary to prepare yourself for applying this knowledge to the design of digital logic. This is analogous to a pianist's practicing chords day after day until the point is reached where the pianist can apply the knowledge and create and play music. We are now ready to solve specific problems; we start by discussing the principles involved.

Suppose a logic network has two inputs. We want to design a logic circuit whose output is HIGH (LOGIC 1) if both inputs are identical and whose output is LOW (LOGIC 0) if both inputs are different. Figure 22.15a illustrates the problem: We are to design the necessary interconnection of gates to produce the output C, under the conditions specified by the problem, given A and B as inputs. This example illustrates the procedural steps necessary to accomplish this goal.

First, list A and B and write all possible combinations under which they can exist. Second, write a 1 for output C when the conditions of the problem specified are met and a 0 for output C when the conditions are not met. Figure 22.15b lists the procedure to this point. Note that C is listed as 1 when A and B are both 0 and again when A and B are both 1. This complies with the conditions that A and B be identical. Under all other conditions, when A and B are different, C is listed as 0.

Third, write the conditions under which C exists; that is, write the conditions under which C has a state of LOGIC 1. Looking at Figure 22.15b, it is clear that C is a LOGIC 1 when A is negated and simultaneously when B is negated. Also observe that C is a LOGIC 1 when A is 1 and simultaneously when B is 1. Written in logic form,

$$C = \overline{A}\,\overline{B} + AB \qquad (22.5)$$

Fourth, construct the Boolean logic described by the statement identifying the conditions of C. Figure 22.15c represents this translation.

If you follow this procedure, you can stipulate any given set of conditions and then design and fabricate the circuitry necessary to accomplish the required result. Note that some simplification procedures, specifically De Morgan's theorem, can be used to reduce the total number of gates (two INVERTERs, two ANDs, and an OR) in Figure 22.15c. Because $\overline{A + B}$ is equal to $\overline{A}\,\overline{B}$, we can rewrite Equation (22.5) as

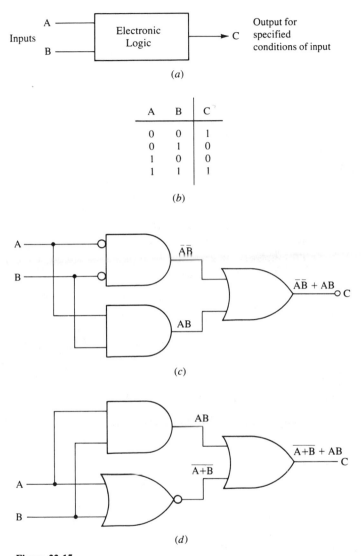

A	B	C
0	0	1
0	1	0
1	0	0
1	1	1

(b)

Figure 22.15

$C = \overline{(A + B)} + AB$. When this new diagram is constructed, Figure 22.15d results. Note that both diagrams produce the same result.

As another example, design a circuit to add two binary digits. This circuit is slightly more complex. There are two inputs. There are also two outputs, a SUM and a CARRY (which goes to the next higher-order position if a string of binary bits are added in parallel).

A	B	Sum	Carry
0	0	0	0
0	1	1	0
1	0	1	0
1	1	0	1

(a)

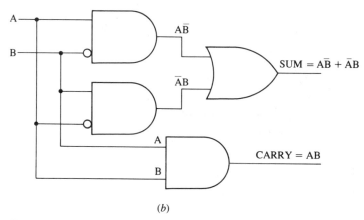

(b)

Figure 22.16

Step 1. Develop the truth table for the problem. See Figure 22.16a.

Step 2. Develop the Boolean equation for the SUM and the CARRY:

$$\text{SUM} = \overline{A}B + A\overline{B}$$
$$\text{CARRY} = AB$$

Step 3. Draw the diagram corresponding to step 2 (see Figure 22.16b).

We made a point of illustrating this "simple adder," often referred to as a *half adder* because it is required to add only 2 bits. More complex adder circuits, called *full adders*, are required to add the two original bits plus the carryover from the adjacent lower-order position. Incidently, the SUM expression $\overline{A}B + A\overline{B}$ occurs so frequently in logic that it is given its own name: the EXCLUSIVE OR. This *special* OR function is symbolized in Figure 22.17. As long as you follow the three procedural steps outlined, you can design any logic function.

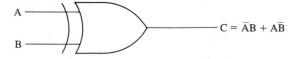

Figure 22.17 EXCLUSIVE OR symbol.

PROBLEMS

Design the necessary logic to perform the tasks in Problems 1 to 4.

1. A three-input circuit producing an output when two and only two inputs are HIGH.

2. A three-input circuit producing an output when two and only two inputs are identical.

3. A three-input circuit producing an output if A and B are different and C is simultaneously HIGH.

4. A three-input circuit producing an output if A and C are different and B is simultaneously LOW.

SUMMARY

There are only three basic logic gates:

1. *AND gate*.   All inputs must be HIGH for the output to be HIGH.
2. *OR gate*.   Any HIGH input produces a HIGH output.
3. *INVERTER gate*.   Changes a HIGH input to a LOW output or a LOW input into a HIGH output.

Two other gates are a combination of the three basic gates:

4. *NAND gate*.   An AND gate that is subsequently INVERTED. If all inputs are HIGH, the output is LOW; otherwise the output is HIGH.
5. *NOR gate*.   An OR gate that is subsequently INVERTED. If all inputs are LOW, the output is HIGH; otherwise the output is LOW.

Using the commutative, associative, and distributive laws, it is possible to convert logic diagrams into Boolean equations or vice versa. However, care must be exercised because although there is some similarity between the mathematical laws of algebra and logic laws, there are also differences caused by the simplification processes of the logic laws.

Any logic circuit can be designed by following a three-step process:

1. Develop a truth table that defines the parameters of the required design.
2. Develop the resultant, simplified Boolean equation.
3. Construct the logic gates defined by the Boolean equation developed in step 2.

REVIEW PROBLEMS, PART V

The following problems are a review of Part V of the text and directly relate to the numbered sections.

Section 21.2, Binary to Decimal

Convert the following binary numbers into their decimal equivalents.

1. 101101

2. 11101101

3. 0.101

4. 0.1101

5. 1110101.101

Section 21.2, Decimal to Binary

Convert the following decimal numbers into binary.

6. 53

7. 125

8. 85.5

9. 781.1875

10. 304.5625

Section 21.3, Decimal to Base N

Convert the following decimal numbers into the base indicated. Do not carry fractional numbers past the fourth decimal place.

11. $5124 = ?_8$

12. $245.621 = ?_8$

13. $4120 = ?_{16}$

14. $541.15 = ?_{16}$

15. $398 = ?_5$

Section 21.3, Base N to Decimal

Use your calculator to convert the following indicated bases into the equivalent decimal number.

16. 754_8

17. 6561_8

18. $3DF_{16}$

19. 321.54_8

20. $C13.A4_{16}$

Section 21.3, Base N to Base M

Make the indicated conversions.

21. $1100110110_2 = ?_8 = ?_{16}$

22. $10011110111.001 = ?_8 = ?_{16}$

23. $5421_8 = ?_2 = ?_{16}$

24. $2A24_{16} = ?_2 = ?_8$

25. $2C1.B_{16} = ?_2 = ?_8$

Section 21.4

For Review Problems 26 to 31, perform the indicated operations by direct application of Tables 21.2 to 21.6. Do not use your calculators.

26. $1111011_2 + 1001101_2 = ?_2$

28. $2DB_{16} - 436_8 = ?_8$

30. $431_8 \times 124_8 = ?_8$

27. $11101011_2 + 536_8 = ?_2$

29. $1011_2 \times 111_2 = ?_2$

31. $327_8 \div 53_8 = ?_8$

Using your calculator, perform the indicated operations on the mixed bases of Review Problems 32 to 35.

32. $431_6 + 516_8 = ?_{10}$

34. $435_6 \times 241_5 = ?_8$

33. $1352_6 + 227_8 = ?_{10}$

35. $721_8 - 521_7 = ?_5$

Section 22.1

Assume an AND, OR, NAND, and NOR gate each have four inputs, respectively referred to as A, B, C, and D. The output is F. For the inputs shown, what is F for each gate?

	A	B	C	D
36.	0	0	0	0
37.	0	1	0	0
38.	0	1	0	1
39.	1	1	1	0
40.	1	1	1	1

Section 22.2

Develop the Boolean equation for the output of the figures in Review Problems 41 to 43. Use the methods discussed in Section 22.2.

41.

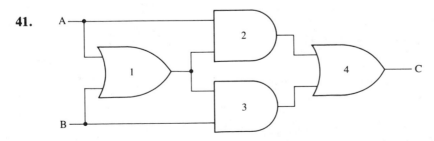

42.

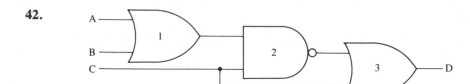

43.

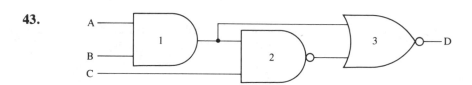

Section 22.3

Convert the Boolean equations in Review Problems 44 to 47 into logic diagrams. If A is the MSD and C is the LSD, define the decimal numbers when output D is high.

44. $\overline{A(\overline{B} + C)} = D$

46. $\overline{AB + \overline{AB} + BC} = D$

45. $\overline{(A + BC) \cdot C} = D$

47. $\overline{(A + \overline{BC})C} = D$

Section 22.4

Simplify the following Boolean expressions.

48. $AB + A\overline{B}$

50. $\overline{(A + B) \cdot \overline{(A + \overline{B})}}$

52. $\overline{(\overline{AB} + C)B}$

49. $(A + B)(A + \overline{B})$

51. $A + \overline{AB} + \overline{A}\,\overline{B}$

Section 22.5

In Review Problems 53 to 55, design the necessary logic to perform the indicated tasks.

53. A three-input circuit producing a HIGH output if all inputs are identical.

54. A three-input circuit producing a HIGH output if only two inputs are identical.

55. A three-input circuit producing a HIGH output if A and C are identical when B is LOW.

appendix _____

Solution of Calculator Examples Using the RPN Method[*]

Section 1.5: Example 1

PRESS	DISPLAY
a. 9.34 ENTER 1.6 CHS × 3.2 ÷	−4.67
b. 3.2 ENTER 1.65 CHS + 13.6 CHS × 2 +	−19.08
c. 3 ENTER 5 + 7 × 42 + 6 ENTER 3 ÷ 5 + 7	
× ÷	2
d. 37.6 ENTER 4.19 ÷ 16.3 ENTER 1.6 CHS ÷ 7.8 + ×	−21.425
e. .2 ENTER 2 + 2 × .2 × 2 ×	1.76
f. 6.3 ENTER 1.7 ÷ 2.1 + 3.21 ENTER 1.6 ENTER 2.3	
× ÷ ÷	6.656

Section 2.2: Example 1

PRESS	DISPLAY
a. 3 ENTER 7 ÷ 6 1/x − 3 ENTER 35 ÷ +	0.3476
7 ENTER 2 × 3 × 5 ×	210
×	73
b. 2 ENTER 21 ÷ 7 ENTER 9 ÷ + 3 ENTER 45 ÷ −	0.8063
3 ENTER 3 × 5 × 7 ×	315
×	254
c. 7 ENTER 32 ÷ 23 ENTER 24 ÷ + 7 ENTER 8 ÷ −	0.3021
2 ENTER 2 × 2 × 2 × 2 × 3 ×	96
×	29

---

*The HP31E keyboard was used.

Section 2.3: Example 1

PRESS	DISPLAY
a. 6 [1/x] [ENTER] 13 [1/x] [+] 8 [1/x] [−]	0.1186
b. 8 [1/x] [ENTER] 12 [1/x] [−]	0.0417
2 [×]	0.0833
[1/x]	12.0000
c. 8 [1/x] [ENTER]	0.1250
3 [1/x] [ENTER] 4 [1/x] [+] 7 [1/x] [−]	0.4405
[÷]	0.2838
d. 12 [1/x] [ENTER] 8 [1/x] [+]	0.2083
6 [1/x] [ENTER] 15 [1/x] [−] 2 [×]	0.2000
[÷]	1.0417
e. 3300 [1/x] [ENTER] 5600 [1/x] [+] 4700 [1/x] [+] 6800 [1/x]	
[+] [1/x]	1188.4579

Section 2.7: Example 1

PRESS	DISPLAY
a. 5 [ENTER] 8 [÷] 2 [+]	2.6250
9 [ENTER] 11 [÷] 1 [+]	1.8182
[÷]	1.4438
b. 7 [ENTER] 8 [÷] 1 [+]	1.8750
5 [ENTER] 6 [÷] 2 [+]	2.8333
[×]	5.3125
3 [1/x] [ENTER] 4 [+]	4.3333
[÷]	1.2260
c. 5 [CHS] [ENTER] 6 [÷]	−0.8333
3 [1/x] [ENTER] 4 [+]	4.3333
[×]	−3.6111
2 [ENTER] 5 [÷]	0.4000
3 [CHS] [ENTER] 4 [÷]	−0.7500
[×]	−0.3000
[÷]	12.0370
d. 2 [ENTER] 5 [÷] 2 [+] 7 [ENTER] 8 [÷] 1 [+] [−] 3.17 [×]	1.6643
7 [ENTER] 9 [÷] 3 [+] [÷]	0.4405
5 [1/x] [ENTER] 3 [+] [−]	−2.7595

PRESS	DISPLAY
e. 9 $\boxed{1/x}$ $\boxed{\text{ENTER}}$ 4 $\boxed{+}$ 3 $\boxed{1/x}$ $\boxed{\text{ENTER}}$ 3 $\boxed{+}$ $\boxed{\times}$	13.7037
2 $\boxed{1/x}$ $\boxed{\text{ENTER}}$ 2 $\boxed{+}$ 3 $\boxed{1/x}$ $\boxed{\text{ENTER}}$ 1 $\boxed{+}$ $\boxed{-}$	1.6667
$\boxed{\times}$	15.9877
6 $\boxed{1/x}$ $\boxed{\text{ENTER}}$ 5 $\boxed{+}$ 7 $\boxed{1/x}$ $\boxed{\text{ENTER}}$ 3 $\boxed{+}$ $\boxed{+}$	8.3095
$\boxed{\div}$	1.9240

Section 3.5: Example 1

PRESS	DISPLAY
a. 3.2 $\boxed{\text{ENTER}}$ 1.6 $\boxed{y^x}$	6.4304
b. 1.84 $\boxed{\text{ENTER}}$ 3 $\boxed{1/x}$ $\boxed{y^x}$	1.2254
c. 3.84 $\boxed{\text{ENTER}}$ 2.61 $\boxed{\text{ENTER}}$ 4 $\boxed{\div}$ $\boxed{y^x}$	2.4059
d. 8.1 $\boxed{\text{ENTER}}$ 1.62 $\boxed{y^x}$ 3.4 $\boxed{\sqrt{}}$ $\boxed{+}$	31.4749
e. 2.32 $\boxed{\text{ENTER}}$ 1.2 $\boxed{y^x}$ 2.64 $\boxed{\text{ENTER}}$ .87 $\boxed{y^x}$ $\boxed{-}$ 1.35 $\boxed{y^x}$	0.3083
f. 6.4 $\boxed{\text{ENTER}}$ 3 $\boxed{\text{ENTER}}$ 7 $\boxed{\div}$ $\boxed{y^x}$	2.2157
1.6 $\boxed{\text{ENTER}}$ 2 $\boxed{\text{ENTER}}$ 5 $\boxed{\div}$ $\boxed{y^x}$	1.2068
$\boxed{\div}$	1.8359
1.4 $\boxed{\text{ENTER}}$ 3 $\boxed{1/x}$ $\boxed{y^x}$	1.1187
3.7 $\boxed{\text{ENTER}}$ 4 $\boxed{1/x}$ $\boxed{y^x}$	1.3869
$\boxed{\div}$ $\boxed{-}$	1.0293

Section 4.4: Example 1

PRESS	DISPLAY	
a. 5.879 $\boxed{\text{EEX}}$ 12	5.879	12
b. 324 $\boxed{\text{CHS}}$ $\boxed{\text{EEX}}$ 14	-32.4	14
c. 1.6 $\boxed{\text{EEX}}$ $\boxed{\text{CHS}}$ 19	1.6	-19
d. 1.6 $\boxed{\text{EEX}}$ 19 $\boxed{\text{CHS}}$	1.6	-19

Section 4.4: Example 2

PRESS	DISPLAY	
a. 8.9 $\boxed{\text{ENTER}}$ 6.02 $\boxed{\text{EEX}}$ 23 $\boxed{\times}$	5.3578	24
63.5 $\boxed{\div}$	8.4375	22

PRESS   **DISPLAY**

b. 9 $\boxed{\text{EEX}}$ 9 $\boxed{\text{ENTER}}$ 1.6 $\boxed{\text{EEX}}$ 19 $\boxed{\text{CHS}}$ $\boxed{\text{ENTER}}$ 2 $\boxed{y^x}$ $\boxed{\times}$ 2.3040 -28

.529 $\boxed{\text{EEX}}$ 10 $\boxed{\text{CHS}}$ $\boxed{\text{ENTER}}$ 2 $\boxed{y^x}$ $\boxed{\div}$ 8.2332 -08

c. 9 $\boxed{\text{EEX}}$ 9 $\boxed{\text{ENTER}}$ 125 $\boxed{\text{EEX}}$ 9 $\boxed{\text{CHS}}$ $\boxed{\div}$ $\boxed{\sqrt{}}$ 1.6 $\boxed{\text{EEX}}$

19 $\boxed{\text{CHS}}$ $\boxed{\times}$ 4.2933 -11

Section 4.4: Example 3

Same as Section 4.4: Example 2, except depress the $\boxed{\text{ENG}}$ (engineering notation key) at the end of the sequence. Note that the HP31E does not have an $\boxed{\text{ENG}}$ key, therefore this function must be done manually.

Section 5.4: Example 1

5.4: Example 1a $(3a + 2b) + (a - 4b) = 4a - 2b$

PRESS   **DISPLAY**

3 $\boxed{\text{ENTER}}$ $\boxed{\text{RCL 1}}$ $\boxed{\times}$ 2 $\boxed{\text{ENTER}}$ $\boxed{\text{RCL 2}}$ $\boxed{\times}$ $\boxed{+}$ 4 $\boxed{\text{ENTER}}$

$\boxed{\text{RCL 2}}$ $\boxed{\times}$ $\boxed{\text{CHS}}$ $\boxed{\text{RCL 1}}$ $\boxed{+}$ $\boxed{+}$ 2.0000

$\boxed{\text{CLX}}$ 0.0000

4 $\boxed{\text{ENTER}}$ $\boxed{\text{RCL 1}}$ $\boxed{\times}$ 2 $\boxed{\text{ENTER}}$ $\boxed{\text{RCL 2}}$ $\boxed{\times}$ $\boxed{-}$ 2.0000

5.3: Example 1c $-(2p + 3.1q - 2.7r) + (4.1p + r) = 2.1p - 3.1q + 3.7r$

PRESS   **DISPLAY**

2 $\boxed{\text{ENTER}}$ $\boxed{\text{RCL 1}}$ $\boxed{\times}$ 3.1 $\boxed{\text{ENTER}}$ $\boxed{\text{RCL 2}}$ $\boxed{\times}$ $\boxed{+}$ 2.7

$\boxed{\text{ENTER}}$ $\boxed{\text{RCL 3}}$ $\boxed{\times}$ $\boxed{-}$ $\boxed{\text{CHS}}$ -2.4750

4.1 $\boxed{\text{ENTER}}$ $\boxed{\text{RCL 1}}$ $\boxed{\times}$ $\boxed{\text{RCL 3}}$ $\boxed{+}$ 10.4250

$\boxed{+}$ 7.9500

$\boxed{\text{CLX}}$ 0.0000

2.1 $\boxed{\text{ENTER}}$ $\boxed{\text{RCL 1}}$ $\boxed{\times}$ 3.1 $\boxed{\text{ENTER}}$ $\boxed{\text{RCL 2}}$ $\boxed{\times}$ $\boxed{-}$ 3.7

$\boxed{\text{ENTER}}$ $\boxed{\text{RCL 3}}$ $\boxed{\times}$ $\boxed{+}$ 7.9500

5.3: Example 1e $-\left(2\frac{1}{5}g - 3\frac{1}{4}k\right) - \left(-1\frac{1}{2}g + 2\frac{1}{5}k\right) = -\frac{7}{10}g + \frac{21}{20}h$

PRESS	DISPLAY
2 [ENTER] 5 [1/x] [+] [RCL 1] [×] 3 [ENTER] 4 [1/x] [+]	
[RCL 2] [×] [−] [CHS]	4.2750
1 [ENTER] 2 [1/x] [+] [CHS] [RCL 1] [×] 2 [ENTER] 5 [1/x] [+]	
[RCL 2] [×] [+]	2.8750
[−]	1.4000
[CLX]	0.0000
7 [ENTER] 10 [÷] [RCL 1] [×] [CHS]	−1.2250
21 [ENTER] 20 [÷] [RCL 2] [×] [+]	1.4000

5.3: Example 2c $X - (Y + Z - (2X + Y) + 3Z) = 3X - 4Z$

PRESS	DISPLAY
2 [ENTER] [RCL 1] [×] [RCL 2] [+] [CHS] 3 [ENTER] [RCL 3]	
[×] [+] [RCL 2] [+] [RCL 3] [+] [CHS] [RCL 1] [+]	−7.7500
[CLX]	0.0000
3 [ENTER] [RCL 1] [×] 4 [ENTER] [RCL 3] [×] [−]	−7.7500

5.3: Example 2d $U - (V - W - (2U - (2V - 2W))) = 3U - 3V + 3W$

PRESS	DISPLAY
2 [ENTER] [RCL 2] [×] 2 [ENTER] [RCL 3] [×] [−] [CHS] 2	
[ENTER] [RCL 1] [×] [+] [CHS] [RCL 3] [−] [RCL 2] [+]	
[CHS] [RCL 1] [+]	7.5000
[CLX]	0.0000
3 [ENTER] [RCL 1] [×] 3 [ENTER] [RCL 2] [×] [−] 3 [ENTER]	
[RCL 3] [×] [+]	7.5000

Section 5.7: Example 1

PRESS	DISPLAY
a. 3.25 [ENTER] [RCL 1] [×] [RCL 2] [1/x] [×] [RCL 3] [ENTER]	
2 [y^x] [×] 1.6 [ENTER] [RCL 1] [ENTER] 2 [y^x] [×] [RCL 2]	
[ENTER] 3 [CHS] [y^x] [×] [RCL 3] [×] [÷]	21.59
[CLX]	0.00
2.03125 [ENTER] [RCL 1] [1/x] [×] [RCL 2] [ENTER] 2 [y^x] [×]	
[RCL 3] [×]	21.59

PRESS	DISPLAY

b. 3.2 [ENTER] [RCL 1] [×] [RCL 2] [×]   10.80

   1.7 [ENTER] [RCL 1] [ENTER] 2 [y^x] [×] [RCL 2] [ENTER] 3 [CHS]

   [y^x] [×]   0.34

   [×] 2.5 [×] [RCL 3] [ENTER] 2 [y^x] [×] [RCL 2] [×]   202.42

   [CLX]   0.00

   13.6 [ENTER] [RCL 1] [ENTER] 3 [y^x] [×] [RCL 2] [$1/x$] [×]

   [RCL 3] [ENTER] 2 [y^x] [×]   202.42

Section 6.1: Example 3

PRESS	DISPLAY

a. 21.3 [ENTER] 3 [$1/x$] [y^x] 17.5 [$\sqrt{}$] [×]   11.5961

b. 117.6 [ENTER] 5 [$1/x$] [y^x] 38.2 [$\sqrt{}$] [CHS] [×] 1 [EEX]

   3 [$\sqrt{}$] [÷]   −0.5071

c. 121.6 [ENTER] 4 [$1/x$] [y^x] 1 [EEX] 4 [ENTER] 5 [$1/x$] [y^x] [÷]   0.5263

   3 [ENTER] 7 [÷] [y^x]   0.7595

d. 11863.24 [ENTER] 15 [$1/x$] [y^x] [CHS]   −1.8690

Section 6.4: Example 5

PRESS	DISPLAY

a. 2.1 [ENTER] 3 [y^x]   9.2610

   2.1 [ENTER] [×] 3 [×] 3.6 [CHS] [×]   −47.6280

   3.6 [ENTER] [×] 3 [×] 2.1 [×] [CHS]   81.6480

   3.6 [CHS] [ENTER] 3 [y^x]   −46.6560

b. π [$\sqrt{}$] [ENTER] 3 [y^x]   5.5683

   π [$\sqrt{}$] [ENTER] [×] 3 [×] 3.29 [CHS] [×]   −31.0075

   3.29 [CHS] [ENTER] [×] π [$\sqrt{}$] [×] 3 [×]   57.5557

   3.29 [CHS] 3 [y^x]   −35.6113

Section 7.6: Example 1

Since the RPN method does not use parentheses, only the second approach is shown.

PRESS	DISPLAY

π [ENTER] [$1/x$] 1 [+] [$1/x$] 1 [+] [$1/x$]   0.5687

Section 7.6: Example 2

PRESS	DISPLAY
a. RCL 0 ENTER 1 + 1/x 1 +	1.3175
RCL 0 CHS ENTER 1 + 1/x CHS 1 +	1.8696
÷	0.7047

To verify:

PRESS	DISPLAY
RCL 0 ENTER × RCL 0 + STO 2 2 −	4.7725
RCL 2 ÷	0.7047
b. RCL 0 ENTER RCL 1 + RCL 0 ENTER RCL 1 − ÷	−3.0476
STO 2	−3.0476
RCL 0 ENTER RCL 1 − RCL 0 ENTER	−0.3281
RCL 1 + ÷ STO 3	−0.3281
RCL 2 ENTER RCL 3 + RCL 2 ENTER RCL 3 − ÷	1.2413

To verify:

PRESS	DISPLAY
RCL 0 ENTER × RCL 1 ENTER × + RCL 0	
ENTER RCL 1 × 2 × ÷	1.2413

Section 9.2: Example 2

PRESS	DISPLAY
4.7 ENTER 1/x 5.6 1/x + 6.8 1/x + 22 ×	11.8447

Section 9.2: Example 3

PRESS	DISPLAY
27 ENTER 2 ÷ 20 ENTER 30 + 1/x 20 × 30 × −	1.5000

Section 9.2: Example 4

PRESS	DISPLAY
25 ENTER 1.25 ENTER 6.8 × − 33 ×	544.5000
6.8 ENTER 33 + 1.25 × 25 −	24.7500
÷	22.0000

Section 9.2: Example 5

PRESS	DISPLAY

2.2 ENTER .47 + 1/x 2.2 × .47 × 3.3 +   3.6873

1.5 1/x ENTER 4.7 1/x + 10 1/x + 1/x +   4.7083

1/x 45 ×   9.5577

Section 9.3: Example 2

PRESS	DISPLAY

6.8 ENTER 33 × 15 ÷ 33 + 6.8 +   54.7600

1/x 25 ×   0.4565

Section 9.3: Example 3

PRESS	DISPLAY

2.2 ENTER .47 + 1/x 2.2 × .47 × STO 0   0.3873

1.5 1/x ENTER 4.7 1/x + 10 1/x + 1/x STO 1   1.0210

3.3 ENTER RCL 0 + RCL 1 + 4.7 ×   22.1288

1/x ENTER RCL 1 × 45 ×   2.0762

Section 9.4: Example 4

PRESS	DISPLAY

2.2 ENTER .47 + 1/x 2.2 × .47 × STO 0   0.3873

1.5 ENTER 1/x 4.7 1/x + 10 1/x + 1/x STO 1   1.0210

3.3 ENTER RCL 0 + RCL 1 + 1/x 45 × STO 2   9.5577

1.5 ENTER 4.7 × 1.5 ENTER 10 × + 4.7 ENTER 10

× +   69.0500

1/x 1.5 × 4.7 × RCL 2 ×   0.9758

Section 9.5: Example 1

PRESS	DISPLAY

2.2 ENTER 5.6 × 2.2 ENTER 5.6 + 1/x × STO 0   1.5795

1.5 ENTER RCL 0 + 1/x 20 × STO 1   6.4946

1.5 ENTER RCL 1 ×   9.7419

5.6 ENTER 2.2 ENTER 5.6 + ÷ RCL 1 × STO 2   4.6628

RCL 1 ENTER RCL 2 −   1.8318

Section 9.5: Example 2

PRESS	DISPLAY
27.5 ENTER 5.6 ÷ STO 0	4.9107
3.3 ENTER 2.2 + 1/x 2.2 × 3.3 × 4.7 + 1/x 27.5	
× STO 1	4.5681
RCL 0 +	9.4788
3.3 ENTER 2.2 + 1/x RCL 1 × 2.2 ×	1.8272
CHS RCL 1 +	2.7409
2.2 ×	6.0299
CHS 27.5 +	21.4701

Section 9.5: Example 3

PRESS	DISPLAY
4.7 ENTER 3.3 + STO 1	0.0000
2.2 ENTER 6.8 + STO 2	9.0000
RCL 1 × RCL 1 ENTER RCL 2 + ÷	4.2353
1.5 + 1 +	6.7353
1/x 30 × STO 0	4.4541
RCL 2 × RCL 1 ENTER RCL 2 + ÷ STO 1	2.3581
CHS RCL 0 + STO 2	2.0961

Section 9.5: Example 4

PRESS	DISPLAY
1.5 ENTER RCL 0 × 2.2 ENTER RCL 2 × +	11.2926
30 ENTER RCL 0 − 4.7 ENTER 3.3 + RCL 1 × −	
2.2 ENTER RCL 2 × +	11.2926

Section 9.5: Example 5

PRESS	DISPLAY
5.6 ENTER 6.8 + STO 1 10 × 10 ENTER RCL 1 +	
÷ 3.3 ⊢ STO 1	8.8357
CLX 1.5 ENTER 1 + STO 2	2.5000
RCL 1 × RCL 1 ENTER RCL 2 + ÷	1.9486
4.7 + 1/x 22.5 × STO 0	3.3841
RCL 2 × RCL 1 ENTER RCL 2 + ÷ STO 1	0.7463
10 × 10 ENTER 5.6 + 6.8 + ÷ STO 3	0.3332
CHS RCL 1 + STO 2	0.4132
RCL 0 ENTER RCL 1 −	2.6378

Section 9.5: Example 6

PRESS   **DISPLAY**

PRESS	DISPLAY
6.8 [ENTER] [RCL 3] [×] [RCL 0] [ENTER] [RCL 1] [−] 1 [ENTER] 1.5 [+] [×] [−] 4.7 [ENTER] [RCL 0] [×] [−]	−20.2343
5.6 [CHS] [ENTER] [RCL 3] [×] 10 [ENTER] [RCL 2] [×] [+] 22.5 [−]	−20.2343

Section 12.3: Example 1

PRESS	DISPLAY
3.4 [ENTER] 1.3 [CHS] [×] .9 [CHS] [×]	3.9780
2.5 [ENTER] 1.1 [CHS] [×] 1.7 [CHS] [×] [+]	8.6530
9.8 [ENTER] 5.6 [×] 7.6 [×] [+]	425.7410
1.7 [CHS] [ENTER] 1.3 [CHS] [×] 9.8 [×] [−]	404.0830
5.6 [ENTER] 2.5 [×] .9 [CHS] [×] [−]	416.6830
3.4 [ENTER] 7.6 [×] 1.1 [CHS] [×] [−]	445.1070

Section 12.3: Example 2

PRESS	DISPLAY
2.3 [ENTER] 3.2 [×] 5.7 [CHS] [×]	−41.9520
1.5 [CHS] [ENTER] 7.1 [×] 3.4 [×] [+]	−78.1620
3.6 [CHS] [ENTER] 1.4 [CHS] [×] 1.5 [×] [+]	−70.6020
3.4 [ENTER] 3.2 [×] 3.6 [CHS] [×] [−]	−31.4340
1.4 [CHS] [ENTER] 1.5 [CHS] [×] 5.7 [CHS] [×] [−]	−19.4640

PRESS	DISPLAY
2.3 [ENTER] 1.5 [×] 7.1 [×] [−] [STO 0]	−43.9590
17.1 [ENTER] 3.2 [×] 5.7 [CHS] [×]	−311.9040
1.5 [CHS] [ENTER] 7.1 [×] 6.8 [×] [+]	−384.3240
3.6 [CHS] [ENTER] 3.6 [CHS] [×] 1.5 [×] [+]	−364.8840
6.8 [ENTER] 3.2 [×] 3.6 [CHS] [×] [−]	−286.5480
3.6 [CHS] [ENTER] 1.5 [CHS] [×] 5.7 [CHS] [×] [−]	−255.7680
17.1 [ENTER] 1.5 [×] 7.1 [×] [−]	−437.8830
[RCL 0] [÷] [STO 1]	9.9612

PRESS	DISPLAY
2.3 [ENTER] 3.6 [CHS] [×] 5.7 [CHS] [×]	47.1960
17.1 [ENTER] 7.1 [×] 3.4 [×] [+]	459.9900
3.6 [CHS] [ENTER] 1.4 [CHS] [×] 6.8 [×] [+]	494.2620

PRESS	DISPLAY
3.4 [ENTER] 3.6 [CHS] [×] 3.6 [CHS] [×] [−]	450.1980
1.4 [CHS] [ENTER] 17.1 [×] 5.7 [CHS] [×] [−]	313.7400
2.3 [ENTER] 6.8 [×] 7.1 [×] [−]	202.6960
[RCL 0] [÷] [STO 2]	−4.6110

PRESS	DISPLAY
2.3 [ENTER] 3.2 [×] 6.8 [×]	50.0480
1.5 [CHS] [ENTER] 3.6 [CHS] [×] 3.4 [×] [+]	68.4080
17.1 [ENTER] 1.4 [CHS] [×] 1.5 [×] [+]	32.4980
3.4 [ENTER] 3.2 [×] 17.1 [×] [−]	−153.5500

PRESS	DISPLAY
1.4 [CHS] [ENTER] 1.5 [CHS] [×] 6.8 [×] [−]	−167.8300
2.3 [ENTER] 1.5 [×] 3.6 [CHS] [×] [−]	−155.4100
[RCL 0] [÷] [STO 3]	3.5353

PRESS	DISPLAY
2.3 [ENTER] [RCL 1] [×] 1.5 [ENTER] [RCL 2] [×] [−] 3.6	
[ENTER] [RCL 3] [×] [−]	17.1000
1.4 [CHS] [ENTER] [RCL 1] [×] 3.2 [ENTER] [RCL 2] [×] [+]	
7.1 [ENTER] [RCL 3] [×] [+]	−3.6000
3.4 [ENTER] [RCL 1] [×] 1.5 [ENTER] [RCL 2] [×] [+] 5.7	
[ENTER] [RCL 3] [×] [−]	6.8000

Section 13.2: Example 1

PRESS	DISPLAY
12.75 [ENTER] .47 [÷] 6.85 [CHS] [ENTER] .68 [÷] [+] 2.9	
[ENTER] .33 [÷] [+]	25.8420
.47 [1/x] [ENTER] .68 [1/x] [+] .33 [1/x] [+]	6.6286
[÷]	3.8986

Section 13.2: Example 2

PRESS	DISPLAY
6.1 [CHS] [ENTER] 1.82 [−]	−7.9200
2.2 [1/x] [ENTER] 4.7 [1/x] [+] 1.82 [×] 12.65 [ENTER] 2.2 [÷]	
[−] 1.05 [ENTER] 4.7 [÷] [−]	−4.7589
[÷] [STO 1]	1.6643

PRESS	DISPLAY
12.65 [ENTER] 2.2 [÷] 1.05 [ENTER] 4.7 [÷] [+] 6.1 [CHS]	
[ENTER] [RCL 1] [÷] [+]	2.3081
2.2 [1/x] [ENTER] 4.7 [1/x] [+] [RCL 1] [1/x] [+]	1.2682
[÷]	1.8200

Chapter 14

The HP31E does not have the D.MS function. However, many RPN calculators have the [→ H.MS] key and the [→ H] key to convert between the Hour (Degree), Minute, Second format and decimal format. Typical examples are:

a. Convert 15.23 hours to hours, minutes, seconds.

PRESS	DISPLAY
15.23	15.23
[→ H.MS]	15.1348

This reads as 15 hours, 13 minutes, 48 seconds.

b. Convert 4 hours, 6 minutes, 12 seconds into decimal format.

PRESS	DISPLAY
4.0612	4.0612
[→ H]	4.1033

This reads as 4.1033 (Fix 4) hours.

Section 15.2: Example 1

PRESS	DISPLAY
60 [SIN]	0.8660
60 [COS]	0.5000
60 [TAN]	1.7321
60 [ENTER] 10 [×] 9 [÷] [STO 1]	66.6667
[GRD]	66.6667
[SIN]	.8660
[RCL 1] [COS]	0.5000
[RCL 1] [TAN]	1.7321
60 [ENTER] π [×] 180 [÷] [STO 1]	1.0472
[RAD]	1.0472
[SIN]	0.8660
[RCL 1] [COS]	0.5000
[RCL 1] [TAN]	1.7321

Section 15.3: Example 1

PRESS	DISPLAY
a. [DEG] .321 [SIN⁻¹]	18.7234
b. No [D.MS] function	
c. [RAD] .936 [COS⁻¹]	0.3597
d. [DEG] 1.432 [1/x] [TAN⁻¹]	34.9275
e. [RAD] 2.155 [1/x] [SIN⁻¹]	0.4825
f. [GRD] 3.862 [1/x] [COS⁻¹]	83.3259

Section 15.6: Example 2

PRESS	DISPLAY
a. [DEG] 28 [SIN] [ENTER] [×]	0.2204
64 [TAN] 3 [yˣ] [+]	8.8394
b. [RAD] 1.13 [COS] [1/x] [√‾] [ENTER]	1.5309
[DEG] 63 [COS] [ENTER] [×] [÷]	7.4279

Section 15.8: Example 1

PRESS	DISPLAY
17.2 [ENTER] 70.1 [SIN] [÷] [STO 1] 68.7 [SIN] [×]	17.0427
[RCL 1] [ENTER] 41.2 [SIN] [×]	12.0489

Section 15.8: Example 2

PRESS	DISPLAY
14.83 [ENTER] 25.14 [SIN] [×] 12.4 [÷] [SIN⁻¹]	30.5363
12.4 [ENTER] 124.324 [SIN] [×] 25.14 [SIN] [÷]	24.1053

Section 15.8: Example 3

PRESS	DISPLAY
15.26 [ENTER] [×] 14.15 [ENTER] [×] [+] 2 [ENTER] 15.26	
[×] 14.15 [×] 19.6 [COS] [×] [−] [√‾]	5.1240
15.26 [ENTER] 19.6 [SIN] [×] 5.124 [÷] [SIN⁻¹]	87.4664

Section 15.8: Example 5

PRESS   **DISPLAY**

12.7 ENTER × 18.112 ENTER × + 29.32 ENTER ×
─   −370.3279

2 ENTER 12.7 × 18.112 × ÷ COS⁻¹   143.6085

Section 16.5: Example 1

PRESS   **DISPLAY**

DEG 25 ENTER 42 COS ×   18.5785
25 ENTER 42 SIN ×   16.7283
17.3 ENTER 105.7 COS ×   −4.6814
17.3 ENTER 105.7 SIN ×   16.6545
RAD 16.2 ENTER 1.15 ENTER π × STO 1 COS ×   −14.4343
16.2 ENTER RCL 1 SIN ×   −7.3545

PRESS   **DISPLAY**

GRAD 22.8 ENTER 37.9 CHS COS ×   18.8775
22.8 ENTER 37.9 CHS SIN ×   −12.7859

Section 16.5: Example 2

PRESS   **DISPLAY**

2.5 ENTER × 3.6 ENTER × + √   4.3829
3.6 ENTER 2.5 ÷ TAN⁻¹   55.2222
1.7 ENTER × .9 ENTER × + √   1.9235
.9 ENTER 1.7 CHS ÷ TAN⁻¹ 180 +   152.1027
5.2 ENTER × 3.6 ENTER × + √   6.3245

PRESS   **DISPLAY**

3.6 CHS ENTER 5.2 CHS ÷ TAN⁻¹ 180 −   −145.3049
3.65 ENTER × 1.42 ENTER × + √   3.9155
1.42 CHS ENTER 3.65 ÷ TAN⁻¹   −21.2581

Section 16.5: Example 3

PRESS	DISPLAY
5.6 $\boxed{\text{ENTER}}$	5.6
3.5	3.5
$\boxed{\rightarrow \text{P}}$	6.6039
$\boxed{x \rightleftarrows y}$	57.9945

Section 16.5: Example 4

PRESS	DISPLAY
151.23 $\boxed{\text{CHS}}$ $\boxed{\text{ENTER}}$	-151.2300
31.62	31.62
$\boxed{\rightarrow \text{R}}$	-27.7169
$\boxed{x \rightleftarrows y}$	-15.2983

Section 19.3: Example 1

PRESS	DISPLAY
1.07034 $\boxed{10^x}$ $\boxed{\text{Fix 6}}$ $\boxed{\text{STO 1}}$	11.758117
$\boxed{\text{LOG}}$ $\boxed{\text{ENTER}}$ 2 $\boxed{\times}$	2.140680
$\boxed{\text{RCL 1}}$ $\boxed{\text{ENTER}}$ 5 $\boxed{\times}$ $\boxed{\text{LOG}}$ $\boxed{+}$	3.909990

Section 19.3: Example 2

PRESS	DISPLAY
.83139 $\boxed{10^x}$ $\boxed{\text{Fix 6}}$ $\boxed{\text{STO 1}}$	6.782347
$\boxed{\text{ENTER}}$ 2 $\boxed{\div}$ $\boxed{\text{LOG}}$ 3 $\boxed{\times}$	1.591080
$\boxed{\text{RCL 1}}$ $\boxed{\text{ENTER}}$ 5 $\boxed{\div}$ $\boxed{\text{LOG}}$ $\boxed{+}$	1.723500

Section 19.3: Example 3

PRESS	DISPLAY
1.6426789 $\boxed{\text{STO 1}}$ $\boxed{\text{Fix 4}}$ 43 $\boxed{\text{ENTER}}$ 2.6 $\boxed{\text{ENTER}}$ $\boxed{\text{RCL 1}}$	
$\boxed{\times}$ $\boxed{y^x}$	507.6001

Section 19.3: Example 4

PRESS	DISPLAY
12.7 $\boxed{\text{ENTER}}$ 5 $\boxed{\text{ENTER}}$ 4 $\boxed{\div}$ $\boxed{y^x}$ $\boxed{\text{Fix 6}}$	23.974781

Section 19.5: Example 1

PRESS	DISPLAY
10 [ENTER] [LOG] 4 [÷] [Fix 6]	1.660964
10 [ENTER] [LN] 4 [LN] [÷]	1.660964

Section 19.5: Example 2

PRESS	DISPLAY
54.1 [ENTER] [LN] 7 [LN] [÷] [STO 1] [Fix 6]	2.050883
7 [ENTER] [RCL 1] [y^x] [Fix 1]	54.1

Section 19.5: Example 3

PRESS	DISPLAY
23.6 [ENTER] 2.41 [y^x] [LN] [LOG] 2.73 [×]	2.4075

Section 19.7: Example 1

PRESS	DISPLAY
1 [ENTER] 1.5496 [CHS] [e^x] [−] 22.5 [×]	17.7225

Section 21.2: Example 2

PRESS	DISPLAY
2 [ENTER] 12 [y^x] 2 [ENTER] 10 [y^x] [+]	5120.0000
2 [ENTER] 9 [y^x] 2 [ENTER] 8 [y^x] [+] 2 [ENTER] 7 [y^x] [+]	6016.0000
2 [ENTER] 5 [y^x] [+] 2 [ENTER] 4 [y^x] [+] 4 [+] 2 [+]	6070.0000

Section 21.2: Example 3

PRESS	DISPLAY
2 [ENTER] 2 [y^x] 2 [ENTER] 0 [y^x] [+]	5.0000
2 [ENTER] 1 [CHS] [y^x] [+]	5.5000
2 [ENTER] 3 [CHS] [y^x] [+]	5.6250
2 [ENTER] 4 [CHS] [y^x] [+]	5.6875

Section 21.2: Example 6

PRESS	DISPLAY
13 ENTER 2 ÷	6.5000
STO 1 6 −	0.5000
2 ×	1.0000
RCL 1 .5 −	6.0000
2 ÷	3.0000
2 ÷	1.5000
STO 1 1 − 2 ×	1.0000
RCL 1 .5 −	1.0000

Section 21.3: Example 2

PRESS	DISPLAY
2694 ENTER 8 ÷	336.7500
STO 1 336 − 8 ×	6.0000
RCL 1 .75 −	336.0000
8 ÷	42.0000
8 ÷	5.2500
STO 1 5 − 8 ×	2.0000
RCL 1 .25 −	5.0000

Section 21.3: Example 5

PRESS	DISPLAY
2919 ENTER 16 ÷	182.4375
STO 1 182 − 16 ×	7.0000
RCL 1 .4375 − 16 ÷	11.3750
STO 1 11 − 16 ×	6.0000
RCL 1 .375 −	11.0000

Section 21.3: Example 6

PRESS	DISPLAY
3 ENTER 8 $1/x$ × 8 ENTER 2 CHS y^x 5 × + 8 ENTER 4 CHS y^x 7 × +	0.4548

Section 21.3: Example 9

PRESS   **DISPLAY**

8 ENTER 3 y^x 7 × 8 ENTER × 3 × + 8 ENTER 5

× + 4 +   3820.0000

8 1/x ENTER 5 × 8 ENTER 2 CHS y^x 4 × + +   3820.6875

Section 21.3: Example 13

PRESS   **DISPLAY**

6 ENTER × 5 × 6 ENTER 4 × + 3 +   207.0000

8 ÷   25.8750

STO 1 25 − 8 ×   7.0000

RCL 1 .875 − 8 ÷   3.1250

STO 1 3 − 8 ×   1.0000

RCL 1 .125 −   3.0000

Section 21.4: Example 9

PRESS   **DISPLAY**

6 ENTER 3 y^x 3 × 6 ENTER × + 6 ENTER 5 ×

+ 4 + STO 1   718.0000

8 ENTER 3 y^x 8 ENTER × 5 × + 8 ENTER 7 ×

+ 6 +   894.0000

RCL 1 + STO 1 9 ENTER × 8 × 9 ENTER 8 ×

+ 5 +   725.0000

CHS RCL 1 + 16 ÷ STO 1   55.4375

55 − 16 ×   7.0000

RCL 1 .4375 − 16 ÷ STO 1   3.4375

3 − 16 ×   7.0000

RCL 1 .4375 −   3.0000

answers to selected problems

Section 1.2

1. 17	7. 244	13. 668	19. 12	25. −2.5
3. −1.1	9. 276.9	15. −1380.5	21. 0.1	
5. 172.9	11. −0.55	17. 61	23. 6.5	

Section 1.3

1. 54   3. −83.92   5. 4.079   7. −1.301   9. 3.974

Section 1.4

1. −2.203   3. 106.640   5. 3.191   7. 20.371

Section 1.5

1. −8.100   3. −19.190   5. −2.174   7. 0.541   9. 19.580

Section 1.6

1. 1031.3   3. 891.9   5. 1183.1   7. 89.4   9. 106.8

Section 2.1

1. 60	7. 3060	13. 10/9	19. 1/5
3. 1200	9. 252	15. −1/10	21. 77/60
5. 56	11. 7/6	17. 9/35	

Section 2.2

1. 1.167, 7/6	5. −0.1, −1/10	9. 0.200, 24/120
3. 1.111, 30/27	7. 0.257, 18/70	11. 1.283, 154/120

Section 2.3

1. 0.589   3. 0.019   5. −1.069   7. −1.271   9. 0.048

Section 2.4

1. 1/9   3. 1/32   5. 4/3   7. 5/8   9. 63/2

Section 2.5

1. 10   3. 1/10   5. 7/10   7. 64/25   9. 21/10

Section 2.6

1. 7/15   3. 75/16   5. −168/285   7. −3125/96   9. 99/25

Section 2.7

1. 0.467   5. −0.589   9. 4.483   13. 18.366   17. −7.094
3. 4.688   7. −32.552   11. 3.860   15. 2.306   19. −0.328

Section 3.1

1. 3^6 5. π^{12} 9. 1.5   13. $6.24^{5/3}$ 17. 2   21. 3
3. 5^5 7. $2.4^{2.25}$ 11. 1.25^{24} 15. 6   19. 1/2   23. 12

Section 3.3

1. 1/7.1   3. 6.7^2 5. $1/6.5^4$ 7. $1.75^{7.5}$

Section 3.4

1. 171   3. −0.1875   5. 274.825   7. 3.4722   9. 0.36

Section 3.5

1. 3.284   5. 3.595   9. 0.311   13. 8.164   15. −1.532   17. 0.414
3. 1.436   7. 15.678   11. −1.931

Section 4.1

1. 1.56×10^{-4} 5. 3.65×10^6 7. 1.2×10^1 9. 1×10^{-1} 11. 1×10^2
3. 1.685×10^3

Section 4.2

1. 1.875 MHz   3. 2.5 mH   5. 1.4 kW   7. 1 MΩ 9. 11.1 kW

Section 4.3

1. 1.22×10^7
3. 2.2386×10^6

5. 2×10^{-4}
7. 9.4657×10^3

9. 5.12×10^{-16}
11. 1

13. 6.4×10^{-23}

15. 1.3940

Section 4.4

1. 523.504×10^{-18}
3. 11.773×10^6

5. 790.599×10^{-18}

7. 3.780×10^6

9. -88.485

Section 4.6

1. $0.000015 \text{ F} = 15 \ \mu\text{F} = 15{,}000{,}000 \text{ pF}$
3. $85.6 \text{ kHz} = 85{,}600 \text{ Hz} = 0.0856 \text{ MHz}$
5. $4650 \text{ nA} = 4.65 \mu\text{A} = 0.00465 \text{ mA}$

7. 96.564 km/h
9. 2787.09 cm^2
11. 9.4633 L

13. 1344168 cm/h
15. 0.6831 m

REVIEW PROBLEMS: PART I

1. 3.1
3. 34305
5. -54.103
7. 47.491
9. 0.3224
11. 0.1591
13. 0.4543
15. -6291
17. 2195.122
19. 170.984
21. 162.162
23. 60
25. 225
27. 252
29. 140
31. 40/39

33. 89/105
35. 31/168
37. 1.5
39. 0.0446
41. 1.7846
43. -1.5873
45. 14/5
47. 7/216
49. 8/21
51. 1/49
53. 9/2
55. $-7/10$
57. $-4/25$
59. 19/7
61. $-34/101$
63. -2.7424

65. 1.8382
67. 15^2
69. 2^{27}
71. $15^{8/3}$
73. $5.6^{5/2}$
75. $0.4^{4/3}$
77. $3.02^{2.35}$
79. 64
81. $1/7^3$
83. 25/0.41
85. 0.4219
87. 0.002930
89. 48.5066
91. 67.4217
93. 0.1138
95. 7.85×10^{-5}

97. 2.3×10^{-1}
99. 10^{-3}
101. 1.39×10^{-4}
103. $10 \times 10^3 \ \Omega$
105. 1.5×10^3 W
107. 1.75×10^6 Hz
109. 1.6147×10^2, 161.47
111. 6.3798×10^{-4}, 637.98×10^{-6}
113. 5.0596×10^5, 505.96×10^3
115. 4.2×10^4, 42×10^3
117. 1.5883×10^9, 1.5883×10^9
119. 120,000 Hz = 0.12 MHz
121. 20,000 mH = 20,000,000 μH
123. 0.000018 F = 18,000,000 pF
125. 37.853 L
127. 1.207 km/min

Section 5.1

1. $3.873 \text{ k}\Omega$
3. 7.826×10^5 (Hz)
5. 6.417
7. 1.4146×10^5 (Hz)

Section 5.2

1. $4.6x - 7.8y$
3. $-4.86u - 4.8w + 9.7v$
5. $\dfrac{5}{14} x - \dfrac{13}{66} y - \dfrac{31}{30} z$
7. $7.2i + 8.7j - 6.8k$
9. $\dfrac{35}{24} p + \dfrac{4}{5} q$

Section 5.3

1. $4.06a - 9$

3. $3.1a - 9.35b - 5.1$

5. $-6.5a + 4.7b - 0.7c + 4.2$

7. $\frac{15}{56} d + \frac{1}{3} e - \frac{17}{30} f$

9. $-\frac{49}{20} u + \frac{33}{8} v - \frac{16}{5} w$

Section 5.4

1. 13.330 for $b = 1.85$, $c = 2.35$
3. 10.805 for $c = 1.85$, $d = 2.35$

5. -4.430 for $s = 1.85$, $t = 2.35$
7. -8.162 for $u = 1.85$, $v = 2.35$, $w = -2.55$

Section 5.5

1. s^3

3. $c^4 d$

5. $e^{-2} f^6$

7. $tu^{-2} v$

9. $e^3 f + e^4 f^2 + e^2 f^2 g^2$

11. $\frac{10}{21} d^{-1} ce^{-1} - \frac{1}{14} c^2 de + d^{-1} c^2 e^{-1}$

Section 5.6

1. $k^3 l^{-1}$

3. $g^3 k l^3$

5. $u^3 w - u^3 vw^2 + uv^{-1} w^3$

7. $3.1183 a^{-3} c^{-1} - 2.4046 a^{-1} c^{-2} + 0.9542$

9. $\frac{13}{14} u^2 v^{-3} w^{-1} + \frac{22}{35} u^3 w^{-2}$

Section 5.7

1. 0.741 for $a = 2.75$, $b = 3.15$, $c = 4.25$
3. 2.281 for $x = 2.75$, $y = 3.15$

5. 1461.389 for $a = 2.75$, $b = 3.15$, $c = 4.25$

Section 6.1

1. $6.25 x^2 y^4 z^2$

3. $\frac{11.56 a^2}{b^2 c^4}$

5. $\frac{25 u^{2.4} t^3}{s^{4.6}}$

7. $\frac{2ab^2}{c^3}$

9. $-11xy^3$

11. $4x^{-1} y^2$

13. $\frac{-3.1913 u^{2/3}}{t^{4/3}}$

15. 11.705

17. 90.036

19. 1906.509

21. 4.398

23. 146.954

25. 0.130

Section 6.2

1. $(xy)^4$

3. $(uv^2)^3$

5. $10\sqrt{3}$

7. $2/3$

9. $30c^2 d^3$

11. $3\sqrt{7} + 21$

13. $\frac{6 + \sqrt{6}}{18}$

15. $\left(\frac{x}{y}\right)^2$

Section 6.3

1. $x^2 + 5x + 6$
3. $14.76s^2 - 27.55s + 10.81$
5. $6i^2 - ij - 2j^2$
7. $-2x^4 - x^2y^3 + 6y^6$

9. $12a^2b^2 + 5abc - 2c^2$
11. $12x^2 - 9xz + 8xy - 6yz$
13. $25.23d^4 - 32.19d^2g + 4.64d^2f - 5.92fg$
15. $5.12wx + 8.64xz - 7.52wy - 12.69yz$

Section 6.4

Case 1

1. $a^2 - 4b^2$ 3. $4x^2y^2 - z^2$ 5. $36u^2v^4 - 25$ 7. $2.25s^{-2} - t^6$ 9. $74.3044\pi^2 - s^{-4}t^{-6}$

Case 2

1. $x^2 - xy - 2y^2$
3. $x^4 + x^2y^2 - 20y^4$
5. $\dfrac{6a^2}{b^4} + \dfrac{10a}{b^2} - 24$

7. $9.52\,x^2y^2 - 10.88xyz - 16.32z^2$
9. $32i^2j^4 + (2\sqrt{30} - 3.2\sqrt{35})ij^2 - \sqrt{42}$

Case 3

1. $x^2 + 4xy + 4y^2$
3. $4x^2y^2 - 4xyz + z^2$

5. $\dfrac{36s^2}{t^2} - 24\dfrac{sv^2}{t} + 4v^4$
7. $23.5225d^2f^{-4} - 9.7de^{-1}f^{-2} + e^{-2}$

9. $\dfrac{5x^2}{y^4} - \dfrac{2\sqrt{46}xz}{y^2}^2 + 9.2z^{-4}$

Case 4

1. $s^3 + 6s^2t + 12st^2 + 8t^3$
3. $27a^3 - 27a^2b + 9ab^2 - b^3$
5. $64x^6 - 240x^4y + 300x^2y^2 - 125y^3$

7. $-15.625d^3 - 28.125d^2e - 16.875de^2 - 3.375e^3$
9. $23\sqrt{23}a^3b^3 - 69\pi a^2b^2 + 3\sqrt{23}\pi^2ab - \pi^3$

Case 5

1. $4a^2 - 4ab + 8ac + b^2 - 4bc + 4c^2$
3. $9s^2 - 6st + 24sv + t^2 - 8tv + 16v^2$
5. $25i^2 - 20ik^2 + 30ij^3 + 4k^4 - 12k^2j^3 + 9j^6$
7. $12.25x^2 + 16.8xy - 25.2xz + 5.76y^2 - 17.28yz + 12.96z^2$
9. $5x^4 - 2\sqrt{15}x^2y + 2\sqrt{10}x^2z + 3y^2 - 2\sqrt{6}yz + 2z^2$

Section 6.5

Case 1

1. $7(R_1 + 3R_2)$

3. $\dfrac{1}{10}\left(I_1 - \dfrac{3I_2}{4}\right)$

5. $5X_L X_C(X_L X_C^2 + 3 - 15X_L^2 X_C)$

7. $1.8R_1 R_2(2 + R_1 + 5R_2)$

9. $4q(q^2 - 7q + 4)$

11. $\left(\dfrac{1}{a_1} + \dfrac{1}{b_1}\right)(d_1 + d_2)$

13. $(s + t)\left(\dfrac{1}{b_1} + \dfrac{1}{b_2}\right)$

15. $(v + 1)(u - w)$

Case 2

1. $(ab - 2)(ab + 2)$

3. $(4 - 2x)(4 + 2x)$

5. $\left(\dfrac{f}{g^2} - \dfrac{h^3}{i^4}\right)\left(\dfrac{f}{g^2} + \dfrac{h^3}{i^4}\right)$

7. $(4 - b)(4 + 5b)$

9. $\left(\dfrac{15}{st^2} - \dfrac{8}{v^3}\right)\left(\dfrac{15}{st^2} + \dfrac{8}{v^3}\right)$

Case 3

1. $(c + 1)^2$

3. $(a + 6b)^2$

5. $\left(\dfrac{a}{3} - \dfrac{b}{2}\right)^2$

7. $9(R_1 - 2R_2)^2$

9. $(\sqrt{5}g + 1)^2$

11. $\left(\dfrac{9u^2}{v} + 4\right)^2$

13. $\left(\dfrac{\sqrt{3}}{5} x^2 y + z\right)^2$

15. $(b^4 + 4b^2 + 4 - 9c)^2$

Case 4

1. $(s + 1)(s + 2)$

3. $(u - 4)(u + 2)$

5. $(2R_1 - 3)(R_1 - 2)$

7. $3(a + 2)(2a - 3)$

9. $(2p - 3q)(p - 2q)$

11. $(e^2 - 5)(e^2 - 7)$

13. $3(x^3 - 2y^3)(x^3 + 7y^3)$

15. $\left(r - \dfrac{1}{2}\right)\left(r + \dfrac{1}{4}\right)$

Section 6.6

1. $a^2 + 2a - 3$

3. $i^3 - 2i^2 + 3i + 5$

5. $3y^3 - 4y^2 - 5y + 2 \quad (R = -7)$

7. $z^3 + z^2 + z + 1$

9. $\dfrac{1}{4}a^2 - \dfrac{1}{6}a + \dfrac{1}{3}$

11. $3a^2 + 4ab - 4b^2 \quad (R = -6b^3)$

13. $3x^2 + 4xy - 5y^2$

15. $\dfrac{1}{4}a - \dfrac{1}{5}b - 1$

17. $3y^2 + 8y + 6$

19. $x - 6 \quad \therefore b = 18$

Section 7.1

1. $\dfrac{1}{23}$ 3. $\dfrac{d^2}{ce}$ 5. $\dfrac{x + 3y}{6xy}$ 7. $\dfrac{3}{s + t}$ 9. $\dfrac{3u - 2v}{2(2u + 3v)}$ 11. $\dfrac{x^2 + 2xy + y^2}{x - y}$

Section 7.2

1. -1 3. $-\dfrac{1}{t + s}$ 5. $-\dfrac{1}{2}$ 7. 1

Section 7.3

1. $\dfrac{7}{12}$ 5. $\dfrac{bc + ac - ab}{abc}$ 9. $\dfrac{y(4y + 4x + 5)}{x^2 - y^2}$ 13. $\dfrac{-\pi^2 - 2\pi + 4}{2\pi^2 + 5\pi - 12}$

3. $\dfrac{3a}{4}$ 7. $\dfrac{s^4 + t^4 - s - t}{st}$ 11. $\dfrac{4 - a}{a^2 + 2a - 15}$ 15. -4

Section 7.4

1. $1/6$ 5. $3cd^2(a + 3)$ 9. 6 13. $\dfrac{1}{R + x}$ 15. y 17. 1 19. 2

3. $\dfrac{c(a - 2b)}{a - b}$ 7. $\dfrac{x + 4}{(x - y)(2x - 3)}$ 11. 1

Section 7.5

1. -20 5. $\dfrac{I(i^2 + 1)}{i(I^2 + 1)}$ 9. $\left(\dfrac{a + b + 1}{a + b}\right)^2$ 13. $\dfrac{t^2 + t + 1}{t}$ 15. $\dfrac{2}{c(c + 2)}$

3. $\dfrac{2}{15}$ 7. $-\dfrac{(b + 2)(1 - b)}{b(1 + b)}$ 11. $\dfrac{v - 6}{v + 4}$

Section 7.6

1. 1.763 3. -7.619 5. -1.380 7. 0.635

Section 8.2

1. $x = 4$ 5. $z = 12$ 9. $x = 10$ 13. $x = 1$ 17. $x = 8$

3. $b = 4$ 7. $c = -2$ 11. $y = -2\dfrac{1}{3}$ 15. $y = 11$ 19. $x = 0$

Section 8.3

1. $i = e/r$

3. $N_0 = \dfrac{nM}{D}$

5. $A = \dfrac{R_1 T_2 - R_2 T_1}{R_2 - R_1}$

7. $R_2 = \dfrac{R_T R_1 R_3}{R_1 R_3 - R_T (R_1 + R_3)}$

9. $R_S = V_S \left(\dfrac{R_L}{P_{RL}}\right)^{1/2} - R_L$

11. $R_a = \dfrac{R_1 (R_b + R_c)}{R_b - R_1}$

13. $B = \left(\dfrac{2\mu F}{A}\right)^{1/2}$

15. $L_2 = \dfrac{L_T L_1 L_3}{L_1 L_3 - L_T L_3 - L_T L_1}$

17. $L = \dfrac{1}{C(2\pi f_R)^2}$

19. $C = \dfrac{L}{(QR)^2}$

21. $Z_b = \dfrac{(Z_a - Z_{in})(Z_c + Z_L)}{Z_{in} - (Z_a + Z_c + Z_L)}$

Section 8.4

1. $s = 31.5$ 5. $y = 1$ 9. $d = 4$ 11. $a = \dfrac{1}{4}$ 13. $\mu = -0.5$ 15. $p = -1/11$

3. $x = -0.5$ 7. $\beta = -3$

Section 8.5

1. $q = \dfrac{\sqrt{t}}{r}$ 5. $x = 72$ 9. $x = 25°$ 13. $\dfrac{10}{16}$ 17. $x = 9$

3. $x = 23$ 7. Side $3 = 80$ cm   11. $x = 98$ 15. $x = 20$

Section 9.2

1. 1.766 mA, 8.302 V   3. 47 kΩ   5. 4.761 mA   7. 34.591 V   9. 20.127 V

Section 9.4

1. 21.265 V   3. $I_1 = 0.686$ mA, $I_6 = 0.237$ mA   5. 1.801 V   7. 0.140 mA   9. 146.110 mA

Section 9.5

1. 6.357 mA   3. 10.289 V   5. 89.03 mA   7. 3.502 V   9. Yes

11. $V_A = -33.5$ V

$V_B = -20.581$ V

$V_C = -3.489$ V

$V_D = -2.346$ V

$V_E = -18.882$ V

$V_F = -16.992$ V

$V_G = -1.133$ V

Section 10.2

1. $\pm\dfrac{5\sqrt{2}}{2}$ 3. ± 6 5. $0, 1$ 7. $0, 9/4$ 9. $0, 2/3$ 11. $-5, 8$ 13. $-6/7, 5$ 15. $-11, 3$

Section 10.3

1. ± 11 3. $0, -5/3$ 5. $-7, 9/2$ 7. $-1, 5/9$ 9. $-1 \pm \sqrt{2}$ 11. $-1.67, 0.36$

Section 11.2

1. $i = -7, h = 3$ 7. $u = 6, v = -4$ 13. $k = 3.1, j = -1.75$
3. $a = 3.1, b = -2.6$ 9. $x = 1.2, y = -1.8$ 15. $m = -1.32, n = 1.96$
5. $c = 1.86, d = 2.45$ 11. $\iota - 3, f - -2$

Section 11.3

1. $a = 2, b = -7, c = 13$ 5. $s = 3.1, t = -2.6, v = -1.7$
3. $\alpha = 1.5, \beta = -3.2, \gamma = 4$ 7. $f = 1.3, g = -5.6, h = -2.5$

Section 12.2

1. a. 24 b. -4 c. 55 d. -747 e. 164 f. -624
3. $-b(di - fg) + e(ai - cg) - h(af - cd)$

Section 12.3

1. $c = -5.4904, d = 2.1538$ 5. $u = -0.5390, v = 0.7301, w = -0.5670$
3. $x = -4.5218, y = -6.3200, z = -4.1749$

Section 12.4

1. $I_{R1}, I_{R4} = 2.591$ mA $I_{R2}, I_{R5} = 1.453$ mA $I_{R3} = 1.138$ mA
3. -25.058 V
5. $I_{R1} = 0.579$ mA from B to C $I_{R3} = 1.408$ mA from F to D $I_{R5} = 0.382$ mA from E to D
   $I_{R2} = 1.790$ mA from D to B $I_{R4} = 0.829$ mA from E to C $I_{R6} = 1.211$ mA from A to E
7. -9.304 V

Section 12.6

Multiplying $1/D$ times each element results in:

1.
$$\begin{bmatrix} 0.2603 & 0.5890 & 0.3562 \\ 0.0822 & -0.2877 & -0.1507 \\ 0.1507 & -0.0274 & -0.1096 \end{bmatrix}$$

3.
$$\begin{bmatrix} 0.0653 & -0.1751 & -0.0740 \\ -0.0102 & -0.0598 & -0.1142 \\ 0.1169 & -0.0168 & -0.0744 \end{bmatrix}$$

5.
$$\begin{bmatrix} 0.1024 & -0.0596 & 0.0055 & 0.0332 \\ 0.1477 & -0.0493 & -0.0877 & 0.0734 \\ -0.0491 & 0.0621 & 0.0489 & -0.0171 \\ -0.1172 & 0.0508 & 0.0474 & 0.0072 \end{bmatrix}$$

7. $x = 7$, $y = -2$, $z = -5$

9. $u = 0.5035$, $v = -1.2180$, $w = -1.788$

11. $I_1 = -5.6818$ mA, $I_2 = 10.1263$ mA, $I_3 = 11.7276$ mA

Section 13.2

1. 24.7 V   3. -60.6 V   5. 3.532 kΩ

Section 13.3

1. $R_1 = 0.755$ kΩ, $R_2 = 1.092$ kΩ, $R_3 = 2.333$ kΩ

5. $\dfrac{5R}{6}$

3. $R_{AB} = 150.8$ kΩ, $R_{BC} = 26.929$ kΩ
   $R_{CA} = 6.855$ kΩ

Section 13.4

1. a. 35.75 kΩ   b. 35.597 kΩ   3. 47 kΩ

REVIEW PROBLEMS: PART II

1. 4.188

3. 166.667

5. 6.059

7. $2.3r - 6.7s$

9. $\dfrac{7}{15}x - \dfrac{1}{6}y - \dfrac{4}{3}z$

11. $-12e + 11d + 23f - 17g$

13. $0.4a$

15. $0.9u - 10.3v - 1.7w$

17. $-4.4r + 9.35s - 3.6t$

19. 16.975

21. 17.943

23. -12.883

25. $3a^6 + 6a^5b - 4a^4b^2$

27. $\dfrac{1}{2}u^4v^4 - \dfrac{1}{6}u^3v^6$

29. $k^5l^{-5}m^{-1}$

31. 7.950

33. 0.2

35. -7.009

37. $9.61a^2b^4c^{-6}$

39. $32.768\ l^6n^9m^{-12}$

41. $4a^2b^{-3}c$

43. $2.93i^2j^3$

45. 50.818

47. 4.270

49. 0.00051

51. $\left(\dfrac{y}{x}\right)^{1/2}$

53. $4\sqrt{3}$

55. $5\sqrt{5}$

57. $5\sqrt{2}$

59. $-6\sqrt{2} - 4$

61. $r^5s^2t^4\sqrt{21s}$

63. $\dfrac{2\sqrt{5} + 5}{5}$

65. $\dfrac{a}{b\sqrt{c}}$

67. $-11.13a^3b^2 + 16.43a^2b + 9.03a^2b^2 - 13.33ab$

69. $2x^4 + 7x^2y^3 - 15y^6$

71. $15x^4y^2 + 58x^2yz - 153z^2$

73. $28xz - 36yz - 14xy + 18y^2$

75. $4a^2 - 16b^2$

77. $\dfrac{1}{2}y^2 + 2y - 6$

79. $x^9 - 12x^6 + 48x^3 - 64$

81. $36r^2 - 84rs + 49s^2$

83. $6a^4b^{-2} + 7a^2b^{-1}c^{-1} - 5c^{-2}$

85. $0.09x^2 - 1$

87. $-6a^2 + 16ab - 8b^2$

89. $9x^2 + 12xy - 6xz + 4y^2 - 4yz + z^2$

91. $216x^3y^6 - 216x^2y^4z^3 + 72xy^2z^6 - 8z^9$

93. $400x^2 - 360x + 81$

95. $9a^2b^4 - 6a^2b^2c + 6ab^3c^2 + a^2c^2 + -2abc^3 + b^2c^4$

97. $12t^4 + 13t^2 - 14$

99. $18x^3 - 33x^2 - 30x$

101. $9r^2 - 12rs + 36rt + 4s^2 - 24st + 36t^2$

103. $12a^3 - 2a^2b - 2ab^2$

105. $a^4 - 625$

107. $2my(1 - 4m)$

109. $x(32 + x)$

111. $(x - 7)(x + 7)$

113. $\pi(r^2 + R^2)$

115. $2(x^2 + 4x + 1)$

117. $(3x - 2)(x - 2)$

119. $(a + 5)^2$

121. $(2c - 3)^2$

123. $(a^2 + 25)(a - 5)(a + 5)$

125. $(y + 9)(y + 1)$

127. $4a^2b^2c^2(2a^2c + 3)$

129. $\left(s - \dfrac{1}{10}\right)\left(s + \dfrac{1}{10}\right)$

131. $(a^2 + 4)(a - 2)(a + 2)$

133. $(2t - 3)(9t + 2)$

135. $(3x + 1)(x + 3)$

137. $(rs - 12)(rs + 12)$

139. $(2a + 3)(a + 4)$

141. $a(a + 5)(a + 2)$

143. $(x - 0.8)(x + 0.8)$

145. $(3u - v)(u - 2v)$

147. $-3 + 6a^{-1} + a^{-3}$

149. $2z^2 - 7z + 4$

151. $x^2 - x + 4$ (remainder $= 16$)

153. $5x - y$ (remainder $= 3y^2$)

155. $\dfrac{3x^3}{4y}$

157. $\dfrac{m}{m - 1}$

159. $\dfrac{a}{a - 2}$

161. $2/3$

163. $\dfrac{7x - 3y}{x + 5y}$

165. $\dfrac{-1}{a + 4}$

167. $\dfrac{2(4n - 1)}{2 - 5n}$

169. $\dfrac{1 - y}{3}$

171. $\dfrac{2 - 4s + 3s^2}{s^3}$

173. $\dfrac{-5a - 96}{24a}$

175. $\dfrac{x - 4}{x^2 - 9}$

177. $\dfrac{4x^2 + 30x - 21}{x^2 + 7x}$

179. $\dfrac{3x - 2y}{12(2x - y)}$

181. $\dfrac{c(a - b)}{2(a + b)}$

183. $\dfrac{y + 2}{y - 2}$

185. $\dfrac{y + 4}{3}$

187. $\dfrac{21}{2(x - 2)}$

189. -27.6

191. $\dfrac{2x + 2}{2x + 9}$

193. $2r^2 - 1$

195. 0.3783

197. 0.7354

199. $y = 5$

201. $v = 13$

203. $d = 20$

205. $x = -1$

207. $a = 5$

209. $B = 3$

211. $R_2 = \dfrac{R_1 v - R_1 v_1}{v_1}$

213. $\alpha = \dfrac{f_\alpha - f_\beta}{f_\alpha}$

215. $V_o = \dfrac{V_i R_F - I_i R_F h_{ie} + V_i h_{ie}}{h_{ie}}$

217. $y = 3$

219. $y = -6$

221. $z = 0$

223. 20 and 32

225. $x = 415$ mi the 3rd day

227. 8 and 40

229. $V_{R1} = 6.318$ V
$V_{R2} = 4.367$ V
$V_{R3} = 3.066$ V

231. 5.202 mA

233. 28.452 kΩ

235. 45.232 kΩ

237. 1.138 kΩ

239. 63.767 mA

241. $I_T = 7.723$ mA
$I_1 = 7.010$ mA
$I_2 = 0.713$ mA
$I_3 = 0.541$ mA
$I_4 = 0.297$ mA
$I_5 = 0.244$ mA
$I_6 = 0.172$ mA

243. $x = 4, 4$

245. $x = 4, -1$

247. $x = 0, 2$

249. $a = -3, -1$

251. $x = 5, 4$

253. $x = 1 + \sqrt{3}, 1 - \sqrt{3}$

255. $\dfrac{5 \pm \sqrt{43}}{2}$

257. $n = -2, m = 5$

259. $s = 2, r = 5$

261. $a = \dfrac{9}{4}, b = \dfrac{7}{2}$

263. $a = \dfrac{1}{2}, b = \dfrac{3}{2}$

265. $m = 2, n = 0$

267. $a = \dfrac{9}{4}, b = \dfrac{7}{2}$

269. $m = 2, n = -1$

271. $x = 8, y = 16$

273. $r = -1, s = 1, t = 2$

275. $a = 0.611, b = -1.50, c = 0.056$

277. a. 28   b. 29   c. 15   d. 0

279. $x = -2.011, y = 0.9377$

281. $a = 1.9931, b = -3.0827, c = -1.8144$

283. $I_1 = 4.202$ mA
$I_2 = -0.363$ mA
$I_3 = 4.521$ mA

285. $I_1 = 12.465$ mA, $I_2 = -7.567$ mA, $I_3 = 9.581$ mA,
$I_4 = -0.779$ mA, $I_{R1} = 12.465$ mA, $I_{R2} = 11.686$ mA,
$I_{R3} = 4.898$ mA, $I_{R4} = 6.788$ mA, $I_{R5} = 7.567$ mA,
$I_{R6} = 2.014$ mA, $I_{R7} = I_{R8} = 9.581$ mA, $I_{R9} = 0.779$ mA

287. $\begin{bmatrix} -0.8333 & 1.3333 & 1.8333 & 0.3333 \\ -1.1667 & 1.6667 & 2.1667 & 0.6667 \\ 0.2917 & -0.1667 & -0.2917 & 0.0833 \\ -0.4583 & 0.8333 & 1.4583 & 0.5833 \end{bmatrix}$

289. a. $x = -1.05, y = -2.2, z = 1.65$
b. $x = 2, y = -1, z = 3, w = 1$

291. -1.549 V

293. 51.66 mA

Section 14.2

1. a. $66°$
   b. $38.4°$
   c. $77°47'$
   d. $52°38'25''$
   e. $59'43''$

3. a. $59°$ c. $334°6'21''$
   b. $56°12'$ d. $201°14'$

5. a. $21.3°$
   b. $75.3567°$
   c. $105.00889°$
   d. $191.01694°$
   e. $51.00194°$
   f. $0.00667°$

7. a. $14.6° = 0.2548$ rad $= 16.2222$ grad
   b. $71°17'21'' = 1.2442$ rad $= 79.2102$ grad
   c. 2.152 rad $= 123°18'2'' = 137$ grad
   d. 0.08 rad $= 4°35'1'' = 5.093$ grad
   e. 41.26 grad $= 0.6481$ rad $= 37°8'2''$
   f. 0.889 grad $= 0.01396$ rad $= 0°48'$

Section 14.3

1. 36.8 cm

Section 15.2

1. a. 7.7
    b. 10.923
    c. −4.734
    d. −5.917

3. a. 0.1534
    b. 1.8492
    c. 0.1098
    d. 0.3280

e. 0.1230
f. 0.8091
g. 0.2439

h. 1.6946
i. 0.9537
j. 0.4842

Section 15.3

1. a. 25.2773°
    b. 0.9728 rad

c. 72.2845 grad
d. 27°26′26″

e. 78.7370°
f. 1.0819 rad

Section 15.4

1. a. $h = 24.1$, $\theta = 42.146°$, $\phi = 47.854°$
    b. $a = 2.86 \times 10^4$, $\phi = 72.142°$, $\phi = 17.858°$
    c. $b = 101.9031$, $\phi = 31.64°$, $\phi = 58.36°$
    d. $a = 20.91$, $b = 6.47$, $\phi = 17.2°$
    e. $b = 69.17$, $h = 99.58$, $\phi = 51.01$ grad

    f. $a = 27.69$, $h = 46.14$, $\phi = 0.644$ rad
    g. $a = 49.5$, $h = 52.61$, $\phi = 70.2°$
    h. $b = 68.72$, $h = 143.49$, $\phi = 31.8$ grad
    i. $a = 3.52 \times 10^5$, $b = 1.58 \times 10^5$,
        $\phi = 0.421$ rad

3. 2086 Hz   5. 263.84

Section 15.5

1. $\csc 0° = \infty$ $\csc 60° = \dfrac{2\sqrt{3}}{3}$ $\csc 30° = 2$ $\csc 90° = 1$ $\csc 45° = \sqrt{2}$

    $\sec 0° = 1$ $\sec 60° = 2$ $\sec 30° = \dfrac{2\sqrt{3}}{3}$ $\sec 90° = \infty$ $\sec 45° = \sqrt{2}$

    $\cot 0° = \infty$ $\cot 60° = \dfrac{\sqrt{3}}{3}$ $\cot 30° = \sqrt{3}$ $\cot 90° = 0$ $\cot 45° = 1$

3. $\csc 120° = \dfrac{2\sqrt{3}}{3}$ $\csc 135° = \sqrt{2}$ $\csc 150° = 2$

    $\sec 120° = -2$ $\sec 135° = -\sqrt{2}$ $\sec 150° = \dfrac{-2\sqrt{3}}{3}$

    $\cot 120° - \dfrac{-\sqrt{3}}{3}$ $\cot 135° = -1$ $\cot 150° = -\sqrt{3}$

5. $\csc 270° = -1$ $\csc 330° = -2$ $\csc 300° = \dfrac{-2\sqrt{3}}{3}$ $\csc 315° = -\sqrt{2}$

    $\sec 270° = \infty$ $\sec 330° = \dfrac{2\sqrt{3}}{3}$ $\sec 300° = 2$ $\sec 315° = \sqrt{2}$

    $\cot 170° = 0$ $\cot 330° = -\sqrt{3}$ $\cot 300° = \dfrac{-\sqrt{3}}{3}$ $\cot 315° = -1$

Section 15.6

1. -0.4919 3. 4.3272   5. 1.7412   7. 86.8619   9. 0.2912

Section 15.7

7. 0.5   9. $\dfrac{\sqrt{2}}{2}$

Section 15.8

1. 47.09°
3. $a = 12.17$, $b = 8.85$
5. 34.05
7. 30.35
9. $c = 121.67$, $B = 27.08°$, $C = 111.42°$
11. $a = 51.19$, $A = 67°43'29''$, $C = 50°41'31''$
13. $a = 4.58$, $B = 32.52°$, $C = 72.88°$
15. $a = 231.38$, $B = 32°27'$, $C = 94°8'12''$

Section 15.9

1. a. $\sin A \cos B - \cos A \sin B$
   b. $\cos \phi \cos \theta - \sin \phi \sin \theta$
   c. $\sin r \cos s + \cos r \sin s$
   d. $\cos x \cos y + \sin x \sin y$
7. $-15.5 \cos \theta$
9. a. $\dfrac{-\sqrt{2}}{2}$
   b. $\dfrac{\sqrt{6} - \sqrt{2}}{4}$

Section 16.1

1. $j11$
3. $j\alpha$
5. $-j7 \, \beta^2$
7. -49
9. $-25\sqrt{6}$
11. -10
13. $-j2.2$
15. $j13$

Section 16.2

1. $4.2 + j4.2$
3. $-7.22 - j6.7$
5. $j20.16$
7. $-28.21 - j36.6$
9. $97.361 - j145.854$
11. -6.48
13. $6.176 - j4.08$
15. $-32.3459 - j38.3673$
17. $-j0.2$
19. $j2.8$
21. $-1.5 - j0.75$
23. $0.7993 - j4.427$
25. $1.113 - j0.839$
27. $-0.424 + j0.913$

Section 16.5

1. $5.1962 + j3$
3. $-15.3174 - j5.2742$
5. $6.1779 - j6.4896$
7. $7.2 \,\underline{/70.96°}$
9. $13.36 \,\underline{/130.75°}$
11. $111.3 \,\underline{/-42.39°}$

Section 16.6

1. $8 - j4.7$; $9.28 \,\underline{/-30.43°}$
3. $-9.21 + j28.3$; $29.76 \,\underline{/108.03°}$
5. $12.98 - j7.47$; $14.98 \,\underline{/-29.92°}$
7. $-3.783 + j3.638$; $5.248 \,\underline{/136.122°}$
9. $75.443 + j60.875$; $96.940 \,\underline{/38.9°}$
11. $10.235 + j12.958$; $16.512 \,\underline{/51.697°}$
13. $6.205 - j2.320$; $6.625 \,\underline{/-20.5°}$
15. $217.504 + j89.726$; $1.040 \,\underline{/217.501°}$
17. $15031.716 - j15674.928$; $21717.639 \,\underline{/-46.2°}$
19. $j4.159$; $4.159 \,\underline{/90°}$

Section 16.7

1. $5, -5$
3. $5, -2.5 + j2.5 \sqrt{3}, -2.5 - j2.5 \sqrt{3}$
5. $2, 0.618 + j1.902, -1.618 + j1.176, -1.618 - j1.76, 0.618 - j1.902$

Section 17.3

1. a. 432,000 rev/h   b. 120 rev/s   c. 0.12 rev/ms
3. a. 0.27 Km/s   b. 134.06 Km   c. 2178.5 Km
5. a. 0.675 rev, 4.2412 rad   b. 364.5 rev, 2290.221 rad   c. 0.1233 rev, 0.7744 rad   d. 534.6×10^3 rev, 3.359×10^6 rad
7. a. 910.28 μs   b. 7423.06 μs   c. 9200.94 μs   d. 14888.79 μs
9. a. 29.94 μs
11. $i = 127 \sin 9000 \pi t$
    $v = 26.128 \sin (9000 \pi t + 51.68°)$
13. 4.032 μs

Section 17.5

1. a. 4.5 A   b. 208 V (line)
3. a. 10.39 A (line)   b. 220 V

Section 17.6

1. $I_{LA} = 15 + j10$
   $I_{LB} = 1.16 - j17.99$
   $I_{LC} = -16.16 + j7.99$
3. $I_{ACl} = 18.48 \,\underline{/121.66°}$
   $I_{CBl} = 18.48 \,\underline{/241.66°}$
   $I_{BAl} = 18.48 \,\underline{/1.66°}$
   $I_{LA} = 32 \,\underline{/151.67°}$
   $I_{LB} = 32 \,\underline{/31.68°}$
   $I_{LC} = 32 \,\underline{/-88.34°}$
5. $V_{ACl} = 154.16 \,\underline{/210.55°}$
   $V_{CBl} = 154.16 \,\underline{/-29.45°}$
   $V_{BAl} = 154.16 \,\underline{/90.55°}$
7. $I_{CA} = 0.48 \,\underline{/-36.87°}$
   $I_{BC} = 1.703 \,\underline{/-162.138°}$
   $I_{AB} = 0.615 \,\underline{/159.806°}$
   $I_{LC} = 2.019 \,\underline{/186.657°}$
   $I_{LA} = 1.083 \,\underline{/-27.488°}$
   $I_{LB} = 1.276 \,\underline{/35.110°}$

Section 18.1

1. $v_R = 13.41 - j11.04; 17.368 \,\underline{/-39.47°}$
   $v_L = 9.09 + j11.04; 14.304 \,\underline{/50.53°}$
3. 4.424 kΩ
5. 17.318 kΩ

Section 18.2

1. $2.349 + j0.463$ kΩ
3. $1.509 - j1.895$

Section 18.3

1. $i_T = 5.191 + j1.248; 5.339 \,\underline{/13.514°}$ (mA)
   $i_1 = 2.781 + j3.287; 4.306 \,\underline{/49.768°}$ (mA)
   $i_2 = 2.410 - j2.039; 3.157 \,\underline{/-40.232°}$ (mA)
3. $i_T = 7.77 \,\underline{/35.16°}$ (mA)
   $i_1 = 5.23 \,\underline{/-12.561°}$
   $i_2 = 5.751 \,\underline{/77.445°}$
   $i_3 = 7.552 \,\underline{/48.791°}$ (mA)
   $i_4 = 8.657 \,\underline{/-41.209°}$ (mA)
   $i_5 = 6.826 \,\underline{/138.791°}$ (mA)
5. $v_{ab} = 64.49 \,\underline{/124.496°}$
   $v_{ac} = 63.581 \,\underline{/81.649°}$
   $v_{bc} = 46.785 \,\underline{/12.036°}$

REVIEW PROBLEMS: PART III

1. a. $72.4°$ b. $37°47'6''$
3. a. $156°$ b. $74°16'$
5. a. $56.7°$ b. $12.086°$
7. a. $94.65°$; 105.2 grad
    b. 1.136 rad; 72.3 grad
    c. $0.56°$; 0.01 rad
9. a. 11.28 in
    b. 14.11 m
11. 12.048
13. 2.605
15. -2.8583
17. 1.1753
19. 3.3311
21. a. $29.3°$
    b. 17.249 grad
    c. $12°17'54''$
    d. 1.4116 rad
23. 16.956
25. $\theta = 1.0818$ rad; $a = 26.925$; $b = 14.327$
27. 294.8 mA

29. 3

31. $\dfrac{-2\sqrt{3}}{3}$

33. $\dfrac{-\sqrt{3}}{3}$

35. 0.2489
37. 0.0048
39. 4.1533
41. cot 115 grad
43. tan $143.2°$
45. 0.5
47. -2
49. a. $36.81°$ b. $109.5°$
51. a. $-\sin 45°$
    b. $\cos 54°$
    c. $\sin 54.5°$
    d. $\sin 103°$

53. $10\left(\dfrac{\sqrt{3}}{2}\sin\theta - \dfrac{1}{2}\cos\theta\right)$

55. $j12$
57. $-j81$
59. -25
61. $4.3 + j11.7$
63. $-19.53 + j41.04$
65. -18.13

67. $35.97 - j15.84$
69. $-j/3$
71. $4.3 - j4.3$
73. $(-3 + j21.4)/13$
75. $(-159.32 - j117.44)/40$
77. $-9.31 - j9.64$
79. $3.67 - j8.87$
81. $6.5\ \underline{/61.2°}$
83. $21.3\ \underline{/211.4°}$
85. $7.0 \times 10^{-4}\ \underline{/149.0°}$
87. $-30.4 + j21.4,\ 37.2\ \underline{/144.86°}$
89. $4.86\ \underline{/234.71°}$
91. $-27.85 + j25.05,\ 37.46\ \underline{/138.03°}$
93. $6.21 + j2.09,\ 6.55\ \underline{/18.6°}$
95. $14351.99 - j26766.36,\ 30371.33\ \underline{/-61.8°}$
97. $-2.84 + j2.84,\ 4.02\ \underline{/135°}$
99. ± 7
101. $5, j5, -5, -j5$
103. $4, j4, -4, -j4$
105. a. 2.97 ms
     b. 4×10^{-7} s
     c. 8 μs
107. a. 45.414
     b. 52.44 V
     c. 55 V
109. a. 21.39 A
     b. 275 V
111. $I_{ACl} = 96.75\ \underline{/120.64°}$; $I_{LA} = 130.20\ \underline{/135.49°}$

     $I_{CBl} = 58.98\ \underline{/302.11°}$; $I_{LB} = 43.67\ \underline{/73.79°}$

     $I_{BAl} = 44.27\ \underline{/-10.45°}$; $I_{LC} = 155.72\ \underline{/-58.80°}$

113. $v_r = 17.082\ \underline{/55.491°}$

     $v_L = 11.647\ \underline{/145.491°}$
     $v_C = 36.491\ \underline{/-34.509°}$
115. $Z_T = 1.671\ \underline{/50.94°}\ \Omega$
     $I_A = 225.33\ \underline{/76.847°}$ mA
     $I_B = 264.72\ \underline{/-10.173°}$ mA
     $I_C = 273.98\ \underline{/-39.06°}$ mA
117. $I_T = 11.375\ \underline{/42.66°}$; $(8.365 + j7.7)$ mA
     $I_1 = 7.652\ \underline{/-5.066°}$; $(7.622 - j0.676)$ mA
     $I_2 = 8.417\ \underline{/84.934°}$; $(0.743 + j8.384)$ mA
     $I_3 = 11.056\ \underline{/56.291°}$; $(6.136 + j9.197)$ mA
     $I_4 = 12.673\ \underline{/-33.709°}$; $(10.542 - j7.033)$ mA
     $I_5 = 9.993\ \underline{/146.291°}$; $(-8.313 + j5.546)$ mA

Section 19.2

1. 1.94077   3. 0.80060   5. 1.13877

Section 19.3

1. 2.53931   3. 202.17   5. 0.5625_{10} 7. 385.9375_{10} 9. 1.7713

Section 19.4

1. Valid   3. Invalid   5. Valid

Section 19.5

3. 4.6021   7. 0.0339   11. 0.2216   15. 0.9343
5. 0.7275   9. 18424.88   13. 4.876   17. 3729.53

Section 19.6

1. a. 13.5539 dB   3. 39.43 μW   7. 53.174 mH
    b. 38.0618 dB   5. 0.1315 μF   9. 2.8486 mm
    c. -33.0103 dB
    d. 50.96 dB
    e. 23.86 dB

Section 19.7

1. 19.25 V   3. 1.146 kΩ 5. 2.190 mA   7. 0.1816 V   9. 0.1601 V

Section 20.1

1. 2   3. 2   5. -2 7. $-3/11$

Section 20.3

1. $(15, -38)$ 3. $(1, -4)$ 5. $(3, 9), (-1, 1)$

REVIEW PROBLEMS: PART IV

1. 1.82745   9. 4.05546×10^{14} 17. 43.558   25. 14.35 dB   33. 4.89 ms
3. 1.99667   11. 3.96×10^{10} 19. 0.4167   27. 57.36 dB   35. -3
5. 0.15484   13. Invalid   21. 2.977   23. 131.785   29. 22.61 W   37. 2.1
7. 1326.96   15. Invalid   23. 131.785   31. 0.276 μs   39. $-23/99$

Section 21.2

Binary to Decimal

1. 13_{10} 3. 1365_{10} 5. 0.5625_{10} 7. 385.9375_{10}

Decimal to Binary

1. 1010110_2 5. 10010111.0101_2 9. 101011111.1111_2
3. 100111010100_2 7. 1101001111.0011_2

Section 21.3

Decimal to Base N

1. 6060_8 5. $F88_{16}$ 9. 2333_5 13. 2513.2524_6
3. 203.4463_8 7. 165.2667_{16} 11. 5040_7

Base N to Decimal

1. 428_{10} 3. 700_{10} 5. 83.5898_{10} 7. 237_{10} 9. 593.1852

Base N to Base M

1. $546_8 = 166_{16}$ 5. $1001010100011_2 = 11243_8$ 9. 14133.0324_5
3. $100101110011_2 = 973_{16}$ 7. 362_8

Section 21.4

1. 1010100_2 7. 1000001_2 13. $2E_{16}$ 19. 1034_6 25. 142_7
3. $1A0_{16}$ 9. 55110_8 15. 11010_2 21. 121343_8
5. 113_8 11. $6E2A_{16}$ 17. 2284_{10} 23. 254551_7

Section 22.1

1. AND $= 0$, OR $= 1$, NAND $= 1$, NOR $= 0$ 5. AND $= 0$, OR $= 1$, NAND $= 1$, NOR $= 0$
3. AND $= 0$, OR $= 1$, NAND $= 1$, NOR $= 0$

Section 22.2

1. $\overline{(A + \overline{B})C}$ 3. $\overline{A + B(\overline{A + B + C})}$

Section 22.3

1. 1, 5, 7   3. 0, 2, 4, 6, 7   5. 0, 1, 2, 3, 5   7. 0, 1, 2, 3, 4, 6

Section 22.4

1. $(\overline{A} + B)\,\overline{C}$ 3. $\overline{A} + \overline{B}$ 5. $A + B$ 7. $A + \overline{B} + \overline{C}$ 9. $\overline{A} + B\overline{C} + BD\overline{E}$ 11. $\overline{AC}$

REVIEW PROBLEMS: PART V

1. 45
3. 0.625
5. 117.625
7. 1111101_2
9. 1100001101.0011_2
11. 12004_8
13. 1018_{16}
15. 3043_5
17. 3441
19. 209.6875
21. $1466_8 = 336_{16}$
23. $1011000100001_2 = B11_{16}$
25. $1011000001.1011_2 = 1301.54_8$

27. 1001001001_2
29. 1001101_2
31. 5_8
33. 507
35. 1310_5
37. AND = 0, OR = 1, NAND = 1, NOR = 0
39. AND − 0, OR − 1, NAND − 1, NOR − 0
41. $\overline{A(A + B) + B(A + B)}$
43. $AB + \overline{ABC}$
45. 1
47. 0, 2, 3, 4, 6
49. A
51. 1

Index

577